Archives of Virology

Supplementum 9

M. A. Brinton, C. H. Calisher, R. Rueckert (eds.)

Positive-Strand RNA Viruses

Springer-Verlag Wien New York

Dr. Margo A. Brinton
Department of Biology, Georgia State University, Atlanta,
Georgia, U.S.A.

Dr. Charles H. Calisher
Arthropod-borne and Infectious Disease Laboratory (AIDL),
Colorado State University, Fort Collins, Colorado, U.S.A.

Dr. Roland Rueckert
Institute for Molecular Virology, University of Wisconsin, Madison,
Wisconsin, U.S.A.

This work is subject to copyright.
All rights are reserved, whether the whole or part of the material is concerned,
specifically those of translation, reprinting, re-use of illustrations, broadcasting,
reproduction by photocopying machines or similar means, and storage in data banks.

© 1994 Springer-Verlag/Wien
Printed in Austria

Typesetting: Best-set Typesetter Ltd., Hong Kong
Printing: Adolf Holzhausens Ngf., A-1070 Wien

Printed on acid-free and chlorine-free bleached paper

With 182 Figures

Library of Congress Cataloging-in-Publication Data
Positive-strand RNA viruses/M.A. Brinton, C.H. Calisher, R. Rueckert (eds.).
p.cm. − (Archives of Virology, Supplementum; 9) Based on the Third International
Symposium on Positive Strand RNA Viruses, held in Clearwater, Fla. in 1992.
ISBN 3-211-82522-3 (alk. paper). − ISBN 0-387-82522-3 (alk. paper)
1. RNA viruses-Congresses. I. Brinton, Margo A. II. Calisher, Charles H. III. Rueckert,
Roland R. IV. International Symposium on Positive Strand RNA Viruses
(3rd : 1992 : Clearwater, Fla.) V. Series.
[DNLM: 1. RNA Viruses-congresses. W1 AR49LA v. 9 1994/QW 168 P855 1994.]
QR395.P66 1994 576′.64 − dc20 DNLM/DLC for Library of Congress 94-3138 CIP

ISSN 0939-1983
ISBN 3-211-82522-3 Springer-Verlag Wien New York
ISBN 0-387-82522-3 Springer-Verlag New York Wien

Preface

This book summarizes plenary lectures from the Third International Symposium on Positive-Strand RNA Viruses. The First International Meeting, organized under the UCLA Symposia Series, was held in Keystone Colorado in April 1986, and the proceedings published in 1987 (Brinton MA, Rueckert RR (eds) Positive Strand RNA Viruses. Alan R. Liss, New York). That meeting, which brought together virologists spanning a broad range of fields from agriculture to medicine, revealed many similarities in the structure and molecular biology of viruses previously regarded as distinct. The success of the first meeting led to unanimous support for a Positive-Strand RNA Virus Symposium Series to be held triennially, alternating between Europe and America. The Second International meeting, held in Vienna, Austria in June 1989, included for the first time virologists from Eastern Europe and the USSR, due to political changes which had opened the borders between Western and Eastern Europe. The proceedings of the second meeting, which were published in 1990 (Brinton MA, Heinz FX (eds) New Aspects of Positive-Strand RNA Viruses. ASM Press, Washington, D.C.), emphasized the emergence of efforts to develop novel antiviral compounds, as well as new insights into the mechanisms of drug resistance and the role of recombination and gene pool mixing in the evolution of RNA viruses. Previously unsuspected evolutionary relationships between viruses were also revealed.

One of the dominant themes emerging from the third meeting held in Clearwater, Florida in 1992 was the application of new knolwedge on the genetics and structure of viruses to genetic engineering and to the design of new antiviral drugs. The analysis of the picornaviral internal ribosome entry site (IRES), the host proteins which interact with the IRES region, and the utilization of the IRES element to express foreign proteins are described. The use of the nodaviral RNA polymerase to amplify transcripts expressed from vaccinia vectors represents a novel and potentially powerful approach to enhancing foreign protein expression in eukaryotic cells. Identification of cellular genes, such as ubiquitin, inserted into RNA genome of pestivirus, bovine diarrheal virus, as the part of the development of pathogenic strains provides a particularly graphic example of how RNA viruses can evolve by capturing cellular genes via copy-choice recombination during a persistent infection in a host. Other notable contributions include discovery of the membrane dependence of RNA-dependent RNA replicase activity, the molecular characterization of several new viruses (Hepatitis C virus, Astroviruses, Calciviruses, and Lelystad virus), the demonstration that RNA viruses can persist in host cells for very long periods as free nucleic acid, the use of an infectious clone as a vaccine, and apoptosis as a mechanism of cell killing in virus-infected cells.

New developments in genetics and X-ray crystallography provide ever more sophisticated insights into virus structure, RNA-protein interactions critical to viral RNA replication and translation, and mechanisms of resistance to antiviral drugs. These insights, in turn, provide new molecular targets for the design of future antiviral drugs specifically targeted to the coat genes, and key regulatory enzymes of pathogenic positive-strand RNA viruses such as human immunodeficiency virus, poliovirus, the common cold viruses (rhinoviruses), and many others.

This book is a tribute to the restless spirit and searching imagination of humankind and to the societal wisdom of sustained financial support for medical research. The Third International Symposium on Positive-Strand RNA Viruses would not have been possible without substantial financial support from: Immuno AG; National Institute of Allergy and Infectious Diseases; National Cancer Institute; National Heart, Lung and Blood Institute; Sterling Drug Company; Institute for Molecular Biologicals, Miles Research Center; Boehringer Ingelheim Pharmaceuticals, Inc.; Eli Lilly and Company; SmithKline Beecham; Merck Corporation; and Pasteur Merieux. We gratefully acknowledge their support.

M.A. Brinton
C.A. Calisher
R. Rueckert

Contents

Keynote address

Strategies for control of virus diseases

Molecular aspects of pathogenesis and virulence

Genome replication and transcription

RNA recombination

RNA-protein interactions and host-virus interactions

Contents   ix

Protein expression and virion maturation

RNA replication

Virus receptors

x   Contents

Virus structure and assembly

Listed in Current Contents

Arch Virol (1994) [Suppl] 9: 1–8

Archives
of
Virology

© Springer-Verlag 1994
Printed in Austria

The importance of antigenic variation in vaccine design

Keynote Address

F. Brown

USDA Plum Island Animal Disease Center, Greenport, New York, U.S.A.

It is an honor for me to give the Keynote Address at the Third International Symposium on Positive Strand RNA Viruses because I am following two major figures in positive strand RNA virus research. The Keynote Addresses at the two previous meetings were given by David Baltimore in 1986 and Paul Kaesberg in 1989. The first of these addresses was a highly detailed description of the genetics of poliovirus. The second was rather more philosophical and historical, pointing out important relationships between the animal and plant viruses coming under the general umbrella of positive strand RNA viruses, but stopping at events before 1975.

My original inclination was to speak about the history of virology as it pertains to this general group of viruses. However, as this seemed to be too similar to what Paul Kaesberg had said at the last meeting in Vienna, I decided that, given the time available, I would rather look forward than backward, and speculate on the uses we will be making of the knowledge and techniques that have accumulated during recent years.

Nevertheless, I think it is appropriate, especially at a positive strand RNA meeting, to mention that exactly 100 years ago, Dimitri Ivanovski [16] described the filterable nature of tobacco mosaic virus. Indeed, the celebration of this event starts in St. Petersburg on September 21, 1992, that is, the day after this address. Although I could argue that Martinius Beijerinck [4] was the true father of virology, when he described what in essence was the molecular nature of viruses, nevertheless Ivanovski made a major contribution to virology.

Following the descriptive period, in which it was firmly established that many of the diseases we encounter are caused by filterable agents, there was a prolonged lull until the impact of chemistry started to be felt, first with Stanley's historic work [28] with tobacco mosaic virus and then with Schlesinger's work with phage [25] in the mid 1930s. Nevertheless the biological implications were not felt until later because in the 1930s the key role of nucleic acids was not appreciated, even though Bawden et al. [2] had demonstrated the presence of RNA in tobacco mosaic virus. However, the key finding by Avery et al. [1] that the transforming principle in pneumococcus is DNA focussed attention on the nucleic acids and soon led to the discovery that the infectivity and information base of micro-organisms is located in their nucleic acid. The experiments of Hershey and Chase [13] with bacteriophage T2 and those of Fraenkel-Conrat and Williams [10] in California, and Gierer and Schramm [12] in Tubingen with the infectious RNA of tobacco mosaic virus were part of the realization that the nucleic acids are central to our understanding, not only of virology, but of biology in general. These key observations have led to a transformation in the way in which biology is studied and have produced a mountain of information on nucleic acids. Sequence data abound, but now we need to interpret what the sequences are trying to tell us. Our task is to come to terms with what is often called the second half of the genetic code.

With positive strand RNA viruses we are in a very favorable position to pursue this task because they have small genomes, coding for a small number of proteins. This allows us to study important fundamental biological problems, such as replication and translation, with simple systems. These studies were catalyzed by the demonstration by Racaniello and Baltimore [23] that the complementary DNA of poliovirus RNA is infectious and can be manipulated in the same way as DNA viruses and plasmids. This discovery is, in my opinion, the single most important event in positive sense RNA virology in the past 20 years.

Viruses were first studied because of the diseases they cause and, indeed, the first attempts at classification were based on the signs or symptoms of the individual diseases (Borrel [5]; Holmes [14]). Such classification schemes resulted in intriguing groupings, such as vesicular stomatitis virus (a rhabdovirus), vesicular exanthema virus (a calicivirus) and foot-and-mouth disease (FMD) virus (a picornavirus), all of which cause vesicular diseases of farm animals. When the characterization of viruses by physico-chemical methods became possible, it was soon apparent that a classification system based on disease signs or symptoms was misleading, to say the least. So the systems devised by Lwoff and Wildy and their colleagues took over (Lwoff and Tournier [18]; Gibbs et al. [11]). Now that more and more sequences are becoming available, interesting relationships are coming to light and those interested in taxonomy and evolution now have a data base which should allow them to make endless speculations on this subject; so light should soon replace hot air when the evolution of viruses is discussed.

But at a more practical level, most of us want to know how viruses invade their hosts and cause disease. This means that at the outset we need to know about host specificity and host-cell interactions. With the increasing knowledge of host cell receptors and their interaction with viruses, a start has been made. But we need to know what happens with the intact animal or plant rather than with the highly artificial systems we use today. It is perhaps worthwhile mentioning that FMD virus, generally regarded as a pathogen for cattle, swine, sheep and goats, can infect and cause disease in humans – although the number of well-documented cases is very small (Hyslop [15]). Those few individuals must have had something different from most of us who have worked with the virus.

The enormous effort that has gone into determining the structures of several plant and animal viruses over the past 15 years is now starting to pay dividends. Moreover, since most of this structural information has been obtained with the small RNA viruses, the scientific groups which have solved these structures can offer leadership in important areas such as drug design. Such studies should also lend insight into the intriguing problem of particle stability which exists within the picornavirus family, in which the structures have been shown to be so similar by X-ray crystallography. Polioviruses and hepatitis A virus are stable in dilute acid (ca. pH2) whereas the rhinoviruses are unstable at pH5 and FMD virus disintegrates below pH7. Structural studies should help us solve this problem and also provide the reason for the stability of an acid resistant mutant of FMD virus, which we have isolated recently (Brown et al. [6]).

At a more immediate practical level, particularly when vaccination is being considered, there is the problem of antigenic variation for the X-ray crystallographers to study. Antigenic variation appears not to be a problem with hepatitis A or the cardioviruses. Even with poliovirus the strains of the virus used to prepare the vaccines 30 to 40 years ago are used in current vaccines and appear to control the disease effectively, especially in the developed countries. The situation with the rhinoviruses and FMD virus is very different. It is difficult to assess the situation with the rhinoviruses because there is no animal model which enables us to relate antigenic differences to protection studies.

However, the variation in FMD viruses has been studied extensively because of the importance of the disease they cause. The multiplicity of antigenic variants of FMD virus is so great that it is a manufacturer's headache, or even nightmare, to produce the appropriate vaccine. Not only are there seven serotypes, between which there is no cross-protection, but even within the serotypes there is sufficient antigenic variation to make it necessary to compare the antigenic properties of the virus causing any new outbreak in the field with those of the vaccine strain in order to predict the efficacy of existing vaccines. Sometimes the antigenic divergence is so great that it is necessary to produce a completely new vaccine to control an outbreak.

This situation is also compounded by the fact that passaging FMD viruses in tissue culture cells can lead to selection of antigenic variants that are neutralized much less well by antiserum against the parent virus than is the parent virus. This selection was demonstrated with viruses belonging to serotypes O, A and Asia 1, and presumably occurs with viruses belonging to the other serotypes (Dinter et al. [9]; Meloen [21]; Martinsen [19]; McVicar and Sutmoller [20]; Cowan et al. [8]). Since the production of vaccines is made using viruses which have been adapted to grow well in cell culture, the potential problem of selecting antigenic variants is clear.

I have been interested in the structural basis of this antigenic diversity for several years, continuing advances in physical-chemical methods allowing us to study the problem in some detail. The basic premise, based on thousands of observations, is that protection against infection is generally correlated with the level of neutralizing antibody at the time of challenge. The virus consists of one molecule of single stranded positive sense RNA, m.wt. 2.5×10^6 and 60 copies of each of four proteins VP1–VP4. Proteins VP1–3 have m.wt. of ca. 24×10^3 and VP4 has a m.wt. of ca. 10×10^6. Treatment of the virus with trypsin, which cleaves capsid protein VP1, drastically reduces its infectivity and in some cases its immunogenicity. This cleavage occurs at amino acids 138 and 154 of VP1 for viruses of serotype O, removing a fragment of 16 amino acids. Synthetic peptides corresponding to this region are highly immunogenic and appear to represent the immunodominant site.

In extending the work to three viruses of other FMD serotypes we found that the same region of VP1 gave good levels of neutralizing antibody with two of these serotypes. But the third, the exception, serves to pinpoint the potential dangers inherent in the way viruses are selected for vaccine production. The antibody elicited by a peptide corresponding to the 141–160 region of a virus of serotype A, isolated from an outbreak in the United Kingdom in 1932, neutralized the parent virus only poorly. This virus had been used in many studies in cattle at the Pirbright Institute in England and was the one used in the early experiments at the Plum Island Animal Disease Center when it was opened in 1953. It was there that Cowan, Martinsen, McVicar and their colleagues (see [7, 19, 20]) had found that, by passage in tissue culture cells, antigenic variants could be selected which were neutralized by antiserum to the parent virus much less well than the parent virus itself.

Table 1. Sequence variation at amino acids 148 and 153 of capsid protein VP1 of viruses (serotype A) isolated from cattle in a single outbreak of FMD

Residue	
148	153
Ser	Leu
Leu	Pro
Ser	Ser
Phe	Pro
Phe	Leu
Phe	Ser[a]
Phe	Gln[a]
Leu	Leu
Ser	Pro
Leu	Gln
Val	Leu

[a] Isolated after passage in presence of monoclonal antibody

The sequence of the virus RNA on which the peptide synthesis had been based had been determined by the Plum Island scientists in collaboration with a group from Genentech [17]. Significantly, the virus used for these studies was neutralized extremely well by the anti-peptide antiserum. Meanwhile, at Pirbright we were encountering problems in sequencing the RNA of the virus at the positions coding for amino acids 148 and 153 of VP1. It was soon established by plaque picking and sequencing, however, that the virus being used at Pirbright, only one passage from the tongue epithelium, was a mixture of at least three viruses differing only at these two positions in the entire capsid protein. Moreover, these viruses could be readily distinguished by serological tests. The amino acid changes are shown in Table 1. Subsequent circular dichroism spectroscopic studies revealed clear structural differences in the peptides, which accounted for the serological differences (Siligardi et al. [26, 27]).

This work has been extended to variants which had been isolated at Plum Island, either by passage in cell culture (Cowan [7]; Moore et al. [22]) or in the presence of monoclonal antibodies against the 141–160 region (Baxt et al. [3]). So far 11 variants have been isolated, all of which differ at amino acids 148 and 153. Serological and structural analyses are still in progress but even at this stage of the work it is possible to draw the tentative conclusion that the presence of Pro at 153 and Leu at 148 has a major influence on the structure of the peptides.

F. Brown

These observations have clear implications for vaccine manufacture. If the virus causing an outbreak is a mixture of significant antigenic variants, as it was with the virus described in this paper, and the antigenic differences are also significant in terms of cross protection, it is important to determine whether the virus used for vaccine production is sufficiently closely related to the outbreak virus to afford protection and does not represent only a small proportion of the diverse population in the initial isolate. In practice, this is usually done by determining whether the antibody elicited by the vaccine neutralizes the outbreak strain, ideally as efficiently as it neutralizes the virus from which the vaccine was produced. But we have found that even quite small changes in culture conditions can lead to a preponderance of different variants in a virus harvest (P. Piatti et al., unpubl.). Consequently each batch of virus to be used for vaccine formulation must be tested for its serological specificity and for its ability to confer protection against the strain responsible for the outbreak.

In conclusion, I would like to speculate on whether it will be possible to construct a peptide which would afford protection to all the serotypes of FMD virus. The 141–160 region of VP1 contains an Arg Gly Asp sequence which is present in the vast majority of FMD viruses that have been sequenced. This tripeptide is involved in attachment of the virus to susceptible cells. If we could construct a peptide allowing this sequence to occupy a dominant position, it is possible that we could use it to elicit extensive cross-reaction between serotypes. Structural studies that define antigenic variation could well provide the information necessary to design such a peptide and, therefore, hasten the day when it will no longer be necessary to make different vaccines against the individual serotypes.

References

1. Avery OT, MacLeod CM, McCarty M (1944) Studies on the chemical nature of the substance inducing transformation of pneumococcal types. Induction of transformation by a desoxyribonucleic acid fraction isolated from Pneumococcus Type III. J Exp Med 79: 137–158
2. Bawden FC, Pirie NW, Bernal JD, Fankuchen I (1936) Liquid crystalline substances from virus-infected plants. Nature 138: 1051–1052
3. Baxt B, Vakharia V, Moore DM, Franke AJ, Morgan DO (1989) Analysis of neutralizing antigenic sites on the surface of type A12 foot-and-mouth disease virus. J Virol 63: 2143–2151
4. Beijerinck MW (1898) Uber ein contagium fluidum als Ursache der Fleckenkrankheit der Tabaksblätter. Verh K Ned Akad Wet 65: 3–21
5. Borrel A (1903) Epithelioses infectieuses et epitheliomas. Ann Inst Pasteur 17: 81–118
6. Brown F (1993) Unpublished data

7. Cowan KM (1969) Immunochemical studies of foot-and-mouth disease virus. V. Antigenic variants of virus demonstrated by immunodiffusion analyses with 19S but not 7S antibodies. J Exp Med 129: 333–350

8. Cowan KM, Erol N, Whitehead AP (1974) Heterogeneity of type Asia 1 foot-and-mouth disease virus and BHK 21 cells and the relationship to vaccine preparation. Bull Off Int Epiz 81: 1271–1298

9. Dinter Z, Philipson L, Wessten T (1959) Properties of foot-and-mouth disease virus in tissue culture. Arch Virusforsch 9: 411–427

10. Fraenkel-Conrat H, Williams RC (1955) Reconstitution of active tobacco mosaic virus from its inactive protein and nucleic acid components. Proc Natl Acad Sci USA 41: 690–698

11. Gibbs AJ, Harrison BD, Watson DH, Wildy P (1966) What's in a virus name? Nature 209: 450–454

12. Gierer A, Schramm G (1956) Infectivity of ribonucleic acid from tobacco mosaic virus. Nature 177: 702–703

13. Hershey AD, Chase M (1952) Independent functions of viral protein and nucleic acid in growth of bacteriophage. J Gen Physiol 36: 39–56

14. Holmes FO (1948) Order virales. The filterable viruses. In: Breed RS (ed) Bergey's manual of determinative bacteriology, 6th edn. Ballière, Tindall and Cox, London, pp 1127–1286

15. Hyslop NStG (1973) Transmission of the virus of foot-and-mouth disease between animals and man. Bull World Health Organ 49: 577–585

16. Ivanovski DJ (1892) Ueber die Mosaikkrankheit der Tabakspflanze. St Petersb Acad Compl Sci Bull 35: Ser 4, 3, 67

17. Kleid DG, Yansura D, Small B, Dowbenko D, Moore DM, Grubman MJ, McKercher PD, Morgan DO, Robertson BH, Bachrach HL (1981) Cloned viral protein vaccine for foot-and-mouth disease; responses in cattle and swine. Science 214: 1125–1129

18. Lwoff A, Tournier P (1966) The classification of viruses. Annu Rev Microbiol 20: 45–74

19. Martinsen JS (1972) Neutralizing activity of sera from guinea pigs inoculated with foot-and-mouth disease virus variants. Res Vet Sci 13: 97–99

20. McVicar JW, Sutmoller P (1972) Three variants of foot-and-mouth disease virus, type O: cell culture characteristics and antigenic differences. Am J Vet Res 33: 1627–1633

21. Meloen RH (1976) Localization on foot-and-mouth disease virus (FMDV) of an antigenic deficiency induced by passage in BHK cells. Arch Virol 59: 299–306

22. Moore DM, Vakharia VN, Morgan DO (1989) Identification of virus neutralizing epitopes on naturally occurring variants of type A12 foot-and-mouth disease virus. Virus Res 14: 281–295

23. Racaniello VR, Baltimore D (1981) Cloned poliovirus complementary DNA is infectious in mammalian cells. Science 214: 916–919

24. Rowlands DJ, Clarke BE, Carroll AR, Brown F, Nicholson BH, Bittle JL, Houghten RA, Lerner RA (1983) Chemical basis of antigenic variation in foot-and-mouth disease virus. Nature 306: 694–697

25. Schlesinger M (1936) The Feulgen reaction of the bacteriophage substance. Nature 138: 508–509

26. Siligardi G, Drake AF, Mascagni P, Rowlands DJ, Brown F, Gibbons WA (1991a) A CD strategy for the study of polypeptide folding/unfolding. Int J Pept Protein Res 38: 519–527

27. Siligardi G, Drake AF, Mascagni P, Rowlands DJ, Brown F, Gibbons WA (1991b) Correlations between the conformations elucidated by CD spectroscopy and the antigenic properties of four peptides of the foot-and-mouth disease virus. Eur J Biochem 99: 445–451

28. Stanley WM (1935) Isolation of a crystalline protein possessing the properties of tobacco mosaic virus. Science 81: 644–645

Author's address: Dr. F. Brown, USDA Plum Island Animal Disease Center, Greenport, NY 11944, U.S.A.

Strategies for control of
virus diseases

Arch Virol (1994) [Suppl] 9: 11–17

Archives
of
Virology
© Springer-Verlag 1994
Printed in Austria

The genetic and functional basis of HIV-1 resistance to nonnucleoside reverse transcriptase inhibitors

E. A. Emini, V. W. Byrnes, J. H. Condra, W. A. Schleif, and **V. V. Sardana**

Merck Research Laboratories, West Point, Pennsylvania, U.S.A.

Summary. The nonnucleoside reverse transcriptase (RT) inhibitors are structurally diverse compounds that are specific inhibitors of the human immunodeficiency virus type 1 RT enzyme. The compounds are largely functionally identical and bind to a common site in the enzyme. HIV-1 variants that exhibit reduced susceptibility to these inhibitors have been derived in cell culture and, more recently, from HIV-1-infected patients undergoing experimental therapy. The variants express amino acid substitutions at RT positions that apparently interact directly with the inhibitors. Effects of specific substitutions at these positions vary among the compounds, suggesting subtle differences in how the compounds physically interact with the enzyme.

Introduction

Human immunodeficiency virus type 1 (HIV-1), a member of the *Lentivirinae* subfamily of the family *Retroviridae*, is the etiologic agent of acquired immunodeficiency syndrome [1, 2]. Upon introduction into the host, the virus establishes a long-term persistent infection that is eventually manifest as clinical disease. The search for therapeutic agents to treat chronically infected patients has focused on the development of compounds that prevent productive virus infection. As with all retroviruses, the productive infection of HIV-1 depends on transcription of the viral genomic RNA into DNA and subsequent integration of viral DNA into the host cell genome.

Transcription of viral RNA is mediated by the virus-encoded enzyme reverse transcriptase (RT). Inhibitors of this enzyme, such as the nucleoside analogs 3′-azido-3′-deoxythymidine (AZT) and dideoxyinosine (ddI), have potent antiviral activity in vitro and are therapeutically useful in infected patients [3, 4]. Unfortunately, long-term use of these compounds is limited by their toxicity and by the emergence of HIV-1 variants that are resistant to them [5–9].

RT1 IC$_{50}$ (μM)

L-697,661   0.052 ± 0.004

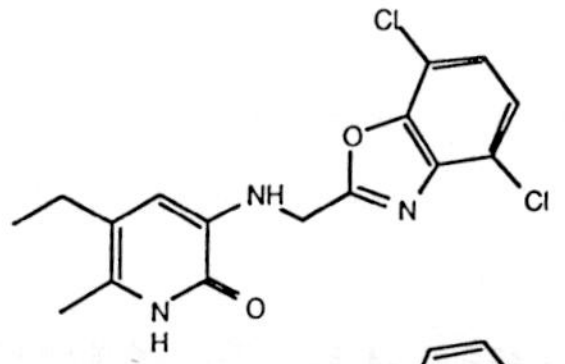

L-696,229   0.039 ± 0.006

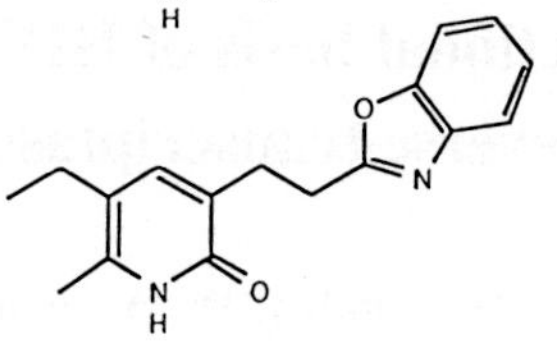

R82913
(Cl-TIBO)   0.381 ± 0.024

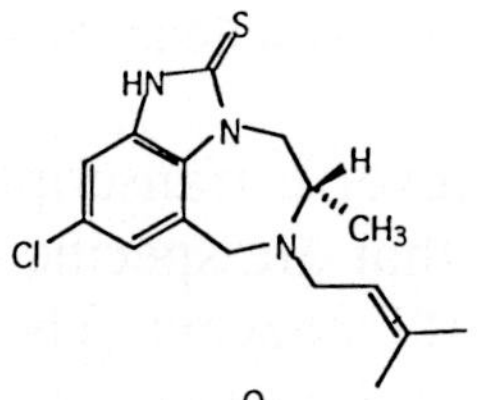

BI-RG-587
(Nevirapine)   0.175 ± 0.014

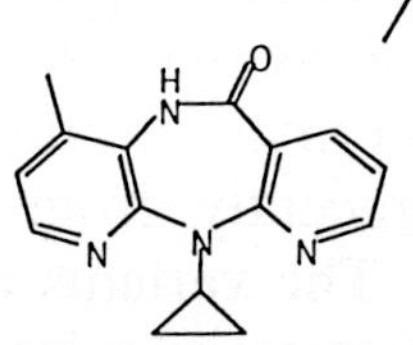

Fig. 1. Structures of the nonnucleoside inhibitors. IC$_{50}$ values, determined using HIV-1 RT, are from Condra et al. [20] and Sardana et al. [21]

Recently, novel classes of nonnucleoside RT inhibitors were described (Fig. 1). These include the pyridinone derivatives L-697,661 and L-696,229 [10, 11], BI-RG-587 (nevirapine, 12) and the TIBO compounds R82150 and R82913 [13, 14]. These inhibitors are structurally distinct, but share functional features. All are non-competitive in their activity and all are specific for the HIV-1 RT; none inhibit the RT from HIV-2. These compounds are active against AZT-resistant variants of HIV-1 and are synergistic for virus inhibition in cell culture when used in combination with AZT or ddI [10, 15].

Viral and RT resistance to nonnucleoside inhibitors

Nunberg et al. [16] first defined virus resistance to nonnucleoside inhibitors by serially passing wild-type sensitive virus in the presence of increasing concentrations of a pyridinone inhibitor. The resulting virus population exhibited a 1000-fold decrease in sensitivity to L-697,661 and L-696,229 as well as to BI-RG-587 and TIBO. These observations provided evidence for the functional identity of the structurally different inhibitors. Subsequently, HIV-1 variants resistant to BI-RG-587 also were obtained [17]. In both cases, the resistant virus variants retained their sensitivity to AZT and ddI.

These initial studies mapped the resistant phenotype to amino acid substitutions at RT residues 181 (tyr→cys) and 103 (lys→asn). In addition, photoaffinity labelling experiments with a derivative of BI-RG-587 demonstrated that the tyr residues at positions 181 and 188 both interact with the inhibitor [18]. The significance of these residues was confirmed by the genetic studies of Shih et al. [19] and Condra et al. [20].

We attempted to further understand the basis of resistance to the nonnucleoside inhibitors by introducing a series of specific amino acid substitutions into recombinantly expressed RT and, subsequently, assessing the effects of these substitutions on the enzyme's sensitivity [21]. Data from these studies are summarized in Table 1. Substitutions

Table 1. Fold-differences in IC_{50} of inhibition resulting from specific amino acid substitutions at positions 103, 181 and 188[a]

	Inhibitor			
	L-697,661	L-696,229	BI-RG-587	R82913
Amino acid at Position 181				
Tyr[b]	1.0	1.0	1.0	1.0
Cys	>200	223.7 (±60.1)	113.2 (±25.3)	13.6 (±3.9)
Ser	33.0 (±10.7)	153.0 (±29.9)	34.8 (±1.2)	11.1 (±0.5)
His	17.8 (±3.5)	31.9 (±12.9)	17.0 (±5.0)	4.8 (±0.0)
Phe	1.0 (±0.1)	0.6 (±0.1)	1.5 (±0.2)	0.9 (±0.3)
Trp	3.7 (±0.5)	16.8 (±5.3)	10.3 (±0.9)	5.4 (±0.7)
Position 188				
Tyr[b]	1.0	1.0	1.0	1.0
Cys	2.2 (±0.4)	31.0 (±6.1)	81.1 (±9.9)	22.5 (±5.0)
His	6.2 (±1.1)	19.8 (±4.1)	1.1 (±0.4)	20.9 (±2.9)
Phe	1.4 (±0.5)	0.7 (±0.2)	0.6 (±0.0)	1.8 (±0.2)
Position 103				
Lys[b]	1.0	1.0	1.0	1.0
Asn	10.4 (±1.7)	15.9 (±3.1)	47.1 (±27.9)	31.6 (±0.9)
Combination Substitutions				
103 (Asn) + 181 (Cys)	>200	571.1 (±79.6)	>1 700	>790
103 (Asn) + 188 (Cys)	39.6 (±6.8)	>1 000	>1 700	>790

[a] Fold differences in IC_{50} are reported as geometric means (± geometric standard errors) of multiple determinations. Data are from Sardana et al. [21]

[b] Wild-type residue

were introduced at RT residues 103, 181 and 188. As would be antic-ipated given the structural diversity of the inhibitors, each substitution exhibited varying effects on the different inhibitors. For example, the tyr→cys replacement at position 181 engendered significantly less re-sistance against the TIBO R82913 compound than it did against the pyridinones. This suggests that each structural class of inhibitor interacts with the RT in a characteristic and specific fashion. Also, different amino acid substitutions yielded different resistance effects. The cys replacement at position 181 had the most noted effect while, in contrast, no effect was seen following replacement of the tyr residue at either 181 or 188 with a phe residue. This latter result suggests that the activity of the inhibitors may depend on aromatic stacking of the amino acid side groups at these two residue positions. Finally, the data suggest functional interdependence of RT residue 103 with the 181 and 188 residues. While the substitution of asn for lys at 103 engenders resistance to the nonnucleoside inhibitors, this substitution, in conjunction with replacements at 181 and 188, resulted in a greater than additive loss of sensitivity.

A further attempt was made to define additional amino acid residues that might influence sensitivity by once again serially passing wild-type virus in increasing concentrations of L-697,661 (according to the method described in [16]). Viral clones were derived following the fourth and sixth passages. These clones exhibited varying degrees of resistance (data not shown). Sequencing of the RT coding regions yielded the presence of two novel amino acid substitutions at residues 100 (leu→ile) and 108 (val→ile). Both substitutions were independently introduced into recombinantly expressed RT and both were indeed found to mediate resistance to the different nonnucleoside inhibitors (Table 2). Again, combinations of substitutions within the same RT molecule generally resulted in a greater than additive degree of resistance.

Following our functional definition of the interdependence of the RT region surrounding residues 103 to 108 and the region encompassing residues 181 and 188, Kohlstaedt et al. [22] published a low-resolution crystallographic structure of the HIV-1 RT. Appropriate crystals for these studies were obtained by incorporating the nonnucleoside inhibitor nevirapine (BI-RG-587) into the RT molecule. The structure indicated that the inhibitor was bound in a pocket located in the so-called "palm" region of the RT. The pocket is formed largely by the interaction of three strands. Two of the strands contain the 181 and 188 residues, respectively, while the third encompasses the region from approximately residues 100 to 110. The three strands appear to interact so as to locate the invariant asp residues at positions 110, 185 and 186 in proximity to each other. These residues may coordinate the Mg^{++} ion that is essential

Table 2. Fold-differences in IC_{50} of inhibition resulting from specific amino acid substitutions at positions 100 and 108[a]

Amino acid substitution	Inhibitor			
	L-697,661	L-696,229	BI-RG-587	R82913
100 (Leu)[b]	1.0	1.0	1.0	1.0
100 (Ile)	13.1 (±2.9)	2.4 (±0.0)	4.5 (±0.6)	41.0 (±8.5)
100 (Ile) + 103 (Asn)	10.8 (±2.6)	20.0 (±4.1)	93.0 (±38.0)	>790
108 (Val)[b]	1.0	1.0	1.0	1.0
108 (Ile)	2.3 (±0.6)	3.4 (±0.0)	1.6 (±0.4)	2.2 (±0.1)
108 (Ile) + 103 (Asn)	16.4 (±4.0)	94.5 (±32.8)	103.8 (±26.4)	196.2 (±22.7)

[a] Fold differences in IC_{50} are reported as geometric means (± geometric standard errors) of multiple determinations. For methods see Sardana et al. [21]

[b] Wild-type residue

for the enzyme's activity. Hence, all of the residues that are defined as mediating resistance to the nonnucleoside inhibitors are located within the apparent inhibitor binding pocket and, therefore, probably mediate their effect by directly altering the interaction of the inhibitor with its binding site.

Finally, L-697,661 was assessed in a human clinical trial for in vivo antiviral activity. The results of the trial suggested that the compound was effective in decreasing the quantity of circulating cell-free virus (unpubl. obs.). However, the effect was transient (lasting only several weeks) due to the rapid selection for resistant virus variants (unpubl. obs.). Sequence analysis showed that resistance was predominantly mediated by the previously defined substitutions at RT residue positions 103 and 181. The in vitro derivation and analysis of inhibitor-resistant virus variants had accurately predicted the nature and degree of resistance that was ultimately observed in the clinic. Indeed, the in vitro studies provided a basis of understanding that resulted in a remarkably fast assessment of resistance selection in vivo.

In conclusion, the nonnucleoside reverse transcriptase inhibitors comprise a structurally diverse class of small compounds that are non-competitive in their activity and that are highly specific for the HIV-1 RT. Resistance to the inhibitors arises readily in cell culture and in treated infected persons. The resistance is mediated by amino acid substitutions within the enzyme's inhibitor-binding site and effects of individual substitutions on the different inhibitors vary. It is hoped that by understanding the structural and functional basis of the resistance eventually we will be able to develop multiple nonnucleoside inhibitors that are defined by non-overlapping resistances. Combination therapy with such a set of inhibitors may result in a long-term, effective anti-HIV-1 therapy.

Acknowledgement

The authors thank D. Wilson for preparation of the manuscript.

References

1. Barre-Sinoussi F, Chermann JC, Rey F, Nugeyre MT, Chamaret S, Gruest J, Dauguet C, Axler-Blin C, Vezinet-Brun F, Rouzioux C, Rozenbaum W, Montagnier L (1983) Isolation of a T-lymphotropic retrovirus from a patient at risk for acquired immunodeficiency syndrome (AIDS). Science 220: 868–871
2. Popovic M, Sarngadharan MG, Read E, Gallo R (1984) Detection, isolation, and continuous production of cytopathic retroviruses (HTLV-III) from patients with AIDS and pre-AIDS. Science 224: 497–500
3. Fischl MA, Richman DD, Grieco MH, Gottlieb MS, Volberding PA, Laskin OL, Leedom JM, Groopman JE, Mildvan D, Schooley RT, Jackson GG, Durack DT, King D, AZT Collaborative Working Group (1987) The efficacy of azidothymidine (AZT) in the treatment of patients with AIDS and AIDS-related complex. N Engl J Med 317: 185–191
4. Butler KM, Husso RN, Balis FM, Brouwers P, Eddy J, El-Amin D, Gress J, Hawkins M, Jarosinski P, Moss H, Poplack D, Santacroce S, Venzon D, Wiener L, Wolters P, Pizzo PA (1991) Dideoxyinosine in children with symptomatic human immunodeficiency virus syndrome. N Engl J Med 324: 137–144
5. Larder BA, Darby G, Richman DD (1989) HIV with reduced sensitivity to Zidovudine (AZT) isolated during prolonged therapy. Science 243: 1731–1734
6. Boucher CAB, Tersmette M, Lange JMA, Kellam P, deGoede REY, Mulder JW, Darby G, Goudsmit J, Larder BA (1990) Zidovudine sensitivity of human immunodeficiency viruses from high-risk, symptom-free individuals during therapy. Lancet 336: 585–590
7. Land S, Treloar G, McPhee D, Birch C, Doherty R, Cooper D, Gust I (1990) Decreased in vitro susceptibility to zidovudine of HIV isolates obtained from patients with AIDS. J Infect Dis 161: 326–329
8. Richmann DD, Grimes JM, Lagakos SW (1990) Effect of stage of disease and drug dose of zidovudine susceptibilities of isolates of human immunodeficiency virus. J Acquir Immune Defic Syndr 3: 743–746
9. St Clair MH, Martin JL, Tudor-Williams G, Bach MC, Vavro CL, King DM, Kellam P, Kemp SD, Larder BA (1991) Resistance to ddI and sensitivity to AZT induced by a mutation in HIV-1 reverse transcriptase. Science 253: 1557–1559
10. Goldman ME, Nunberg JH, O'Brien JA, Quintero JC, Schleif WA, Freund KF, Gaul SL, Saari WS, Wai JS, Hoffman JM, Anderson PS, Hupe DJ, Emini EA, Stern AM (1991) Pyridinone derivatives: specific human immunodeficiency virus type 1 reverse transcriptase inhibitors with antiviral activity. Proc Natl Acad Sci USA 88: 6863–6867
11. Saari WS, Hoffman JM, Wai JS, Fisher TE, Rooney CS, Smith AM, Thomas CM, Goldman ME, O'Brien JA, Nunberg JH, Quintero JC, Schleif WA, Emini EA, Stern AM, Anderson PS (1991) 2-pyridinone derivatives: a new class of

nonnucleoside, HIV-1 specific reverse transcriptase inhibitors. J Med Chem 34: 2922–2925

12. Merluzzi VJ, Hargrave KD, Labadia M, Grozinger K, Skoog M, Wu JC, Shih C-K, Eckner K, Hattox S, Adams J, Rosenthal AS, Faanes R, Eckner RJ, Koup RA, Sullivan JL (1990) Inhibition of HIV-1 replication by a nonnucleoside reverse transcriptase inhibitor. Science 250: 1411–1413

13. Pauwels R, Andries K, Desmyter J, Schols D, Kukla MJ, Breslin HJ, Raeymaeckers A, van Gelder J, Woestenborghs R, Heykants J, Schellekens K, Janssen MAC, DeClerq E, Jannsen PA (1990) Potent and selective inhibition of HIV-1 replication in vitro by a novel series of TIBO derivatives. Nature 343: 470–474

14. White EL, Buckheit RW, Ross LJ, Germany JM, Andries K, Pauwels R, Janssen PAJ, Shannon WM, Chirigos MA (1991) A TIBO derivative, R82193, is a potent inhibitor of HIV-1 reverse transcriptase with heteropolymer templates. Antiviral Res 16: 257–266

15. Richman D, Rosenthal AS, Skoog M, Eckner RJ, Chou T-C, Sabo JP, Merluzzi VJ (1991) BI-RG-587 is active against zidovudine-resistant human immunodeficiency virus type 1 and synergistic with zidovudine. Antimicrob Agents Chemother 35: 305–308

16. Nunberg JH, Schleif WA, Boots EJ, O'Brien JA, Quintero JC, Hoffman JM, Emini EA, Goldman ME (1991) Viral resistance to human immunodeficiency virus type 1-specific pyridinone reverse transcriptase inhibitors. J Virol 65: 4887–4892

17. Richman D, Shih C-K, Lowy I, Rose J, Prodanovich P, Goff S, Griffin J (1991) Human immunodeficiency virus type 1 mutants resistant to nonnucleoside inhibitors of reverse transcriptase arise in tissue culture. Proc Natl Acad Sci USA 88: 11241–11245

18. Cohen KA, Hopkins J, Ingraham RH, Pargellis C, Wu JC, Palladino DEH, Kinkade P, Warren TC, Rogers S, Adams J, Farina PR, Grob PM (1991) Characterization of the binding site for nevirapine (BI-RG-587), a nonnucleoside inhibitor of human immunodeficiency virus type-1 reverse transcriptase. J Biol Chem 22: 14670–14674

19. Shih C-K, Rose JM, Hansen GL, Wu JC, Bacolla A, Griffin JA (1991) Chimeric human immunodeficiency virus type 1/type 2 reverse transcriptase display reversed sensitivity to nonnucleoside analog inhibitors. Proc Natl Acad Sci USA 88: 9878–9882

20. Condra JH, Emini EA, Gotlib L, Graham DJ, Schlabach AJ, Wolfgang JA, Colonno RJ, Sardana VV (1992) Identification of the human immunodeficiency virus reverse transcriptase residues that contribute to the activity of diverse nonnucleoside inhibitors. Antimicrob Agents Chemother 36: 1441–1446

21. Sardana VV, Emini EA, Gotlib L, Graham DJ, Lineberger DW, Long WJ, Schlabach AJ, Wolfgang JA, Condra JH (1992) Functional analysis of HIV-1 reverse transcriptase amino acids involved in resistance to multiple nonnucleoside inhibitors. J Biol Chem 267: 17526–17530

22. Kohlstaedt LA, Wang J, Friedman JM, Rice PA, Steitz TA (1992) Crystal structure at 3.5A resolution of HIV-1 reverse transcriptase complexed with an inhibitor. Science 256: 1783–1790

Authors' address: Dr. E. A. Emini, Merck Research Laboratories, WP16-225, West Point, PA 19486, U.S.A.

Arch Virol (1994) [Suppl] 9: 19–29

Archives
of
Virology

© Springer-Verlag 1994
Printed in Austria

Structure-based design of symmetric inhibitors of HIV-1 protease

J. Erickson[1] and **D. Kempf**[2]

[1] Structural Biochemistry Program, Frederick Biomedical Supercomputing Center, Program Resources, Inc., National Cancer Institute, Frederick Cancer Research and Development Center, Frederick, Maryland
[2] Pharmaceutical Products Division, Abbott Laboratories, Abbott Park, Illinois, U.S.A.

Summary. HIV-1, the causative agent of AIDS, encodes a protease that processes the viral polyproteins into the structural proteins and replicative enzymes found in mature virions. Protease activity has been shown to be essential for the proper assembly and maturation of fully infectious HIV-1. Thus, the HIV-1 protease (HIV PR) has become an important target for the design of antiviral agents for AIDS. Analysis of the three-dimensional structures of related aspartic proteinases, and later of Rous sarcoma virus protease, indicated that the active site and extended substrate binding cleft exhibits two-fold (C2) symmetry at the atomic level. We therefore set out to test whether compounds that contained a C2 axis of symmetry, and that were structurally complementary to the active site region, could be potent and selective inhibitors of HIV PR. Two novel classes of C2 or pseudo-C2 symmetric inhibitors were designed, synthesized and shown to display potent inhibitory activity towards HIV PR, and one of these, A-77003, recently entered clinical trials. The structure of the complex with A-74704 was solved using X-ray crystallographic methods and revealed a highly symmetric mode of binding, confirming our initial design principles. These studies demonstrate that relatively simple symmetry considerations can give rise to novel compound designs, allowing access to imaginative new templates for synthesis that can be translated into experimental therapeutic agents.

Introduction

One of the most active areas of antiviral research concerns efforts to inhibit the human immunodeficiency virus type 1 (HIV), the causative agent of AIDS. An explosion of information about the molecular virology of HIV has resulted in drug design strategies that target virtually

HIV-1 GENOME

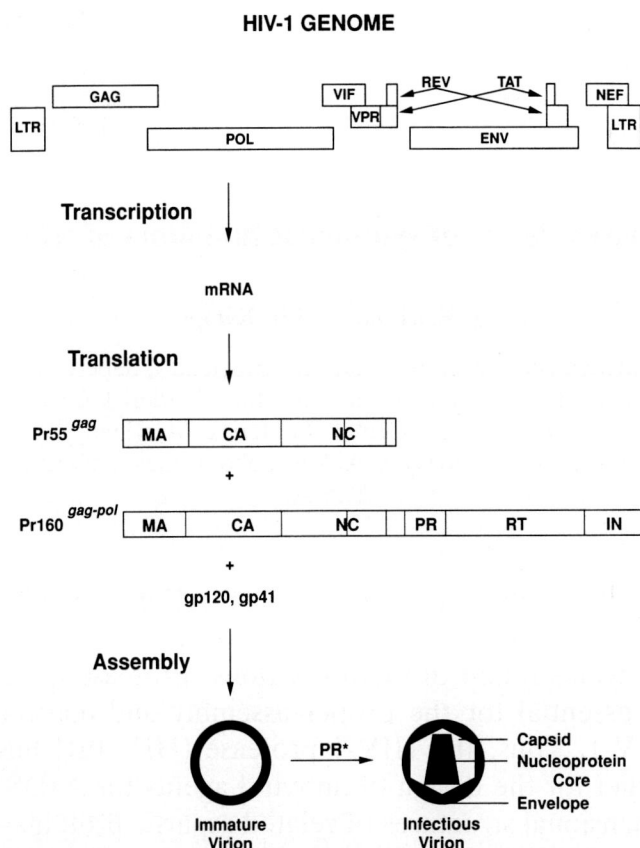

Fig. 1. Schematic diagram of the role played by HIV protease in viral maturation. Expression of the HIV genome during the acute stage of infection leads to the production of Pr55 gag and Pr160 gag-pol precursors. These polyproteins are packaged with viral RNA into immature virions. Maturation of newly-assembled particles into fully-infectious virions requires processing, or cleavage, of Pr55 gag and Pr160 gag-pol by the viral protease at discrete sites as indicated by the vertical, stippled lines

every aspect of the viral life cycle [1]. The HIV genome encodes a protease (HIV PR) that cleaves, or processes, viral *gag* and *gag-pol* protein precursors during virus assembly and maturation (Fig. 1). In 1988, the late I. Segal and co-workers observed that deletion mutagenesis of the HIV PR gene resulted in the production of non-infectious, immature virus particles [2]. This experiment demonstrated that HIV PR performs an essential function in the life cycle of HIV and thus makes this enzyme an important target for the design of specific antiviral agents for AIDS.

Structural considerations

The structure of HIV protease and its relationship to the aspartic protease family of enzymes has been reviewed recently [3]. Unlike the latter

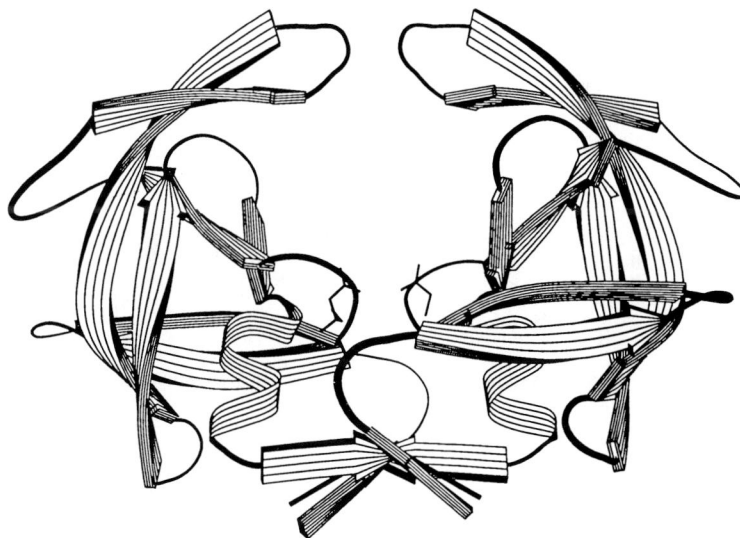

Fig. 2. Ribbon drawing of the secondary structural elements of the HIV protease dimer

single chain, bilobed enzymes, whose active sites are formed at the interface of the N- and C-domains, HIV PR exists as a dimer that can exhibit exact crystallographic, two-fold rotational (C_2) symmetry (Fig. 2). The active site is formed by the dimer interface and is composed of nearly equivalent contributions of residues from each subunit. The substrate binding cleft is bound on one side by the active site aspartic acids, Asp25 and Asp125, and on the other by a pair of two-fold related, anti-parallel β hairpin structures, or "flaps". In HIV PR, crystal packing forces maintain the flap in a conformation that is unsuitable for substrate binding [4].

Intensive drug design efforts for renin, an aspartic protease that plays a key role in hypertension in humans, have led to the development of substrate-based approaches to inhibitor design in which the scissile amide bond of a peptide substrate is replaced by a non-cleavable, transition state analogue or isostere (Fig. 3) (reviewed in [5]). The understanding that HIV PR may be a distant relative of the cellular aspartic proteases led, in turn, to the immediate application of similar strategies to design a variety of highly potent HIV PR inhibitors (reviewed in [6–8]). However, the usefulness of peptidomimetics as drug candidates has been hampered by their generally poor pharmacologic properties of oral bioavailability, metabolic stability, and pharmacokinetics. In an effort to explore alternative approaches to HIV PR inhibitor design, we utilized our knowledge of aspartic protease structure and function to conceptualize several novel structural classes of inhibitors based on the premise of active site symmetry.

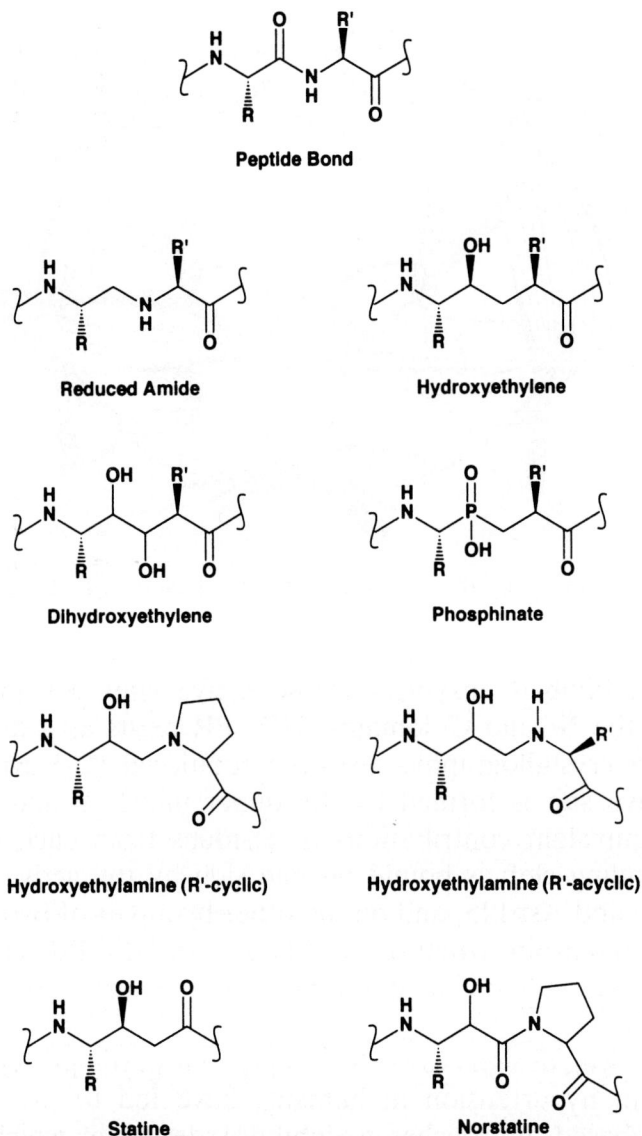

Fig. 3. Non-cleavable transition-state isosteres employed for the synthesis of HIV protease inhibitors. Reprinted from [13] with permission

Symmetry-based inhibitor design

Our design strategy had two requirements: first, that the inhibitor possess the same C_2 symmetry as the enzyme; and, second, that the symmetry elements of the inhibitor and enzyme approximately superimpose when the inhibitor is bound in the active site [9, 10]. We initially designed a C_2 symmetric diaminoalcohol in which the pseudo-C_2 axis passed through the alcohol carbon atom and bisected the O-C-H angle (Fig. 4A). Each side of the diaminoalcohol resembled a phenylalanine

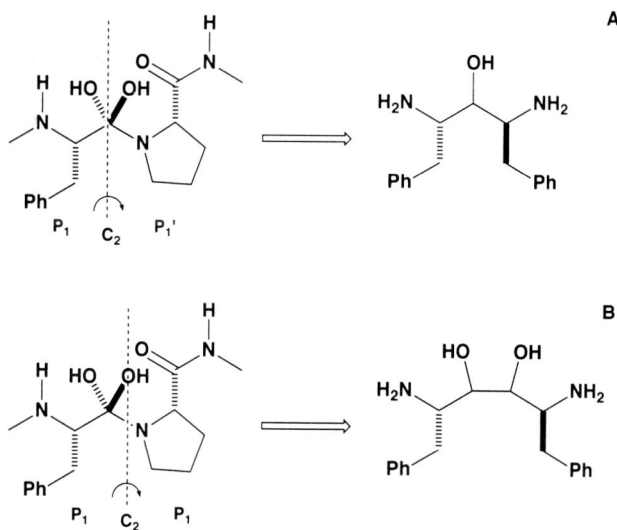

Fig. 4. Design of C_2 symmetric inhibitors of HIV protease. **A** Placement of the C_2-axis through the carbon atom produces the diaminoalcohol. **B** Placement of the C_2-axis through the midpoint of the C–N peptide bond produces the diaminodiol. Adapted from [10]

moiety which is a common P1 substituent for HIV PR substrates. This compound obviously satisfied the first constraint.

In order to determine whether the second requirement would be met by our design, we performed a modeling experiment using the crystal structure of a reduced peptide inhibitor complexed with rhizopuspepsin [11], a fungal aspartic protease, and the crystal structure of Rous sarcoma virus protease (RSV PR) which had just become available to us [12]. The active site regions of RSV PR and rhizopuspepsin were superimposed in order to "dock" the rhizopuspepsin-bound inhibitor into the active site of RSV PR. The C-terminal portion of the inhibitor was deleted beyond the CH_2-group of the reduced amide moiety, and the N-terminal half was rotated by the enzyme two-fold axis to produce a pseudo-C_2 symmetric inhibitor. The deviations from ideal geometry for the computer-generated inhibitor were small enough to suggest that the corresponding diaminoalcohol might bind favorably in the orientation as modeled in this experiment.

The prototype compound I was synthesized based on the fact that aromatic amino acid side chains are prevalent in the P1 position of naturally-occurring substrates for HIV PR (Fig. 5). This molecule, which closely resembles the central "core" structure of the computer-modeled inhibitor, is essentially C_2 symmetric except for the secondary OH group on the central carbon atom. Compound I was a weak inhibitor of HIV-1

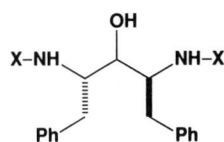

No.	X	IC$_{50}$ (nM)
I	H-	>10,000
II	Ac-	>10,000
III	Boc-	3,000
IV	H-Val	590
V	Ac-Val-	12
VI	Ac-Val-Val-	10
VII	Cbz-Val-	3

Fig. 5. Structure-activity relationships for C$_2$ symmetry-based diaminoalcohols. Adapted from [10]

protease (IC$_{50}$ > 200 µM) and did not exhibit significant anti-HIV activity in vitro as measured by HIV p24 antigen production in H9 cells [9].

The next stage in our design entailed increasing the binding potency to HIV protease. The usefulness of the modeled RSV PR/inhibitor complex for more sophisticated design was limited by the absence of ordered flap in the RSV PR crystal structure [12]. However, careful examination of the substrate binding site region of this structure indicated that the P1 and P2 substituents would be buried in subsites in the enzyme, and that the P3 residues were more likely to be exposed. These considerations led us to extend I by the symmetric addition of NH$_2$ blocked amino acids. The inhibitory potency for a series of diaminoalcohols of increasing size ranged from >10 000 nM for the core structure, I, to 3 nM for the bifunctionalized, Cbz-Val compound, VII (Fig. 5) [10]. The most potent HIV PR inhibitor of this series, compound VII, also designated A-74704, exhibited measurable anti-HIV activity in vitro with an IC$_{50}$ ≤ 1 µM. A-74704 also demonstrated good specificity for HIV PR over human renin (>10 000 : 1), low cellular toxicity (EC$_{50}$: TC$_{50}$ = 500 : 1), and was resistant to proteolytic degradation in a renal cortex homogenate at 37°C (t$_{1/2}$ ≫ 3 h).

Crystal structure of A-74704/HIV PR complex

To validate the proposed symmetry-based mode of binding, we cocrystallized A-74704 with recombinant HIV PR and solved the 2.8 Å crystal structure of the complex in the hexagonal space group, P6$_1$ [9]. The inhibitor formed a fairly symmetric pattern of hydrogen-bonding interactions with the enzyme and includes a buried water molecule that

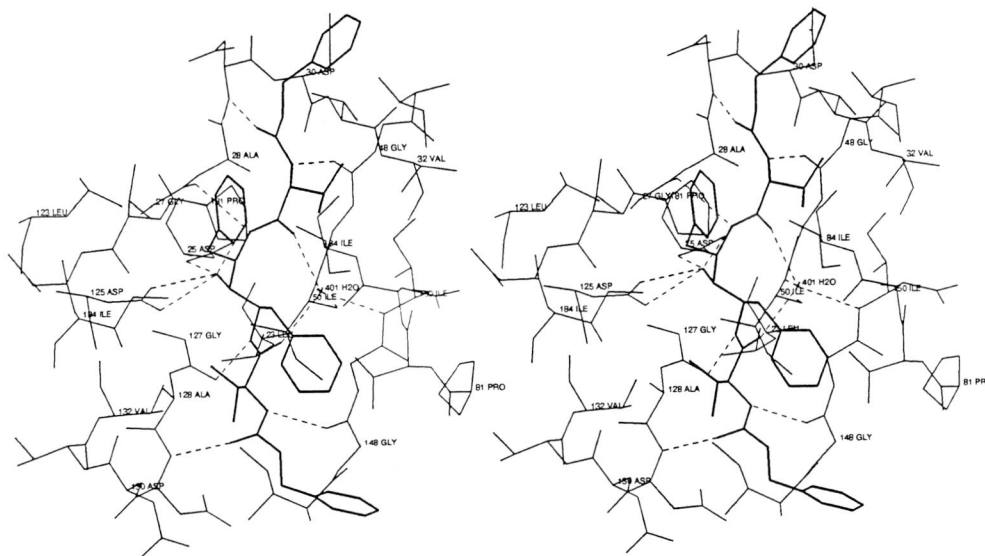

Fig. 6. Stereo view of hydrogen-bonding interactions (dashed lines) between A-74704 (thick lines) and atoms of the protease (thin lines). The buried water molecule is shown along with the Cα backbone. Residues of the enzyme making contacts of less than 4.2 Å to the inhibitor are illustrated. From [9] with permission

makes bridging hydrogen bonds between the inhibitor P2 and P2′ CO groups and the enzyme IIe50 and IIe50′ NH groups on the flaps (Fig. 6). The water molecule is located within 0.2 Å of the enzyme pseudo-C_2 axis and exhibits approximate tetrahedral coordination. The inhibitor and enzyme pseudo-C_2 axes pass within 0.2 Å of each other and make an angle of approximately 6°. The symmetry of the inhibitor is nearly exact: 20 nonhydrogen atom pairs from both halves of the inhibitor superimpose to with 0.36 Å rms by an approximate dyad (177.9°). The backbone Cα atoms of the two monomers of HIV PR could be superimposed to within 0.42 Å rms after a rotation of 179.9°.

Design of symmetry-based diols

Analysis of the crystal structure of the A-74704/HIV PR complex revealed that the short spacing between the P1 and P1′ NH groups resulted in hydrogen bonds with poor geometry between the inhibitor P1 and P1′ amides and the carbonyl groups of Gly27 and Gly127. The replacement of an ethyl alcohol by a glycol isostere has been shown to result in potent renin inhibitors [13]. Thus, a second series of symmetry-based HIV PR inhibitors was designed by application of a C_2 axis placed at the midpoint of the scissile bond (Fig. 4B). The resulting diol compounds were generally 10- to 50-fold more potent than the symmetry-based

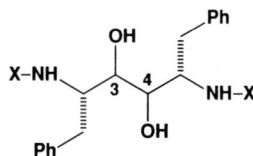

No.	X	Conf.	IC$_{50}$ (nM)
VIII	Boc	3R,4R	40
IX	Boc	3R,4S	12
X	Boc	3S,4S	280
XI	Cbz-Val	3R,4R	0.22
XII	Cbz-Val	3R,4S	0.22
XIII	Cbz-Val	3S,4S	0.38

Fig. 7. Structure-activity relationship for C$_2$ symmetry-based diaminodiols. Adapted from [10]

alcohols [10]. The additional hydroxymethyl group leads to three distinct diastereoisomers for the diol analogues. The short, Boc-protected diols, VIII-X, were moderately potent (Fig. 7). Replacement of Boc with Cbz-Val led to a more potent series, XI-XIII, in which the stereochemistry of the two hydroxy groups exhibited surprisingly little effect on inhibitor potency in sharp contrast to the case for substrate-based, hydroxy-containing peptidomimetic inhibitors (reviewed in [13]).

The potential usefulness of compounds XI–XIII were limited by their poor solubility. Efforts to improve the solubility of this series were aided by examination of the solvent-accessible surface of the A-74704/HIV PR complex which indicated that the terminal portions of the inhibitor were exposed to solvent (Fig. 8). Thus, solubility enhancement efforts were directed primarily towards making modifications at the termini of the inhibitor. This strategy led to a new series of compounds with markedly improved solubilities without sacrificing either enzyme inhibitor or antiviral potency (Fig. 9). Compound XVIII, designated A-77003, was chosen as a clinical candidate and is currently being evaluated in Phase I/II clinical trials [14].

Summary and conclusions

We have used structural and mechanistic considerations, and knowledge of substrate preferences to conceptualize novel, symmetry-based inhibitors of HIV PR. Two classes of compounds – a pseudo-C$_2$ symmetric diaminoalcohol and a diastereomeric set of diaminodiols – were designed based on the concept of C$_2$ active site symmetry. The structure of A-74704 complexed with HIV PR confirmed the proposed symmetric mode of binding that was based on our initial modeling studies, and also

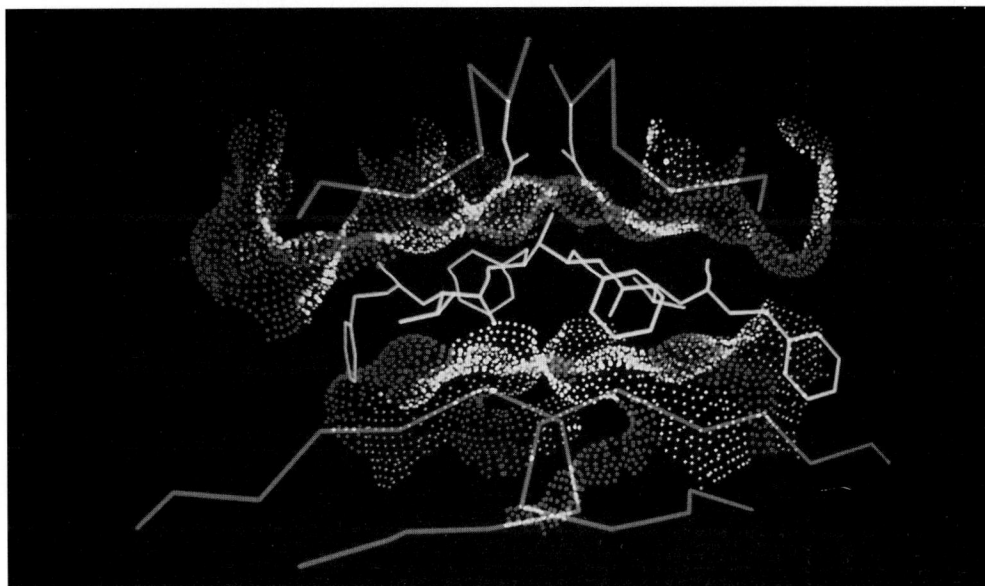

Fig. 8. View of a 10–15 Å thick section of the active site of the A-74704/HIV PR complex. A solvent-accessible surface (dots) of the active site region was computed after removal of the inhibitor

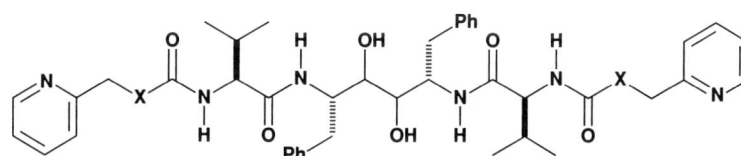

No.	X	Conf.	K_I(nM)	EC_{50} (μM)	Solubility (pH 7.4, μM)
XIV	O	3R,4R	0.16	0.12-0.67	6.5
XV	O	3R,4S	0.09	0.02-0.14	3.1
XVI	O	3S,4S	0.19	0.05-0.18	0.27
XVII	NCH₃	3R,4R	1.66	0.28-1.5	292
XVIII	NCH₃	3R,4S	0.15	0.07-0.20	256
XIX	NCH₃	3S,4S	0.18	0.06-0.17	4.7

Fig. 9. Structure-activity relationships and solubilities for diaminodiols with different end-group substituents. Adapted from [14]

proved to be useful in subsequent efforts to improve the solubility of the more potent diol series.

It is still unclear as to why the diols are consistently more potent than the diaminoalcohols. Preliminary crystallographic studies indicate that the modes of interaction of the *R,R* and *S,S* analogues of A-77003 are quite distinct from that described above [15]. A recent review of the crystal structures of numerous HIV protease inhibitor complexes with structurally diverse compounds has revealed some general features of binding that may be important for future design strategies of both symmetric and non-symmetric inhibitors [3]. However, it is not yet possible to extrapolate from structural data to binding potency in a rigorously predictive fashion. Despite the current theoretical limitations at predicting and analyzing free energies of binding, it is clear that structure-based approaches to inhibitor design are mature enough to contribute to the conceptualization of medicinal chemistry strategies that can lead to useful clinical candidates for AIDS, cancer and other conditions and diseases [16].

Acknowledgements

The authors wish to acknowledge the collaborative efforts and support of numerous colleagues at Abbott Laboratories and at the NCI-Frederick Cancer Research and Development Center. The content of this publication does not necessarily reflect the views or policies of the Department of Health and Human Services, nor does mention of trade names, commercial products, or organization imply endorsement by the U.S. Government.

References

1. Mitsuya H, Yarchoan R, Broder S (1990) Molecular targets for AIDS therapy. Science 249: 1533–1544
2. Kohl NE, Emini EA, Schleif WA, Davis LJ, Heimbach JC, Dixon RAF, Scolnick EM, Sigal IS (1988) Active human immunodeficiency virus protease is required for viral infectivity. Proc Natl Acad Sci USA 85: 4686–4690
3. Wlodawer A, Erickson JW (1993) Structure-based inhibitors of HIV-1 protease. Annu Rev Biochem 62: 543–585
4. York DM, Darden TA, Pedersen LG, Anderson MW (1993) Molecular dynamics simulation of HIV-1 protease in a crystalline environment and in solution. Biochemistry 32: 1443–1453
5. Greenlee WJ (1990) Renin inhibitors. Med Res Rev 10: 173–236
6. Meek TD (1992) Inhibitors of HIV-1 protease. J Enzym Inhib 6: 65–98
7. Huff JR (1991) HIV protease: a novel chemotherapeutic target for AIDS. J Med Chem 34: 2305–2314
8. Norbeck DW, Kempf DJ (1991) HIV protease inhibitors. In: Bristol JA (ed) Annual reports in medicinal chemistry, vol 26. Academic Press, Harcourt Brace Jovanovich, San Diego, pp 141–149
9. Erickson J, Neidhart DJ, VanDrie J, Kempf DJ, Wang XC, Norbeck DW, Plattner JJ, Rittenhouse JW, Turon M, Wideburg N, Kohlbrenner WE, Simmer R, Helfrich R, Paul DA, Knigge M (1990) Design, activity and 2.8 Å crystal structure of a C_2 symmetric inhibitor complexed to HIV-1 protease. Science 249: 527–533
10. Kempf DJ, Codacovi L, Wang XC, Norbeck DW, Kohlbrenner WE, Wideburg NE, Paul DA, Knigge MF, Vasavanonda S, Craig-Kennard A, Saldivar A,

Rosenbrook Jr W, Clement JJ, Plattner JJ, Erickson J (1990) Structure-based, C_2 symmetric inhibitors of HIV protease. J Med Chem 33: 2687–2689

11. Suguna K, Padlan EA, Smith CW, Carlson WD, Davies DR (1987) Binding of a reduced peptide inhibitor to the aspartic proteinase from *Rhizopus chinensis*: implications for a mechanism of action. Proc Natl Acad Sci USA 84: 7009–7013

12. Miller M, Jaskolski M, Rao JKM, Leis J, Wlodawer A (1989) Crystal structure of a retroviral protease proves relationship to aspartic protease family. Nature 337: 576–579

13. Luly JR, BaMaung N, Soderquist J, Fung AKL, Stein H, Kleinert HD, Marcotte PA, Egan D, Bopp B, Merits I, Bolis G, Greer J, Perun T, Plattner JJ (1988) Renin inhibitors. Dipeptide analogues of angio-tensinogen utilizing a dihydroxye-thylene transition-state mimic at the scissile bond to impart greater inhibitory potency. J Med Chem 31: 2264–2276

14. Kempf DJ, March K, Paul MF, Knigge DW, Norbeck WE, Kohlbrenner WE, Codacovi L, Vasavanonda S, Bryant P, Wang XC, Wideburg NE, Clement JJ, Plattner JJ, Erickson J (1991) Antiviral and pharmacokinetic properties of C_2 symmetric inhibitors of HIV-1 protease. Antimicrob Agents Chemother 35: 2209–2214

15. Hosur MV, Bhat TN, Gulnik S, Wideburg NE, Norbeck DW, Appelt K, Baldwin E, Kempf D, Liu B, Erickson J (1994) Influence of stereochemistry on activity and binding modes for C_2 symmetry-based diol inhibitors of HIV-1 protease. J Am Chem Soc, in press

16. Erickson JW, Fesik SW (1992) Macromolecular X-ray crystallography and NMR as tools for structure-based drug design. Ann Rep Med Chem 727: 271–189

Authors' address: Dr. J. Erickson, Structural Biochemistry Program, Frederick Biomedical Supercomputing Center, Program Resources, Inc., National Cancer Institute, Frederick Cancer Research and Development Center, Frederick, MD 21702–1201, U.S.A.

Arch Virol (1994) [Suppl] 9: 31–39

Archives
of
Virology
© Springer-Verlag 1994
Printed in Austria

Age-dependent susceptibility to fatal encephalitis: alphavirus infection of neurons

D. E. Griffin, B. Levine, W. R. Tyor, P. C. Tucker, and **J. M. Hardwick**

Johns Hopkins University School of Medicine, Baltimore, Maryland, U.S.A.

Summary. Sindbis virus encephalitis in mice provides a model for studying age-dependent susceptibility to acute viral encephalitis. The AR339 strain of SV causes fatal encephalitis in newborn mice, but weanling mice recover uneventfully. Increased virulence for older mice is associated with a single amino acid change from Gln to His at position 55 of the E2 glycoprotein. Weanling mice with normal immune systems clear infectious virus from neurons through an antibody-mediated mechanism. This does not happen in newborn mice because the infected neurons die soon after they are infected. Death in immature neurons, as well as most other mammalian cells infected with Sindbis virus, occurs by induction of apoptosis. This can be prevented by cellular expression of *bcl-2*, an inhibitor of apoptosis, which is expressed by mature neurons in culture. We conclude that mature neurons are resistant to induction of apoptosis after infection with SV through expression of cellular inhibitors of apoptosis. This provides the opportunity for antibody to clear virus by a noncytolytic mechanism.

Introduction

Newborn mice are uniquely susceptible to a number of virus infections and alphaviruses have frequently been isolated by inoculation of infected mosquito pools or infected tissue into suckling mice. Many of the same viruses that are lethal for newborn mice cause no disease or only mild disease in older mice. Sindbis virus (SINV) is the prototype alphavirus (family *Togaviridae*) and serves as an excellent model for studies of age-dependent susceptibility to virus infection. AR339, the original wildtype strain of SINV caused fatal encephalitis in suckling mice [1] and was shown to be avirulent for weanling mice [2]. Although it is clear that maturation of the host is of prime importance for age-dependent susceptibility, studies have not revealed an important role for maturational changes in induction of interferon or in virus-specific humoral or cellular immune responses that could explain this decrease in susceptibility [3–5]. The target cell within the central nervous system in both immature and mature mice is the neuron [4, 6]. Therefore, maturation of this cell may be an important determinant for the outcome of infection.

Maturation of the host is not the only determinant of severity of disease however, because it is also clear that viruses selected by serial passage, either in tissue culture cells or in animals, can cause mild disease in newborn mice or fatal encephalitis in weanling mice [4, 7]. To investigate the host cell and the viral determinants of age-dependent susceptibility to alphavirus encephalitis we have used this model system of SINV infection in mice.

Viral determinants of age-dependent virulence

Neurovirulence of different strains of SINV is determined largely by amino acid changes in the E1 and E2 surface glycoproteins; the effects of different amino acid substitutions have been studied primarily in newborn mice [8–11]. NSV, adapted by serial intracerebral passage of AR339, is a neurovirulent strain of SINV that causes fatal encephalitis in weanling mice [7]. NSV differs from AR339 at only 2 residues in E1 and 2 residues in E2 [12] (Table 1). When these sequences were compared with HRSP, a laboratory-adapted strain of SINV the substitution of His for Gln at E2–55 was noted to be unique for NSV. Therefore, we investigated the role of the Gln to His substitution at E2–55 with respect to virulence for older mice by preparing (in collaboration with R. Kuhn, J. H. Strauss and E. G. Strauss, California Institute of Technology) recombinant viruses differing only at this position [13]. When these viruses were tested for neurovirulence and replication in mice of different ages, the His substitution was associated with greater mortality in 1 and 2 week-old, but not in 1 day-old mice (Table 2). With His at E2–55 amounts of virus produced within 24 h after intracerebral

Table 1. Amino acid sequence differences in 3 strains of SINV differing in passage history and neurovirulence for suckling mice (*SM*) and weanling mice (*WM*)

Strain	Virulence		E2					E1		
	SM	WM	3	23	55	172	209	72	237	313
NSV	+	+	T	E	H	G	G	A	A	D
AR339	+	−	T	E	Q	G	R	V	A	G
HRSP	±	−	I	V	Q	R	G	A	S	G

Table 2. Effect of amino acid at E2–55 on mortality and SINV growth in the brains of mice of different ages inoculated intracerebrally with 1 000 pfu of recombinant virus

Age	% Mortality		Virus titer (24 h) ($\log_{10}$ pfu/gm)	
	E2–55 His	Gln	E2–55 His	Gln
1 day	100	100	8.5	8.5
1 week	100	18	10	8.1
2 weeks	88	0	ND	ND

ND Not determined

inoculation into 1 week-old mice were also higher (Table 2). The data suggest that this amino acid change increases the efficiency of the replication of SINV in the brains of mature mice. The mechanism by which this enhancement occurs is not yet clear.

Clearance of infection from the nervous system of weanling mice

Newborn mice infected subcutaneously or intracerebrally with the AR339 strain of SV die within 4 days of infection without mounting a virus-specific immune response [4, 14]. Histopathology of the brains of infected animals shows neuronal infection and cell death. Weanling (3–4 week old) mice infected intracerebrally with AR339 develop antiviral antibody, inflammation, and other histopathologic evidence of encephalitis, but remain asymptomatic and clear infectious virus within 7–8 days of infection (Fig. 1). Mice unable to mount an immune response to SINV, such as those with severe combined immunodeficiency (*scid*), develop persistent, asymptomatic infection (Fig. 1) with no readily apparent functional or histopathologic evidence of neuronal injury [15].

Clearance of virus from tissue has generally been assumed to occur by lysis and, thus, elimination of infected cells by virus infection *per se* or by virus-specific cytotoxic T cells. Major histocompatibility complex (MHC) class I-restricted CD8 T cells and class II-restricted CD4 T cells can recognize virus-infected cells through appropriate presentation of viral peptides in the context of MHC antigens on the cell surface and can eliminate these cells. Although this mechanism appears to be important in many tissues, its importance in the nervous system is in question because neurons are incapable of expressing either MHC class I or class II antigens and elimination of this nonrenewable cell population would appear to be self-defeating. *Scid* mice, which are incapable of generating an antibody or T cell response due to a deficiency in a recombinase

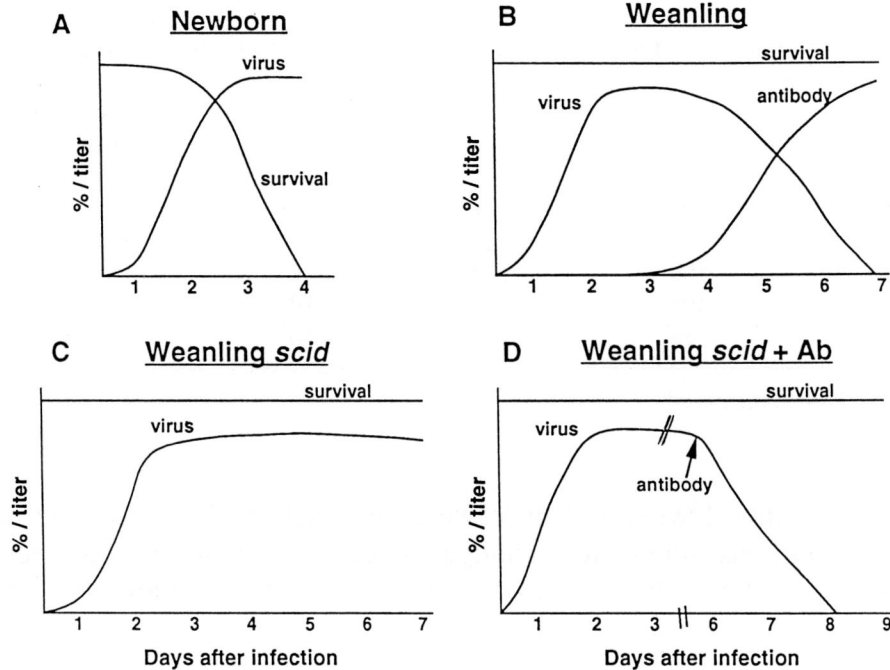

Fig. 1. Schematic representations of the effects on survival and induction of antibody secretion after intracerebral inoculation of the AR339 strain of SINV into (**A**) immunocompetent newborn (1–2 days old) mice, (**B**) immunocompetent weanling (3–4 weeks old) mice, and (**C**) immunodeficient weanling *scid* mice. The effect of antibody treatment on brain virus in *scid* mice is also shown (**D**)

necessary for production of mature rearranged antibody and T cell receptor genes, were used to determine the immunologic mechanism by which SINV is cleared from the brains of weanling mice.

Hyperimmune serum and immune T cells were passively transferred into persistently infected *scid* mice [15]. Infection was cleared from the brains of mice receiving immune serum (Fig. 1), but not from the brains of mice receiving immune T cells. Experiments using monoclonal antibodies determined that the antibodies capable of clearing virus were specific for the E2 glycoprotein. This effect could be replicated in vitro using cultures of differentiated rat neurons. Neither complement nor monocytes were necessary for antibody-mediated clearance of SINV from neurons and there was no evidence in vitro or in vivo of neuronal destruction associated with viral clearance.

The dose of E2 antibody is important because approximately 40% of mice relapsed with infectious virus 30–90 days after treatment with hyperimmune serum or low doses of monoclonal antibody to E2. Interestingly, the viruses recovered from these mice, as well as from persistently infected *scid* mice that had not been treated with antibody,

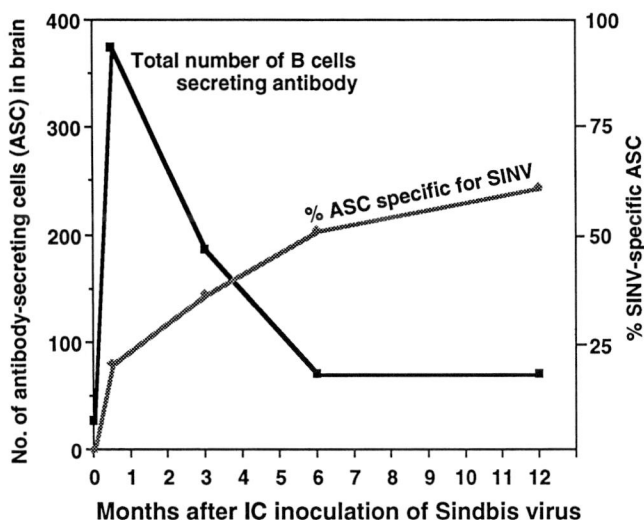

Fig. 2. Appearance and persistence of antibody-secreting cells (*ASC*) in the brains of immunocompetent weanling mice infected intracerebrally (*IC*) with SINV. Initially, B cells enter the brain with other inflammatory cells and secrete antibody, but many of these cells are not secreting antibody specific for SINV. SINV antibody-secreting cells are progressively enriched with time after apparent recovery from infection

frequently had His rather that the original Gln at E2–55 further suggesting the importance of this amino acid change for optimal replication of SINV in mature neurons [15a]. Reappearance of infectious virus in these mice suggested that viral RNA may persist in neurons, even though RNA could not be detected in the brains of treated mice by in situ hybridization 20 days after transfer of antibody [15].

The possibility of persistence of low levels of viral RNA was investigated using reverse transcriptase-polymerase chain reaction (RT-PCR). Viral RNA could be detected using primers to amplify the E2 gene present in the abundant subgenomic RNA as well as using primers to amplify the NSP2 gene, present only in full length RNA, for months to years after recovery from infection [16, 17]. This PCR-amplifiable RNA was present not only in *scid* mice treated with antibody, but also in immunocompetent mice that had cleared infection. Analysis of the brains of immunocompetent mice for as long as one year after infection showed persistence of intraparenchymal B cells secreting antibody specific for SINV (Fig. 2) [17]. These data suggest that one consequence of a nonlytic mechanism for eliminating virus from tissue is failure to eliminate all viral RNA from the infected cell. If the cells are terminally differentiated and are not replaced then long term control (i.e., continued secretion of antibody) may be important in preventing reactivation of infection.

Immature   **Mature**

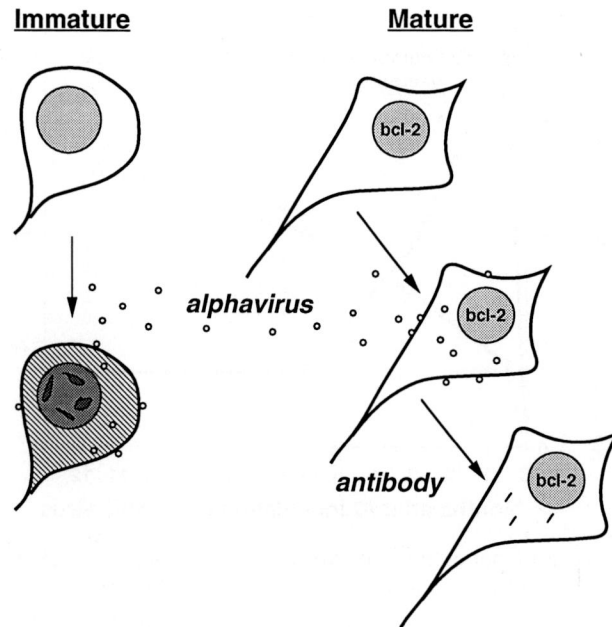

Fig. 3. Schematic diagram of the outcome of SINV infection of neurons of immature and mature mice and the immune response to infection. Immature neurons are induced to undergo apoptotic cell death as a consequence of infection. Mature neurons express *bcl-2* and do not die in response to infection. Antibody to the virus downregulates virus replication, but does not completely eliminate virus, resulting in the persistence of small amounts of viral RNA in the nervous system

Response of immature and mature neurons to SINV infection

The age-dependent susceptibility of neurons to SINV infection can be observed in vitro. Neuronal cultures from embryonic rodents are lysed by SINV if infected within the first several days after explantation. If the neurons are allowed to mature in culture for several weeks they are not lysed, but become persistently infected. Therefore, we sought to determine how SINV usually kills mammalian cells and the maturational changes that prevent cell death and lead to persistent neuronal infection.

Mouse neuroblastoma cells (N18 clone) [18] were used as a model for immature neurons. Two distinctive mechanisms of cell death, necrosis and apoptosis, are generally recognized. Continuous observation by videomicroscopy of N18 cells infected with SINV showed active blebbing of the cell membrane consistent with death by apoptosis or programmed cell death. Electron microscopy confirmed the characteristic condensation of chromatin and fragmentation of nuclei. Endonucleolytic cleavage of DNA in these and other mammalian cell lines (BHK, AT3) infected with SINV confirmed that SINV induces cell lysis by activating the cellular mechanism for programmed cell death [19].

A cellular oncogene, *bcl-2*, can prevent induction of apoptosis in many cells, including neurons [20–22], and has been postulated to play a role in development of the nervous system. To determine whether expression of the *bcl-2* protein might play a role in the resistance of mature neurons to SINV-induced death, expression of *bcl-2* mRNA was measured in explanted neurons early and late in differentiation. Newly explanted neurons have almost no detectable mRNA for *bcl-2*, while mature neurons have substantial amounts. To determine whether expression of this gene could influence the outcome of SINV infection, lines of AT3 cells expressing *bcl-2* were produced and infected with SINV. Control AT3 cells, lacking *bcl-2* expression, were all dead within 48 h post infection while *bcl-2*-expressing AT3 cells survived and became persistently infected with SINV. Therefore, we postulate that mature neurons resist alphavirus-induced death, and may resist the lethal effects of other viruses, by upregulating expression of *bcl-2* or related genes during differentiation.

Conclusion

Studies of age-dependent susceptibility to alphavirus encephalitis have used SINV infection of mice as a model system. Neurons of immature (newborn) mice are lysed by SINV infection through induction of apoptosis leading to rapidly fatal disease. Neurons of mature (weanling) mice express genes, such as *bcl-2*, that render them resistant to induction of apoptosis and these cells become persistently infected by SINV. Virus can be cleared from these persistently infected neurons by treatment with antibody, but one consequence of this mechanism of virus clearance is persistence of viral RNA (Fig. 3). Therefore, longterm downregulation of virus replication requires continued administration of antibody or local production of antibody within the brain, a well-recognized consequence of previous central nervous system infection. The price the host pays for preservation of the nonrenewable neuron after virus infection is failure to completely eliminate the viral RNA. This creates the necessity for continued, long-term local production of antiviral antibody to prevent renewed virus replication. Strains of SINV that are virulent for mature mice may be able to overcome neuronal resistance to lytic infection through specific mutations in the E2 glycoprotein.

References

1. Taylor RM, Hurlbut HS (1953) The isolation of Coxsackie-like viruses from mosquitoes. J Egypt Med Assoc 36: 489–494
2. Taylor RM, Hurlbut HS, Work TH, Kingsbury JR, Frothingham TE (1955) Sindbis virus: a newly recognized arthropod-transmitted virus. Am J Trop Med Hyg 4: 844–846

3. Griffin DE (1976) Role of the immune response in age-dependent resistance of mice to encephalitis due to Sindbis virus. J Infect Dis 133: 456–464

4. Sherman LA, Griffin DE (1990) Pathogenesis of encephalitis induced in newborn mice by virulent and avirulent strains of Sindbis virus. J Virol 64: 2041–2046

5. Reinarz ABG, Broome MG, Sagik BP (1971) Age dependent resistance of mice to Sindbis virus infection: viral replication as a function of host age. Infect Immun 3: 268–273

6. Jackson AC, Moench TR, Trapp BD, Griffin DE (1988) Basis of neurovirulence in Sindbis virus encephalomyelitis of mice. Lab Invest 58: 503–509

7. Griffin DE, Johnson RT (1977) Role of the immune response in recovery from Sindbis virus encephalitis in mice. J Immunol 118: 1070–1075

8. Davis NL, Fuller FJ, Dougherty WG, Olmsted RA, Johnston RE (1986) A single nucleotide change in the E2 glycoprotein gene of Sindbis virus affects penetration rate in cell culture and virulence in neonatal mice. Proc Natl Acad Sci USA 83: 6771–6775

9. Tucker PC, Griffin DE (1991) The mechanism of altered Sindbis virus neurovirulence associated with a single amino acid change in the E2 glycoprotein. J Virol 65: 1551–1557

10. Polo JM, Davis NL, Rice CM, Huang HV, Johnston RE (1988) Molecular analysis of Sindbis virus pathogenesis in neonatal mice by using virus recombinants constructed in vitro. J Virol 62: 2124–2133

11. Polo JM, Johnston RE (1990) Attenuating mutations in glycoproteins E1 and E2 of Sindbis virus produced a highly attenuated strain when combined in vitro. J Virol 64: 4438–4444

12. Lustig S, Jackson AC, Hahn CS, Griffin DE, Strauss EG, Strauss JH (1988) The molecular basis of Sindbis virus neurovirulence in mice. J Virol 62: 2329–2336

13. Tucker PC, Strauss EG, Kuhn RJ, Strauss JH, Griffin DE (1993) Viral determinants of age-dependent virulence of Sindbis virus for mice. J Virol 67:4605–4610

14. Johnson RT, McFarland HF, Levy SE (1972) Age-dependent resistance to viral encephalitis: studies of infections due to Sindbis virus in mice. J Infect Dis 125: 257–262

15. Levine B, Hardwick JM, Trapp BD, Crawford TO, Bollinger RC, Griffin DE (1991) Antibody-mediated clearance of alphavirus infection from neurons. Science 254: 856–860

15a. Levine B, Griffin DE (1993) Molecular analysis of neurovirulent strains of Sindbis virus that evolve during persistent infection of scid mice. J Virol 67:6872–6875

16. Levine B, Griffin DE (1992) Persistence of viral RNA in mouse brains after recovery from acute alphavirus encephalitis. J Virol 66: 6429–6435

17. Tyor WR, Wesselingh S, Levine B, Griffin DE (1992) Long term intraparenchymal Ig secretion after acute viral encephalitis in mice. J Immunol 149: 4016–4020

18. Amano T, Richelson E, Nirenberg M (1972) Neurotransmitter synthesis by neuroblastoma clones. Proc Natl Acad Sci USA 69: 258–263

19. Levine B, Huang Q, Isaacs JT, Reed JC, Griffin DE, Hardwick JM (1993) Conversion of lytic to persistent alphavirus infection by the bcl-2 cellular oncogene. Nature 361: 739–742

20. Hockenbery DM, Zutter M, Hickey W, Nahm M, Korsmeyer SJ (1991) Bcl-2 protein is topographically restricted in tissues characterized by apoptotic cell death. Proc Natl Acad Sci USA 88: 6961–6965

21. Garcia I, Martinou I, Tsujimoto Y, Martinou J-C (1992) Prevention of program-
    med cell death of sympathetic neurons by the bcl-2 proto-oncogene. Science 258:
    302–304
22. Korsmeyer SJ (1992) Bcl-2 initiates a new category of oncogenes: regulators of cell
    death. Blood 80: 879–886

Authors' address: Dr. D. E. Griffin, Meyer 6-181, Johns Hopkins Hospital, 600 N.
Wolfe Street, Baltimore, MD 21287-7681, U.S.A.

Archives

V̲of
irology

© Springer-Verlag 1994
Printed in Austria

Principles and background for the construction of transgenic plants displaying multiple virus resistance

E. Truve[1], **M. Kelve**[1], **A. Aaspôllu**[1], **A. Kuusksalu**[1], **P. Seppänen**[2], and **M. Saarma**[1,3]

[1] Institute of Chemical Physics and Biophysics, Estonian Academy of Sciences, Tallinn, Estonia
[2] Kemira OY, Espoo Research Centre, Espoo, [3] Institute of Biotechnology, University of Helsinki, Helsinki, Finland

Summary. We investigated the possibility of reconstructing the $2'-5'$ oligoadenylate (2–5A) pathway into the plant kingdom to achieve multiple virus resistance. Differently phosphorylated 2–5A trimers and tetramers inhibited TMV RNA translation in cell-free systems. In wheat germ extracts the most potent inhibitors were nonphosphorylated forms of 2–5A. Triphosphorylated forms of 2–5A were deposphorylated and hydrolysed in plant extracts. Since we could not detect homologous DNA to mammalian 2–5A synthetase cDNA in tobacco or potato, we cloned rat 2–5A synthetase cDNA and transformed it by the *Agrobacterium*-mediated mechanism into tobacco and potato. Transformed tobacco plants were resistant to PVS infection and propagation of PVX was reduced. In transgenic potatoes tolerance to PVX and, in one transgenic clone, also to PVY was observed.

Introduction

The $2'-5'$ oligoadenylate (2–5A) pathway is part of the mammalian antiviral response system induced by interferons. The key enzyme of the pathway, the 2–5A synthetase, polymerizes ATP to a family of oligonucleotides, the 2–5A. Virus replication is inhibited due to the degradation of viral RNA by the specific 2–5A-activated ribonuclease, RNase L. Recently we have demonstrated that exogenously added 2–5A molecules can inhibit protein synthesis in plant cell-free extracts and protoplasts [21]. This inhibition was accompanied by the degradation of translated RNA. Thus, we concluded that 2–5A can activate a putative plant RNase. In this paper, the effects of exogenous 2–5A on protein synthesis in plant systems, the cloning of a novel mammalian 2–5A synthetase cDNA, and the properties of transgenic tobacco and potato plants expressing 2–5A synthetase are discussed.

Approaches to engineered resistance against plant viruses

The following approaches have successfully been taken to make plants virus-resistant by means of genetic engineering: (1) expression of the viral coat protein (CP) gene [1]; (2) expression of viral nonstructural proteins [10]; (3) use of artificial antisense genes [2]; (4) expression of nucleic acid sequences encoding viral satellite RNAs [8, 11]. The main drawback of all approaches here mentioned for the construction of virus-resistant transgenic plants is that resulting plants are resistant to only one specific virus. There is some evidence for a broad-spectrum protection being raised by viral CP expression [15, 17] but nevertheless this is limited to a closely related group of viruses.

2–5A system in mammals

The 2–5A synthetase polymerizes ATP in the presence of double-stranded RNA (dsRNA, for example replicative intermediates of RNA viruses) to produce a family of oligonucleotides with the general structure $ppp(A2'p5')_nA$ with $n \geq 2$, abbreviated 2–5A. These oligonucleotides possess $2'-5'$ phosphodiester bonds that are unusual in comparison with ordinary $3'-5'$ links in the nucleic acids. Two other enzymes involved in the 2–5A system are: (i) $2'-5'$ phosphodiesterase which degrades 2–5A, and (ii) 2–5A-dependent ribonuclease (RNase L). The 2–5A synthetase is expressed as an inactive enzyme. For its activation, the presence of dsRNA is required. Only dsRNA molecules at least about 50 base pairs and with no more than one mismatch per 45 nucleotides can activate the synthesis of 2–5A [16]. Viral RNA has been shown to be a very potent activator of the 2–5A pathway [14]. Actually, 2–5A is not a single compound but a mixture of oligoadenylates with different chain lengths and states of phosphorylation. Oligomers with at least three residues are required to activate RNase L. Another requirement for 2–5A activation of RNase L in mammalian cells is a $5'$ di- or triphosphate group. The existence of nonphosphorylated "core" 2–5A molecules in cells also has been reported, but they neither bind to nor activate the RNase L [12]. Activation of the RNase L in mammalian cell extracts is observed already at nanomolar concentrations of 2–5A [12]. Due to the activity of $2'-5'$ phosphodiesterase in cells, the activation of the 2–5A-dependent RNase is transient without the persistent de novo synthesis of 2–5A.

Components of the 2−5A pathway in plants

In plants ATP polymerizing activity was reported from tobacco mosaic virus (TMV)-infected or "antiviral factor" treated *Nicotiana glutinosa* leaves, which mediated the discharge of histidinyl-TMV-RNA [6]. The polymerization product was synthesized in vitro using a partially purified enzyme from *N. glutinosa* or *N. tabacum* immobilized on a poly(rI:rC) column [7]. An inhibitory effect on TMV replication was demonstrated as well with the exogenous 2−5A trimer "core" at a concentration of 100−200 nM in *N. glutinosa* leaf discs as at 10 nM in *N. tabacum* protoplasts [5]. It was suggested that plant oligoadenylates differ from the mammalian analogs because they did not compete with $(2'-5')$ $pppA_4[^{32}P]pCp$ for binding on the RNase L [7]. Furthermore, no 2−5A-binding proteins were detected in plant extracts [7], despite the fact that 2−5A was able to activate the discharging factor [6] referred above. The absence of $(2'-5')pppA_nA$-binding proteins in plants has been independently demonstrated [4]. From metabolic stability assays, plants were thought to lack the $2'-5'$ phosphodiesterase activity [5]. A probe of the human 2−5A synthetase was shown to hybridize to tobacco genomic DNA and mRNA from TMV-infected tobacco [19] although 2−5A synthetase activity had not been detected in tobacco earlier [4]. The partially purified ATP-polymerizing plant enzyme also reacted with antibodies to human 2−5A synthetase [19]. Recently, exogeneous nonphosphorylated 2−5A molecules longer than trimers were demonstrated to induce both increased cytokinin activity and the synthesis of pathogenesis-related and heat shock proteins in tobacco and wheat [13].

Detection of 2−5A-dependent ribonuclease activity and 2−5A degradation in plant system

We have followed the in vitro translation rate of TMV RNA in wheat germ extract and rabbit reticulocyte lysate in the presence and absence of different 2−5A forms [21]. In the case of rabbit reticulocyte lysate, di- and triphosphorylated forms of 2−5A trimers and tetramers were the most potent inhibitors of in vitro translations (Table 1). These results fit well with those obtained by several other groups [12]. We also showed a clear inhibition of TMV RNA translation by 2−5A in wheat germ extract. However, in contrast with a mammalian cell-free system, in wheat germ extract nonphosphorylated forms of 2−5A were the best inhibitors of protein synthesis, whereas $5'$ phosphorylated compounds had much weaker inhibitory effects on the translation efficiency (Table 1). Addition of the 2−5A "core" resulted in the greatest reduction of the in vitro translation rate, consistent with the in vitro data obtained by Devash et al. [5].

Table 1. Effect of differently phosphorylated forms of 2–5A trimers and tetramers on TMV RNA in vitro translation in wheat germ extract and rabbit reticulocyte lysate[a]

Compound	Incorporation of the [³H]-Leu (%)	
	wheat germ extract	rabbit reticulocyte lysate
Control without 2–5A	100	100
1 μM A₃ "core"	19.4	84.4
1 μM pA₃	84.3	90.2
1 μM ppA₃	42.6	73.8
1 μM pppA₃	70.4	54.9
200 nM A₄ "core"	61.5	69.7
200 nM pA₄	67.9	53.8
200 nM ppA₄	68.8	40.9
200 nM pppA₄	118.8	40.2

[a] 1 μg of TMV RNA was translated in the cell-free extract in the presence of [³H]-labelled leucine; the rate of in vitro translation was measured by TCA precipitation of the translation products on Whatman GF/C filters and counting the radioactivity of the filters [21]

The ability of the 2–5A trimer "core" to inhibit protein synthesis was examined in tobacco mesophyll protoplasts. Addition of the 2–5A trimer "core" resulted in an at least two-fold reduction in protoplast protein synthesis compared to that of 2–5A-untreated control cells [21].

When TMV RNA was isolated from the wheat germ in vitro translation mix with and without the 2–5A "core", a three to seven-fold faster degradation of TMV RNA in the mix with the 2–5A "core" was observed [21]. This means that a rapid degradation of RNA is induced by 2–5A in the plant cell-free extracts. This is similar to the activity found in mammalian cells, in which 2–5A-dependent RNase L is responsible for the cleavage of cytoplasmic RNAs, thus inhibiting the protein synthesis. Using chemical crosslinking method, we detected a 70 kD plant protein from potato leaf extracts that specifically bind (2′–5′)A₄[³²P]pCp [21]. This observation is contrary to that of Cayley et al. [4] and Devash et al. [5] who could not detect 2–5A-binding protein in plants. The reason for such a difference may be the use of radioactively labeled triphosphorylated 2–5A forms instead of "core" molecules by the other groups. Because 2–5A inhibits protein synthesis and stimulates the hydrolysis of RNA in wheat germ extract, the 2–5A-binding 70 kD plant protein may be the plant 2–5A-dependent RNase, analogous to the mammalian RNase L. We emphasize that the putative plant 2–5A-dependent RNase differs from its mammalian analogs in that the plant

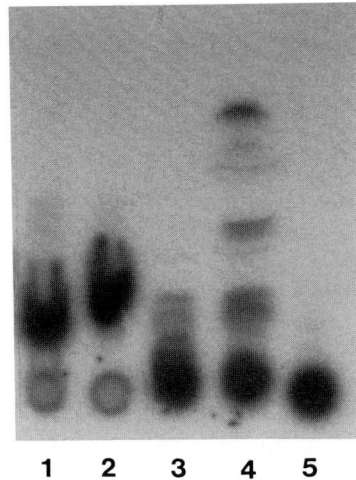

Fig. 1. Separation of 2–5A degradation products analyzed by silica gel thin layer chromatography. *1* pppA$_4$, incubated 1 h at 37°C with 0.5 units of calf intestine alkaline phosphatase; *2* A$_3$ "core", incubated 1 h at 37°C with 0.5 units of calf intestine alkaline phosphatase; *3* pppA$_4$, incubated overnight at 25°C with mouse L cell extract; *4* pppA$_4$, incubated overnight at 25°C with tobacco leaf extract; *5* pppA$_4$, untreated

enzyme seems to be activated preferentially by the 2–5A "core" whereas in mammals "cores" neither bind nor activate RNase L [12].

Thin layer chromatography on silica gel plates revealed that (2'–5') pppA$_4$[^{32}P]pCp was efficiently degraded in tobacco leaf extracts (Fig. 1). As triphosphorylated forms of 2–5A remain in the start on silica gel plates (Fig. 1, lane 5), our results show that enzymes that both dephosphorylate 2–5A and hydrolyze 2'–5' phosphodiester bonds are present in plant extracts. These data contradict those reported by Devash et al. [5].

In plant extracts we have found enzyme activities resembling 2–5A-dependent RNase and 2'–5' phosphodiesterase activities of mammalian cells. However, we have been unable to detect 2–5A synthetase activity in healthy or virus-infected plants. We did not detect neither tobacco or potato genomic DNA sequences nor mRNAs from virus-infected tobacco or potato hybridizing to the murine 2–5A synthetase cDNA probe. Therefore we assume that plants do not exhibit an enzyme activity similar to that of mammalian 2–5A synthetase.

As plants do contain a 2–5A-dependent ribonuclease, we assumed that expression of the mammalian 2–5A synthetase gene in transgenic plants might simultaneously protect against infection by many different RNA viruses. Mammalian 2–5A synthetase is normally expressed in an inactive form, so the pathway in transgenic plants, as we assume, would be switched on only after the appearance of dsRNA, i.e. after viral infection. As the product of the 2–5A synthetase, 2–5A is degraded by plant enzymes, the effect would be transient and the reconstructed pathway is switched off after the disappearance of dsRNA, i.e. after the degradation of viral RNA.

Cloning of a novel 2–5A synthetase cDNA from rat

At least three major forms of 2–5A synthetase have been reported in mammalian cells: 40–46 kD, 69 kD, and 100 kD. The cDNA sequences encoding the small form of the 2–5A synthetase have been identified from human [3] and mouse [9, 18]. We have used the mouse L3 2–5A synthetase cDNA, which is derived from the unique mouse genetic locus not present in human and which encodes a 40 kD 2–5A synthetase [18]. Using L3 cDNA as a probe, we isolated 2–5A synthetase cDNA from a rat hippocampus cDNA library. We have cloned and sequenced this cDNA (EMBL Acc. No. Z18877), which is 1.5 kb long and contains a 1 074 bp open reading frame. The nucleotide sequence showed 86% homology to the mouse L3 2–5A synthetase cDNA. Homology to the mouse synthetase L2 cDNA was even higher (89%). The L2 cDNA is characterized by an additional 600 bp untranslated sequence at its 3′-end, which is not present in the L3 cDNA [18]. Since in our cloned rat 2–5A synthetase cDNA the starting point for the poly(A) tail was nearly at the same position as in the mouse L3 cDNA (−3 bp compared to L3) we believe that the cloned sequence is the rat analog of mouse 2–5A synthetase L3 cDNA. The amino acid sequence is 82% identical to the corresponding mouse 2–5A synthetase. The main difference between rat and mouse 40 kD 2–5A synthetases was the nine amino acid long deletion at the C-terminal region of the rat enzyme [20].

Construction of plants simultaneously tolerant to different plant viruses

We have used an *Agrobacterium*-mediated transfer to obtain transgenic tobacco and potato plants expressing the rat 2–5A synthetase. Plants expressing detectable amounts of the 2–5A synthetase mRNA were used for the virus infection experiments. In a series of independent experiments transgenic tobacco plants were infected with potato virus X (PVX), potato virus S (PVS), and potato virus Y (PVY), belonging to the potex-, carla- and potyvirus groups, respectively. The propagation of the viruses was followed for a month after infection. The results revealed that all transgenic tobacco clones were resistant to infection by PVS and in some clones the propagation rate of PVX was considerably reduced. However, protection to PVY was not observed (Table 2).

Leaf disc experiments using both PVX and TMV, a member of tobamovirus group, were done with transgenic tobacco. Again, some

Table 2. Propagation of four different plant viruses in leaf discs and intact transgenic tobacco plants expressing 2–5A synthetase[a]

Virus	Intact plants	Leaf discs
Potato virus X	+/−	+/−
Potato virus S	−	N.D.
Potato virus Y	+	N.D.
Tobacco mosaic virus	N.D.	+/−

[a] Virus concentration was measured a week (leaf discs) or a month (intact plants) after infection by homogenizing the leaf material and detecting the virus by ordinary ELISA or time-resolved fluoroimmunoassay (TRFIA) measurement [20]. Specific monoclonal antibodies against PVX, PVS and PVY were used, TMV was detected by polyclonal antibodies

−: virus propagation is inhibited in all tested transgenic lines

+/−: virus propagation is inhibited in some transgenic lines, but not in all

+: virus propagating as in nontransgenic control tobacco plants

N.D. Not determined

tobacco clones markedly reduced the infection rate of both viruses, as determined one week following infection (Table 2).

Transgenic potato plants were infected with PVX and PVY, using sap from infected plant leaves [20]. Some clones exhibited remarkable tolerance, not only to PVX infection as with tobacco, but one potato clone also showed slight tolerance to PVY infection, which was not the case in transgenic tobaccos (data not shown). Infection experiments with PVS in potato are now in progress.

We conclude that the expression of the mammalian 2–5A synthetase in transgenic plants leads to reconstruction of a pathway analogous to that of the mammalian 2–5A system. As plants contain endogenous 2–5A-dependent ribonuclease as well as the 2–5A degrading enzyme activity, we propose a model for the reconstituted 2–5A pathway (Fig. 2). As revealed in the infection experiments, transgenic plants expressing 2–5A synthetase exhibit the broadest spectrum of protection against virus infections ever reported. Recently a patent has been filled for this novel approach.

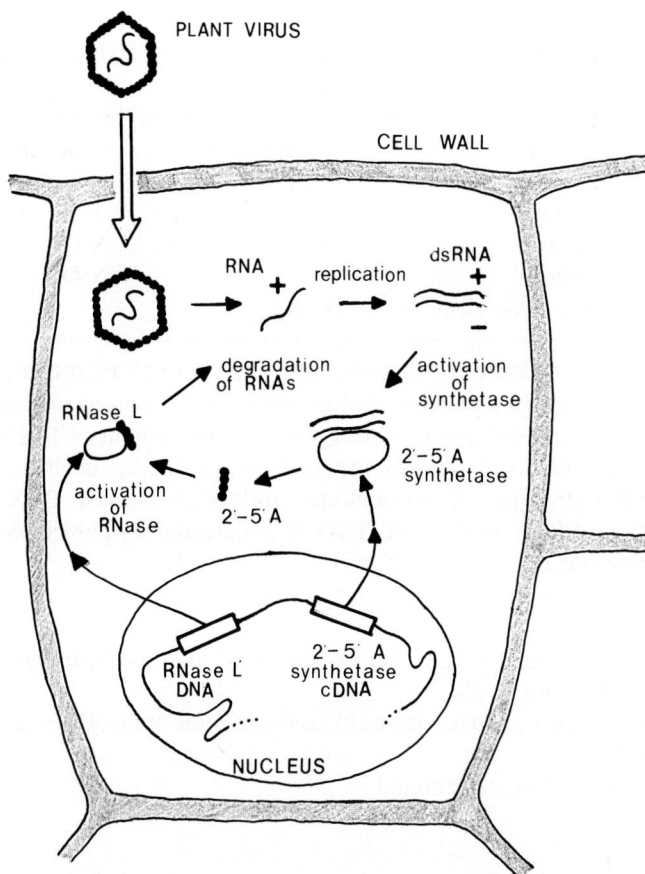

Fig. 2. A model for the reconstructed 2–5A pathway in transgenic plants expressing the mammalian 2–5A synthetase. The putative plant 2–5A-dependent RNase is marked as RNase L′

Acknowledgements

We greatfully acknowledge A. H. Shulman for useful comments and critical reading of the manuscript, I. A. Mikhailopulo for supplying us chemically synthesized 2–5A, H. Persson for a rat hippocampus cDNA library, Y. Sokawa for the mouse 2–5A synthetase cDNA, T. H. Teeri for the plant expression vector pHTT202, and Y. Varitsev for the virus strains. This work was partly supported by grants from the Finnish Ministry of Agriculture and Forestry and by a grant to E.T. from Finnish Centre of International Mobility.

References

1. Beachy RN, Loesch-Fries S, Tumer N (1990) Coat protein-mediated resistance against virus infection. Annu Rev Phytopathol 28: 451–474
2. Bejarano ER, Lichtenstein CP (1992) Prospects for engineering virus resistance in plants with antisense RNA. Trends Biotechn 10: 383–388

3. Benech P, Mory Y, Revel M, Chebath J (1985) Structure of two forms of the interferon-induced $(2'-5')$ oligo A synthetase of human cells based on cDNAs and gene sequences. EMBO J 4: 2249–2256

4. Cayley PJ, White RF, Antoniw JF, Walesby NJ, Kerr IM (1982) Distribution of the $ppp(A2'p)_nA$-binding protein and interferon-related enzymes in animals, plants, and lower organisms. Biochem Biophys Res Commun 108: 1243–1250

5. Devash Y, Gera A, Willis DH, Reichman M, Pfleiderer W, Charubala R, Sela I, Suhadolnik RJ (1984) 5'-dephosphorylated 2',5'-adenylate trimer and its analogs. Replication in tobacco mosaic virus-infected leaf discs, protoplasts, and intact tobacco plants. J Biol Chem 259: 3482–3486

6. Devash Y, Hauschner A, Sela I, Chakraburtty K (1981) The antiviral factor (AVF) from virus-infected plants induces discharge of histidinyl-TMV-RNA. Virology 111: 103–112

7. Devash Y, Reichman M, Sela I, Reichenbach NL, Suhadolnik RJ (1985) Plant oligoadenylates: enzymatic synthesis, isolation, and biological activities. Biochemistry 24: 593–599

8. Gerlach WL, Llewellyn D, Haseloff J (1987) Construction of a plant disease resistance gene from the satellite RNA of tobacco ringspot virus. Nature 328: 802–805

9. Ghosh SK, Kusari J, Bandyopadhyay SK, Samanta H, Kumar R, Sen GC (1991) Cloning, sequencing and expression of two murine 2'-5'-oligoadenylate synthetases. J Biol Chem 266: 15293–15299

10. Golemboski DB, Lomonossoff GP, Zaitlin M (1990) Plants transformed with a tobacco mosaic virus nonstructural gene sequence are resistant to the virus. Proc Natl Acad Sci USA 87: 6311–6315

11. Harrison BD, Mayo MA, Baulcombe DC (1987) Virus resistance in transgenic plants that express cucumber mosaic virus satellite RNA. Nature 328: 799–802

12. Kerr IM (1987) The 2–5A system: a personal view. J Interferon Res 7: 505–510

13. Kulaeva ON, Fedina AB, Burkhanova EA, Karavaiko NN, Karpeisky MYa, Kaplan IB, Taliansky ME, Atabekov JG (1992) Biological activities of human interferon and 2'–5' oligoadenylates in plants. Plant Mol Biol 20: 383–393

14. Lengyel P (1987) Double-stranded RNA and interferon action. J Interferon Res 7: 511–519

15. Ling K, Namba S, Gonsalves D (1991) Protection against detrimental effects of potyvirus infection in transgenic tobacco plants expressing the papaya ringspot virus coat protein gene. Bio/Technology 9: 752–758

16. Minks MA, West DK, Benvin S, Baglioni C (1979) Structural requirements of double-stranded RNA for the activation of 2',5'-oligo (A) polymerase and protein kinase of interferon-treated HeLa cells. J Biol Chem 254: 10180–10183

17. Neijdat A, Clark WG, Beachy RN (1990) Engineered resistance against plant virus diseases. Physiol Plant 80: 662–668

18. Rutherford MN, Kumar A, Nissim A, Chebath J, Williams BRG (1991) The murine 2–5A synthetase locus: three distinct transcripts from two linked genes. Nucleic Acids Res 19: 1919–1924

19. Sela I, Grafi G, Sher N, Edelbaum O, Yagev H, Gerassi E (1987) Resistance systems related to the *N* gene and their comparison with interferon. In: Evered D, Harnett S (eds) Plant resistance to viruses. J. Wiley, Chichester, pp 109–119

20. Truve E, Aaspõllu A, Honkanen J, Puska R, Mehto M, Hassi A, Teeri TH, Kelve M, Seppänen P, Saarma M (1993) Transgenic potato plants expressing mammalian

$2'-5'$ oligoadenylate synthetase are protected from potato virus X infection under field conditions. Bio/Technology 11: 1048–1052

21. Truve E, Kelve M, Teeri T, Saarma M (1993) The effects of 2–5A on protein synthesis in wheat germ extracts and tobacco proplasts (submitted)

Authors' address: Dr. E. Truve, Department of Molecular Genetics, Institute of Chemical Physics and Biophysics, Akadeemia tee 23, EE0026 Tallinn, Estonia.

Arch Virol (1994) [Suppl] 9: 51–58

Archives
of
Virology
© Springer-Verlag 1994
Printed in Austria

The structure of an immunodominant loop on foot and mouth disease virus, serotype O1, determined under reducing conditions

D. Rowlands[3], D. Logan[4], R. Abu-Ghazaleh[2], W. Blakemore[2], S. Curry[2], T. Jackson[2], A. King[2], S. Lea[1], R. Lewis[1], J. Newman[2], N. Parry[3], D. Stuart[1], and E. Fry[1]

[1] Laboratory of Molecular Biophysics, University of Oxford, Oxford
[2] AFRC Institute for Animal Health, Pirbright, Surrey
[3] Wellcome Research Laboratories, Beckenham, Kent, U.K.
[4] U.P.R. de Biologie Structurale, I.B.M.C. du C.N.R.S., Strasbourg Cedex, France

Summary. Residues 136-159 of VPI of foot and mouth disease virus (FMDV) comprise the G-H loop of the protein and form a prominent feature on the surface of virus particles. This sequence contains an immunodominant neutralizing epitope, which can be mimicked with synthetic peptides, and includes an Arg, Gly, Asp motif which has been implicated in the binding of the virus to cellular receptors. Crystallographic analysis of native virus particles failed to resolve the structure of this region due to its disordered state. However, reduction of a disulphide bond between cysteine residues 134 of VP1 and 130 of VP2 caused the G-H loop to collapse onto the surface of the virus particle and allowed its conformation to be determined.

Introduction

Foot and mouth disease viruses (FMDVs) comprise the *Aphthovirus* genus of the family *Picornaviridae* and are the causative agents of one of the most important infections of domestic livestock [1]. The disease can affect all cloven hoofed animals and is endemic in many parts of the world, especially in tropical and less developed countries. Rigorous import controls and hygiene methods are maintained to keep non-endemic areas free from disease, while vaccination and restrictions on stock movement are used to control spread of disease in endemic regions. Vaccines in current use consist primarily of tissue culture grown virus which is chemically inactivated. Protection is afforded by circulating virus neutralizing antibodies and it is usually necessary to vaccinate two or three times per year to maintain sufficiently high levels of antibody.

Control by vaccination is complicated by, amongst other things, the antigenic diversity displayed by the virus. There are seven distinct serotypes of the virus distributed unequally throughout the world and there is significant antigenic variation within the serotypes. Consequently, it is of interest and importance to understand the molecular basis of this variation.

An important antigenic feature of the virus is located on the VP1 protein approximately in the region of amino acids 140–160 and synthetic or biosynthetic peptides representing this sequence are extraordinarily effective in inducing virus neutralizing antibodies and can afford protection against live virus challenge in a range of species [2–6]. The sequence of this region of VP1 is highly variable between different virus serotypes and between virus isolates within serotypes and there is evidence to suggest that it is an important and immunodominant antigenic feature of the virus. In contrast to the general variability displayed by this sequence, it contains a highly conserved triplet of the amino acids Arg, Gly, Asp located at positions 145–147 (sequence numbering of type O1 virus). Several lines of evidence had suggested that the VP1 140–160 region is involved in binding of the virus to its cell receptors and these were further supported by the demonstration that synthetic peptides containing the Arg, Gly, Asp motif can competitively inhibit the binding of virus to susceptible cells [7–9]. Arg, Gly, Asp is the hallmark of many integrin receptor binding proteins and it has been proposed that the cellular receptor(s) for FMDV belong to the integrin family [7–10]. Recently it has been shown that certain other picornaviruses use receptors of this family [11, 12].

Structure of the virus under oxidizing conditions

X-ray crystallographic analysis of serotype O1 virus showed that the general organization of the particle was similar to that of other picornaviruses, as expected [13]. There were, however, many intriguing differences of detail. For example, the shortening of the structural proteins VP1–3 relative to other picornaviruses has resulted in a generally thinner and smoother protein capsid due primarily to truncation of the surface oriented loops linking the β strand components of the β barrel motifs of each protein. Consequently, FMDV particles lack the groove or canyon of the entero and rhinoviruses [14, 15] or series of pits in the cardioviruses [16], which encircle the axes of five fold symmetry in those viruses and are thought to contain the cell receptor binding domains and shield them from immune surveillance [17]. However, the smooth surface of FMDV is interrupted by a large excursion of VP1 sequence between residues ca. 136–159, i.e. that part of the protein which had been shown to contain an important antigenic determinant and to be involved in cell receptor binding. Unfortunately, nothing could be deduced concerning the structure of this feature, the G-H loop of VP1, apart from its surface location, because it was too disordered in the virus crystals to determine electron density. The nature of this disorder is, of course, conjectural since it could be a consequence of flexibility along the whole length of the G-H loop sequence or due to the lack of fixed orientation between the structured loop and the remainder of the virus particle. Supportive evidence for the latter possibility was provided by crystallographic analyses of mutant viruses selected under pressure of neutralizing antibodies [18]. These mutants displayed antigenic characteristics compatible with alterations of the VP1 G-H loop sequence but

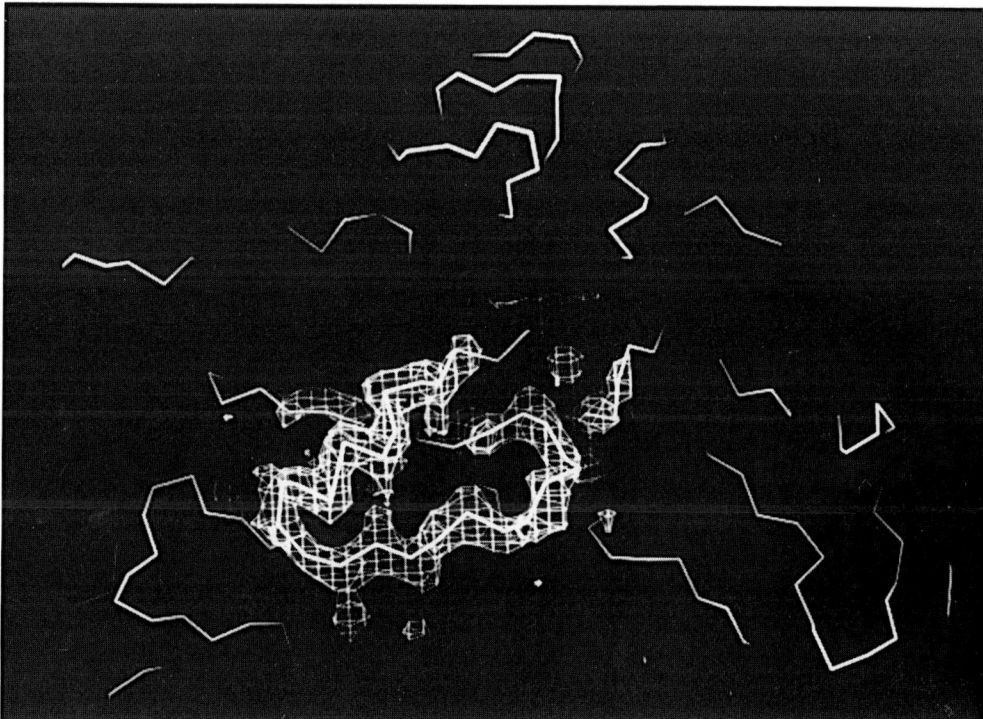

Fig. 1. A difference electron density map showing the continuous density defining the whole VP1 G-H loop in the reduced virus structure. In the native virus electron density representing this loop was diffuse or absent between residues 134 and 156. The electron density map is superimposed on the α carbon tracings of the surface oriented loops of a single asymmetric unit of the virus and is viewed from the outside of the capsid. VP1 occupies the top portion of the triangular unit, pointing towards the 5 fold axis, while VP2 and VP3 occupy the bottom left and right hand portions, respectively. An electron density map with coefficients $2F_{reduced} - F_{native}$ and native phases showed clear side chain density correlating exactly with the sequence for the loop residues and allowed a model to be built (reproduced from [19])

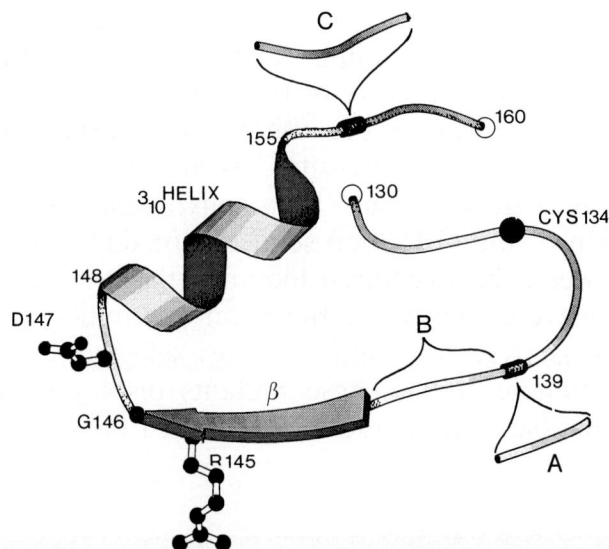

Fig. 2. A molscript ribbon diagram of the VP1 G-H loop with side chains shown for the Arg-Gly-Asp triplet. *A*, *B*, and *C* are positions at which insertions or deletions are found in viruses of other serotypes. *A* As many as three residues inserted in SAT serotypes; *B* as many as four residues deleted in C, A and Asia 1 serotypes; *C* as many as four residues inserted in SAT serotypes (reproduced from [19])

contained amino acid substitutions, not in the G-H loop, but in the adjacent B-C loop. In the structural analysis of these viruses it was found that there was considerably more electron density attributable to the G-H loop, overlying part of VP2, and a change in distribution between the two alternative densities attributable to a disulphide bond between cysteines 130 of VP2 and 134 of VP1, i.e. at the start of the G-H loop. This suggested that the B-C loop substitutions in these mutant viruses caused a defined G-H loop structure, which is linked to the virus particle via a flexible "stem", to preferentially orient in a "southerly" position, i.e. towards the 3-fold axis of symmetry, overlying VP2.

Structure of the virus under reducing conditions

These observations, and in particular the changes in orientation of the disulphide bond between VP1 134 and VP2 130, led us to investigate the X-ray crystallographic structure of the virus under reducing conditions. Virus crystals were soaked in a 10 mM solution of dithiothreitol prior to exposure to the high energy X-ray beam of the Synchrotron Radiation Source at Daresbury, U.K. Analysis of the data showed that under these conditions the G-H loop is ordered, although it is still one of the most mobile regions of the virus, and lies snugly against the surface

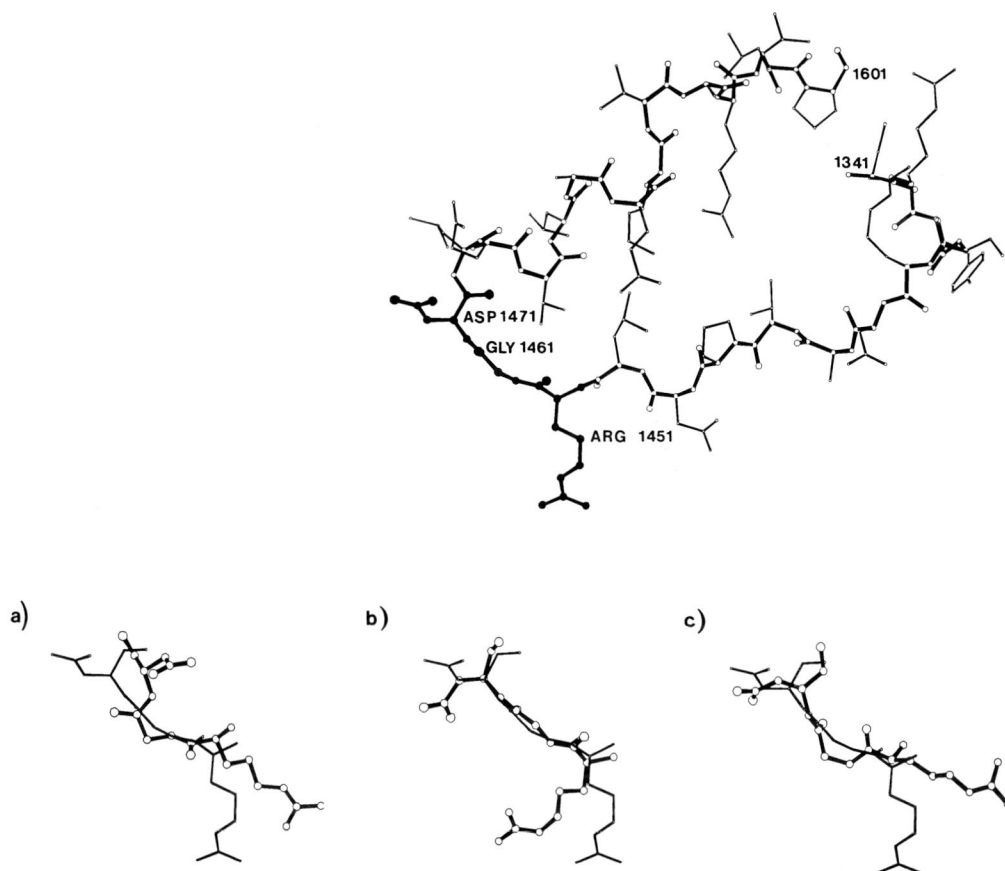

Fig. 3. An all atom trace of residues 134–160 of VP1 G-H loop. Spheres representing the atoms of the Arg-Gly-Asp residues are filled in. Superimpositions of the Arg-Gly-Asp of FMDV (narrow lines) with those of **a** α-lytic protease, **b** γ-II crystallin, and **c** thermolysin (reproduced from [19])

of the particle (Fig. 1) [19]. The part of the VP2 chain containing the Cys 130 was moved by ca. 2Å and VP1 Cys 134 had shifted by ca. 12Å as compared to the oxidized virus structure. These relatively massive rearrangements suggest that the VP2 130, VP1 134 disulphide bond is under considerable strain in the oxidized state.

The structural features of the VP1 G-H loop are shown diagramatically in Fig. 2, from which it can be seen that the Arg, Gly, Asp triplet occurs at a turn where the loop, having traversed towards the 3-fold axis, returns as a 3_{10} helix towards the 5-fold axis. As outlined above, there is circumstantial evidence that the Arg, Gly, Asp sequence is involved in receptor binding, possibly to an integrin. It was, therefore, of interest to compare the conformation of these residues with those in other proteins in which they occur in the protein Data Bank [20]. The structures are known for three proteins that have been tested for integrin binding activity, γ-II-crystallin [21], α-lytic protease [22] and thermolysin [23]. The Arg, Gly, Asp of FMDV conformed most closely to that in γ-II-crystallin (Fig. 3), the only one of the three active in integrin binding, further suggesting that the FMDV sequence functions in binding to an integrin receptor. Of six other proteins that contain Arg-Gly-Asp and have been structurally resolved but which have not been tested for their ability to bind to integrin receptors, the FMDV conformation most closely resembles that seen in the VP1 protein of human rhinovirus 1A [24].

In both FMDV and γ-II crystallin the acidic and basic side chains of the Arg-Gly-Asp are well separated in the "open" conformation and nuclear magnetic resonance studies with peptide inhibitors of integrins suggest that such an arrangement enhances their potency [25, 26]. Furthermore, recent studies of the disintegrins, a family of polypeptide integrin inhibitors, have shown that the Arg, Gly, Asp motif that they bear is located in a highly mobile loop [25, 26] as in the VP1 G-H loop of FMDV.

Reoxidation of the G-H loop

On returning reduced crystals to an oxidizing environment, the rate of reformation of the VP2–VP3 disulphide bond was very slow, with a half time of several days. This, however, may be an artefact caused by the conditions within the crystal. Virus freshly released from infected cells is in the reduced state and formation of the disulphide bond appears to occur more rapidly during storage than was observed with virus within crystals.

Conclusions

For those FMDV sequences that have been determined, only serotype O1 viruses have the potential to form a disulphide bond between VP2 and the G-H loop of VP1. It is therefore of interest to examine the state of order or disorder of the VP1 G-H loop of viruses of other serotypes. Such analyses are being performed with viruses of types A and C and preliminary evidence suggests that in these viruses the VP1 G-H loop is in a disordered state, similar to that of oxidized O1 viruses. This suggests the intriguing conclusion that the VP2 130 – VP1 134 disulphide bond in O1 viruses serves to induce disorder in the VP1 G-H loop. The biological relevance of the highly disordered state of the VP1 G-H loop in oxidized O1 viruses and in viruses of other serotypes is unclear, because the infectivity of the virus is not altered by exposure to reducing conditions (Sangar and Rowlands, unpubl. results, [18]).

However, the kinetics of attachment of reduced or oxidized virus to susceptible cells have not been compared and it is conceivable that this may be influenced by the state of disorder of the VP1 G-H loop.

References

1. Brooksby JB (1982) Portraits of viruses: foot and mouth disease virus. Intervirology 18: 1
2. Bittle JL, Houghten RA, Alexander H, Schimick TM, Sutcliffe JG, Lerner RA, Rowlands DJ, Brown F (1982) Protection against foot and mouth disease by immunisation with a chemically synthesised peptide predicted from the viral nucleotide sequence. Nature 298: 30–33
3. Pfaff E, Mussgay M, Bohm HO, Schalz GE, Schaller H (1982) Antibodies against a preselected peptide recognise and neutralize foot and mouth disease virus. EMBO J 1: 869–874
4. Dimarchi R, Brooke G, Gale C, Cracknell V, Doel T, Mowat N (1986) Protection of cattle against foot and mouth disease by a synthetic peptide. Science 232: 639–641
5. Brockhuijsen MP, Van Rijn JMM, Blom AJM, Pouwels PH, Enger-Valk BE, Brown F, Francis MJ (1987) Fusion proteins with multiple copies of the major antigenic determinant of foot and mouth disease protect both the natural host and laboratory animals. J Gen Virol 68: 3137–3143
6. Morgan DO, Moore DM (1990) Protection of cattle and swine against foot and mouth disease using biosynthetic peptide vaccines. Am J Vet Res 51: 40–45
7. Baxt B, Becker Y (1990) The effect of peptides containing the arginine, glycine, aspartic acid sequence on the adsorbtion of foot and mouth disease virus to tissue culture cells. Virus Genes 4: 74–83
8. Fox G, Parry NR, Barnett PV, McGinn B, Rowlands DJ, Brown F (1989) The cell attachment site on foot and mouth disease virus includes the amino acid sequence RGD (Arginine-Glycine-Aspartic Acid). J Gen Virol 70: 625–637
9. Surovoi AY, Ivanov VT, Cherpurkin AV, Ivanyushckenkov VN, Dryagalin NN (1989) Is the Arg-Gly-Asp sequence the binding site of foot and mouth disease virus with the cell receptor? Sov J Bio-Org Chem 14: 572–580
10. Geysen HM, Barteling SJ, Meloen RH (1985) Small peptides induce antibodies with a sequence and structural requirement for binding antigen comparable to antibodies raised against the native protein. Proc Natl Acad Sci USA 82: 178–182
11. Chang KH, Day C, Walker J, Hyypia T, Stanway G (1992) The nucleotide sequences of wild type coxsackie virus strains imply that an RGD motif in VP1 is functionally significant. J Gen Virol 73: 621–626
12. Bergelson JM, Shepley MP, Chan BMC, Hemler ME, Finberg RW (1992) Identification of the integrin VLA-2 as a receptor for echovirus 1. Science 255: 1718–1720
13. Acharya R, Fry E, Stuart D, Fox G, Rowlands D, Brown F (1989) The three-dimensional structure of foot and mouth disease virus at 2.9Å resolution. Nature 327: 709–716
14. Hogle JM, Chow M, Filman DJ (1985) The three dimensional structure of poliovirus at 2.9Å resolution. Science 229: 1358–1365
15. Rossmann MG, Arnold E, Erickson JW, Frankenberger EA, Griffith JP, Hecht H-J, Johnson E, Kramer G, Luo M, Mosser AG, Reuckert RR, Sherry B, Vriend G (1985) Structure of a human common cold virus and functional relationship to other picornaviruses. Nature 317: 145–153

16. Luo M, Vriend G, Kamer G, Minor I, Arnold E, Rossmann MG, Boege U, Scraba DG, Duke GM, Palmenberg AC (1987) The atomic structure of Mengovirus at 3.0Å resolution. Science 235: 182–191

17. Rossmann MG (1989) The canyon hypothesis. Virol Immunol 2: 143–161

18. Parry N, Fox G, Rowlands D, Brown F, Fry E, Acharya R, Logan D, Stuart D (1990) Structural and serological evidence for a novel mechanism of antigenic variation in foot and mouth disease virus. Nature 347: 569–572

19. Logan D, Abu-Ghazaleh R, Blakemore W, Curry S, Jackson T, King A, Lea S, Lewis R, Newman J, Parry N, Rowlands D, Stuart D, Fry E (1993) The structure of a major immunogenic site on foot and mouth disease virus. Nature 362: 566–568

20. Bernstein FC, Koetzle TF, Williams GJB, Meyer EF, Brice MD, Rodgers JR, Kennard O, Shimanouchi T, Tasumi M (1977) The protein data bank: a computer based archival file for macromolecular structures. J Mol Biol 112: 535–542

21. Wistow G, Turnell B, Summers L, Slingsby C, Moss D, Miller L, Lindley P, Blundell T (1983) X-ray analysis of the eye lens protein γ-II crystallin at 1.9Å resolution. J Mol Biol 170: 175–202

22. Fujinaga M, Delbaere LTJ, Brayer GD, James MNG (1985) Refined structure of α-lytic protease at 1.7Å resolution. Analysis of hydrogen bonding and solvent structure. J Mol Biol 183: 479–502

23. Holmes MA, Matthews BW (1982) Structure of thermolysin refined at 1.6Å resolution. J Mol Biol 160: 623–639

24. Kim S, Smith TJ, Chapman MS, Rossmann HG, Pevear DC, Dutko FJ, Felock PJ, Diana GD, McKinlay MA (1989) Crystal structure of human rhinovirus serotype 1A (HRV1A). J Mol Biol 210: 91–111

25. Aumailley M, Gurrath M, Muller G, Calvete J, Timpl R, Kessler H (1991) Arg-Gly-Asp constrained within cyclic pentapeptides. Strong and selective inhibitors of cell adhesion to vitronectin and laminin fragment P1. FEBS Lett 291: 50–54

26. Reed J, Hull WE, von der Lieth C, Kubler D, Suhai S, Kinzel V (1988) Secondary structure of the Arg-Gly-Asp recognition site in proteins involved in cell-surface adhesion. Evidence for the occurrence of nested β-bends in the model hexapeptide GRGDSP. Eur J Biochem 178: 141–154

27. Saudek V, Atkinson RA, Pelton JT (1991) Three dimensional structure of echistatin, the smallest active RGD protein. Biochemistry 30: 7369–7372

28. Adler M, Lazaraus RA, Dennis MS, Wagner G (1991) Solution structure of kistrin, a potent platelet aggregation inhibitor and gp IIb-IIIa antagonist. Science 253: 445–448

Authors' address: Dr. D.J. Rowlands, Wellcome Research Laboratories, Beckenham, Ken BR3 3BS, UK.

Arch Virol (1994) [Suppl] 9: 59–64

Archives
of
Virology
© Springer-Verlag 1994
Printed in Austria

Immunopathologic mechanisms of dengue hemorrhagic fever and dengue shock syndrome

I. Kurane[1], **A. L. Rothman**[1], **P. G. Livingston**[1], **S. Green**[1], **S. J. Gagnon**[1], **J. Janus**[1], **B. L. Innis**[2], **S. Nimmannitya**[3], **A. Nisalak**[2], and **F. A. Ennis**[1]

[1] Division of Infectious Diseases and Immunology, Department of Medicine, University of Massachusetts Medical Center, Worcester, Massachusetts, U.S.A.
[2] Department of Virology, Armed Forces Research Institute of Medical Sciences, Bangkok
[3] Children's Hospital, Bangkok, Thailand

Summary. Dengue virus infections are a major cause of morbidity and mortality in tropical and subtropical areas of the world. The immuno-pathological mechanisms that result in severe complications of dengue virus infection, i.e. dengue hemorrhagic fever (DHF), are important to determine. Primary dengue virus infections induce serotype-specific and serotype-cross-reactive, CD4+ and CD8+ memory cytotoxic T lymphocytes (CTL). In secondary infections with a virus of a different serotype from that which caused primary infections, the presence of cross-reactive non-neutralizing antibodies results in an increased number of infected monocytes by dengue virus – antibody complexes. This in turn results in marked activation of serotype cross-reactive CD4+ and CD8 ı memory CTL. We hypothesize that the rapid release of cytokines and chemical mediators caused by T cell activation and by CTL-mediated lysis of dengue virus-infected monocytes triggers the plasma leakage and hemorrhage that occurs in DHF.

Introduction

Dengue viruses (genus *Flavivirus*, family *Flaviviridae*) are known as four serotypes, named dengue-1, dengue-2, dengue-3 and dengue-4 [1]. Infection with any of these viruses can be asymptomatic or can cause either dengue fever (DF) or dengue hemorrhagic fever (DHF) [2]. DF is a self-limited febrile disease. DHF is a life-threatening syndrome characterized by plasma leakage into interstitial spaces, thrombocytopenia and hemorrhage. When plasma leakage is so profound that shock occurs, the illness is called dengue shock syndrome (DSS). The pathogenesis of DHF/DSS is not understood. There are no good animal models to analyze it and, therefore, analyses of human materials are important to elucidate its pathogenesis.

Epidemiological studies in Thailand and Cuba have shown that DHF/DSS is much more commonly observed during secondary infection with dengue virus serotypes different from those that caused primary infections [3, 4]. It has been reported that in Thailand as many as 99% of cases of DHF/DSS occur in children who have antibody to dengue virus before dengue virus infection that causes DHF/DSS [2]. It is also known that antibodies to any of the four dengue viruses augment dengue virus infection of $Fc\gamma$ receptor $(Fc\gamma R)$-positive cells such as monocytes [5]. Based on these epidemiological and laboratory observations, it has been speculated that serotype cross-reactive immune responses, including enhancing antibodies, contribute to the pathogenesis of DHF/DSS [6].

The role of T lymphocytes in dengue virus infections is not understood. It is generally accepted that virus-specific T lymphocytes play a critical role in recovery from virus infections. On the other hand, the adoptive transfer of virus-specific cytotoxic T lymphocytes (CTL) has resulted in development of severe symptoms during infections with lymphocytic choriomeningitis virus or respiratory syncytial virus [7, 8]. These observations suggest that virus-specific CTL contribute to recovery from virus infections, but may also induce immunopathology in certain situations.

From analyses of T cell responses to dengue viruses we hypothesize that massive activation of dengue virus-specific $CD4^+$ and $CD8^+$ T lymphocytes triggers these complications.

Analysis of dengue virus-specific T lymphocytes in vitro

We analyzed dengue virus-specific, memory $CD4^+$ T lymphocytes induced by primary infections [9, 10]. In bulk culture proliferation assays, $CD4^+$ T lymphocytes showed the highest responses to the serotype that had caused primary infections, but also responded, albeit at lower levels, to other serotypes [10]. We then established dengue virus-specific $CD4^+$ $CD8^-$ T lymphocyte clones from dengue virus-immune donors. These $CD4^+$ T cell clones included both serotype-specific and serotype-cross-reactive clones with various patterns of cross-reactivities [11]. Most of the $CD4^+$ T cell clones were cytotoxic, and lysed dengue virus-infected or non-infectious dengue antigen-pulsed target cells by HLA class II-restriction. HLA DP, DQ and DR were all used as restriction elements by the $CD4^+$ T cell clones. Furthermore, these $CD4^+$ T cell clones produced $IFN\gamma$ and IL-2 upon stimulation with dengue antigens [10, 12]. NS3 protein was recognized by most of the $CD4^+$ T cell clones [11] and envelope (E) protein was recognized by certain clones.

Dengue virus-specific memory CD8$^+$ cytotoxic T lymphocytes (CTL) induced by primary infection were also serotype-cross-reactive in bulk cultures [13]. The CD8$^+$ CTL in bulk culture recognized non-structural protein(s) and E protein. We then established CD8$^+$ CD4$^-$ CTL clones. The dengue virus-specific CD8$^+$ CTL clones also included serotype-specific and serotype-cross-reactive clones, and lysed dengue virus-infected target cells by HLA class I-restriction.

The results obtained in bulk culture and with the clones were consistent, indicating that dengue virus-specific CD4$^+$ and CD8$^+$ memory T lymphocytes include serotype-cross-reactive clones as well as serotype-specific clones.

Activation of T lymphocytes in vivo

Speculating that strong, systemic T cell responses would occur in patients with DHF/DSS, we examined coded serum samples of Thai children with DHF/DSS or DF and serum samples of healthy, age matched, Thai children for levels of soluble IL-2 receptor (sIL-2R), soluble CD4 (sCD4), soluble CD8 (sCD8), IL-2 and IFNγ [14]. SIL-2R, sCD4, sCD8, IL-2 and IFNγ are released or secreted by activated T cells. We believe that serum levels of these T cell activation markers reflect the levels of T cell activation in vivo. The levels of sIL-2R, sCD4, IL-2 and IFNγ were significantly higher in samples from patients with DHF/DSS or DF than in samples from healthy children, and the levels of sCD8 were higher in patients with DHF/DSS than in healthy children. The levels of sIL-2R, sCD4 and sCD8 in patients with DHF/DSS is significantly higher than those in patients with DF. These results indicate that a) T lymphocytes are activated and produce IFNγ and IL-2 in vivo, during DHF/DSS and DF, b) CD4$^+$ T lymphocytes are activated both in DF and DHF, and the levels of activation is significantly higher in DHF than in DF, and c) CD8$^+$ T lymphocytes are significantly activated in DHF/DSS, but not in uncomplicated DF.

Possible mechanisms of DHF/DSS

Based on analysis of T cell responses to dengue viruses in vitro and in vivo, we offer a model of T cell-mediated immunopathology underlying complications of severe dengue virus infections, DHF/DSS (Fig. 1) [15].

In secondary infections with a dengue virus serotype different from that which caused primary infections, enhancing antibodies increase the number of virus-infected monocytes, which activate serotype-cross-reactive CD4$^+$ memory T cells induced by primary infection. The activated crossreactive CD4$^+$ T cells produce lymphokines such as IFNγ and IL-2. IFNγ upregulates the expression of FcγR [16] and HLA class I and class

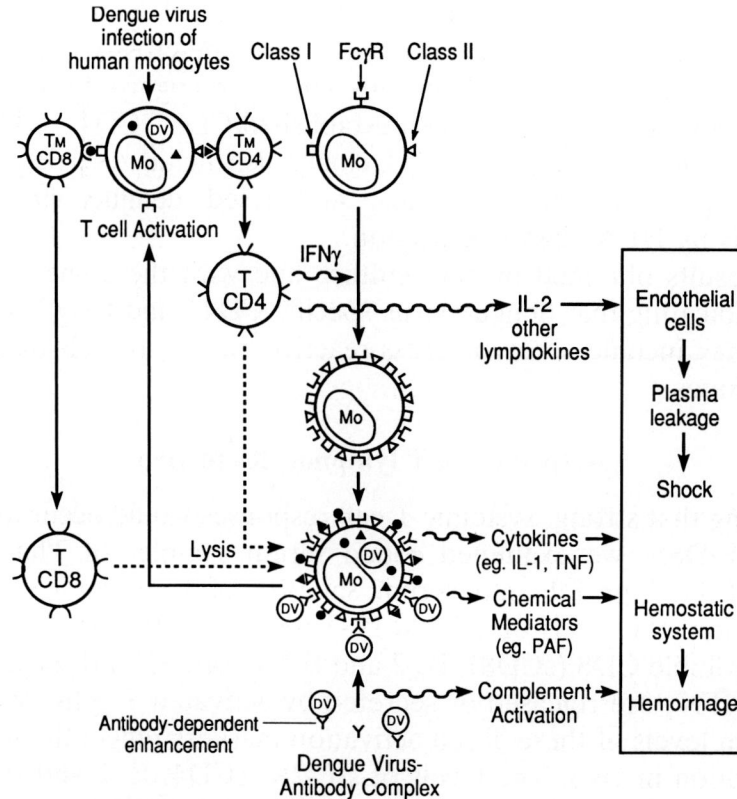

Fig. 1. Immunopathology of DHF/DSS induced by dengue serotype-cross-reactive antibodies and T lymphocytes (from [15])

II molecules [17]. Increase in the number of FcγR further augments infection of monocytes by dengue virus-antibody complexes [18], and upregulation of HLA molecules facilitates recognition of dengue virus antigens by CD4$^+$ and CD8$^+$ CTL. Further activation of CD4$^+$ T cells may increase the plasma levels of IL-2. Lysis of dengue virus-infected monocytes by these cytotoxic T lymphocytes (CTL) may release monokines or chemical mediators. The complement system may be activated by dengue virus-antibody complexes, and C3a or C5a will be also produced. Rapid increase of lymphokines, monokines, chemical mediators and complement activation products will cause plasma leakage, derangement of coagulation and hemorrhage.

There are, however, some patients with DHF/DSS who do not have preexisting antibody and show primary immune responses [19]. Therefore, it is clear that other factors such as virulence of viruses and human genetics may contribute to the pathogenesis of DHF/DSS.

Acknowledgements

This work was supported by grants from the U.S. Army Medical Research and Development Command (DAMD 17-86-C-6208), and the National Institutes of Health (NIH-RO1-AI30624, NIH-T32-AI07272). The opinions contained herein are those of the authors and should not be construed as representing the official policies of the Department of the Army or the Department of Defense of the USA.

References

1. Henchal EA, Putnak JR (1990) The dengue viruses. Clin Microbiol Rev 3: 376–396
2. Halstead SB (1980) Immunological parameters of togavirus disease syndromes. In: Schlesinger RW (ed) The togaviruses: biology, structure, replication. Academic Press, New York, pp 107–173
3. Halstead SB (1981) Dengue hemorrhagic fever – a public health problem and a field for research. Bull World Health Organ 58: 1–21
4. Kouri G, Guzman G, Bravo J (1986) Hemorrhagic dengue in Cuba: history of an epidemic. Pan Am Health Organ Bull 20: 24–30
5. Halstead SB, O'Rourke EJ (1977) Dengue viruses and mononuclear phagocytes. I. Infection enhancement by non-neutralizing antibody. J Exp Med 146: 201–217
6. Halstead SB (1988) Pathogenesis of dengue. Challenge to molecular biology. Science 239: 476–481
7. Baenzinger J, Hengartner H, Zinkernagal RM, Cole GA (1986) Induction or prevention of immunopathological disease by cloned cytotoxic T cell lines specific for lymphocytic choriomeningitis virus. Eur J Immunol 16: 387–393
8. Cannon MJ, Openshaw PJM, Askonas BA (1988) Cytotoxic T cells clear virus but augment lung pathology in mice infected with respiratory syncytial virus. J Exp Med 168: 1163–1168
9. Kurane I, Innis B, Nisalak A, Hoke C, Nimmannitya S, Meager A, Ennis FA (1989) Human T cell responses to dengue virus antigens. Proliferative responses and interferon gamma production. J Clin Invest 83: 506–513
10. Kurane I, Meager A, Ennis FA (1989) Dengue virus-specific human T cell clones: serotype cross-reactive proliferation, interferon gamma production and cytotoxic activity. J Exp Med 170: 763–775
11. Kurane I, Brinton MA, Samson AL, Ennis FA (1991) Dengue virus-specific, human $CD4^+$ $CD8^-$ cytotoxic T cell clones: multiple patterns of virus crossreactivity recognized by NS3-specific T cell clones. J Virol 65: 1823–1828
12. Kurane I, Rothman AL, Bukowski JF, Kontny U, Janus J, Innis BL, Nisalak A, Nimmannitya S, Meager A, Ennis FA (1990) T-lymphocyte responses to dengue viruses. In: Brinton MA, Heinz FX (eds) New aspects of positive-strand RNA viruses. American Society for Microbiology, Washington, pp 301–304
13. Bukowski JF, Kurane I, Lai C-J, Bray M, Falgout B, Ennis FA (1989) Dengue virus-specific cross-reactive $CD8^+$ human cytotoxic T lymphocytes. J Virol 63: 5086–5091
14. Kurane I, Innis BL, Nimmannitya S, Nisalak A, Meager A, Janus J, Ennis FA (1991) Activation of T lymphocytes in dengue virus infections: high levels of soluble interleukin 2 receptor, soluble CD4, soluble CD8, interleukin 2 and interferon gamma in sera of children with dengue. J Clin Invest 88: 1473–1480
15. Kurane I, Ennis FA (1992) Immunity and immunopathology in dengue virus infections. Semin Immunol 4: 121–127

16. Guyre PM, Gorganerli P, Miller R (1983) Recombinant immune interferon increases immunoglobulin G Fc receptors on cultured human mononuclear phagocytes. J Clin Invest 72: 393–397
17. Kelley VE, Fiers W, Strom TB (1983) Cloned human interferon-γ but not interferon-β or -α, induces experession of HLA-DR determinants by fetal monocytes and myeloid leukemic cell lines. J Immunol 132: 240–245
18. Kontny U, Kurane I, Ennis FA (1988) Gamma interferon augments Fcγ receptor-mediated dengue virus infection of human monocytic cell. J Virol 62: 3928–3933
19. Scott RM, Nimmannitya S, Bancroft WH, Mansuwan P (1976) Shock syndrome in primary dengue infections. Am J Trop Med Hyg 26: 337–343

Authors' address: Dr. I. Kurane, Division of Infectious Diseases and Immunology, Department of Medicine, University of Massachusetts Medical Center, 55 Lake Avenue North, Worcester, MA 01655, U.S.A.

Molecular aspects of pathogenesis
and virulence

Arch Virol (1994) [Suppl] 9: 67–77

Archives
Virology
of
© Springer-Verlag 1994
Printed in Austria

Cardioviral poly(C) tracts and viral pathogenesis

A. C. Palmenberg and **J. E. Osorio**

Institute for Molecular Virology, Department of Animal Health and Biomedical Sciences, University of Wisconsin Madison, Wisconsin, U.S.A.

Summary. Mengovirus is a prototypical member of the cardiovirus genus of the family *Picornaviridae*. The positive-strand RNA genome is 7761 bases in length and encodes a polyprotein of 2293 amino acids. The 5′ non-coding region (758 bases) contains an unusual homopolymeric poly(C) tract, which in the wild-type virus, has a sequence of $C_{50}UC_{10}$. We have discovered through genetic engineering that truncation or deletion of this poly(C) sequence yields infectious virus isolates that grow well in cell culture, but are 10^6 to 10^9 fold less pathogenic to mice than the wild type strain. Animals receiving sublethal doses of the short poly(C) strains characteristically develop high levels of neutralizing antibodies and acquire lifelong protective immunity against challenge with wild type virus. Effectively, the genetically engineered strains are superb vaccines against cardiovirus disease. Moreover, their potential is not limited to murine hosts. Pigs and sub-human primates have also been protectively vaccinated with short poly(C) tract Mengoviruses. The molecular mechanism of poly(C)-mediated pathogenesis is currently under study. Most hypotheses link the activity to induction of the antiviral cytokine, interferon.

Introduction

The pathogenic expression of disease during natural virus infection is dependent upon intrinsic genetic properties of the infecting agent and also upon cellular or humoral responses, induced or inherent, in the infected host. Nonetheless, even for picornaviruses (family *Picornaviridae*), whose genomes and virion structures can be probed at the molecular level, very little is known about how or why a specific viral infection brings about a particular course of disease within its natural host. Central to these processes is a molecular examination of all facets of a virus life cycle, especially those ensuing within the infected cell. Such studies are significant because they bring to light the (exploitable) individualities of particular viral phenotypes, and also serve as critical probes to the inner workings of elementary biological phenomena whose disruption through infection can mediate the expression of disease.

The medical and agricultural importance of picornavirus-caused disease has long targeted these agents for intensive laboratory study. The family includes a diverse variety of highly virulent agents. Among the better known are polioviruses, rhinoviruses, hepatitis A virus and foot-and-mouth disease viruses. Others include the coxsackie viruses, ECHO (enteric cytopathic human orphan) viruses, bovine enteroviruses, and cardioviruses (genus *Cardiovirus*) such as Mengo, encephalomyocarditis virus and Theiler's murine encephalomyelitis viruses. The family members' small genomic size and rapid replication cycle make for easy experimental manipulation. Yet, with the exception of polio vaccines and foot-and-mouth disease vaccines, there are still very few prophylactic or preventive intervention therapies that are uniquely effective against picornaviral diseases. The worldwide economic impact and long term historic interest in viruses of this family make it essential to develop additional intercession methods.

Recent advances in biotechnology, recombinant engineering, and X-ray crystallography now are providing powerful avenues of investigation and hope. Despite the widespread host ranges and disparate medical afflictions caused by individual picornaviruses, members of the family have remarkable similarities in particle structure and genome organization. The molecular mechanisms of pathogenesis slowly are being unraveled for several viruses, and there is every expectation that seminal discoveries in one virus system will find immediate applications in others. This commentary will summarize a recent set of findings with cardioviruses that unexpectedly implicate a peculiar segment of the viral noncoding region, a homopolymeric poly(C) tract, in the pathogenic expression of disease. The development is now being exploited in the preparation of new cardiovirus vaccines.

The viruses

Picornavirus genomes are positive sense, monocistronic RNAs, and encode a single long open reading frame that can be translated to form polyprotein. The RNAs vary in length from 7100 to 8300 bases, depending upon the strain. The polyprotein reading frame generally occupies 85–90% of the theoretical coding capacity of the genome, with the remainder of the bases distributed between 5' (610–1300 bases) and 3' (42–126 bases) noncoding regions. As is characteristic of most eucaryotic mRNAs, there is also a 3' polyadenylate tail of 50–150 bases (see [14] for review of genome structure). Post-translational proteolytic

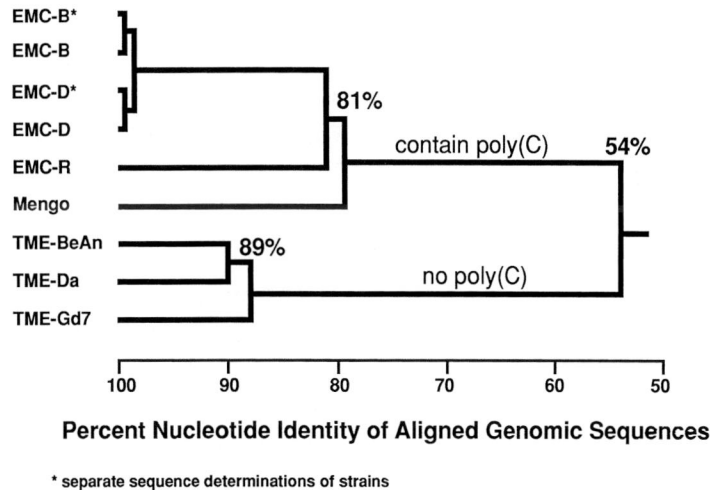

EMC-B*
EMC-B
EMC-D*
EMC-D
EMC-R
Mengo
TME-BeAn
TME-Da
TME-Gd7

81%

contain poly(C)   54%

89%

no poly(C)

100   90   80   70   60   50

Percent Nucleotide Identity of Aligned Genomic Sequences

* separate sequence determinations of strains

Fig. 1. The complete nucleotide sequences of seven strains of cardiovirus were aligned according to the methods and sequences in [15], and their relationships are presented graphically (percent identity). EMC-B and EMC-D were independently sequenced by different laboratories [1, 4], and both determinations are included for completeness

processing of the polyprotein is a distinguishing feature of the viral life cycle. It occurs through a remarkable cascade of primary, secondary and maturational cleavages that provide and regulate the entire spectrum of virus proteins necessary for productive infection.

Our research efforts center on the cardioviruses, a genus of *Picornaviridae* containing two related, but distinct virus groupings (Fig. 1). The Theiler's murine encephalomyelitis viruses (TMEV) and the encephalomyocarditis-like viruses (i.e. EMCV, Mengo, Columbia SK, and Maus Elberfeld) both have murine host ranges, although EMCV was originally isolated from a chimpanzee, and similar viruses have been obtained from several human and non-human primates as well as from domestic pigs and other mammals [10, 11]. All cardioviruses are closely related, and share at least 54% amino acid identity along their polyprotein lengths. Nucleotide identity between genomes from within either group is typically greater than 80%. Yet despite this overall similarity, a special genetic feature, peculiar to the EMCV-like viruses, allows easy discrimination between the TMEV and EMCV group members.

The EMCV-like viruses (genus *Cardiovirus*) and the aphthoviruses (genus *Aphthovirus*), including foot-and-mouth disease viruses, are distinguished among picornaviruses, and indeed among all positive-strand RNA viruses, by the notable presence within their 5′ noncoding sequences of long, homopolymeric polyribocytidylate segments. Typically, the poly(C) regions are about 150 (cardioviruses) to 350 (aphthoviruses) nucleotides from the 5′ end of the genome and may contain as many as

60 to 420 pyrimidine residues in a row [2, 8]. Natural viral isolates are characterized by the distinctive lengths of their poly(C) tracts and also by specific sequence discontinuities, such as uridine residues, which sometimes disrupt the homopolymer [5]. All cardioviruses except the TMEV contain these unusual segments.

Our plasmids

During the past several years, we have attempted to develop a series of cDNA plasmids containing genomic segments from EMCV and Mengo so we could explore the role of the poly(C) in the viral life cycle. Construction of bacterial plasmids with picornavirus sequences is, of course, not very novel these days. Infectious polio, rhino, coxsackie and hepatitis-A constructs have been characterized and reported (e.g. [3, 16]). However, none of these viruses contain poly(C). As we and others discovered, extended poly(dC):poly(dG) sequences, engineered from the natural tracts, are frustratingly difficult to clone into viable bacterial vectors. Nevertheless, we were eventually able to patch together several cDNAs that contained full-length cardioviral sequences, albeit at first, only sequences with discretely shorter poly(C) tracts than were present in the original parental viruses ($C_{50}UC_{10}$ in Mengo, and $C_{115}UCUC_3UC_{10}$ in EMCV, Rueckert strain). Tracts of C_8, C_{12} and $C_{13}UC_{10}$ were initially isolated.

Because, by virtue of their existence, the long parental poly(C) tracts were assumed to play an essential role in the life cycle of viruses that contain them, we at first doubted whether our truncated poly(C)s would have any biological value. To our surprise, RNA transcripts derived from these clones proved completely infectious for cells in culture. The progeny virus resulting from transfections could be propagated in HeLa cells in a manner indistinguishable from parental isolates (with long poly(C)s). Moreover, the short tract viruses steadfastly maintained their engineered, deleted sequences during serial tissue culture passage [7]. Why then should the enigmatic longer tracts be so faithfully carried in wildtype strains when they are apparently dispensable for reproductive functions of the virus?

Poly(C)-mediated attenuation

During further attempts to identify altered growth characteristics peculiar to the long poly(C) tracts, we began inoculation experiments with mice, the natural host for cardioviruses. Both Mengo and EMCV normally cause a rapid and lethal meningoencephalomyelitis in animals inoculated intraperitoneally (i.p.) or intracerebrally (i.c.). Remarkably, we discovered that all cDNA-derived mengoviruses containing artificially shortened

Mengo Viral Genome

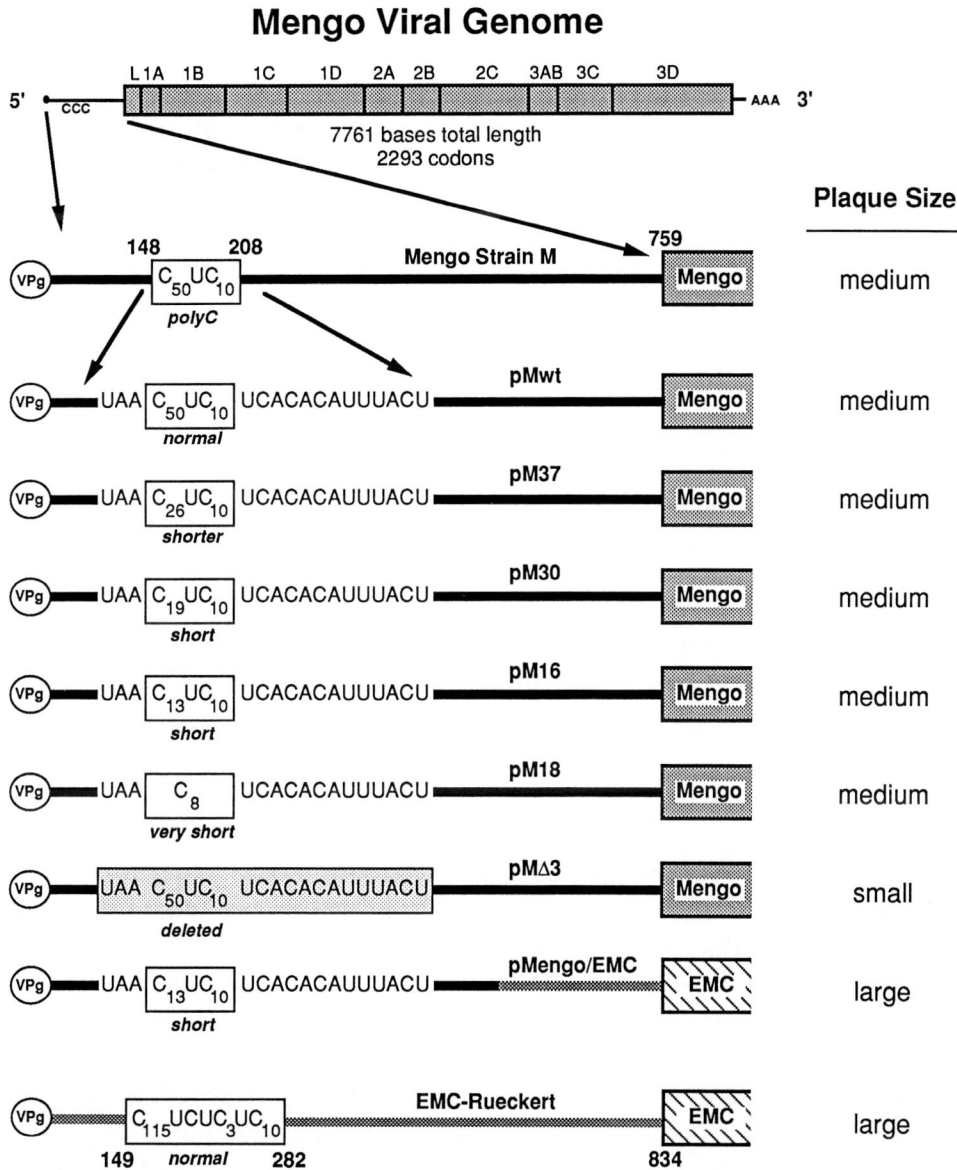

Fig. 2. The schematic diagram shows the poly(C) region and flanking sequences from a representative panel of natural and genetically engineered cardioviruses. The characteristic plaque phenotype of each strain (on Hela cells) is indicated

poly(C)s had dramatically attenuated pathogenicity when compared to wildtype virus or when compared to a newly engineered progeny virus (pMwt) containing poly(C) of wildtype length. Viruses isolated from the brains of sacrificed animals were attenuated progeny that only replicated to a limited extent following inoculation, and clearly did not induce the devastating pathological effects of infection with wildtype virus. Rather, animals receiving the short tract strains uniformly developed life-long protective immunity against normally lethal challenge with wildtype cardioviruses [6]. Effectively, our truncated poly(C) mutations produced new attenuated virus strains with superb vaccine potential for mice.

To date, we have constructed and characterized more than 25 cDNA plasmids containing full-length copies of Mengo (strain M) and Mengo/-EMCV chimeric genomes. A representative panel is shown in Fig. 2. Progeny viruses derived from the pM plasmids are identical in sequence except for their poly(C) tracts, which range from wild-type length (C_{50}-UC_{10}) to C_0. We have also developed viruses that are missing short segments (5′ and 3′) of the heteropolymeric sequences that flank the poly(C) (e.g. pMΔ3). All strains are viable for growth in HeLa cells, and all except those with 3′ flanking deletions have the medium-sized plaque phenotype characteristic of the parental strain. Not for lack of trying, we have still been unable to clone the long EMCV-Rueckert poly(C) tract ($C_{115}UCUC_3UC_{10}$) [5]. Instead, we have derived a parallel series of viable chimeric constructions that link the 5′ portions of the Mengo plasmids and short poly(C)s to the polyprotein region from EMCV. The chimeric viruses (pMengoEMC) have the large plaque phenotype of the EMCV parent from which most of the genome was derived [7].

Poly(C) length requirement

The extensive panel of poly(C) tract viruses has enabled experimentation with the length requirements for viral pathogenicity. Our functional definition of pathogenicity is the ability of a virus to cause mortality within 14 days following i.p. or i.c. inoculation. Wildtype Mengo or EMCV-inoculated mice generally show clear-cut signs of neurological degeneration within the first week after inoculation (hind leg paralysis, hunched posture and ataxia), and virtually all mice are dead by day 10, even from i.p. inoculations. Of those animals that have died after receiving a short tract progeny virus, none have been observed to exhibit overt signs of disease until nearly 10 days after injection. Routinely, all animals are monitored for at least 4 weeks, and scored as "ill" if they show any sign of disease.

Figure 3 and Table 1 summarize representative results. When tested in 4-week old mice inoculated i.c. the LD_{50} for pM16 virus (8×10^6 pfu), and pM18 virus (10^7 pfu), were measured as more than 10^6 fold higher than for Mengo (9 pfu), pMwt (9 pfu), or EMCV-R (<1 pfu). Within the window of C_{37} to C_{24} (including discontinuities), the viral LD_{50} increased by about one $\log_{10}$ for every 3 Cs that were removed. Longer poly(C)s (>C_{37}) and shorter poly(C)s (<C_{24}) were more pathogenic or more attenuated, respectively, but the steepest portion of the curve appears to

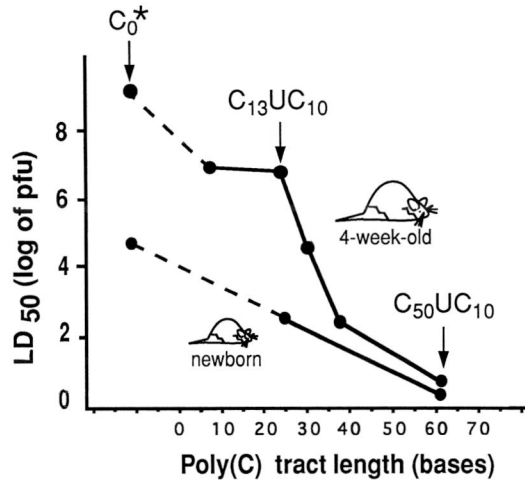

Fig. 3. The LD_{50} values (dose causing 50% mortality within 14 days post-inoculation) for representative natural and genetically engineered cardioviruses are presented graphically. All experiments were carried out in ICR Swiss mice (4-week old females, or 12-h old newborns). Each determination represents cumulative data from at least 50 animals. Virus titer was quantitated by plaque forming units (pfu) in HeLa cell assays, and verified immediately before inoculation (i.c. or i.p.). Co^*, contains flanking deletions

Table 1. LD_{50} of representative viruses (p.f.u.)

Virus	Route	i.c.	i.c.	i.p.
		4-week-old	newborn	4-week-old
EMC-R	$C_{115}UCUC_3UC_{10}$	<1	nd	$<10^2$
Mengo	$C_{50}UC_{10}$	9	nd	10^5
pMwt	$C_{50}UC_{10}$	9	6	10^5
pM37	$C_{26}UC_{10}$	7×10^2	nd	nd
pM30	$C_{19}UC_{10}$	6×10^4	nd	nd
pM16	$C_{13}UC_{10}$	8×10^6	3×10^2	$>10^{11}$
pM18	C_8	10^7	nd	$>10^{10}$
pMΔ3	C_0[a]	$>10^9$	4×10^4	nd

[a] Flanking deletions, too
nd Not determined

be centered on a length of about C_{30}. The poly(C)-mediated attenuation effects in i.p. experiments are even more dramatic, and we have been unable to kill 4-week old animals with as much as 50 µg of live, short-tract virus (10^{11} pfu, i.p.). Though newborn mice are known to be particularly susceptible to cardiovirus infection, even these animals proved quite resistant to viral strains that completely lacked poly(C). Some newborn mice, for example survived i.c. doses of up to 10^6 pfu of pMΔ3 virus without showing illness or disease (LD_{50} is about 10^4 pfu).

Vaccine potential

All animals receiving a sublethal dose of any pM or pMengoEMC short tract virus, rapidly seroconvert with high titers of neutralizing antibodies and are effectively protected against challenge with EMCV or Mengo [6]. A few inoculated animals have been observed long term, and more than two years after receiving their initial dose (10^4 pfu pM16), they still appear healthy. Current experiments are exploring various routes and doses for optimum inoculation, but it is already clear that very small amounts of short tract virus (e.g. 10^2 pfu pM16), administered orally or by subcutaneous injection, are sufficient to elicit a reliable protective response (manuscript in prep.). Given the high LD_{50} of the short tract viruses and the potency of the immune reaction to inoculation, it is obvious that one of our experimental objectives must be to characterize the molecular basis for poly(C)-mediated attenuation, with the eventual goal of exploiting this phenomenon and the principles to be learned from it, so that we can develop new and effective, live picornavirus vaccines.

While most experiments are carried out in ICR Swiss mice, we have also learned that poly(C) attenuation is not exclusive for this strain. Results similar to those in Fig. 3 have been obtained in Balb-C mice and nude mice with Balb-C backgrounds (not shown), and we have no reason to suspect that other mouse strains will behave differently. Perhaps of greater interest, however, are recent results with non-murine animals. As mentioned above, EMCV and Mengo are known to infect human and non-human primates as well as pigs and other animals. In 1991, a spontaneous EMCV epizootic in colony of baboons used for biomedical research provided important impetus to test a short tract virus (i.e. pM16) in primate hosts [12]. The results of these trials (and other collaborative experiments with *Rhesus* macaques and domestic pigs), will be published in detail elsewhere, but the data can easily be summarized by stating that the poly(C)-mediated attenuation phenomenon is not restricted to murine hosts. In all tested animals, pM16 virus proved a safe, efficacious vaccine, and induced long-lasting protective immunity against the lethal, circulating strains of EMCV.

Preliminary genetic stability experiments also suggest that the reversion potential of short poly(C) tract viruses is minimum, a property that could add to their value as putative vaccines. Our engineered viruses passage with fidelity in tissue culture [7], and we have been unable to detect a pathogenic revertant after multiple serial brain passages in 4-week old mice (experiment now in progress). The reasons for such stability are presently unclear. One might expect that polymerase slippage mechanisms or RNA recombination could easily restore the missing Cs to a pathogenic tract length during an infection. That we have not observed this suggests, surprisingly, that the attenuated viruses may be phenotypically dominant, and their elicited protective response is sufficient to overwhelm a pathogenic variant, should it arise. In this light, it is tempting to speculate that the extraordinary length of the natural poly(C) tracts (up to 420 bases in foot-and-mouth disease viruses) [8] might, conversely, be a safeguard against significant accidental deletion. Should inadvertent rearrangements leave a tract length shorter than C_{30}, the virus would effectively vaccinate the host, and probably be cleared before it could passage further. Our initial clumsy attempts to clone long poly(C) tracts, which resulted instead in unnatural short sequences, now seem fortuitous indeed, as we apparently stumbled upon a genome configuration that would be strongly rejected by natural selection.

How does poly(C) work?

We do not know why a simple truncation of a non-coding homopolymeric sequence should have such profound effects on viral pathogenesis, but there are clear hypothetical possibilities. Either, (a) there is a host cell-type (or tissue) that distinctively restricts replication of short tract virus during early infection, or (b) both short and long tract viruses replicate equivalently in all cells but induce differential host responses during primary infection, which in consequence set alternate pathogenic directions. Specific hypotheses abound, and there are many potential mechanisms that singly or in concert could contribute to an attenuated phenotype. Whether the short tract strains are simply more effective at tripping important protective elements within the host's defense system (e.g. NK cells, or T-cell recognition?) or whether they are defective in some critical replication step within a specific target tissue are only conjecture at the present time. Perhaps it may be more realistic to regard the long poly(C) viruses as having enhanced pathogenic effects, rather than to consider the short tract viruses as relatively attenuated. Conceivably, elements from the host's own immune system, amplified or up-regulated by long poly(C)s, facilitate the rapid establishment of infection in animals inoculated with wild-type viruses. This view is putatively bolstered by experiments showing that drug-induced immunosuppression temporarily protects mice from cardioviruses that are normally pathogenic and lethal [18].

One clear candidate for this type of agent is the antiviral cytokine, interferon. Links between interferon induction and cardiovirus pathogenicity are already well established. Moreover, synthetic poly(I) : poly(C), which is chemically quite similar to our enigmatic viral tract, is known to be a potent inducer of cellular interferon [9]. The in vivo properties of interferon sensitive (*is*) and resistant (*ir*) mutants of Mengo have been described [17]. Unfortunately, these isolates have not been characterized genetically, and we do not know if they differ in poly(C) length. Likewise, studies with the closely related plaque variants of EMCV, EMC-B and EMC-D, have revealed differences in the interferon response associated with cardioviral infection. It has been proposed that the greater ability of EMC-B than of EMC-D to induce interferon in certain inbred mouse strains is the direct cause of the "attenuated" diabetogenic phenotype of EMC-B [9, 13]. Though we still do not understand these phenomena at the mechanistic level, in view of our Mengo results with engineered short tract viruses, we are beginning to look more closely at the natural cause and effect connections between interferon, pathogenicity and viral poly(C). Mice are difficult test tubes, but unraveling at least some of these relevant processes seems key to understanding the molecular basis of poly(C)-mediated attenuation.

Acknowledgements

This work was supported by Public Health Service grant AI-30566 from the NIH.

References

1. Bae YS, Eun HM, Yoon JW (1989) Molecular identification of a diabetogenic viral gene. Diabetes 38: 316–320
2. Brown F, Newman JFE, Stott EJ, Porter AG, Frisby D, Newton D, Carey N, Fellner P (1974) Poly C in animal viral RNAs. Nature 251: 342–344
3. Callahan PL, Mizutani S, Colonno RJ (1985) Molecular cloning and complete sequence determination of RNA genome of human rhinovirus type 14. Proc Natl Acad Sci USA 82: 732–736
4. Cohen SH, Naviaux RK, Brink KMV, Jordan GW (1988) Comparison of the nucleotide sequences of diabetogenic and nondiabetogenic encephalomyocarditis virus. Virology 166: 603–607
5. Duke GM, Hoffman MA, Palmenberg AC (1992) Sequence and structural elements that contribute to efficient encephalomyocarditis viral RNA translation. J Virol 66: 1602–1609
6. Duke GM, Osorio JE, Palmenberg AC (1990) Attenuation of Mengovirus through genetic engineering of the 5' noncoding poly(C) tract. Nature 343: 474–476
7. Duke GM, Palmenberg AC (1989) Cloning and synthesis of infectious cardiovirus RNAs containing short, discrete poly(C) tracts. J Virol 63: 1822–1826
8. Escarmis C, Toja M, Medina M, Domingo E (1992) Modifications of the 5' untranslated region of foot-and-mouth disease virus after prolonged persistence in cell culture. Virus Res 26: 113–125
9. Giron DJ, Agostini HJ, Thomas DC (1988) Effect of interferons and poly(I) : poly(C) on the pathogenesis of the diabetogenic variant of encephalomyocarditis virus in different mouse strains. J Interferon Res 8: 745–753

10. Grainer JH (1961) Studies on the natural and experimental infection of animals in Florida with the encephalomyocarditis virus. Proc US Livestock San A: 556–572

11. Helwig FC, Schmidt ECH (1945) A filter passing agent producing intersticial myocarditis in anthropoid apes and small animals. Science 102: 31–33

12. Hubbard GB, Soike KF, Butler TM, Carey KD, Davis H, Butcher WI, Gauntt CJ (1992) An encephalomyocarditis virus epizootic in a baboon colony. Lab Anim Sci 42: 233–239

13. Jordan GW, Cohen SH (1987) Encephalomyocarditis virus-induced diabetes mellitus in mice: Model of viral pathogenesis. Rev Infect Dis 9: 917–924

14. Palmenberg AC (1987) Genome organization, translation and processing in picornaviruses. In: Rolands DJ, Mahy BWJ, Mayo M (eds) The molecular biology of positive strand RNA viruses. Academic Press, London, pp 1–15

15. Palmenberg AC (1989) Sequence alignments of picornaviral capsid proteins. In: Semler B, Semler BL, Ehrenfeld E (eds) Molecular aspects of picornavirus infection and detection. ASM Publications, Washington, pp 211–241

16. Racaniello VR, Baltimore D (1981) Cloned poliovirus complementary DNA is infectious in mammalian cells. Science 214: 916–918

17. Simon EH, Kung S, Koh TT, Brandman P (1976) Interferon sensitive mutants of Mengovirus. I. Isolation and biological characterization. Virology 69: 727–736

18. Zschiesche W, Veckenstedt A (1979) Pathogenicity of Mengo virus to mice. III. Potentiation of infection by immunosuppressants. Exp Pathol 17: 387–393

Authors' address: Dr. A. C. Palmenberg, Institute for Molecular Virology, Department of Animal Health and Biomedical Sciences, 1655 Linden Drive, University of Wisconsin, Madison, WI 53706, U.S.A.

Arch Virol (1994) [Suppl] 9: 79–86

Archives
Virology
of

© Springer-Verlag 1994
Printed in Austria

Transgenic mice and the pathogenesis of poliomyelitis

V. R. Racaniello and **R. Ren**

Department of Microbiology, Columbia University College of Physicians
and Surgeons, New York, New York, U.S.A.

Summary. Transgenic mice expressing the cell receptor for poliovirus have been generated and are susceptible to poliovirus infection. TgPVR mice have been used to answer questions about the pathogenesis of poliovirus infection. Despite the widespread pattern of PVR expression, poliovirus infection in TgPVR mice is restricted to only a few sites, indicating that poliovirus tropism is not controlled solely by the ability of cells to bind virus. After intramuscular inoculation, poliovirus travels to the spinal cord by axonal transport. This route of entry into the central nervous system may play a role in the pathogenesis of poliovirus infections in humans.

Introduction

In the past decade, studies on the molecular biology, structure and genetics of polioviruses have made these among the best understood viruses of eukaryotic cells. Because a convenient animal model for poliomyelitis has not been available, our understanding of the pathogenesis of this disease has not kept up with progress in elucidating other aspects of poliovirus replication. The recent establishment of transgenic mice expressing the cell receptor for poliovirus (TgPVR mice) should now lead to renewed studies on poliovirus pathogenesis [1, 2]. We have used TgPVR mice to investigate two unanswered questions about the pathogenesis of poliomyelitis: the basis for the restricted tropism of the virus, and the mechanism by which the virus enters the central nervous system (CNS) from its initial sites of infection.

The cell receptor for polioviruses

Poliovirus replication begins when the virus attaches to a cell surface receptor, which has been identified as a novel member of the immunoglobulin (Ig) superfamily [3]. The poliovirus receptor (PVR) is composed of three extracellular immunoglobulin (Ig)-like domains, a transmembrane helix and a cytoplasmic tail. Other virus receptors have been identified as members of the Ig superfamily, including those for HIV-1 [4], the majority of rhinovirus serotypes [5–7], and mouse hepatitis virus [8].

Cultured mouse cells are resistant to poliovirus infection because they do not express PVR. However, if PVR is expressed in these cells, they become susceptible to poliovirus infection [3]. Mice are resistant to infection with most strains of poliovirus because they lack a receptor for the virus (the exceptions are certain mouse-adapted strains of poliovirus, which appear to use a different receptor in mice). Transgenic mice that express the human PVR gene are susceptible to infection with all three poliovirus serotypes, and develop a disease that clinically and histopathologically resembles human poliomyelitis [2]. PVR is therefore the cellular determinant of the host range of poliovirus in mice.

Poliovirus tissue tropism

Human poliovirus infections begin when virus is ingested and replicates in the oropharyngeal and intestinal mucosa. This primary replication leads to the establishment of a viremia, which enables the virus to spread to many other tissues; however, in the animal subsequent poliovirus replication is limited to a few sites: neurons of the brain and spinal cord, and an undefined extraneural site. The basis for the restriction of poliovirus replication to so few sites has not been determined. Early approaches to this question included studies of the ability of polioviruses to bind to tissue homogenates, to determine whether virus binding correlated with tissue sensitivity. In some studies, poliovirus binding activity was identified only in susceptible sites, such as CNS tissue and intestine [11]. In other studies, virus binding was more extensive [12–14]. There are clear difficulties in interpreting the results of binding assays conducted with tissue homogenates. For example, absence of virus binding activity may be an experimental artifact resulting from lability of virus receptors. Nevertheless, it is often concluded that poliovirus tissue tropism is controlled at the level of receptor expression [13, 15].

Studies on PVR RNA expression in human and TgPVR tissues suggest that susceptibility to poliovirus infection may not be determined simply by expression of virus-binding sites. In TgPVR mice, the patterns of PVR RNA expression and poliovirus replication were determined by Northern blot hybridization analysis and by in situ hybridization [2, 16]. PVR RNA is expressed in virtually all TgPVR mouse tissues, including spleen, muscle, lung, heart, small intestine, liver, kidney, thymus, brain and spinal cord, as determined by Northern blot hybridization. The results of in situ hybridization show that PVR RNA is expressed at high levels in neurons of the central and peripheral nervous system, developing T lymphocytes in the thymus, epithelial cells of Bowman's capsule and tubules in the kidney, alveolar cells in the lung, and endocrine cells in the adrenal cortex, and at low levels in intestine, spleen, skeletal muscle and brown fat. The cell type specificity of PVR expression is striking. For example, within the CNS and PNS, PVR RNA was detected only in neurons. This restricted pattern of expression may be related to the cell function of PVR.

Despite the widespread expression of PVR RNA in TgPVR mice, poliovirus replication was detected only at a few locations, including neurons of the brain and spinal cord, skeletal muscle, and occasionally in brown adipose fat. For example, after intraperitoneal inoculation of type 1 poliovirus, infectious virus could be detected in the kidney within 24 h, but no viral replication ensued. Viral replication was also absent in a variety of other tissues from the same animals. Thus, despite expression of PVR RNA, many TgPVR tissues remain refractory to poliovirus infection.

One possible explanation for these findings is that despite expression of PVR RNA, receptor protein may not be expressed on the cell surface. Immunohistochemical analysis of TgPVR tissues with anti-PVR mAbs is currently being used to address this question. However, other evidence suggests that in at least some TgPVR tissues, PVR protein does reach the cell surface. Cultured thymocytes and freshly dispersed kidney cells from TgPVR mice can bind polioviruses, yet these cells are resistant to infection [16; unpubl. res.]. In these cells, the block to poliovirus infection clearly is not at the level of binding to PVR.

If thymocytes and kidney cells can bind polioviruses, what other obstacles might prevent infection? Inability of virus to reach cells expressing PVR, such as T lymphocytes of the thymus, might preclude poliovirus infection, but this explanation cannot account for the resistance to infection of tubular epithelial cells in the kidney. Poliovirus might bind to many tissues, but might not be able to deliver its RNA genome into the cell. If other factors are required for poliovirus entry and/or uncoating, these factors might only be expressed in susceptible tissues. Finally, certain cell types may be unable to support other aspects of poliovirus replication, such as translation and replication of poliovirus RNA, and assembly of new virus particles. The basis for the restriction of poliovirus replication in different tissues may vary.

In human tissues, Northern and Western blot analyses have demonstrated widespread expression of PVR RNA and protein, suggesting that PVR expression may not be the primary determinant of poliovirus tropism [3, 17]. However, it is not possible to determine from these results whether poliovirus-binding sites are expressed on the cell surface. Studies to examine the expression of PVR on specific cells by immuno-histochemistry, using anti-PVR mAbs, are in progress. It will also be informative to purify PVR from the membranes of poliovirus-susceptible and -resistant human tissues, and determine its ability to bind polio-viruses. This approach would circumvent the difficulties associated with binding assays using crude tissue homogenates. Although such studies will provide information on the ability of human tissues to bind polioviruses, it will also be necessary to assess the permissivity of other steps in poliovirus replication. Many human tissues develop susceptibility to poliovirus infection when explanted into cell culture, and thus cannot be used for such studies. It will therefore be necessary to identify human tissues that retain resistance to poliovirus infection during growth in cell culture.

Spread of poliovirus to the central nervous system

The route by which polioviruses reach the CNS has been a subject of much discussion. It has been suggested that virus may enter the CNS from the blood across the blood-brain barrier, or may enter a peripheral nerve and be transmitted to the CNS [18–20]. In support of the hypo-thesis that poliovirus spreads to the CNS from the blood, it appears that viremia is necessary for spread to the CNS and precedes paralytic infection, and that the presence of antiviral antibodies in the blood halts viral spread and prevents invasion of the CNS [21, 22].

The neural spread hypothesis is also supported by several observa-tions. Following intramuscular injection of monkeys with the highly neurotropic poliovirus type 2 MV strain, localization of initial paralysis in the injected limb occurred at high frequency, and freezing the sciatic nerve blocked spread of this virus from muscle to the CNS [23]. In the Cutter incident, in which children received incompletely inactivated poliovirus vaccine, a high frequency of initial paralysis was observed in the inoculated limb [24].

The amount of virus causing paralysis in 50% of inoculated TgPVR mice is similar for intramuscular and intracerebral inoculation, suggest-ing that polioviruses may reach the CNS directly after intramuscular inoculation. Paralysis is initially observed in the injected limb in TgPVR mice but not in non-transgenic mice inoculated intramuscularly with P2/Lansing, a mouse-adapted poliovirus type 2 strain. Following intra-muscular injection, poliovirus spreads first to the inferior segment of the spinal cord, then to the superior spinal cord, and then to the brain. Finally, development of CNS disease after inoculation in the hindlimb footpad is blocked by sciatic nerve transection. These results directly demonstrate that poliovirus spreads from muscle to the CNS through nerve pathways [25].

Based on our studies of poliovirus pathogenesis in transgenic mice, and the observations of others on the disease in humans and monkeys [26, 27], we suggest a revised description of the pathogenesis of poliomyelitis in humans. After ingestion, poliovirus first replicates in the intestinal (and occasionally oropharyngeal) mucosa, although the specific cell type in which replication occurs has not been identified. Virus produced in the mucosa is shed into the gut lumen and the blood, leading to virus in the feces and viremia, respectively. Disseminated virus then replicates in skeletal muscle cells, enters peripheral nerves and spreads by axonal transport to the CNS. Virus replication in skeletal muscle maintains a persisting viremia, which may disseminate infection to multiple sites from which virus may also enter the CNS. The persisting viremia that precedes paralytic infection is important for virus spread to the CNS [21], an observation that has been used as evidence in support of the hypothesis that virus enters the CNS from the blood. However, persisting viremia may be the result of successful replication of poliovirus in skeletal muscle, from which virus can spread to the CNS and to the bloodstream.

Poliovirus replication is occasionally detected in brown adipose tissues and neurons of peripheral ganglia of TgPVR mice, consistent with observations made in monkeys (unpubl. res.). Brown adipose tissues might be another extraneural tissue that supports poliovirus replication and maintains the persisting viremia, and transmission of virus along nerve fibers from peripheral ganglia may provide an additional route for entry into the CNS. However, because poliovirus replication can be more readily detected in skeletal muscle, viral replication in skeletal muscle and subsequent spread along nerves to the CNS may still be the major route for virus entry into the CNS.

There are other pathways that polioviruses might use to gain entry to the CNS. Because the initial site of poliovirus replication is the alimentary tract [27], it is possible that polioviruses may spread from the intestinal lumen to the CNS through vagal autonomic nerve fibers, as found for a reovirus [28]. The observation that 5–30% of poliovirus infections involve the brain stem is consistent with this route of spread. However, in most paralytic infections in humans, virus appears to initially infect the lower motor neurons of the spinal cord, which is consistent with the hypothesis that poliovirus spreads from the muscle to the CNS.

Because passive or active immunization against poliovirus blocks viremia and prevents CNS infection, it has been assumed that poliovirus enters the CNS through the blood-brain barrier [21, 22]. However, studies on other virus infections suggest that this observation may be consistent with neural spread. During infection of mice with reovirus serotype 3, the virus spreads by nerves and not by the bloodstream to the CNS, despite the presence of viremia [29, 30]. Antiviral antibody decreases viremia and prevents appearance of virus in the CNS after inoculation of reovirus serotype 3 in the hindlimb footpad [31]. Administration of anti-viral antibody can result in clearance of Sindbis virus infection from mouse neurons by restricting viral gene expression [32]. These studies demonstrate that blocking virus entry to the CNS by the blood-brain barrier is not the only mechanism by which antibody prevents CNS infection. The mechanism by which anti-poliovirus antibody prevents CNS infection therefore remains to be determined.

Entry of polioviruses into the CNS is a rare event, occurring only in approximately 1 in 100 infections. Polioviruses may therefore be viewed as inhabitants of the gut which occasionally, and perhaps by chance, replicate in neurons. Why cells of the gut and the CNS are both susceptible to infection with polioviruses, and how the virus travels from the gut to the CNS, are questions which can now be addressed by studying poliomyelitis in TgPVR mice.

References

1. Koike S, Taya C, Kurata T, Abe S, Ise I, Yonekawa H, Nomoto A (1991) Transgenic mice susceptible to poliovirus. Proc Natl Acad Sci USA 88: 951–955
2. Ren R, Costantini FC, Gorgacz EJ, Lee JJ, Racaniello VR (1990) Transgenic mice expressing a human poliovirus receptor: A new model for poliomyelitis. Cell 63: 353–362
3. Mendelsohn C, Wimmer E, Racaniello VR (1989) Cellular receptor for poliovirus: molecular cloning, nucleotide sequence and expression of a new member of the immunoglobulin superfamily. Cell 56: 855–865
4. Maddon PJ, Dalgleish AG, McDougal JS, Clapham PR, Weiss RA, Axel R (1986) The T4 gene encodes the AIDS virus receptor and is expressed in the immune system and the brain. Cell 47: 333–348
5. Greve JM, Davis G, Meyer AM, Forte CP, Yost SC, Marlor CW, Kamarck ME, McClelland A (1989) The major human rhinovirus receptor is ICAM-1. Cell 56: 839–847
6. Staunton DE, Merluzzi VJ, Rothlein R, Barton R, Marlin SD, Springer TA (1989) A cell adhesion molecule, ICAM-1, is the major surface receptor for rhinoviruses. Cell 56: 849–853
7. Tomassini JE, Graham D, DeWitt CM, Lineberger DW, Rodkey JA, Colonno RJ (1989) cDNA cloning reveals that the major group rhinovirus receptor on HeLa cells is intercellular adhesion molecule 1. Proc Natl Acad Sci USA 86: 4907–4911
8. Williams RK, Jiang G-S, Holmes KV (1991) Receptor for mouse hepatitis virus is a member of the carcinoembryonic antigen family of glycoproteins. Proc Natl Acad Sci USA 88: 5533–5536
9. Kaplan G, Freistadt MS, Racaniello VR (1990) Neutralization of poliovirus by cell receptors expressed in insect cells. J Virol 64: 4697–4702

10. Kaplan G, Racaniello VR (1991) Down regulation of poliovirus receptor RNA in HeLa cells resistant to poliovirus infection. J Virol 65: 1829–1835

11. Evans CA, Byatt PH, Chambers VC, Smith WM (1954) Growth of neurotropic viruses in extraneural tissues VI. Absence of in vivo multiplication of poliomyelitis virus, types I and II, after intratesticular inoculation of monkeys and other animals. J Immunol 72: 348–352

12. Brown RH, Johnson D, Ogonowski M, Weiner HL (1987) Type 1 human poliovirus binds to human synaptosomes. Ann Neurol 21: 64–70

13. Holland JJ (1961) Receptor affinities as major determinants of enterovirus tissue tropisms in humans. Virology 15: 312–326

14. Kunin CM, Jordan WS (1961) In vitro adsorption of poliovirus by noncultured tissues. Effect of species, age and malignancy. Am J Hyg 73: 245–257

15. Crowell RL, Landau BJ (1983) Receptors in the initiation of picornavirus infections. In: Fraenkel-Conrat H, Wagner RR (eds) Comprehensive virology. Academic Press, New York, pp 1–42

16. Ren R, Racaniello V (1992) Human poliovirus receptor gene expression and poliovirus tissue tropism in transgenic mice. J Virol 66: 296–304

17. Freistadt MF, Kaplan G, Racaniello VR (1990) Heterogeneous expression of poliovirus receptor-related proteins in human cells and tissues. Mol Cell Biol 10: 5700–5706

18. Bodian D (1959) Poliomyelitis: pathogenesis and histopathology. In: Rivers TM, Horsfall FL (eds) Viral and rickettsial infections of man. Lippincott, Philadelphia, pp 479–498

19. Morrison LA, Fields BN (1991) Parallel mechanisms in the neuropathogenesis of enteric virus infections. J Virol 65: 2767–2772

20. Sabin AB (1957) Properties of attenuated polioviruses and their behavior in human beings In: Rivers TM (ed) Cellular biology, nucleic acids and viruses. New York Academy of Science, New York, pp 113–133

21. Bodian D, Horstmann DH (1965) Polioviruses. In: Horsfall FL, Tamm I (eds) Viral and rickettsial infections of man. Lippincott, Philadelphia, pp 430–473

22. Melnick JL (1985) Enteroviruses: polioviruses, coxsackieviruses, echoviruses and newer enteroviruses. In: Fields BN, Knipe DM, Chanock RM, Melnick JL, Roizman B, Shope RE (eds) Virology. Raven Press, New York, pp 705–738

23. Nathanson N, Bodian D (1961) Experimental poliomyelitis following intramuscular virus injection. 1. The effect of neural block on a neurotropic and a pantropic strain. Bull Johns Hopkins Hosp 108: 308–319

24. Nathanson N, Langmuir A (1963) The Cutter incident: poliomyelitis following formaldehyde-inactivated poliovirus vaccination in the United States during the spring of 1955. III. Comparison of the clinical character of vaccinated and contact cases occurring after use of high rate lots of Cutter vaccine. Am J Hyg 78: 61–81

25. Ren R, Racaniello VR (1992) Poliovirus spreads from muscle to the central nervous system by neural pathways. J Infect Dis 166: 635–654

26. Bodian D (1955) Emerging concept of poliomyelitis infection. Science 12: 105–108

27. Sabin AB (1956) Pathogenesis of Poliomyelitis: reappraisal in light of new data. Science 123: 1151–1157

28. Morrison LA, Sidman RL, Fields BN (1991) Direct spread of reovirus from the intestinal lumen to the central nervous system through vagal autonomic nerve fibers. Proc Natl Acad Sci USA 88: 3852–3856

29. Flamand A, Gagner J, Morrison LA, Fields BN (1991) Penetration of the nervous systems of suckling mice by mammalian reoviruses. J Virol 65: 123–131

30. Tyler KL, McPhee DA, Fields BN (1986) Distinct pathways of viral spread in the host determined by reovirus S1 gene segment. Science 233: 770–774
31. Tyler KL, Virgin IVth HW, Bassel-Duby R, Fields BN (1989) Antibody inhibits defined stages in the pathogenesis of reovirus serotype 3 infection of the central nervous system. J Exp Med 170: 887–900
32. Levine B, Hardwick JM, Trapp BD, Crawford TO, Bollinger RC, Griffin DE (1991) Antibody-mediated clearance of alphavirus infection from neurons. Science 254: 856–859
33. Morrison ME, Racaniello VR (1992) Molecular cloning and expression of a murine homolog of the human poliovirus receptor gene. J Virol 66: 2807–2813

Authors' address: Dr. V. R. Racaniello, Department of Microbiology, Columbia University College of Physicians and Surgeons, 701 West 168th Street, New York, NY 10032, U.S.A.

Arch Virol (1994) [Suppl] 9: 87–97

Archives
Virology
© Springer-Verlag 1994
Printed in Austria

Adaptation of positive-strand RNA viruses to plants

R. Goldbach[1], **J. Wellink**[2], **J. Verver**[1], **A. van Kammen**[2], **D. Kasteel**[1], and **J. van Lent**[1]

Departments of [1] Virology and [2] Molecular Biology, Agricultural University
Wageningen, Wageningen, The Netherlands

Summary. The vast majority of positive-strand RNA viruses (more than 500 species) are adapted to infection of plant hosts. Genome sequence comparisons of these plant RNA viruses have revealed that most of them are genetically related to animal cell-infecting counterparts; this led to the concept of "superfamilies". Comparison of genetic maps of representative plant and animal viruses belonging to the same superfamily (e.g. cowpea mosaic virus [CPMV] versus picornaviruses and tobacco mosaic virus versus alphaviruses) have revealed genes in the plant viral genomes that appear to be essential adaptations needed for successful invasion and spread through their plant hosts. The best studied example represents the "movement protein" gene that is actively involved in cell-to-cell spread of plant viruses, thereby playing a key role in virulence and pathogenesis. In this paper the host adaptations of a number of plant viruses will be discussed, with special emphasis on the cell-to-cell movement mechanism of comovirus CPMV.

The plant virus infection cycle

The infection cycle of plant RNA viruses includes the following steps:

1. *Penetration* of plant cells (primary infection), either mechanically (some viruses, like potato virus X and tobacco mosaic virus [TMV]) or with the aid of a biological vector (most viruses), e.g.
   — insects, especially aphids (many viruses)
   — nematodes (some viruses, such as nepo- and tobraviruses)
   — fungi (rarely, for instance beet necrotic yellow vein virus and some potyviruses).
2. *Uncoating and translation* of the viral genome.
   Removal of the viral coat, probably by the binding of ribosomes that start translation at the 5′ end ("co-translational disassembly", [5]).

3. *Replication* of (and – for some of the plant viruses – concurrent mRNA *transcription* from) the viral genome. The viral replicase, involved in this process, is probably constructed by both virus-encoded and host-encoded subunits.
4. Further *translation* of the newly produced viral plus-strand RNAs to produce more replicase, coat protein (CP), and the transport or "movement" protein.
5. *Cell-to-cell movement* of virus through plasmodesmata. For this process the movement protein is required. The mechanism(s) of cell-to-cell movement are still unknown but should involve modification of the plasmodesmata to allow passage. Some viruses (e.g. cowpea mosaic virus [CPMV]) move as complete viral particles, some as non-encapsidated genomes (e.g. TMV).
6. *Systemic spread* through the infected plant. This requires, in addition to cell-to-cell movement, long-distance transport through the vascular system (mostly phloem) of the plant.
7. *Transmission* to other, non-infected plants. Since the plants themselves do not move, this often requires the aid of a biological vector (see step 1).

Striking differences with animal-infecting RNA viruses seem not to occur on the level of translation and RNA synthesis (steps 2–4) but are apparent in the other steps.

Transmission of plant RNA viruses

Transmission has been most extensively studied for (the many) aphid-transmitted viruses. Aphid transmission may be either persistent (the virus circulates non-propagatively in the insect and is transmitted persistently after a certain latency period) or non-persistent (the virus is acquired immediately and transmitted only during a very short period, i.e. a few minutes). For some non-persistently transmitted viruses, a specific viral gene product has been shown to be essential for successful transmission. For potyviruses this protein (approx. 55K) was originally called the "helper component" (HC). Later this (two-domain) protein was also shown to possess proteolytic activity and therefore has been renamed HC-PRO. As visualized in the genome comparisons in Fig. 1 this protein is encoded by one of the extra genes in the potyviral genome, and is not present in the genomes of the other picorna-like viruses.

Penetration of plant cells

Because plant cells have rigid cell walls, consisting of cellulose, pectin and other macromolecular components, plant viruses cannot penetrate

FAMILY GENUS GENETIC MAP
(SPECIES)

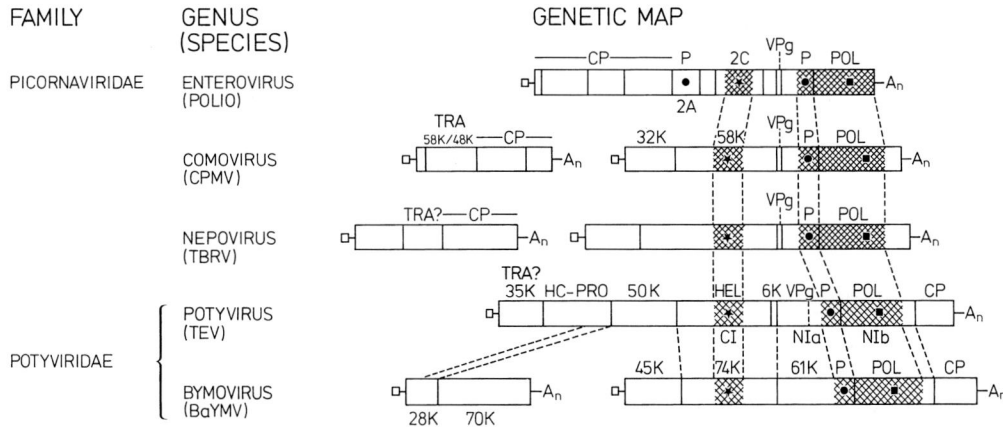

Fig. 1. Comparison of some members of the "supergroup" [1–4] of picorna-like viruses. The figure shows the genetic maps of poliovirus (*Picornaviridae*), cowpea mosaic virus (CPMV, comovirus) tomato black ring virus (TBRV, nepovirus), tobacco etch virus (TEV, genus potyvirus, family *Potyviridae*) and barley yellow mosaic virus (BaYMV, genus baymovirus, family *Potyviridae*). It is anticipated that comovirus and nepovirus groups will be considered genera. Coding regions in the genomes are indicated as open bars; regions of amino acid sequence homology in the gene products are indicated by similar shading. □ VPg; A_n, poly(A) tail; CP, coat protein(s); TRA, transport or "movement" protein; HEL, helicase; P, proteinase, POL, polymerase; * NTP motif; ● cysteine proteinase motif; ■ polymerase motif. Comparison of the various genetic maps indicate close genetic interrelationships between the picornaviruses and the "picorna-like" viruses of plants. At least part of the extra genes in the plant viral genomes reflect adaptations to plants, i.e. the TRA genes and the HC-PRO gene

via receptor-mediated endocytosis. Instead, they must be introduced mechanically or with the aid of a biological vector, mostly plant-parasitizing insects, nematodes or fungi. As a consequence, plant viruses enter their host cell via (transient) lesions, where apparently intact virus particles become internalized. Uncoating which is, due to binding of ribosomes by a mechanism referred to as co-translational disassembly, as observed for some positive strand RNA viruses in vitro [5–7], has been proposed to occur in vivo as well. For TMV, this hypothesis is supported by the observation of translationally active virus-ribosome complexes in epidermal cells of tobacco leaves soon after infection [8, 9]. However, whether co-translational disassembly represents a general mechanism of uncoating during plant virus infections, still remains unknown. If it is, plant virus particles may have ribosome binding sites, rather than cell receptor binding sites.

Cell-to-cell movement

After the elucidation of the genomic sequences of the most important plant RNA viruses and the assignment of their basic genetic properties (related to translation and replication) the mechanism of cell-to-cell transport has gained more and more attention. Though plant cells are surrounded by rigid cell walls they are interconnected by channels through these cell walls, called plasmodesmata. A plasmodesma can be regarded as the plant analog of the gap junctions in animal cells. Basically a plasmodesma is a cytoplasmic connection, surrounded by the plasma-membrane and containing a desmotubule, derived from and connecting the endoplasmic reticulum of two neighbouring cells. Plasmodesmata have not yet been biochemically characterized and structural models are mainly based on electron microscopic studies (Fig. 2A). Viruses may use plasmodesmata for their spread through plant tissue, though they permit only the passage of small molecules. Their effective diameter is only 3 nm (10) and their M_r exclusion limits ($\approx$1K) indicate that they must be modified to allow virus particles or even viral nucleic acid (average diameter 10 nm) to pass through.

For several plant viruses the active involvement of virally encoded proteins in plasmodesma modification has been demonstrated (TMV: the 30K protein; CPMV: the 58K/48K protein pair, see Fig. 1) or suggested (e.g. the 35K protein of potyvirus tobacco etch virus [TEV], Fig. 1). The

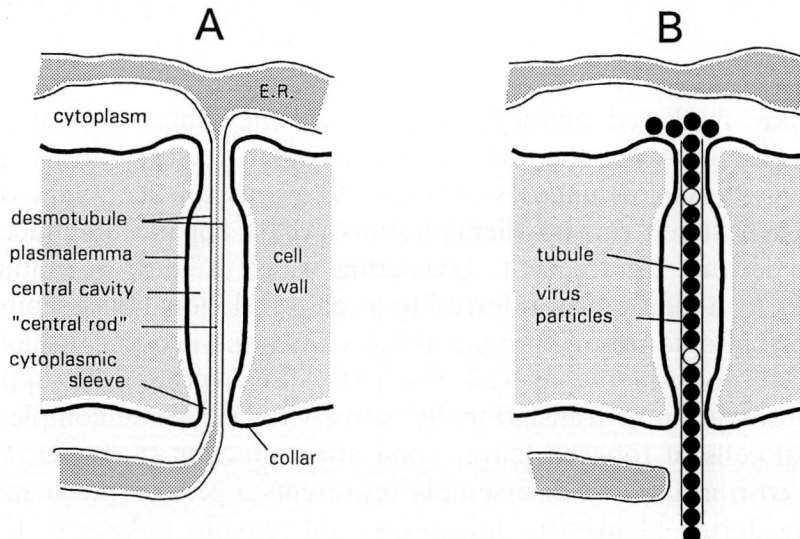

Fig. 2. Schematic representation of a plasmodesma before (**A**) and after (**B**) modification due to CPMV infection. It is suggested that after removal of the desmotubule a rigid tubular structure is formed, along which viral particles move to the neighbouring cell

evidence obtained so far indicates that there are at least two different cell-to-cell viral movement mechanisms, i.e. the movement of non-encapsidated RNA genomes through slightly modified plasmodesmata (exemplified by TMV, for review see [11]) and the movement of whole virus particles through considerably modified plasmodesmata, as exemplified by comovirus CPMV [12].

Cell-to-cell movement of CPMV

Electron microscopy of CPMV-infected cowpea cells reveals that the plasmodesmata are very conspiciously modified, containing long tubes extending into the cytoplasm and filled with virions (Fig. 3). The diameter of these tubes is approximately 28 nm, i.e. the size of the virus particles. Obviously virus particles are moving along these tubular structures to reach the neighbouring cells. A schematic interpretation of a plasmodesma modified upon CPMV infection is presented in Fig. 2B. It is proposed that the desmotubule is removed to allow tubule formation.

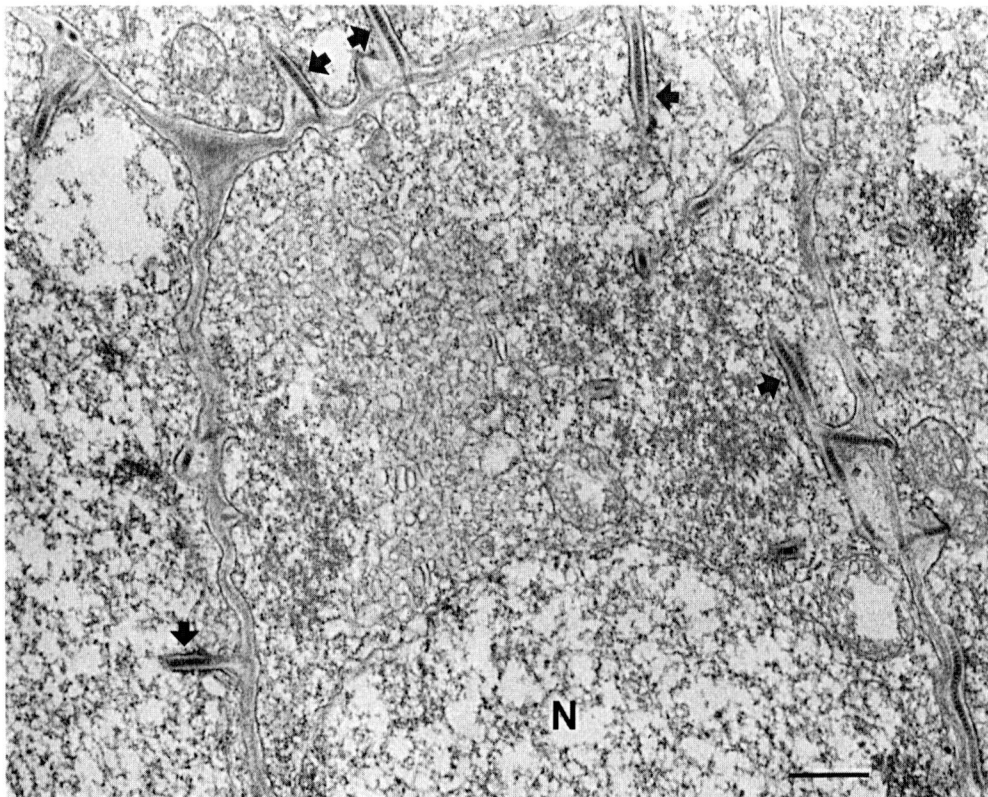

Fig. 3. Electron micrograph of osmium-fixed CPMV-infected cowpea tissue, revealing several tubular structures (arrows) in the cell walls. *N* Nucleus. The bar represents 0.5 μm

Because (the larger) B-RNA of CPMV is able to replicate independently in single cowpea protoplasts but needs (the shorter) M-RNA for systemic spread through plant tissues, it was postulated that M-RNA encodes a protein(s) essential for cell-to-cell transport of virus [13, 14]. Using infectious transcripts from cDNA clones of M-RNA, it was demonstrated that systemic invasion requires both the M-RNA-coded coat proteins and the 58K/48K protein pair, which are derived from the N-terminal parts of the two (105K and 95K) overlapping M polyproteins [15]. In concert with this, immunogold labelling of thin sections of infected tissue localized the 58K/48K proteins to the plasmodesma-linked tubules [16]. Because an antiserum directed against their common C-terminus was used, no differentiation between (the location of) these proteins could be made. Recent cell fractionation studies, however, have provided a first indication that these two proteins are differently targeted (58K: cytoplasm; 48K: membrane fraction). Furthermore, deletion studies have demonstrated that the 58K protein is a prerequisite for M-RNA amplification, implicating a function in RNA replication rather than in cell-to-cell movement (see also Wellink and Van Kammen [15]).

Tubular structures are also formed in infected cowpea protoplasts

A break-through for further analysis of the tubular structures and their involvement in cell-to-cell movement has been the recent finding [17] that they are also formed in cowpea protoplasts, from which the cell wall and plasmodesmata are absent (Fig. 4A).

Between 12 and 21 h post-inoculation, tubule formation starts at the periphery of the protoplast near the plasma membrane. Upon assembly, the virus-containing tubule is enveloped by the plasma membrane and extends into the culture medium. This suggests that the tubule has functional polarity and makes it likely that a tubule "grows" into a neighbouring cell in vivo. On average, 75% of infected protoplasts possess tubular structures extending from their surface. The tubule wall is 3 to 4 nm thick and as long as 20 μm, as shown by fluorescent light microscopy and negative staining electron microscopy. By analogy to infected plant cells, both the viral 58K/48K movement and capsid proteins were located in these tubules, as determined by immunofluorescent staining and immunogold labelling using specific antisera against these proteins (Figs. 4B and C). These results demonstrate that the formation of tubules is not necessarily dependent on the presence of plasmodesmata or the cell wall, and that they are composed, at least in part, of virus-encoded components.

Interestingly, mutant viruses containing deletions in the coat protein-coding region of the M RNA (e.g. mutant M58S, Δ5, Fig. 5) are still

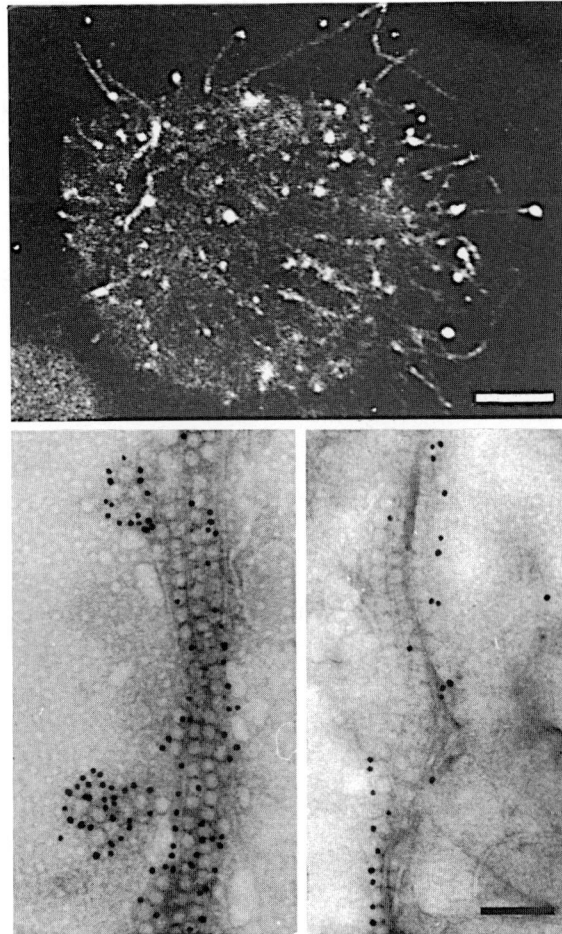

Fig. 4. Tubule formation in CPMV-infected cowpea protoplasts. Upper panel: immunofluorescene of an infected protoplast, stained with anti-58K/48K serum and visualized by confocal laser scanning microscopy; The bar represents 5 µm. Left below: immunogold labeling of a tubule using anti-CPMV serum; Right below: idem, using anti-58K/48K serum. The bar represents 0.15 µm

able to induce tubule-formation in protoplasts, but these tubules remain empty, because no mature virus particles can be formed. The diameter of the empty tubules is identical to that of virus particles-containing tubules, indicating that the presence of virus is not crucial for assembly of a tubule of the appropriate diameter. Mutant viruses with deletion in the 58K/48K-coding region (e.g. mutant MΔP, Fig. 5) are not able to induce tubules, the mature particles accumulating in the cytoplasm. This finding confirms that either or both the 58K and the 48K protein, are the trigger for tubule-formation. Because mutant MΔAUG 2/3 (Fig. 5), though showing almost wild type-levels of replication, does not induce

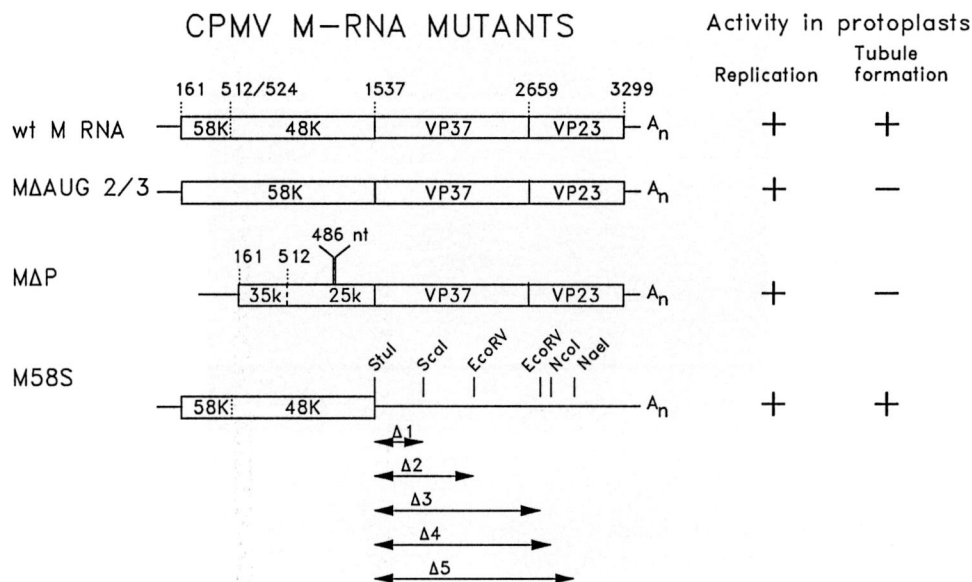

Fig. 5. CPMV M-RNA mutants and their activity in cowpea protoplasts

tubules in infected protoplasts, it is tempting to assume that only the 48K protein, but not the 58K protein, is responsible for tubule formation.

Expression of the CPMV 48K protein is sufficient for tubule formation

To determine the minimal viral information able to induce the tubular structures, the 58K and 48K proteins have been cloned and expressed from a transient expression vector pMon999 (a kind gift of C. Hemenway, Monsanto). Construct pMM58/48 contains the unaltered viral cistrons encoding the overlapping 58K/48K protein pair, while in construct pMM48 the ATG start codon at position 161 of the original M-RNA sequence has been deleted, thereby permitting expression of only the 48K protein (Fig. 6).

Transfection of cowpea protoplasts with either of these vectors leads to comparable rates of tubule induction in up to 10% of the protoplasts. This result demonstrates that the expression of only the 48K protein, even in the absence of virus replication, is sufficient for the induction of tubules (which, of course, remain empty due to the lack of virus particles). This finding also substantiates previous indications, based on deletion analysis (see above), that the 58K protein (or at least its unique N-terminus) has a function in M-RNA synthesis rather than in tubule formation.

CPMV M—RNA

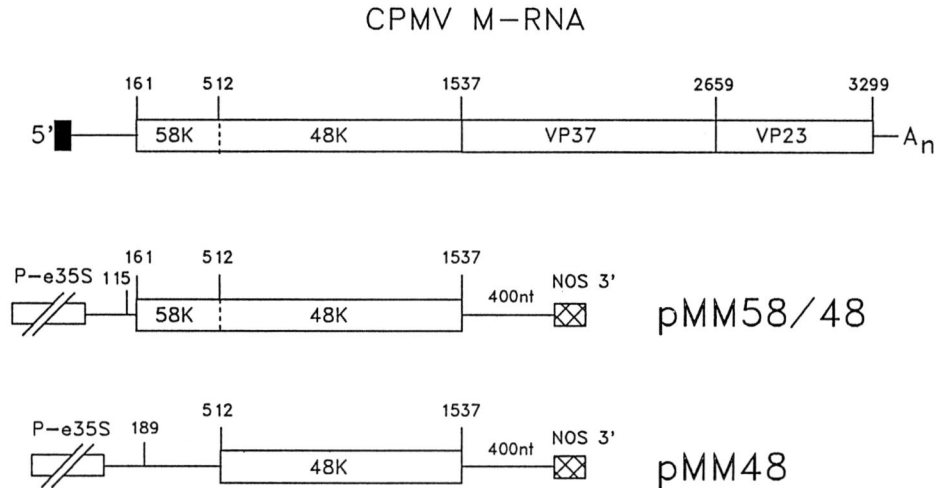

Fig. 6. Transient expression vector constructs containing the CPMV 58K and/or 48K protein-encoding region. The gene cassettes contain an enhanced cauliflower mosaic virus promoter (P-e35S) and a nopaline synthase (*nos*) transcription termination sequence

Tubule formation in non-host protoplasts

For several plant-virus combinations it has been shown that virus replication is fully achieved in protoplasts of non-host plants. Thus it has been shown that CPMV can replicate in protoplasts of tobacco [18] while bromoviruses BMV (restricted to monocotyledons such as barley) and CCMV (restricted to dicotyledons such as cowpea) can replicate in protoplasts of each other hosts [19, 20]. To investigate whether the (in)capability of the CPMV 48K gene to induce tubules in a given plant species determines the pathogenicity and host range of the virus, protoplasts from a series of plant species were tested for their susceptibility for CPMV infection and for tubule induction by the 48K gene. Transfection with CPMV RNA resulted in high levels of infection (more than 25% of the cells infected) in protoplasts of pea and tomato, and lower to poor levels of infection (less than 10% of the cells infected) in protoplasts of barley, *Arabidopsis* and carrot. The results obtained upon transfection with vector pMM48 were in parallel with these infectivity data, with efficient induction of tubules in protoplasts of pea and tomato (to 10% of protoplasts showing tubules) and poor (but unambiguous!) tubule formation in protoplasts of barley, *Arabidopsis* and carrot. Even transfection of protoplasts of a succulent (cactus) revealed induction of tubules directed by the CPMV 48K gene. These results indicate that the host range and pathogenicity of CPMV is not defined by the (in)capability of the viral 48K gene to induce tubules. The conclusion must therefore be drawn that successful systemic invasion of CPMV through a given plant depends not only on the activity of the 48K movement protein, but that effective cell-to-cell transport requires more than the potential to induce tubules.

Conclusion

Our data show that during CPMV infection of cowpea tissue tubular structures are induced within the plasmodesmata, in which intact virus particles move to neighbouring cells.

Deletion studies of the viral genome as well as transient expression studies indicate that the M-RNA-coded 48K protein is involved in the induction of these tubules and therefore represents the viral transport or "movement" protein. On the other hand, the M-RNA-coded 58K protein, though encompassing the complete 48K protein sequence, appears not to be involved in viral cell-to-cell movement but its function seems essential for M-RNA amplification. Because the 48K protein expressed from a transient expression vector is capable of inducing tubules in protoplasts from a series of non-host plant species, we conclude that this gene product, though essential for cell-to-cell movement of the virus, is not the only viral key function in virulence and pathogenesis.

Acknowledgements

The authors thank M. Storms and J. Groenewegen for help in the immuno-gold analysis and T. Sijen and H. Bloksma for protoplasts preparation. This research was partly supported by the Netherlands Foundation for Biological Research (BION) with financial aid from the Netherlands Organization for the Advancement of Pure Research (NWO).

References

1. Goldbach RW (1986) Molecular evolution of plant RNA viruses. Annu Rev Phytopathol 24: 289–310
2. Goldbach R, Wellink J (1988) Evolution of plus-strand RNA viruses. Intervirology 29: 260–267
3. Strauss EG, Strauss JH, Levine AJ (1990) Virus evolution. In: Fields BN, Knipe DM, Chanock RM (eds) Virology. Raven Press, New York, pp 167–190
4. Koonin EV (1991) The phylogeny of RNA-dependent RNA polymerases of positive-strand RNA viruses. J Gen Virol 72: 2197–2206
5. Wilson TMA (1984) Cotranslational disassembly of tobacco mosaic virus in vitro. Virology 137: 255–265
6. Brisco MJ, Hull R, Wilson TMA (1986) Swelling of isometric and of bacilliform plant virus nucleocapsids is required for virus-specific protein synthesis in vitro. Virology 148: 210–217
7. Roenhorst JW, Verduin BJM, Goldbach RW (1989) Virus-ribosome complexes from cell-free translation systems supplemented with cowpea chlorotic mottle virus particles. Virology 168: 138–146

8. Shaw JG, Plaskitt KA, Wilson TMA (1986) Evidence that tobacco mosaic virus particles disassemble cotranslationally in vivo. Virology 148: 326–336

9. Plaskitt KA, Watkins PAC, Sleat DE, Gallie DR, Shaw JG, Wilson TMA (1987) Immunogold labeling locates the site of disassembly and transient gene expression of tobacco mosaic virus pseudovirus particles in vivo. Mol Plant Microbe Int 1: 10–16

10. Terry BR, Robards AW (1987) Hydrodynamic radius alone governs the mobility of molecules through plasmodesmata. Planta 171: 145–157

11. Deom CM, Lapidot M, Beachy RN (1992) Plant virus movement proteins. Cell 69: 221–224

12. Goldbach R, Eggen R, De Jager C, Van Kammen A, Van Lent J, Rezelman G, Wellink J (1990) Genetic organization, evolution and expression of plant viral genomes. In: Fraser RSS (ed) Recognition and response in plant-virus interactions. Springer, Berlin Heidelberg New York Tokyo, pp 147–162

13. Goldbach R, Rezelman G, Van Kammen A (1980) Independent replication and expression of B-component RNA of cowpea mosaic virus. Nature 286: 297–300

14. Rezelman G, Franssen HJ, Goldbach RW, Le TS, Van Kammen A (1982) Limits to the independence of bottom component RNA of cowpea mosaic virus. J Gen Virol 60: 335–342

15. Wellink J, Van Kammen A (1989) Cell-to-cell transport of cowpea mosaic virus requires both the 58K/48K proteins and the capsid proteins. J Gen Virol 70: 2279–2286

16. Van Lent J, Wellink J, Goldbach R (1990) Evidence for the involvement of the 58K and 48K proteins of intercellular movement of cowpea mosaic virus. J Gen Virol 71: 219–223

17. Van Lent J, Storms M, Van der Meer F, Wellink J, Goldbach R (1991) Tubular structures involved in movement of cowpea mosaic virus are also formed in infected cowpea protoplasts. J Gen Virol 72: 2615–2623

18. Huber R, Hontelez J, Van Kammen A (1977) Cowpea mosaic virus infection of protoplasts from Samsun tobacco leaves. J Gen Virol 34: 315–323

19. De Jong W, Ahlquist P (1991) Bromovirus host specifity and systemic infection. Semin Virol 2: 97–105

20. Roenhorst JW, Van Lent JWM, Verduin BJM (1988) Binding of cowpea chlorotic mottle virus to cowpea protoplasts and relation of binding to virus entry and infection. Virology 164: 91–98

Authors' address: Dr. R. Goldbach, Department of Virology, Agricultural University Wageningen, Binnenhaven 11, 6709 PD Wageningen, The Netherlands.

Arch Virol (1994) [Suppl] 9: 99–109

Archives
Virology

© Springer-Verlag 1994
Printed in Austria

A molecular genetic approach to the study of Venezuelan equine encephalitis virus pathogenesis

N. L. Davis[1]**, F. B. Grieder**[1]**, J. F. Smith**[2]**, G. F. Greenwald**[1]**, M. L. Valenski**[1]**,
D. C. Sellon**[1]**, P. C. Charles**[1]**, and R. E. Johnston**[1]

[1] University of North Carolina, Chapel Hill, North Carolina
[2] U.S. Army Research Institute of Infectious Diseases, Frederick, Maryland, U.S.A.

Summary. Viral pathogenesis can be described as a series of steps, analogous to a biochemical pathway, whose endpoint is disease of the infected host. Distinct viral functions may be critical at each required step. Our genetic approach is to use Venezuelan equine encephalitis virus (VEE) mutants blocked at different steps to delineate the process of pathogenesis. A full-length cDNA clone of a virulent strain of VEE was used as a template for in vitro mutagenesis to produce attenuated single-site mutants. The spread of molecularly cloned parent or mutant viruses in the mouse was monitored by infectivity, immunocytochemistry, in situ hybridization and histopathology. Virulent VEE spread through the lymphatic system, produced viremia and replicated in several visceral organs. As virus was being cleared from these sites, it began to appear in the brain, frequently beginning in the olfactory tracts. A single-site mutant in the E2 glycoprotein appeared to block pathogenesis at a very early step, and required a reversion mutation to spread beyond the site of inoculation. The feasibility of combining attenuating mutations to produce a stable VEE vaccine strain has been demonstrated using three E2 mutations.

Introduction

Venezuelan equine encephalitis virus (VEE), a member of the *Alphavirus* genus of the family *Togaviridae*, was first isolated from a diseased horse during an epidemic of equine encephalitis in Venezuela in 1938 [1]. Subtypes and varieties of this mosquito-borne virus are enzootic in many areas of South and Central America and in Florida, and some isolates have caused widespread epizootics of equine encephalitis with a significant incidence of human disease [2].

VEE virion structure and pattern of gene expression are similar to those of Sindbis virus, the prototype of the genus [3]. The virion is enveloped and spherical, containing an icosahedral nucleocapsid, which in turn encases a nonsegmented single-stranded RNA genome. The spikes that cover the outer surface of the virion contain two closely associated integral membrane glycoproteins, E1 and E2 [4]. These surface proteins carry domains involved in initial interactions with target cells and with cells of the immune system of the animal host [5]. The complete genomic sequence of the virulent Trinidad donkey (TRD) strain of VEE has been determined by Kinney et al. [6, 7] and shares all features of Sindbis virus.

VEE does differ from Sindbis virus, however, in its pathogenic phenotype. Infection with VEE in its natural rodent and equine hosts proceeds in two stages, an initial lymphotropic phase and a final neurotropic phase [8]. We have taken a molecular genetic approach to the study of host and viral determinants involved in this complex disease process. Our aim is to delineate some of the mechanisms underlying both tissue and organ tropism and invasion of the central nervous system by neurotropic viruses. In addition, we are applying these results to the design of a genetically engineered live virus vaccine.

A molecular genetic approach

The process of VEE infection in mice may be considered as a series of steps, analogous to a biochemical pathway. The steps show a characteristic order and time course and may involve distinct intermediates in the form of tissue-adapted virus variants. The endpoint of the pathway is fatal encephalitis in this host. The elements in a genetic study of this process are single-site virus mutants that fail to induce fatal encephalitis because they are blocked at different steps in the pathogenesis pathway. The few viruses that progress beyond that step are likely to have reverted with respect to the block. Therefore, the point of restriction can be inferred from both a decrease in virus replication and an increase in the frequency of revertants isolated at steps subsequent to the block encoded by the primary mutation. The stage at which a given mutant is blocked can be identified, and a differentiated cell culture system that mimics the restriction of that mutant can be adapted or established. A study of the restriction can then be pursued in vitro at the molecular level. By combining information obtained with several mutants, the virus-target cell interactions that make up the pathway can be defined in detail.

The genetic system consists of a full-length cDNA clone, pV3000, produced from the virulent TRD strain of VEE, and placed directly downstream from a T7 bacteriophage promoter. Infectious RNA genomes can be transcribed from the linearized plasmid and used to transfect susceptible tissue culture cells [9]. The resulting virus progeny display a phenotype that is determined by the DNA sequence in the full-length clone. Virus produced from pV3000 is indistinguishable from wild type TRD strain VEE in terms of growth properties, virulence and pathogenesis ([9], and Grieder, Davis, Sellon and Johnston, unpubl. res.). Two aspects of this system are important in the experiments outlined here. First, the cDNA clone is a stable form of the virus genome and a consistent source of genetically pure wild-type virus. Second, site-directed mutagenesis can be used to produce single-site mutants at specific loci.

Infection of mice with molecularly cloned virulent VEE

The starting point for a study of single-site attenuated mutants is a description of the steps in the replication and pathogenesis of the molecularly cloned virulent parent virus, V3000, in adult mice. CD-1 mice were injected in the left rear footpad with 10^3 plaque forming units (pfu) of V3000. At various times after infection two mice were sacrificed, and samples of serum and fifteen organs were collected. A 10% homogenate was made of a portion of each organ in phosphate-buffered saline containing 1% calf serum. Anatomic sites of viral replication were detected by plaque assay titration on BHK-21 cells of clarified homogenates, immunocytochemistry and in situ hybridization [10], and parallel tissue samples were examined for histopathological lesions.

Figure 1 shows a sampling of the titration results. Virus spread began very soon after inoculation. Increased titer was seen by 6 h post-infection (pi) in the draining lymph node, which became the site of extensive virus replication. Sections of this organ were analyzed using a radiolabeled RNA probe whose sequence is complementary to both viral genome RNA and subgenomic RNA. Probe-specific hybridization confirmed the titration results, indicating multifocal viral replication within the lymph node. Serial sections stained with hematoxylin and eosin showed scattered pyknotic lymphocytes.

VEE entered the bloodstream to initiate replication in several organs, primarily those of the lymphoid system. Results of titration of spleen and right popliteal lymph node are shown in Fig. 1. Significant virus replication and accompanying histologic lesions also were seen in the thymus, pancreas and heart by 24 h pi. A high titer viremia was produced during this lymphotropic phase (12 to 72 h pi) which has been described previously for wild-type strains of VEE in mice [8]. In organs shown to have significant and consistent titers of infectious virus, hybridization with the radiolabeled RNA probe delineated areas of viral

VEE Titer Left Popliteal Ln

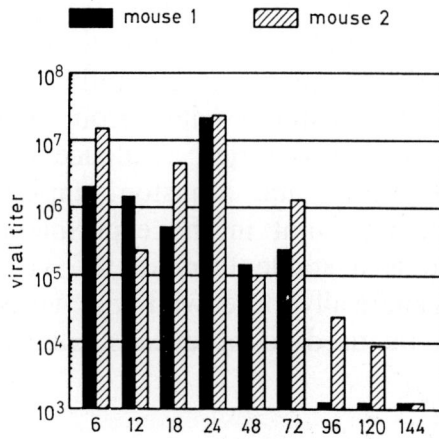

VEE Titer Serum

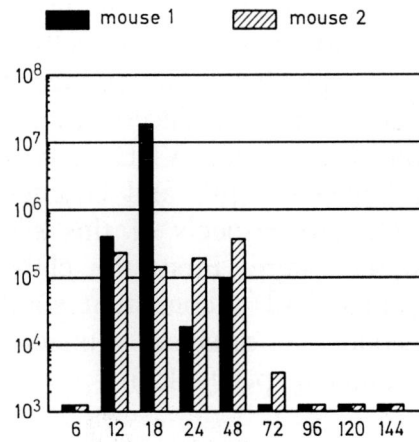

VEE Titer Right Popliteal Ln

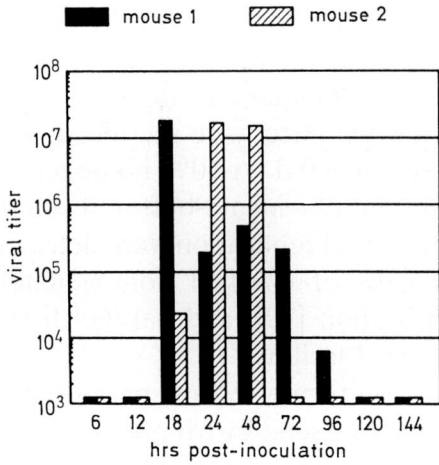

VEE Titer Spleen

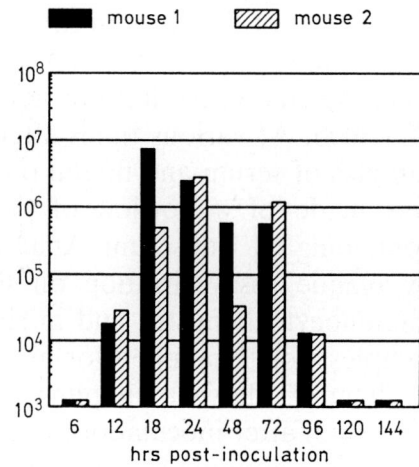

VEE Titer Brain

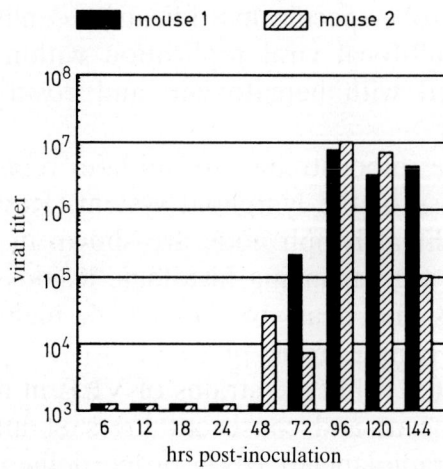

replication, and histopathological examination showed cell degeneration and inflammation.

The next stage of infection, beginning at about 72 h pi, involved clearance of the virus from the serum and the visceral organs and the repopulation of these organs with cells. The host mechanisms responsible for clearance have not yet been defined. However, mice with severe combined immunodeficiency (*scid*) did not clear virus from these organs at this stage. Moreover, many more tissues of *scid* mice have been shown to be active sites of VEE replication (Charles and Johnston, unpubl. res.).

Infectious virus was detected in brain homogenates beginning at two days pi. Most often, virus replication was first detected in the region of the lateral olfactory tracts using in situ hybridization. At two to three days pi, pathological changes began to appear in the brain, by five days there was extensive perivascular infiltration of small lymphocytes, gliosis and neuronal degeneration.

Entry of virulent VEE into the brain

An important step in the pathogenesis pathway of VEE is entry into the central nervous system (CNS). Neurotropic viruses have been shown to enter the CNS by axonal transport through peripheral nerves, or from the bloodstream [11], and both these routes were considered for VEE. Results of our experiments suggest that the major path of entry of VEE into the CNS does not involve transport of VEE through the sciatic nerve following footpad inoculation, infection of (or transport through) endothelial cells lining blood vessels in the brain, or replication of VEE in cells of the choroid plexus.

Transendothelial passage of neurotropic viruses also occurs via diapedesis of infected lymphoid or myeloid cells. Although the early, diffuse panencephalitis associated with this route is not seen with VEE, our experiments do not rule out the existence of such VEE-infected "Trojan horses".

An alternative route for serum-borne virus to gain access to the brain is through the nerves of the olfactory neuroepithelium, as has been shown by Monath et al. for St. Louis encephalitis virus [12]. The neurons

Fig. 1. Spread of virulent VEE, V3000, in the mouse following injection into the left rear footpad. At various times after inoculation, two mice were sacrificed, and samples of serum, thymus, heart, lung, liver, spleen, pancreas, duodenum, kidney, adrenal gland, salivary gland, eye, brain, skeletal muscle, and left and right popliteal lymph nodes were harvested. Results are presented as pfu per gram tissue, or pfu per ml serum. At each time point in the five panels, the two bars represent results from the same two mice

of the neuroepithelium make a direct connection with neurons in the olfactory bulb of the brain and also are in close contact with fenestrated capillaries. Initial studies of brain sections using in situ hybridization showed that the first detection of VEE in the brain was most often associated with olfactory tracts, suggesting that olfactory nerves may serve as a portal of entry for VEE. Further analysis of sections of decalcified heads at 36 h pi by in situ hybridization showed that replication of virus was consistently detected in neurons of the olfactory neuroepithelium and in the olfactory bulb at this time, when a positive signal was seen only occasionally in other parts of the brain (Charles and Johnston, unpubl. res.). At 48 h pi, the same type of analysis showed an intense signal in the lateral olfactory tracts. Therefore, our present hypothesis is that the olfactory neuroepithelium is a major route used by VEE to move from the serum to the brain.

It has been shown that VEE is present in nasal secretions following peripheral inoculation [3]. However, when virus was instilled intranasally in CD-1 mice at a dose of 10^4 pfu, only 83% of the animals died from VEE infection (Charles and Johnston, unpubl. res.). This suggests that infection of olfactory nerves directly from the nasal mucosa can occur, but that efficient and rapid infection of the same nerves that we observed following footpad inoculation occurs from the bloodstream.

Infection with a single-site attenuated mutant of VEE

We have used site-directed mutagenesis to introduce specific single nucleotide changes into the full-length VEE clone [9]. These are attenuating mutations, originally identified in biologically selected mutants, that may be blocked at different steps in the pathogenic process. Characterization of these mutants will be the key to understanding those steps. Molecularly cloned viruses can be defined at the nucleotide level, which not only allows pathogenesis phenotypes to be linked to particular mutations, but also supplies a marker for tracing the fate of the infecting genotype in the host animal.

VEE strain V3010, a molecularly cloned mutant that carries a Lys at E2 position 76 in place of the Glu in V3000, is avirulent in adult mice infected by a peripheral route [9], and gives 0–22% mortality when injected intracerebrally. The spread of V3010 in the mouse after injection of 10^3 pfu into the left rear footpad was followed using the same techniques described above for the virulent parent. The most striking difference between this mutant and its parent was the delay in appearance of mutant virus replication. Virus first appeared in the serum at 48 h, and only at 72 h in the draining lymph node, while the virulent parent was seen in the draining lymph node by 6 h pi, and in the serum at 12 h pi.

Table 1. Virulence phenotypes of viruses recovered from E2 Lys 76 mutant-infected mice[a]

	V3000 Parent	Recovered viruses			
		B1	Sp1	L48	L42
Virulence phenotype	virulent	virulent	avirulent	attenuated	avirulent
Residue at E2 76	Glu	Glu	Glu	Lys	Lys

[a] *B1* Virus recovered from the brain; *Sp1* virus recovered from the spleen; *L48, L42* viruses recovered from the left popliteal lymph node. *Virulent* 100% mortality; *attenuated* between 0% and 100% mortality with a significantly extended average survival time; *avirulent* 0% mortality

Also, the results with this mutant did not suggest a consistent pattern of spread, as seen for the virulent parent. Rather, virus replication was sporadic, occurring in a few tissues in some mice.

The slower time course of spread and the sporadic nature of replication of this mutant suggest that its spread in the mouse may require a reversion event. In addition, although the infecting mutant produced a small plaque in cultured baby hamster kidney (BHK-21) cells, a significant fraction of the virus recovered from the lymph nodes, spleen and brain produced large plaques.

To determine whether revertant viruses were present in mice infected with the attenuated V3010 mutant, viruses were plaque-purified from homogenates of the organs from mutant-infected mice. These recovered viruses were tested for virulence in adult female CD-1 mice and their genomes were analyzed by direct RNA sequencing to identify the E2 76 codon. From the results we were able to divide the recovered viruses into four groups, representatives of which are shown in Table 1. Seven of eighteen recovered viruses tested were probably same site revertants (e.g. recovered virus B1), with Glu at E2 76 and the same mortality and average survival time as the virulent parent virus. A second type of isolate, e.g. Sp1, showed a same-site reversion to the wild type Glu at E2 76, but was avirulent. This virus must have sustained at least two mutation events in the mouse, a same-site reversion and a second attenuating mutation. A third type of recovered virus, represented by virus L48, carried the original Lys at E2 76, but was significantly less attenuated than the infecting mutant. Finally, three of the eighteen isolates tested, including virus L42, appeared to have the original genotype, Lys at 76, and the avirulent phenotype of the infecting virus. However, when spread to the draining lymph node and serum following footpad inoculation was measured for two of these isolates, it was much more rapid than that of the infecting mutant. These viruses must carry a second-site mutation that enabled them to spread beyond the site of inoculation.

All recovered viruses examined showed evidence of some type of mutation event during growth in the mouse, suggesting that the mutant V3010 is blocked at a very early step after infection and probably requires a reversion mutation to spread beyond the site of inoculation. In terms of the analysis of the pathogenesis pathway, the restriction caused by the mutation at E2 76 may define the very first step. Preliminary experiments with another single-site attenuating mutation, at E2 209, showed normal spread to the draining lymph node, but there was no viremia or invasion of the CNS. Further study of these primary mutations and mapping of their second-site reversion mutations presents an opportunity to define the functions required to complete later steps in the pathway.

Identification of attenuating mutations

An important part of our approach is the identification of additional attenuating mutations that may affect different steps in the pathogenesis pathway. Specific alterations in conserved regions of the genome, which probably perform required functions in the virus life cycle, may block virus growth in a particular target cell of the host. One such conserved region lies between positions 74 and 108 of the E1 glycoprotein and includes a stretch of uncharged and hydrophobic residues [6]. This region has been proposed to be involved in membrane fusion during virus entry. Previous work with Sindbis virus [13] showed that a mutation at E1 75 caused attenuation of virulence in neonatal mice. Figure 2 shows a series of random amino acid substitutions between codons 80 and 93 of E1 made using a modification of the Kunkel method for site-directed mutagenesis [14]. The specific infectivity of radiolabeled RNA transcribed from each mutant clone was tested by quantitating the number of plaques initiated by the RNA on transfected BHK-21 cells. All but one of the amino acid changes were lethal for VEE, supporting the idea that this region performs a vital function in viral replication. The one viable mutant, ile for phe at position 81, was avirulent in mice. This mutant, and possibly other attenuating, viable substitutions at this site, will be studied to determine the stage of pathogenesis at which they are blocked.

A second target was the signal for processing of PE2, the precursor of the mature E2 glycoprotein. PE2 is cleaved by a host enzyme that recognizes the basic-X-basic-basic amino acid sequence at the −1 to −4 position relative to the amino terminus of E2 [15, 16]. This signal is

E1 glycoprotein

	val	phe	thr	gly	val	tyr	pro	phe	met	trp	gly	gly	ala	tyr
VEE wild type	80	81	82	83	84	85	86	87	88	89	90	91	92	93
site-directed mutants	ala	val	ile	trp	phe	asn			val	arg	val	ala	ser	
		tyr	pro	ala						gly	asp			
		ile												

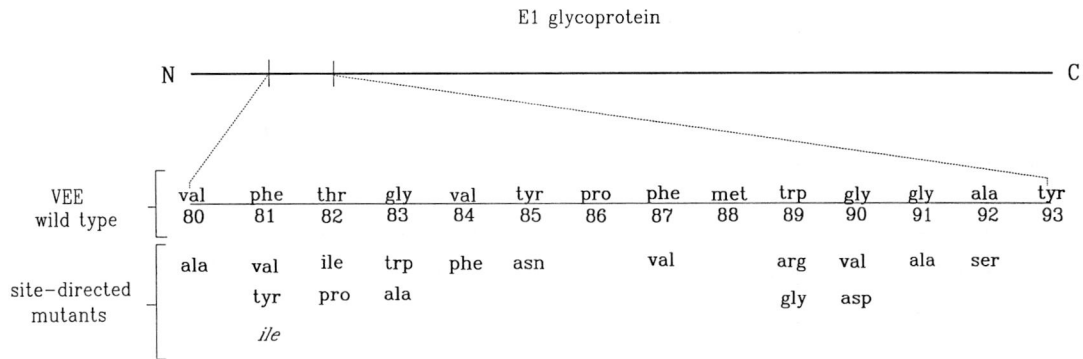

Fig. 2. Single-site mutants in the conserved hydrophobic region of the E1 glycoprotein. The 17 amino acid substitutions shown were identified by sequence analysis of putative mutant clones, and placed individually into full-length VEE cDNA clones. The single viable substitution is shown in italics

found in all alphavirus PE2 glycoproteins and in glycoprotein precursors of several other virus groups [16]. A mutation that blocked PE2 cleavage in the S.A.AR86 strain produced viable, attenuated mutants that incorporated PE2 into virions [17]. Site-directed mutations of VEE included a deletion of the entire four amino acid cleavage signal and various amino acid substitutions at the −1 position relative to the cleavage site. All of these mutations were lethal in VEE. However, a few plaques were formed following transfection of cultured cells, and these viruses proved to be second-site revertants of the original lethal mutation, containing not only the original cleavage site mutation engineered into the full-length clone, but also a second mutation, either in E2 or in E1. Three double mutant full-length clones, each containing a lethal cleavage site mutation coupled with a second-site change, were produced and transcribed into fully infectious RNA genomes, indicating that the second site mutations restored viability to the mutants carrying altered PE2 cleavage sites. In addition, we have shown that two of the double mutants produced virions containing PE2, and were attenuated in mice. These "resuscitated" mutants will be used in our study of VEE pathogenesis.

A genetically engineered live virus vaccine for VEE

Certain live virus vaccines have been shown to be very effective in preventing virus disease, but some have presented the very real problem of reversion to a more virulent phenotype. However, live virus vaccine strains containing multiple independently attenuating mutations are less likely to cause side effects associated with reversion of the attenuated phenotype [18]. The additive or synergistic effect of multiple attenuating mutations has been demonstrated directly in the Sindbis virus-neonatal mouse system [13]. We are engineering a multiply attenuated strain of VEE by altering a full-length clone at a small number of loci to attenuate virulence without severely impairing the replication or immunogenicity of the virus.

Preliminary studies have demonstrated that this strategy will be useful for VEE. A mortality study in mice of different ages showed that combining independently attenuating mutations in pairs often increased attenuation in one-week-old and two-week-old mice, while three mutations appeared to allow the greatest percentage survival [9]. The triple mutant, containing attenuating mutations at positions 76, 120 and 209 of the E2 glycoprotein, was also tested in horses (Smith et al., unpubl. res.). Although the parent molecularly cloned virus caused severe neurological symptoms in these animals, the triple mutant caused no clinical signs of disease and produced solid protection against challenge with a highly virulent strain of VEE. New attenuating mutations will be identified and tested with the aim of placing the most effective combination into a new vaccine strain for VEE.

Acknowledgements

This work was supported by U.S. Public Health Service grants AI22186 and NS26681 and by the U.S. Army Medical Research and Development Command (DAMD 17-91-C-1092)

References

1. Kubes V, Rios FA (1939) The causative agent of infectious equine encephalomyelitis in Venezuela. Science 90: 20–21
2. Young N, Johnson KM (1969) Antigenic variants of Venezuelan equine encephalitis virus: their geographic distribution and epidemiologic significance. Am J Epidemiol 89: 286–307
3. Peters CJ, Dalrymple JM (1990) Alphaviruses. In: Fields BN, Knipe DM (eds) Virology. Raven Press, New York, pp 713–761
4. Pedersen CE, Eddy GA (1974) Separation, isolation, and immunological studies of the structural proteins of Venezuelan equine encephalomyelitis virus. J Virol 14: 740–744
5. Roehrig JT, Matthews JH (1985) The neutralization site on the E2 glycoprotein of Venezuelan equine encephalomyelitis (TC-83) virus is composed of multiple conformationally stable epitopes. Virology 142: 347–356
6. Kinney RM, Johnson BJB, Brown VL, Trent DW (1986) Nucleotide sequence of the 26S mRNA of the virulent Trinidad donkey strain of Venezuelan equine encephalitis virus and deduced sequence of the encoded structural protein. Virology 152: 400–413
7. Kinney RM, Johnson BJB, Welch JB, Tsuchiya KR, Trent DW (1989) The full-length nucleotide sequence of the virulent Trinidad donkey strain of Venezuelan

equine encephalitis virus and its attenuated derivative, strain TC-83. Virology 170: 19–30

8. Gleiser CA, Gochenour Jr ES, Berge TO, Tigertt WD (1962) The comparative pathology of experimental Venezuelan equine encephalitis infection in different animal hosts. J Infect Dis 110: 80–97

9. Davis NL, Powell N, Greenwald GF, Willis LV, Johnson BJB, Smith JF, Johnston RE (1991) Attenuating mutations in the E2 glycoprotein gene of Venezuelan equine encephalitis virus: Construction of single and multiple mutants in a full-length cDNA clone. Virology 183: 20–31

10. Clabough DL, Gebhard D, Flaherty MT, Whetter LE, Perry ST, Coggins L, Fuller FJ (1991) Immune-mediated thrombocytopenia in horses infected with equine infectious anemia virus. J Virol 65: 6242–6251

11. Tyler KL, Fields BN (1990) Pathogenesis of viral infections. In: Fields BN, Knipe DM (eds) Virology. Raven Press, New York, pp 191–239

12. Monath TP, Cropp CB, Harrison A (1983) Mode of entry of a neurotropic arbovirus into the central nervous system: reinvestigation of an old controversy. Lab Invest 48: 399–410

13. Polo JM, Johnston RE (1990) Attenuating mutations in glycoproteins E1 and E2 of Sindbis virus produce a highly attenuated strain when combined in vitro. J Virol 64: 4438–4444

14. Hutchinson III CA, Nordeen SK, Vogt K, Edgell MH (1986) A complete library of point substitution mutations in the glucocorticoid response element of mouse mammary tumor virus. Proc Natl Acad Sci USA 83: 710–714

15. Strauss JH, Strauss EG, Hahn CS, Hahn YS, Galler R, Hardy WR, Rice CM (1987) Replication of alphaviruses and flaviviruses: proteolytic processing of polyproteins. In: Brinton MA, Rucckert RR (eds) Positive strand RNA viruses. Alan R. Liss, New York, pp 209–225

16. Hosaka M, Nagahama M, Kim W-S, Watanabe T, Hatsuzawa K, Ikemizu J, Murakami K, Nakayama K (1991) Arg-X-Lys/Arg-Arg motif as a signal for precursor cleavage catalyzed by furin within the constitutive secretory pathway. J Biol Chem 266: 12127–12130

17. Russell DR, Dalrymple JM, Johnston RE (1989) Sindbis virus mutations which coordinately affect glycoprotein processing, penetration, and virulence in mice. J Virol 63: 1619–1629

18. Almond JW (1987) The attenuation of poliovirus neurovirulence. Annu Rev Microbiol 41: 153–180

Authors' address: Dr. N. L. Davis, Department of Microbiology and Immunology, School of Medicine, CB No. 7290, University of North Carolina, Chapel Hill, NC 27599, U.S.A.

Arch Virol (1994) [Suppl] 9: 111–119

Archives
of
Virology
© Springer-Verlag 1994
Printed in Austria

Use of drug-resistance mutants to identify functional regions in picornavirus capsid proteins

A. G. Mosser, D. A. Shepard, and **R. R. Rueckert**

University of Wisconsin, Institute for Molecular Virology, Madison, Wisconsin, U.S.A.

Summary. The WIN drugs and similar hydrophobic compounds that insert into the capsid of picornaviruses have been shown to block viral uncoating. In some of the human rhinoviruses they also block attachment of virus to cells. Spontaneously occurring drug-resistant mutants of human rhinovirus 14 and poliovirus type 3 were selected for their ability to make plaques in the presence of the selecting drug. The HRV-14 mutants either prevented drug binding or allowed the virus to attach to cells in the presence of drug. About two thirds of the poliovirus mutants were dependent on the presence of drug for plaque formation. In single cycle growth curves, drug was not required for the formation of drug-dependent progeny virus. However, progeny virus grown without drug never accumulated outside of cells, thus making the formation of plaques impossible. This behavior was apparently caused by the extreme thermolability of these mutants. In the absence of drug, heating to 37°C rapidly converted them to non-infectious particles with a sedimentation coefficient of 135S.

Introduction

The capsids of picornaviruses contain 60 copies each of 4 different proteins. Three of these, VP1, VP2 and VP3, are roughly the same size, and include a characteristic eight-stranded antiparallel beta barrel as their core structure [1, 2]. They form the protein shell, with the smallest protein, VP4, on the virus interior. On the surfaces of the rhinoviruses and polioviruses, there is a star-shaped plateau at the five-fold axes, surrounded by a 15–25 Å-deep canyon. For human rhinovirus 14 (HRV-14), the canyon has been shown to be the location of the viral acceptor, that part of the virus surface which interacts with the cellular receptor molecule [3]. The HRV-14 cellular receptor is ICAM-1, an intercellular adhesion molecule of the immunoglobulin superfamily [4, 5].

Within the canyon of polio- and rhinoviruses there is a space between the amino acids of capsid protein VP1 which forms a channel into the hydrophobic interior of the VP1 beta barrel. Electron density has been observed in the interior of the VP1 beta barrel of several viruses, among them poliovirus-1 (Mahoney) [6], poliovirus-3 (Sabin) and HRV-1A [7], but not native HRV-14 [7]. This electron density has been attributed to hydrophobic hydrocarbons termed pocket factor.

Because the interior of the beta barrel connects to the surface, it is possible to exchange the pocket factor with other hydrophobic compounds, or to diffuse such compounds into the "empty pocket" of HRV-14 [8, 9]. Among these are the WIN drugs [10, 11] and several other families of drugs. In the presence of drugs of this type, viruses are stabilized against degradation due to exposure to heat or extremes of pH [12–14]. Capsid stabilization probably is responsible for the main antiviral action of these drugs: as a blocker of viral uncoating [13, 15, 16].

Drug escape mutants of human rhinovirus type 14

Using several of the WIN drugs, Heinz et al. [17] selected about 70 drug-escape mutants of HRV-14. For HRV-14 and other rhinoviruses of the major receptor group, these drugs block attachment of viruses to their cellular receptors [18, 19]. Insertion of the drug deforms the floor of the canyon [8, 9]. Since this is the site of the viral acceptor, this deformation leads to a strong block to the attachment of HRV-14 to cells.

When HRV-14 drug escape mutants were characterized, two types of mutations were found [19, 20]. Some mutants were unable to bind drug, which could be demonstrated as the inability of drug to stabilize these mutants against thermodegradation. Such drug-*exclusion* mutants represent the simplest way in which these viruses can escape the effects of drug. Exclusion mutants have substituted bulky amino acids into the lining of their VP1 beta barrel, thus preventing access to the pocket. Mutations were found at VP1 residues 188 and 199 as shown in Fig. 1 (top panels).

A second class of mutants still bind drug but have improved their attachment to cells in the presence of drug [20]. These *compensation* mutants have mutations on the floor of the canyon, in the virus acceptor region (Fig. 1, bottom panels).

Thus the HRV14 drug-escape mutants identified two functional regions of the viral capsid: (a) the drug-binding pocket, and (b) the deformable part of the canyon, a region important for attachment to cells. Since these mutants could be used to identify sites of drug action, we were encouraged that a similar study with poliovirus should provide

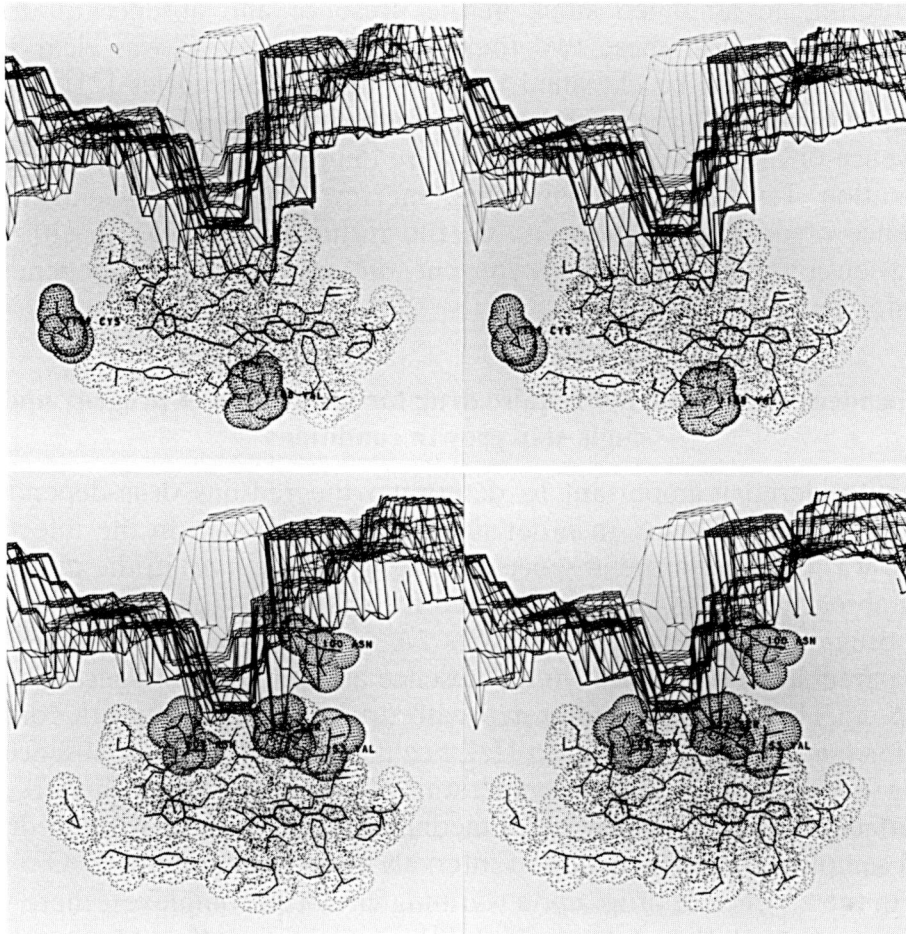

Fig. 1. Stereoscopic views of a portion of the HRV-14 capsid showing the amino acid residues altered in drug-escape mutants. The outer capsid surface is at the top, showing the canyon. Atoms of the amino acids lining the drug-binding pocket are depicted as dotted spheres. Upper panels: amino acid residues altered in drug-exclusion mutants are highlighted. Lower panels: amino acid residues altered in drug-compensation mutants are highlighted

information about capsid regions important for uncoating. For poliovirus-1 and -2, these drugs block uncoating but do not affect attachment [13, 15, 16]. We elected to use poliovirus-3 (Sabin) because its structure is known [6] and it is more sensitive to the drug WIN 51711 than is poliovirus-1 [11, 21].

Drug escape mutants of poliovirus-3

We selected 22 mutants of poliovirus-3 by picking plaques formed in the presence of 2 μg/ml of WIN 51711. Each mutant was characterized by titering an amplified stock in the presence and absence of drug. When the ratios of these two titers were compared, it was clear that the mutants could be classified into two distinct categories [21]. Eight mutants produced about the same number of plaques in either the presence or absence of drug and 14 were dependent on drug for plaque formation. For these latter mutants, the few plaques produced in the absence of drug were shown to be the result of reversion to either a drug-sensitive or a drug-resistant but, in any event, non-dependent phenotype [21].

Dependent mutants do not require drug for production of progeny under single-step growth conditions

We considered it important to determine the reasons drug-dependent mutants required drug. In order to identify which step in the infection cycle was blocked in the absence of drug, growth curves in the presence and absence of drug were constructed for wild type virus and for one of the drug-dependent mutants. Figures 2A and B show growth curves for the parental wild type virus in the absence and presence, respectively, of WIN 51711. Virus preequilibrated with drug-containing or with control buffer was allowed to attach to HeLa cells in the presence or absence of drug for 30 minutes at room temperature. Cells were then washed free of unattached virus, resuspended in medium at 37°C with or without drug, and sampled immediately and at intervals after incubating at 37°C.

In the absence of drug, upon warming virus-cell complexes, there was an immediate decline (about 500-fold) in infectivity, the eclipse phase. This was followed by a period of virus increase beginning at about 2½ h after infection, reaching a plateau at about 6 h. When cell associated and released virus were quantitated separately, it was observed that at 6 h post attachment about 1% of infective virus had been released into the medium, the proportion increasing steadily thereafter until slightly more than half had been released at 12½ h. In the presence of drug (Fig. 2B), the eclipse period was almost completely eliminated, and only at about 4 h after attachment was the beginning of a slow increase in infectivity seen.

The comparable growth curves of mutant 8, a drug-dependent mutant, are presented in Figs. 2C and D. Comparison of the first time points suggests that attachment was impaired in the absence of durg. However, another interpretation, more likely in view of the final virus yield, was that almost all the apparent difference observed was due to partial eclipse of this mutant during the attachment period in the absence of drug. After warming, the disappearance of infectivity was very rapid and dramatic (Fig. 2C), but to our surprise, the time course of virus synthesis

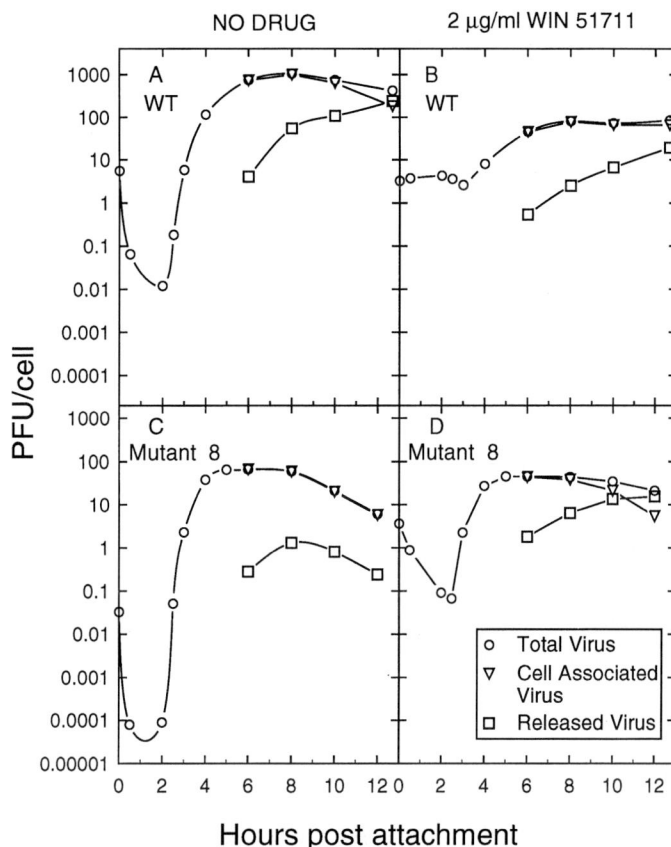

Fig. 2. Single-cycle growth curves of wild-type poliovirus-3 and drug-dependent mutant 8 in the presence and absence of $2\,\mu g/ml$ WIN 51711. Virus was pretreated at room temperature for 1 h and overnight at 4°C with or without WIN 51711, and was allowed to attach to HeLa cells for 30 min at room temperature. Multiplicities of infection were 5 PFU/cell for wild type virus and 10 PFU/cell for mutant 8. Cells were washed free of unattached virus, resuspended in warm medium with or without drug, maintained at 37°C and sampled periodically by quick-freezing 0.5 ml samples in dry ice-ethanol. Infectivity was determined by plaque assay in the presence (mutant 8) or absence (wild type) of drug

was similar to that for wild type virus. New infectivity was detected at about 2½ h and cell-associated infectivity peaked at about 5–6 h and at approximately 50–60 PFU/cell, regardless of the presence or absence of drug. Therefore, drug was not required for any of the functions leading to the production of virus.

When released virus was quantitated separately, infectivity steadily accumulated in the presence of drug (Fig. 2D). However, without drug only about 1% of virus infectivity was detectable outside the cells and infectious virus did not accumulate over time. This suggests that infectivity was lost after virus was released from cells.

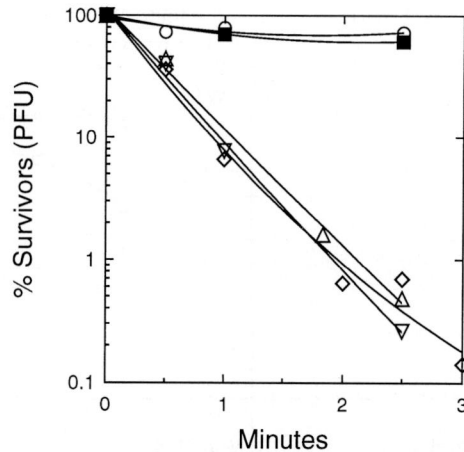

Fig. 3. Kinetics of thermal inactivation of poliovirus-3 and several drug-resistant mutants. Virus (150 µl) was diluted into plaque assay medium and pipetted to warmed glass tubes in a 37°C waterbath. After appropriate intervals, the tubes were transferred to an ice bath. Surviving infectivity was determined by plaque assay in the absence (wild type virus) or presence (mutants) of 2 µg/ml of WIN 51711. ■ Wild type virus; ○ non-drug-dependent mutant 5; ◇ drug-dependent mutants 8, ▽ 13 and △ 22

Drug-dependent mutants are extremely thermolabile

We hypothesized that loss of infectivity in the extracellular medium was the the result of thermolability at 37°C. Figure 3 summarizes results of tests with wild type and mutant viruses. These were diluted in medium used for plaque assays and then pipetted into glass tubes preequilibrated in a 37°C waterbath. The wild type virus, and mutant 5, one of the one-dependent mutants, survived under these conditions, but three of the drug-dependent mutants lost about 90% of residual infectivity every minute.

This result was shown to be general: when the mutants were tested by heating for 2½ min, all drug-dependent mutants were about equally thermolabile. The non-dependent mutants were considerably more variable [21]. We conclude that, in a plaque assay, drug is not required for attachment, uncoating or virus synthesis, but rather to stabilize the hyperlabile mutants long enough to survive the interval between release and infection of new cells.

How do these mutants lose their infectivity? Figure 4A shows that drug-dependent mutant 13 cosedimented with native wild type virus when unheated. When it was heated to 37°C in the absence of drug (Fig. 4B), it was converted to a particle sedimenting at approximately 135S. Neither wild type virus (Fig. 4C) nor mutant 13 treated with drug (Fig. 4D) was converted to 135S particles when heated to 37°C.

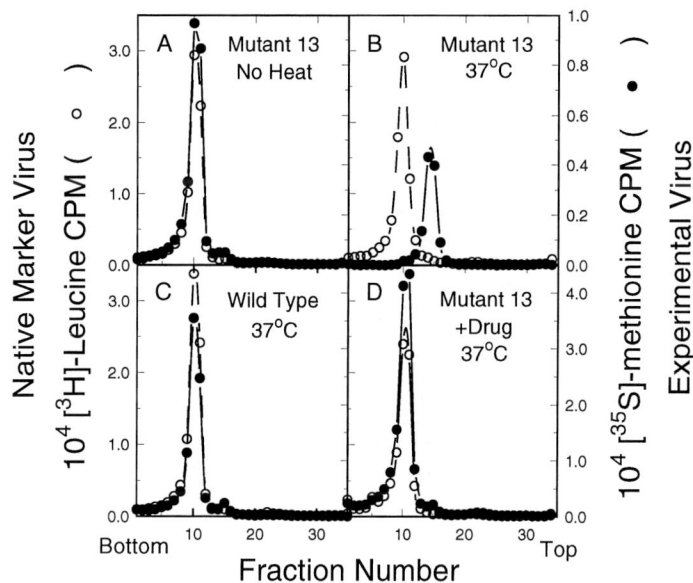

Fig. 4. Sedimentation of virus on sucrose gradients. [^{35}S] methionine-labeled virus samples (●) were either left unheated (**A**) or heated for 2.5 min as described in the legend to Fig. 3. They were then mixed with [^{3}H] leucine-labeled native wild type poliovirus-3 (○) and sedimented through a 5–20% (w/w) sucrose gradient. Fractions were collected from the bottom and radioactivity of samples was determined by scintillation counting. [^{35}S] methionine-labeled experimental virus was (**A**) unheated mutant 13, (**B**) mutant 13 heated to 37°C, (**C**) wild type virus heated to 37°C and (**D**) mutant 13 heated to 37°C after incubating in 2 µg/ml of WIN 51711 at room temperature for 1 h

The noninfectious 135S particles had all the characteristics of A-particles, the initial product of virions that have interacted with cellular receptors at 37°C [23–25]. In addition to sedimentation at 135S, they were hydrophobic, as manifested by poor recovery in the absence of detergents, they had lost capsid protein VP4, and protein VP1 was subject to proteolytic degradation. We speculate that modifications of the viral capsid that allow drug-dependent mutants to uncoat, even in the presence of drug, involve the acquisition of a "hair trigger" for the first step in uncoating. These particles undergo the first step of uncoating, loss of VP4 and formation of noninfectious A particles, when warmed to 37°C, even in the absence of cell receptor interaction. In a plaque assay, rapid decay of the infectivity of free virus released from the initially infected cells slows the rate of cell-to-cell spread to the point where plaques do not form.

Sequencing studies are underway to identify the amino acid substitutions in these drug-dependent mutants. Based on the successful use of mutations to identify functional regions involved in attachment of HRV-14, we expect that these drug-resistant mutants will provide information about capsid structures involved in triggering the conversion of native virus to the 135S particle.

Acknowledgements

We thank J. Y. Sgro for help with the preparation of Fig. 1, and R. Rueckert for excellent technical assistance. This work was supported by National Institutes of Health grants AI24939 and AI31960 to RRR.

References

1. Hogle J, Chow M, Filman DJ (1985) Three-dimensional structure of poliovirus at 2.9 A resolution. Science 229: 1358–1365
2. Rossmann MG, Arnold E, Erickson JW, Frankenberger EA, Griffith JP, Hecht H-J, Johnson JE, Kamer G, Luo M, Mosser AG, Rueckert RR, Sherry B, Vriend G (1985) Structure of a human common cold virus and functional relationship to other picornaviruses. Nature 317: 145–153
3. Olson NH, Kolatkar PR, Oliveira MA, Cheng RH, Greve JM, McClelland A, Baker TS, Rossmann MG (1993) Structure of a human rhinovirus complexed with its receptor molecule. Proc Natl Acad Sci USA 90: 507–511
4. Greve JM, Davis G, Meyer AM, Forte CP, Yost SC, Marlor CW, Kamarck ME, McClelland A (1989) The major human rhinovirus receptor is ICAM-1. Cell 56: 839–847
5. Staunton DE, Merluzzi VJ, Rothlein R, Barton R, Marlin SD, Springer TA (1989) A cell adhesion molecule, IACM-1, is the major surface receptor for rhinoviruses. Cell 56: 849–853
6. Filman DJ, Syed R, Chow M, Macadam AJ, Minor PD, Hogle JM (1989) Structural factors that control conformational transitions and serotype specificity in type 3 poliovirus. EMBO J 8: 1567–1579
7. Kim S, Smith TJ, Chapman MS, Rossmann MG, Pevear DC, Dutko FJ, Felock PJ, Diana GD, McKinlay MA (1989) Crystal structure of human rhinovirus serotype 1A (HRV1A). J Mol Biol 210: 91–111
8. Smith TJ, Kremer MJ, Luo M, Vriend G, Arnold E, Kamer G, Rossmann MG, McKinlay MA, Diana GD, Otto MJ (1986) The site of attachment in human rhinovirus 14 for antiviral agents that inhibit uncoating. Science 233: 1286–1293
9. Badger J, Minor I, Oliveira MJ, Smith TJ, Griffith JP, Guerin DMA, Krishnaswamy S, Luo M, Rossmann MG, McKinlay MA, Diana GD, Dutko FJ, Fancher M, Rueckert RR, Heinz BA (1988) Structural analysis of a series of antiviral agents complexed with human rhinovirus 14. Proc Natl Acad Sci USA 85: 3304–3308
10. McKinlay M (1985) WIN 51711, a new systematically active broad-spectrum antipicornavirus agent. J Antimicrob Chemother 16: 284–286
11. Otto MJ, Fox MP, Fancher MJ, Kuhrt MF, Diana GD, McKinlay MA (1985) In vitro activity of WIN 51711: a new broad-spectrum antipicornavirus drug. Antimicrob Agents Chemother 27: 883–886
12. Caliguiri LA, McSharry JJ, Lawrence GW (1980) Effect of arildone on modifications of poliovirus in vitro. Virology 105: 86–93
13. Fox MP, Otto MJ, McKinlay MA (1986) The prevention of rhinovirus and poliovirus uncoating by WIN 51711: a new antiviral drug. Antimicrob Agents Chemother 30: 110–116

14. Gruenberger M, Pevear D, Diana GD, Kuechler E, Blaas D (1991) Stabilization of human rhinovirus serotpye 2 against pH induced conformational change by antiviral compounds. J Gen Virol 72: 431–433

15. McSharry JJ, Caliguiri LA, Eggers HJ (1979) Inhibition of uncoating of poliovirus by arildone, a new antiviral drug. Virology 97: 307–315

16. Zeichhardt H, Otto MJ, McKinlay MA, Willingmann P, Habermehl K-O (1987) Inhibition of poliovirus uncoating by disoxaril (WIN 51711). Virology 160: 281–285

17. Heinz BA, Rueckert RR, Shepard DA, Dutko FJ, McKinlay MA, Fancher M, Rossmann MG, Badger J, Smith T (1989) Genetic and molecular analysis of spontaneous mutants of human rhinovirus 14 that are resistant to an antiviral compound. J Virol 63: 2476–2485

18. Pevear DC, Fancher MJ, Felock PJ, Rossmann MG, Miller MS, Diana G, Treasurywala AM, McKinlay MA, Dutko FJ (1989) Conformational change in the floor of the human rhinovirus canyon blocks adsorption to HeLa cell receptors. J Virol 63: 2002–2007

19. Heinz BA, Shepard DA, Rueckert RR (1990) Escape mutant analysis of a drug-binding site can be used to map functions in the rhinovirus capsid. In: Laver G, Air G (eds) Use of X-ray crystallography in the design of antiviral agents. Academic Press, New York, pp 173–186

20. Shepard DA, Heinz BA, Rueckert RR (1993) WIN 52035-2 inhibits both attachment and eclipse of human rhinovirus 14. J Virol 67: 2245–2254

21. Mosser AG, Rueckert RR (1993) WIN 51711-dependent mutants of type 3 poliovirus: evidence that virions decay after release from cells unless drug is present. J Virol 67: 1246–1254

22. DeSena J, Mandell B (1977) Studies on the in vitro uncoating of poliovirus. II. Characterization of the membrane-modified particle. Virology 78: 554–566

23. Fricks CE, Hogle JM (1990) Cell-induced conformational changes of poliovirus: externalization of the amino terminus of VP1 is responsible for liposome binding. J Virol 64: 1934–1945

24. Kaplan G, Freistadt MS, Racaniello VR (1990) Neutralization of poliovirus by cell receptors expressed in insect cells. J Virol 64: 4697–4702

Authors' address: Dr. A. Mosser, Institute for Molecular Virology, University of Wisconsin, 1525 Linden Drive, Madison, WI 53706, U.S.A.

Arch Virol (1994) [Suppl] 9: 121–132

_Archives_____
V̇irology
of

© Springer-Verlag 1994
Printed in Austria

Flock house virus: a simple model for studying persistent infection in cultured *Drosophila* cells

R. Dasgupta, B. Selling*, and **R. Rueckert**

Institute for Molecular Virology and Department of Biochemistry, Graduate School and College of Agriculture and Life Sciences, University of Wisconsin, Madison, Wisconsin, U.S.A.

Summary. Flock house virus (FHV), isolated from twenty *Drosophila melanogaster* cell lines, persistently infected with the virus, were examined during successive serial passages by plaque assay and sequence analysis. No phenotypic or genotypic changes in the virus were observed during the establishment of persistent infection, suggesting that it was a cellular modification that led to the first step in establishing the persistent state. Once this state was initiated, the virus was relieved of the need for a functional coat protein to propagate itself and mutations began to accumulate selectively in RNA2, the gene for the coat protein. These changes were manifested by a gradual drift to a smaller plaque population. The replicase activity, coded by RNA1, remained unaltered.

Introduction

Flock house virus (FHV) is a member of the family of small, non-enveloped, isometric insect viruses called *Nodaviridae*. Nodaviruses are the only known messenger sense, bipartite genome RNA viruses of higher or lower animals (for a review see [1]). Plant bipartite viruses are known, but their genome parts are encapsidated in separate virions. The family name *Nodaviridae* was derived from the type member Nodamura virus (NOV), which was named after a village in Japan. The natural host of nodaviruses are insect species including mosquitoes, ticks, honey bees and moth larvae but NOV can also replicate in suckling mice and possibly in pigs. That these non-enveloped viruses have the ability to replicate in both insects and animals aroused much interest and many of the physical properties of these viruses were elucidated in the 1970s. Subsequently, tissue culture methods were developed for nodaviruses

---

* Present address: Wyeth – Ayerst Laboratories, Radnor, Pennsylvania, U.S.A.

such as FHV, black beetle virus (BBV) and Boolarra virus (BOV). It was found that the latter three viruses replicate vigorously in *Drosophila melanogaster* cells and can be quantitatively assayed by plaque formation. This has resulted in a number of significant advances in the molecular biology of nodaviruses: the elucidation of the genome strategy, complete sequence of the genomic RNAs of BBV and FHV, development of infectious RNA transcripts from cDNA clones of FHV RNAs, and determination of the atomic structures of BBV and FHV. A recent study showed that FHV can also replicate in plant protoplasts as well as in some whole plants [2].

Here we report studies on the ability of FHV to establish and maintain persistent infection in cultured cells. Persistent infection (PI) occurs with many viruses but the precise mechanism(s) by which it happens is not well understood at the molecular level (for reviews see [3–5]). However, studies on PI with viruses in cell culture reveal a number of common features: (a) PI cells are generally resistant to superinfection by related viruses; (b) PI is associated with a change in the plaque size; and (c) coevolution of virus and host cell occurs with continued maintenance of PI cells [6, 7].

That nodaviruses might produce PI was first suggested by Friesen et al. [8]. The first experimental reconstruction of such a PI was demonstrated by Selling [9] who found that about 1% of cultured *Drosophila melanogaster* cells survive infection with FHV and produce virus in yields only a few percent those of a normal lytic infection. The small size of the FHV genome and the natural segregation of its coat and replicase genes on separate RNA segments (Fig. 1) greatly facilitate experimental analysis of mutants associated with PI.

The results of our studies on 20 PI cell lines indicate that the establishment of PI is not due to a mutation in the virus genome. This conclusion is based on two key observations: (1) sequence analysis revealed no mutation in the coat protein and (2) the polymerase activity remained unchanged, as judged by the synthesis of FHV RNAs in fresh cells infected with viruses isolated from PI cells. These findings suggest the possibility that the first event in establishment of PI is a change in the expression of one or more cellular genes. During maintenance of PI, that is after repeated passaging of these PI cells into fresh medium, there is an accumulation of mutations in the coat protein, but not in the viral coded polymerase, resulting in small plaque phenotype.

Genomic organization of nodaviruses

The genome strategy of FHV is shown in Fig. 1. The genome, with a total of 4506 bases in two segments, is among the smallest known

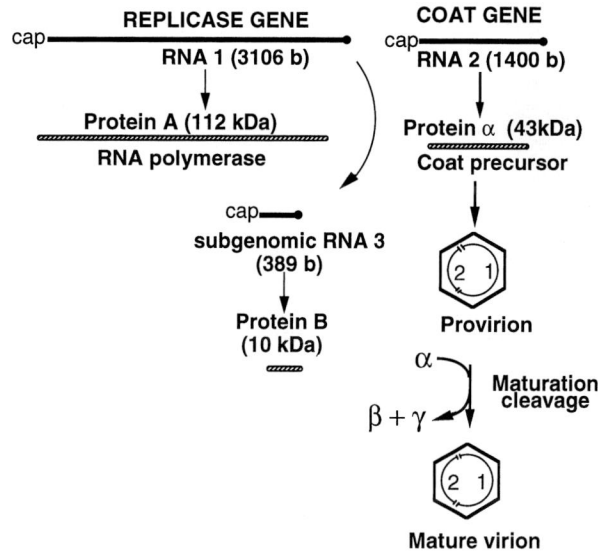

Fig. 1. Genomic organization of FHV. See text for details. Solid circles at the 3′-termini indicate that these ends are masked or blocked

eukaryotic RNA viruses. The messenger sense RNAs, called RNA1 and RNA2, appear to be encapsidated within a single virion. FHV RNA1 (3 106 b) codes for a protein A of mass 112 kDa (calculated from the known sequence) which is responsible for the polymerase activity. FHV RNA2 (1 400 b) directs the synthesis of a 43 kDa virion capsid precursor α which is processed into mature coat protein β (38 kDa) and a small peptide γ (5 kDa). Cells infected with FHV produce an additional messenger, RNA3 (389 b), coding for a protein called B of mass 10 kDa. RNA3 is a subgenomic messenger RNA derived from the 3′-end of RNA1 and is not encapsidated into the virion. Protein B, expressed only from RNA3, appears to be involved in polymerase function but its exact role is unknown. All the nodaviral RNAs are capped (m7GpppGp) at their 5′-ends. The 3′-ends of nodaviral RNAs are not polyadenylated nor do they have a free−OH group; rather they appear to be masked or blocked, as evidenced by their resistance to modification by standard enzymatic procedures even in denaturing conditions [10].

Establishment and maintenance of FHV persistence in *Drosophila* cells

We have shown earlier that infection of *Drosophila* cells with FHV causes extensive lysis within 3 days. However, about 1% cells survived which grew into colonies. The proportion of cells surviving infection was independent of the multiplicity of infection (m.o.i = 1 to 100, [9]). Within two weeks, the medium changed from pink to yellow (due to

Establishment : Passage 1

| **Day 1**
Monolayer of cells
infected w/FHV | **Day 3**
99% Cells lysed | **Day 7 - 10**
Surviving cells. |

Maintenance : Passage 2 and onwards

1 to 100 dilution in fresh medium

Allowed to grow for 2 weeks

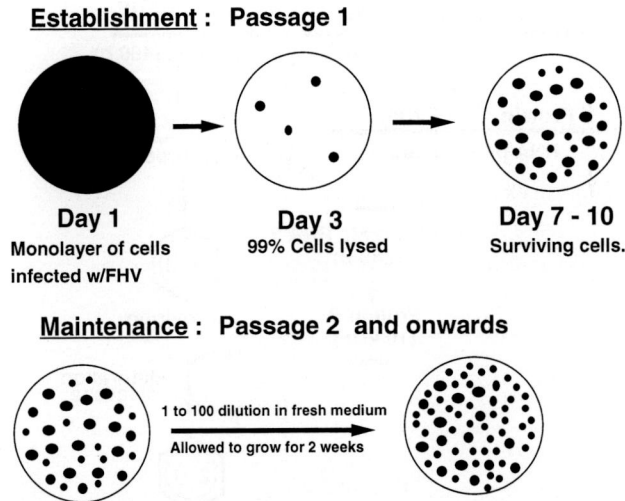

Fig. 2. Development of *Drosophila* cell lines persistently infected with FHV. Establishment; Conditions for the growth of *Drosophila melanogaster* cells and infection with FHV have been described [9, 13]. Briefly, cells (10^6) in 5 ml of complete growth medium (CGM) were pipetted into each well of six well (35 mm diameter) microtiter plates (Falcon Primaria tissue culture plates). These cells were infected with FHV (10000 pfu per well; m.o.i = 0.01) grown from a single plaque and incubated at 26°C. Extensive cpe was apparent after 3 days. Within 5 days, however, colonies (black spots) could be detected by microscopy. By 10 days to two weeks post infection, the colonies showed sufficient growth by acidification of the medium (color changed from pink to yellow). These cells were designated as passage 1. Maintenance: cells from passage 1 were diluted 100 fold (0.05 µl to 5 µl of fresh CGM), at two-week intervals (passage 2 through 26). During initial passages cells grew as clumps which decreased slowly at later passages. Cells from each passage were saved in liquid nitrogen in presence of 10 percent dimethyl sulfoxide (DMSO). The cells were revived, when necessary, by transferring into fresh medium after removal of DMSO (by pelleting the cells and washing with fresh medium)

acidification) showing sufficient growth. Based on this observation, we developed 20 resistant *Drosophila melanogaster* cell lines infected with FHV isolated from individual plaques (Fig. 2). Plaque assay with aliquots removed from each culture showed infectivity in all 20 cultures. These cells were, therefore, persistently infected in a non-lytic fashion. They were resistant to superinfection by closely related but serologically distinguishable nodaviruses BBV and BOV but they were not resistant to superinfection by unrelated insect viruses Cricket paralysis virus and Drosophila X virus (data not shown).

PI cell lines were maintained by subpassaging in fresh medium at two week intervals (Fig. 2). An initial tendency of PI cells to clump together diminished after the first few passages.

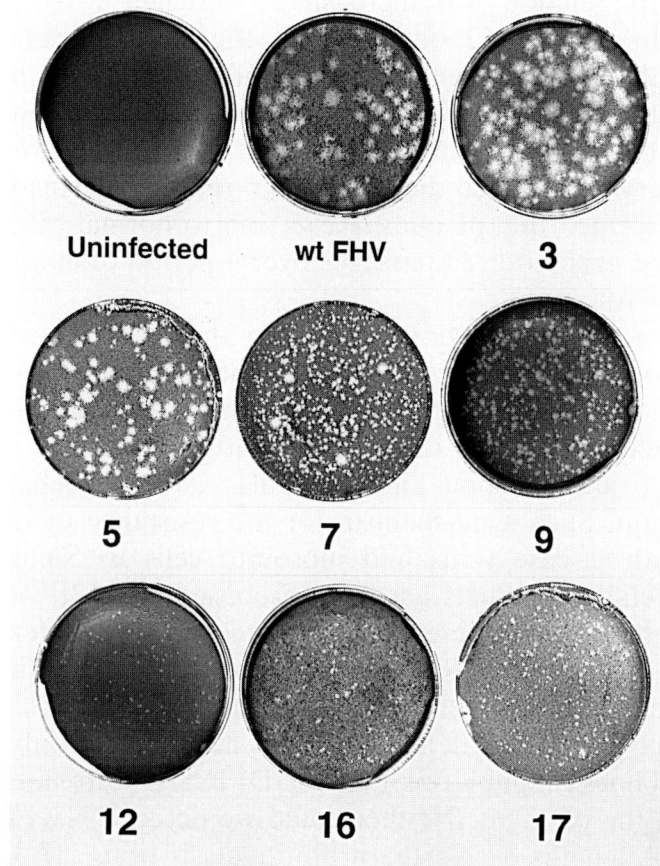

Fig. 3. Plaque assay of viruses from PI cells at ealy and late passages. Virus was released from PI cells by freezing and thawing 3 times and assayed on monolayers of *Drosophila* cells as described [13]. The numbers at the bottom of the plates indicate the number of passages. All assays were done with the same PI cell line (cell line 6). Note the transition to small plaque phenotype starting from passage 7

PI viruses always produced normal size plaques at early passages but the size of the plaques decreased during later passages

Persistent infections with animal viruses generally are associated with a decrease in plaque size [3–5]. However, the basis for this correlation has been unclear. By examining the plaque size of FHV isolates from several PI cell lines at early and late passages, we show that establishment of PI and appearance of small plaque formers are separate events. Figure 3 shows the results of a plaque assay with PI cell line 5 at passages 3, 5, 7, 9, 12, 16 and 17. There were no changes in the plaque size (compared to wild type FHV) until passage 5 when we observed a mixture of small and normal size plaques (approximately 50% of each type). At passages

7 and 9 the population was increasingly enriched with small plaque formers. From passage 12 onwards, all the plaques were tiny. This implied that the establishment of PI was independent of change in the plaque size; rather the decreased plaque size was a secondary event, taking place during the maintenance stage, typically between passages 5–9. At a very late stage, the PI viruses reached a steady state and continuously formed tiny plaques. Reversion to normal plaque size was never observed even after 30 passages over a period of almost two years (data not shown).

Absence of any phenotypic changes in the virus during the establishment period indicates that FHV persistence is initiated by the modification of the virus-cell relationship. The nature of this modification is not known. The tendency of the cells to clump together in first few passages may reflect activation of some kind of cellular defense mechanism against virus replication. Such a mechanism for the resistance of BHK cells to foot-and-mouth disease virus and mosquito cells to Sindbis virus by trans-acting cellular products has been proposed [11, 12]. Alternatively, the virus might generate defective templates which interfere with viral polymerase, a mechanism suggested for persistent infection in many animal viruses [3]. However, this seems unlikely in FHV persistence because we have observed defective RNAs in virions of only four out of twenty PI cell lines (unpubl. obs.). These DI RNAs were not consistently present in all the passages. Neither could we detect RNAs shorter than genomic RNA lengths by Northern blot analysis using ^{32}P labelled plus and minus RNA1 and RNA2 transcripts as probe (data not shown).

Genomic RNA2 is responsible for small plaque phenotype

We tested the hypothesis that small plaque phenotype, characteristic of PI viruses at high passage number, could be mapped to one of the two genomic RNAs. Advantage was taken of the fact that RNAs 1 and 2 are physically separable by sucrose density gradient or gel electrophoresis. Plaque assays were done after transfecting fresh *Drosophila* cells with reassorted RNAs purified from wild type FHV and PI viruses isolated from different passages. The results with passage 26 of cell line 15 are shown in Fig. 4. When RNA1 from either wild type FHV or PI viruses plus RNA2 from wild type FHV were used, normal size plaques were obtained. By contrast, when RNA2 of PI viruses were mixed with RNA1 from either source, it resulted in small plaques. Reassortment plaque assay with all 20 PI cell lines from passage 9 onwards (small plaque formers) produced the same results. RNA2 of PI viruses from passages 1 through 5 (normal size plaque formers; establishment and early maintenance period), as expected, produced normal size plaques. These reas-

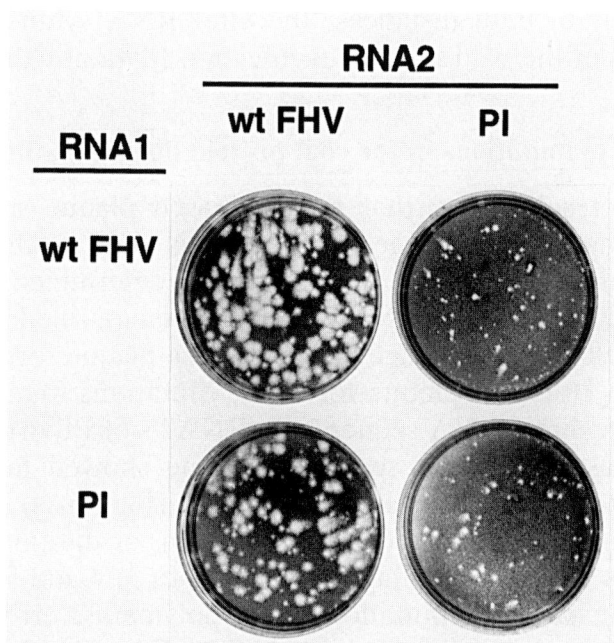

Fig. 4. Reassortment analysis demonstrating that RNA2 is responsible for the small plaque phenotype of PI viruses. Purification of FHV from normal infection and persistent infection has been described [9]. RNA1 and RNA2 were separated on a 1.2% low melting point agarose (BRL) gel and recovered by phenol extraction and ethanol precipitation. Cells were transfected [14] using lipofectin (1 mg/ml, BRL) with mixtures of RNA 1 (250 ng) and RNA2 (60 ng) purified from wt FHV and FHV isolated from PI cell lines. Assay with passage 26 of PI cell line 15 is shown here. Viral RNAs from all the twenty PI cell lines after passage 9 (small plaque phenotypes) produced same results

sortment experiments showed that mutations in RNA2, the coat protein gene, were invariably responsible for the small plaque trait of PI viruses, while RNA1, the replicase gene, played no role in the change of plaque morphology.

Evidence that the viral polymerase activity is not impaired in PI

It has been shown earlier [9] that FHV multiplies in PI cells at a rate 1–10% of that of the wild type virus in fresh *Drosophila* cells. However, when fresh cells were infected with PI viruses, a yield comparable to that of the wild type FHV infection was obtained. This indicated that the function of RNA1 (RNA synthesis) in PI viruses remained unchanged. This was confirmed by measuring the synthesis of ^{3}H-uridine labelled RNA in presence of actinomycin D after infection of cells with PI viruses from passage 3 (normal size plaque former) and with passage 16 (small plaque former). In both instances, the viral RNA synthesis was comparable to that of the wild type virus infection (data not shown).

Accumulation of mutations in the coat protein during maintenance of PI

The foregoing results show that the decreased plaque size of viruses produced during serial passage of PI cells is due to changes in the RNA2, the coat protein cistron. We, therefore, determined the complete sequence of c-DNA clones of RNA2's isolated from purified FHV progeny of several PI cell lines at passage 3 (normal size plaque former), passage 16 and passage 26 (small plaque formers). Mutations found in c-DNAs were verified by direct RNA sequencing. RNA2's of PI viruses from five cell lines at passage 3 were sequenced; none showed mutations that would lead to a change in the amino acid (Fig. 5A). There were, however, one or two silent mutations (mutations at the third position of the codons, not shown) in passage 3 PI cell lines 3, 4 and 5. These were apparently "hot spots" for mutation due to polymerase error and had no effect on the plaque sizes (see Fig. 3, plate 3). One RNA2 isolate (PI 6) had no mutation at all. This meant that the establishment of PI was not accompanied by mutation in the coat gene nor was the ability of RNA1 to replicate impaired when PI viruses were transferred to fresh cells. Apparently then, the establishment of PI is caused by a change in the cellular environment, rather than by a modification of the viral genome.

Analyses of RNA2's of PI viruses isolated from passages 16 and 26 (maintenance period, small plaque formers), however, showed accumulation of mutations in the coat proteins. A total of twelve amino acids were mutated in the coat proteins of viruses from five PI cell lines at passage 16 (an average of 2–3 mutations in each, Fig. 5B). By the 26th passage, the coat proteins had accumulated a total of 65 mutations in viruses isolated from six PI cell lines (an average of 10–11 mutations in each, Fig. 5C). Most mutations occurred randomly over the genomic RNA2's. However, codons for four amino acids mutated more frequently than the others. Table 1 shows these common mutations in passage 16 and passage 26 combined. The most frequent mutation was in Ala at position 265 changing to Valine (Ala265Val, corresponding to base 816 highlighted by asterisks in Figs. 5B and 5C), found in 9 of 11 PI viruses. Ser156Ala, Val195Ileu and Val288Ileu mutations each occurred in 5 of 11 PI viruses. Analysis of these mutations in PI viruses in terms of hydrophobicity and charge showed that 95% were conservative changes.

Interestingly, the Ala265Val mutation was present in the coat proteins of viruses from all five PI cell lines at passage 16 and from four of six PI cell lines at passage 26 (Figs. 5B and 5C). Because this mutation was not present in viruses at passage 3 (Fig. 5A), it must have occurred

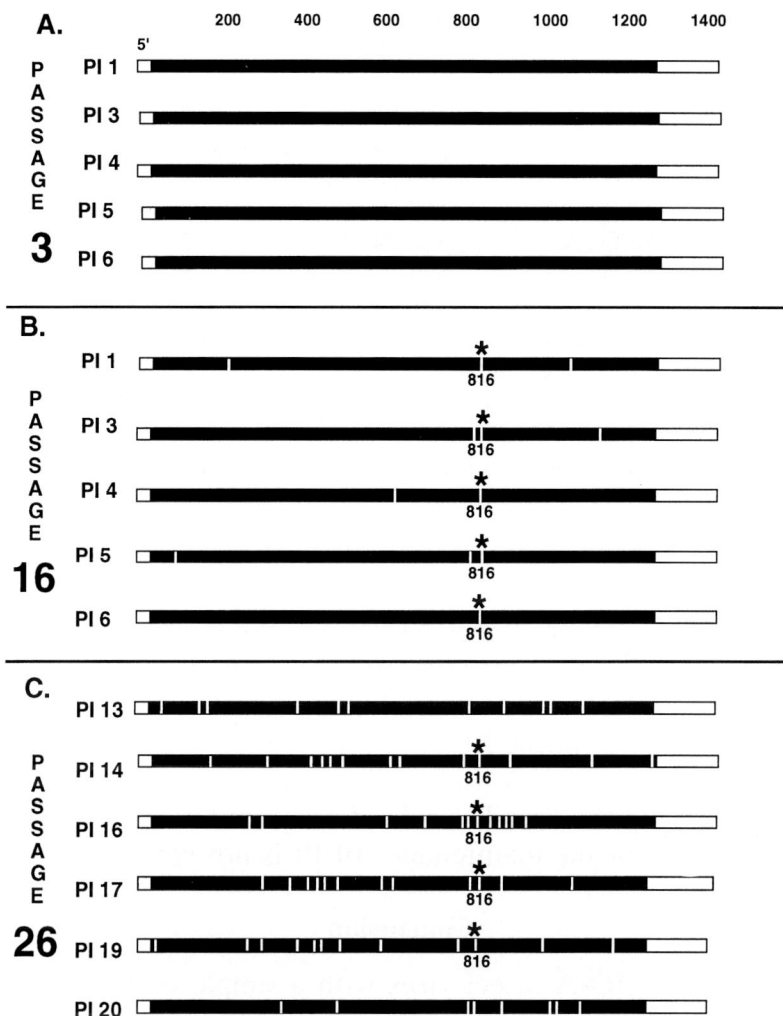

Fig. 5. Mutations accumulating in the coat proteins of PI viruses. c-DNA copies of RNA2's purified from PI viruses were inserted into Pstl-Xbal sites of the plasmid PUC118 and in pBluescriptllKS+ (Stratagene) [15]. Complete sequence analyses [16] of these c-DNA clones were done using oligonucleotide primers (synthesized at the University of Wisconsin-Madison, Biotechnology Center) complementary to known FHV RNA2 sequence [17] at an interval of 200 bases. Sequences were assembled and compared with known RNA2 sequence of FHV [17] using GCG programs [18]. Passage numbers and PI cell lines (Pln) from which viruses were isolated are indicated. Solid black bars indicate regions coding for FHV coat proteins. White bars indicate non-coding regions. Each vertical white line indicates a single mutation in the coat protein. Scale at the top shows the positions of nucleotides. The asterisk highlights base 816 (corresponding to Ala265Val mutation in the FHV coat protein) which mutated at a high frequency (see Table 1). **A** RNA2 from five PI virus isolates at passage 3 (normal size plaque formers) showing no mutation in the coat protein; **B, C** RNA2s from PI virus isolates at passage 16 and 26 (both are small plaque phenotypes, maintenance of PI), showing that mutations accumulate in the coat protein at late stages of PI and they increase with number of passages

Table 1. Frequently mutated amino acids in the coat proteins of PI viruses

Codon change[a]	aa Mutation[b]	Frequency[c]
GCU 816 GUU	Ala 265 Val	9
GUA 884 AUA	Val 288 Ileu	5
UCA 488 GCA	Ser 156 Ala	5
GUA 605 AUA	Val 195 Ileu	5

[a] The number indicates the position of the nucleotide, the change is underlined

[b] For example, Ala 265 Val indicates that Ala at position 265 mutated to Valine

[c] Number of times the mutation occurred in 11 PI RNA2 isolates (five from passage 16 and six from passage 26) sequenced

independently in all nine instances. Moreover, this was the only mutation in PI cell line 6 (PI 6) at passage 16, suggesting that this mutation alone could be responsible for the small plaque phenotype (see Fig. 3, plate 16). This has been confirmed by appropriate site directed mutagenesis at base 816 in the c-DNA followed by infectivity assay with RNA transcripts (data not shown). The role of other mutations in the reduction of plaque size and/or the maintenance of PI is not yet clear.

Conclusion

We used FHV, an RNA insect virus with a simple bipartite genome to study PI in cultured *Drosophila melanogaster* cells. The ease with which PI develops in these cells, in combination with the known molecular biology of FHV makes this an attractive system for studying the molecular basis of PI. Our results strongly suggest that the establishment of PI in cultured *Drosophila* cells requires no mutation in the virus. It appears that the modification, at this point, is due to an unidentified change in the virus-cell relationship. With continued passage of the PI cells (maintenance phase), however, mutations begin to accumulate selectively in the coat proteins of PI-viruses. This is manifested by a shift in the virus population to a smaller plaque size. RNA1 does not contribute to the small plaque phenotype at all, nor is its ability to replicate in fresh cells impaired. We propose that this preferential accumulation of mutations in RNA2 is due to loss of selective pressure on the coat protein, which in PI is no longer needed for infection of fresh host cells. Even in PI cells however, RNA1 must replicate to survive because failure to do so might "cure" the infection and thereby make the cured cells again susceptible to killing by PI-viruses.

Our studies change the focus of attention in PI from mutation(s) in the virus population to modification(s) in the infected cells. Particularly at the establishment step. It now becomes important to identify the nature of the modification(s) that enables the virus to co-exist with the host.

Acknowledgements

We thank Prof. P. Kaesberg for support and valuable suggestions and N. Dreckschmidt and G.-Q. Zhang for expert technical assistance. This investigation was supported by NIH grants AI23742 and AI22813.

References

1. Hendry DA (1991) Nodaviridae of invertebrates. In: Kurstak E (ed) Viruses of invertebrates. Marcel Dekker, New York, pp 227–276
2. Selling BH, Allison RF, Kaesberg P (1990) Genomic RNA of an insect virus directs synthesis of infectious virions in plants. Proc Natl Acad Sci USA 87: 434–438
3. Ahmed RA, Stevens JG (1989) Viral persistence. In: Fields BN (ed) Virology. Raven Press, New York, pp 241–266
4. Mahy BWJ (1985) Strategies of virus persistence. Br Med Bull 41: 50–55
5. Oldstone MBA (1989) Viral persistence. Cell 56: 517–520
6. Ahmed R, Canning WM, Kauffman RS, Sharpe AH, Hallum JV, Fields BN (1981) Role of the host cell in persistent viral infection: coevolution of L cells and reovirus during persistent infection. Cell 25: 325–332
7. de la Torre JC, Martinez-Salas E, Diez J, Villverde A, Gebauer F, Rocha E, Davila M, Domingo E (1988) Coevolution of cells in a persistent infection of foot and mouth disease virus in cell culture. J Virol 62: 2050–2058
8. Friesen P, Scotti P, Longworth J, Rueckert R (1980) Black Beetle Virus: Propagation in Drosophila line 1 cells and an infection-resistant subline carrying endogenous Black Beetle Virus-related particles. J Virol 35: 741–747
9. Selling B (1986) Infectivity of Black Beetle Virus in cultured *Drosophila* cells. Ph.D. thesis, University of Wisconsin-Madison, pp 88–135
10. Dasgupta R, Ghosh A, Dasmahapatra B, Guarino LA, Kaesberg P (1984) Primary and secondary structure of black beetle virus RNA2, the genomic messenger for BBV coat protein. Nucleic Acids Res 12: 7215–7223
11. de la Torre JC, de la Luna S, Diez J, Domingo E (1989) Resistance to foot and mouth disease virus mediated by *trans*-acting cellular products. J Virol 63: 2385–2387
12. Riedel B, Brown DT (1979) Novel antiviral activity found in the media of Sindbis virus persistently infected mosquito (*Aedes albopictus*) cell cultures. J Virol 29: 51–60
13. Selling BH, Rueckert RR (1984) Plaque assay for black beetle virus. J Virol 51: 251–253
14. Zhong W, Dasgupta R, Rueckert R (1992) Evidence that the packaging signal for nodaviral RNA2 is a bulged stem-loop. Proc Natl Acad Sci USA 89: 11146–11150

15. Dasmahapatra B, Dasgupta R, Saunders K, Selling B, Gallagher T, Kaesberg P (1986) Infectious RNA derived by transcription from cloned cDNA copies of the genomic RNA of an insect virus. Proc Natl Acad Sci USA 83: 63–66

16. Kraft R, Tardiff J, Krauter KS, Leinwand L (1988) Using mini-prep plasmid DNA for sequencing double stranded templates with sequenase. Biotechniques 6: 544–549

17. Dasgupta R, Sgro JY (1989) Nucleotide sequence of three nodavirus RNA2's: the messengers for their coat protein precursors. Nucleic Acids Res 17: 7525–7526

18. Devereaux J, Haeberli P, Smithies O (1984) A comprehensive set of sequence analysis programs for the VAX. Nucleic Acids Res 12: 387–395

Authors' address: Dr. R. Dasgupta, Institute for Molecular Virology, Department of Biochemistry, Graduate School and College of Agriculture and Life Sciences, University of Wisconsin, Madison, WI 53706, U.S.A.

Genome replication and transcription

Archives
Virology
© Springer-Verlag 1994
Printed in Austria

Protein-protein interactions and glycerophospholipids in bromovirus and nodavirus RNA replication

P. Ahlquist[1,2], **S.-X. Wu**[1,3], **P. Kaesberg**[1,3], **C. C. Kao**[1,2], **R. Quadt**[1,2], **W. DeJong**[1,2], and **R. Hershberger**[1,2]

[1] Institute for Molecular Virology and Departments of [2] Plant Pathology and [3] Biochemistry, University of Wisconsin-Madison, Madison, Wisconsin, U.S.A.

Summary. The plant bromoviruses and animal nodaviruses are distinct groups of positive strand RNA viruses that have proven to be useful models for RNA replication studies. Bromoviruses encode two large proteins required for RNA replication: 1a contains domains implicated in helicase and capping functions, and 2a contains a central polymerase-like domain. Using immunoprecipitation and far-western blotting, we have now shown that 1a and 2a form a specific complex in vitro and have mapped the interacting domains. Molecular genetic data implicate the 1a–2a complex in RNA replication and suggest that it supports coordinate action of the putative helicase, polymerase, and capping domains. The locations of the interacting 1a and 2a domains have implications for replication models and the evolution of virus genomes bearing homologous replication genes in fused vs. divided forms. For the nodavirus Flock house virus (FHV), a true RNA replicase has been isolated that carries out complete, highly active replication of added FHV RNA, producing newly synthesized positive strand RNA in predominantly ssRNA form. Positive strand RNA synthesis in this FHV cell-free system is strongly dependent on the addition of any of several glycerophospholipids. Positive strand RNA synthesis depends on the complete glycerophospholipid structure, including the polar head group and diacyl glycerol lipid portion, and is strongly influenced by acyl chain length.

Introduction

RNA-dependent RNA synthesis by positive strand RNA viruses is an intricate process that involves the combined function of multiple components including viral proteins, cis-acting recognition and regulatory sequences on the viral RNA, and likely several distinct types of host components such as proteins and membranes. Elucidating the nature, function, and interaction of the various participants in RNA replication and transcription remains a challenging frontier in the study of positive strand RNA viruses. This article reviews some recent results on RNA replication in two different virus systems: the demonstration and mapping of interactions between the helicase-like and polymerase-like RNA replication proteins of the bromoviruses, and the important role of glycerophospholipids in a cell-free system that catalyzes the complete in vitro replication of nodavirus RNA.

Bromoviruses

The bromoviruses are a group of icosahedral, positive strand RNA plant viruses [7]. Bromovirus genomes are divided among three capped RNAs, designated RNAs 1–3. RNA3 encodes two proteins involved in the spread of infection: the 3a noncapsid "movement" protein is translated directly from RNA3 and is required for initial cell-to-cell spread of infection, while coat protein is expressed via a subgenomic mRNA and is required for long range, vascular spread of infection ([3, 5]; Mise and Ahlquist, unpubl. res.). Recent experiments show that the 3a gene of a cowpea-adapted bromovirus can be functionally replaced with the nonhomologous movement gene from a very different positive strand RNA virus, i.e. a rod-shaped, single component tobamovirus that also infects cowpeas [5]. The ability of this hybrid virus to spread between cells and even achieve systemic infection indicates that the tobamovirus movement protein functions independently of sequence-specific interactions with other virus factors or RNA sequences. These results also indicate that the function of bromovirus coat protein in vascular spread is independent of specific interaction with the 3a protein, and provide direct experimental support for the concept of modular evolution in RNA viruses [11].

RNA1 (3.2 kb) and RNA2 (2.9 kb) encode RNA replication factors 1a (109 kDa) and 2a (94 kDa), respectively [1]. 2a contains a central polymerase-like domain and additional N- and C-terminal flanking regions [26]. 1a contains a C-terminal domain with extended sequence similarity to helicases [9, 13] and an N-terminal domain with sequence similarity to the Sindbis virus nsP1 protein, which has methyltransferase and possibly guanylyltransferase activities thought to be involved in viral RNA capping [18, 23]. All three domains are conserved among all members of the "alphavirus-like superfamily" of positive strand RNA viruses, which includes diverse plant viruses and the animal alphaviruses [8, 11]. Mutational studies of bromovirus 1a and 2a show that all three conserved domains are required for RNA replication [1, 16, 17, 25, 26].

Interaction of bromovirus RNA replication proteins

The best-studied bromoviruses are brome mosaic virus (BMV) and cowpea chlorotic mottle virus (CCMV). RNA reassortment [2] and gene exchange experiments between BMV and CCMV [29] show that heterologous combinations of the 1a and 2a genes are unable to support replication of BMV or CCMV RNA3, even though either homologous gene combination will do so. Thus, the functions of the 1a and 2a proteins are not independent; successful RNA replication involves some form of direct or indirect 1a–2a interaction at one or more steps of RNA synthesis.

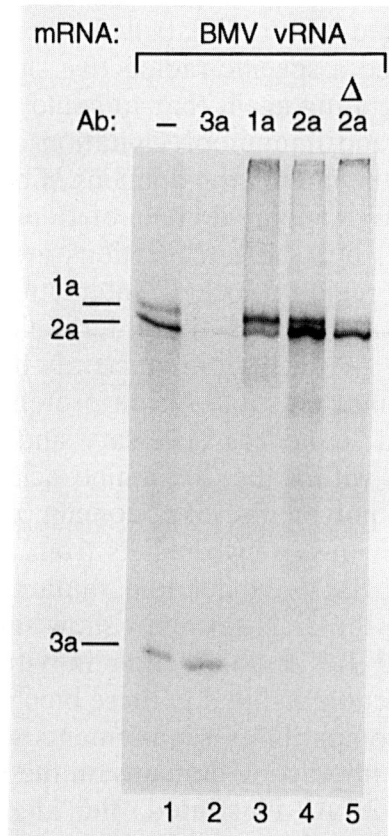

Fig. 1. Co-immunoprecipitation of in vitro-translated BMV 1a and 2a proteins. BMV virion RNA was used translated in a rabbit reticulocyte lysate and a portion of the products was reserved for direct electrophoresis (*1*). The remaining translation products were pre-treated with protein A-expressing *S. aureus* to remove any proteins interacting nonspecifically with protein A, aliquoted into 4 tubes, and then immunoprecipitated with anti-3a serum (*2*), anti-1a serum (*3*), or anti-2a serum (*4, 5*) before electrophoresis on an SDS-polyacrylamide gel. The sample in *5*, marked with a Δ, was incubated for 5 min at 42°C prior to addition of antiserum (from [15])

To further explore the 1a–2a association revealed by the above experiments, we initiated biochemical studies to examine the potential for the BMV 1a and 2a proteins to interact in vitro [15]. The first series of experiments showed that radiolabelled, in vitro-translated 1a and 2a both could be specifically co-precipitated by either of two non-cross-reacting antisera specific for 1a or 2a (Fig. 1). 1a–2a interaction was more efficient when both were co-translated in the same reaction, but complex formation also occurred when 1a and 2a were separately translated and mixed prior to immunoprecipitation.

As a separate test of the ability of 1a and 2a to interact in vitro, an antibody-independent "far-western blot" assay was developed. BMV 2a protein was expressed in insect cells via a baculovirus vector, partially purified, electrophoresed on an SDS-polyacrylamide gel, transferred to a membrane, and incubated with radiolabelled 1a protein. After washing, autoradiography revealed a specific radioactive signal at the position of the 2a protein, demonstrating again that 1a could bind to 2a [15].

In vitro translation and immunoprecipitation of modified 1a and 2a were then used to map the interacting domains in both proteins [14]. The entire N-terminal putative capping domain of 1a protein was dispensable for interaction with 2a, but short truncations at either end of the C-proximal helicase-like region blocked interaction. Interestingly, the 1a helicase-like region alone, with the N-terminal 502 1a amino acids deleted, interacted with 2a much more efficiently than the full 1a protein [14]. For 2a, similar truncations and fusion proteins showed that amino acids 25–140 contained sequences necessary and sufficient to interact with 1a. This region is within the 200 amino acid N-terminal segment preceding the central polymerase-like domain of 2a. This mapping correlates well with previous in vivo results from BMV/CCMV 2a gene hybrids [25] and from BMV 2a deletion mutants [26]. In particular, progressive deletions in the 2a N-terminus show similar effects on RNA replication in vivo and 1a–2a interaction in vitro, providing further evidence that the interaction mapped in these biochemical experiments is relevant to the 1a–2a compatibility requirements seen in in vivo.

The positions of the interacting domains on the linear maps of 1a and 2a are shown in Fig. 2. As illustrated, the alignment of 1a and 2a domains that results from this mapping is remarkably similar to the arrangement of the analogous conserved domains in the TMV p183 readthrough protein. Thus, as we have previously suggested, a bromovirus 1a–2a complex may be a functional analog of the TMV p183 read-through protein [1, 25]. These results also partially explain the absence in p183 of any region corresponding to the N-terminal segment of the bromovirus 2a protein. At least one purpose of this bromovirus 2a segment is to mediate noncovalent interaction of the polymerase-like 2a

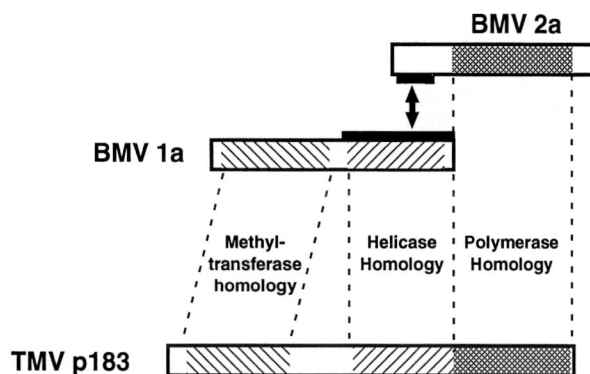

Fig. 2. Position of the interacting domains in the linear maps of BMV 1a and 2a proteins, and their relationship to the p183 readthrough protein of TMV. The portions of 1a and 2a indicated by the solid black bars contain sequences that are necessary and sufficient for their mutual interaction in vitro. The intervirally-conserved polymerase-, helicase-, and methyltransferase-like domains of the BMV 1a and 2a polypeptides and TMV p183 are represented by differential shading, as indicated (from [14])

core with 1a (Fig. 2). In TMV p183, the covalent fusion of domains analogous to 1a and 2a apparently renders this function unnecessary. However, N-terminal extensions do precede the conserved polymerase-like cores of the independent polymerase-like proteins expressed by many other positive strand RNA viruses besides the bromoviruses, including the animal alphaviruses and many plant viruses. Just as in BMV 2a, such flanking segments might contribute to interaction with the separate helicase-like protein that is also expressed by each of these viruses.

Possible integration of functions in a 1a–2a complex

Why, in mechanistic terms, do 1a and 2a interact for viral RNA synthesis? One possibility is that the association of 1a and 2a in a single complex allows the three intervirally-conserved domains of these proteins to integrate and coordinate their actions to make RNA replication a more continuous and efficient process [1]. This possibility is further supported by genetic tests showing that no complementation occurs between any of a number of 1a mutants in the helicase-like domain and other 1a mutants in the methyltransferase-like domain [17]. Thus, the operation of the two 1a domains also appears to be interdependent or cooperative.

   A simplified model to illustrate some of the possibilities for integration of functions in a 1a–2a complex is shown in Fig. 3. In such a complex, the helicase-like domain might promote initiation of RNA synthesis by unwinding the substantial secondary and tertiary structure at the ends of the viral RNAs, providing the polymerase access to its initiation sites.

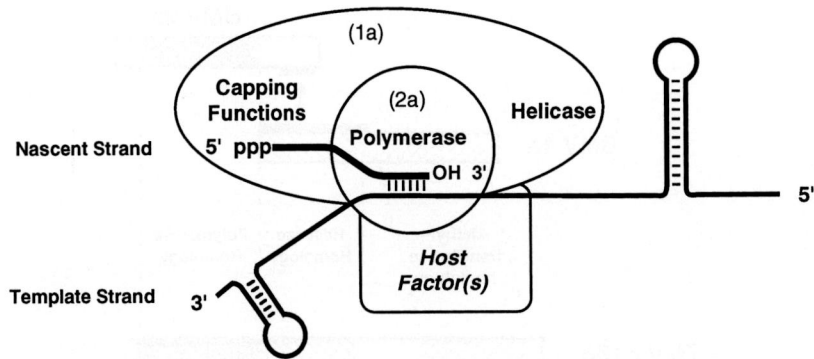

Fig. 3. Highly schematic model of a 1a–2a complex to illustrate the possible coordination of helicase, polymerase, and capping functions in viral RNA replication and transcription. See text for additional discussion

This could explain why 1a appears to influence some aspects of template specificity [25]. Alternatively or in addition, the putative 1a helicase domain might expedite elongation by removing internal template secondary structure just in front of the polymerase or assist in releasing the product strand from the template. Possible capping functions in the 1a N-terminus may be poised to cap the 5′ ends of new positive strands immediately after their synthesis, as appears to occur in the vaccinia virus transcription complex [4].

The model of Fig. 3 and its discussion here is, of course, deliberately oversimplified and meant to illustrate only some aspects of viral RNA synthesis. This discussion does not address the stoichiometry of 1a and 2a in the replication complex or whether 1a or 2a have independent functions outside of a complex as well as in a complex. As suggested in Fig. 3, several results also imply that host proteins are associated with the bromovirus RNA replication complex [12, 22]. Finally, it is not yet known whether different steps in RNA replication might be carried out by compositionally distinct replication complexes.

Nodaviruses and a nodavirus RNA replicase

Flock house virus (FHV) is a member of the *Nodaviridae* family of positive strand RNA viruses [7]. All known nodaviruses infect insects and at least the type member, Nodamura virus, can also infect vertebrates. Nodaviruses have an isometric virion that co-encapsidates two genomic RNAs, RNA1 (3.1 kb) and RNA2 (1.4 kb). RNA1 encodes at least two proteins, A and B, that are involved in RNA replication while RNA2 encodes the coat protein precursor.

Despite extended efforts, template-dependent, cell-free systems capable of completely replicating an exogenously supplied eukaryotic viral RNA have been obtained only recently and for only two to three positive strand RNA viruses [12, 19, 28]. Among these is a highly active FHV RNA replicase extract [28]. Complete replication of added FHV RNA in the originally reported system was achieved by presenting the template RNA together with lipofectin, a reagent commonly used to introduce nucleic acids into membrane compartments [6]. In the unsupplemented system, added FHV positive strand RNA was copied into negative strand RNA, yielding a dsRNA product, but no new positive strands were synthesized. However, when FHV RNA templates were preincubated with lipofectin in an attempt to stimulate or redirect template RNA delivery to the membrane-associated FHV RNA polymerase, the enzyme additionally synthesized new positive strands in primarily ssRNA form.

Glycerophospholipid activates FHV positive strand RNA synthesis in vitro

Lipofectin is a mixture of a cationic lipid and phosphatidylethanolamine, a neutral glycerophospholipid (GPL). In recent experiments we have explored the role of these individual components and found that the activation of FHV single strand RNA synthesis is solely due to the phosphatidylethanolamine [27]. The cationic lipid is actually inhibitory to FHV RNA synthesis in this in vitro system. Since the cationic lipid is the essential component for transfecting nucleic acids into membrane compartments [6], this suggests that the activation of positive strand RNA synthesis is not due to a liposome-mediated pathway of template delivery, as originally suspected.

Further tests showed that other GPLs will also support full FHV RNA replication in vitro. At varying levels, phosphatidylcholine (PC), phosphatidylglycerol, phosphatidylinositol, and phosphatidylserine will all support single strand RNA synthesis (Fig. 4). The polarity of the radiolabelled strands in RNA products synthesized from unlabelled positive strand FHV RNA templates in the presence and absence of PC was determined by ribonuclease T1 analysis [27]. In the absence of PC or any other GPL, the dsRNA product was exclusively labelled in the negative strand. In the presence of PC, however, the dsRNA product contained radiolabelled positive and negative strand RNA in approximately equal proportions and the radiolabelled ssRNA was exclusively positive strand RNA. This confirms that PC addition supports the full replication of FHV RNA in this in vitro system. In addition, the approximately equal labelling of positive and negative strand RNA in dsRNA synthesized in the presence of PC is of interest. This indicates that nearly all negative strands were used as templates for at least one round of positive strand RNA synthesis, and that these negative strands remained preferentially associated with newly synthesized, product positive strands.

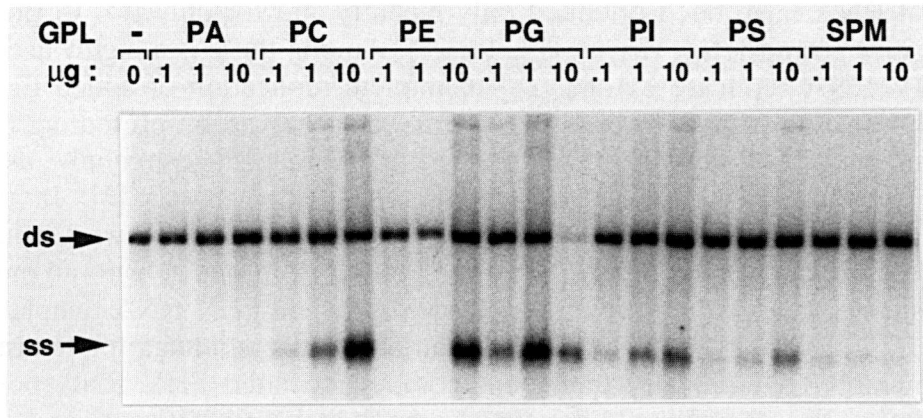

Fig. 4. Effect of different phospholipids on in vitro RNA synthesis in FHV replicase reactions initiated by addition of purified FHV RNA2, the smaller of the two FHV genomic RNAs, and incubated for 3 h at 30°C. *1* contains the products of a control reaction without any added GPL. The type and amount (µg per 12.5 µl reaction) of phospholipid added to each of the other reactions is shown above the corresponding lanes, using the following abbreviations: *PA* Phosphatidic acid; *PC* phosphatidylcholine; *PE* dioleoyl phosphatidylethanolamine; *PG* dioleoyl phosphatidylglycerol; *PI* phosphatidylinositol; *PS* phosphatidylserine; *SPM* sphingomyelin (from [27]). *ds* Double strand RNA, *ss* single strand RNA

Activation of FHV positive strand RNA synthesis in vitro is GPL-specific

In experiments to date, activation of positive strand synthesis in the FHV cell-free system has been specific to GPLs [27]. No detectable stimulation of positive strand synthesis above background was seen with either sphingomyelin or cholesterol, which are common membrane lipids that are not GPLs (Fig. 4). A variety of smaller compounds derived from active GPLs have also proven inactive. For example, neither the polar, phosphorylated head group portions nor the diacyl glycerol lipid portions of several active GPLs were active. Phosphatidic acid, which lacks an additional alcohol esterified to the head group phosphate, was similarly inactive (Fig. 4). Thus it appears that the complete GPL structure, including both the lipid and the full polar head group, is required to activate positive strand RNA synthesis in this FHV system.

Stimulation of positive strand RNA synthesis was dramatically affected by the length of the fatty acid chains within GPL [27]. PC varieties with fatty acid chains of 12 or fewer carbons were inactive, while those with 14- to 18-carbon acyl chains were highly active. Positive strand RNA synthesis was also affected to a lesser degree by acyl chain saturation, so that PC varieties having 18-carbon chains with one or two double bonds were more active than PC with saturated 18-carbon chains.

Possible implications of the GPL effects

It is not yet known why added GPL is required to stimulate positive strand RNA synthesis in the FHV in vitro RNA replication system. However, when the levels of FHV replicase and PC were varied, it was found that the threshold concentration of PC required to induce FHV RNA replication increased as more replicase was added [27]. This suggests that GPL interacts directly with one or more components of the replicase extract. For example, GPL might activate the enzymatic function of a replicase component in a manner similar to the recycling of *E. coli* DNA replication factor dnaA by interaction with diphosphatidyl glycerol [24] or the activation of protein kinase C by phosphatidylserine and diacyl glycerol [20]. Alternatively, since the FHV replicase that was used was membrane-associated [27, 28], the added GPL might support a membrane assembly or modification process required for positive strand RNA synthesis. This possibility might be consistent with the inhibition of picornavirus and alphavirus RNA synthesis by the lipid biosynthesis inhibitor cerulenin, which has been interpreted to indicate that continuous lipid synthesis or membrane formation is required for viral RNA replication in vivo [10, 21].

Acknowledgements

We thank S. Dinant, M. Janda, K. Mise, N. Chua, and E. Smirnyagina for helpful discussions at many stages of this work. This research was supported by NIH grants GM35072 and AI23742 and NSF grant DMB-9004385.

References

1. Ahlquist P (1992) Bromovirus RNA replication and transcription. Curr Opin Gen Dev 2: 71–76

2. Allison R, Janda M, Ahlquist P (1988) Infectious in vitro transcripts from cowpea chlorotic mottle virus cDNA clones and exchange of individual components with brome mosaic virus. J Virol 62: 3581–3588

3. Allison RF, Thompson C, Ahlquist P (1989) Regeneration of a functional RNA virus genome by recombination between deletion mutants and requirements for cowpea chlorotic mottle virus 3a and coat genes for systemic infection. Proc Natl Acad Sci USA 87: 1820–1824

4. Broyles S, Moss B (1987) Sedimentation of an RNA polymerase complex from vaccinia virus that specifically initiates and terminates transcription. Mol Cell Biol 7: 7–14

5. DeJong W, Ahlquist P (1992) A hybrid plant virus made by transferring the noncapsid movement protein from a rod-shaped to an icosahedral virus is competent for systemic infection. Proc Natl Acad Sci USA 89: 6808–6812

6. Felgner PL, Gadek TR, Holm M, Roman R, Chan HW, Wenz M, Northrop JP, Ringold GM, Danielsen M (1987) Lipofection: a highly efficient, lipid-mediated DNA-transfection procedure. Proc Natl Acad Sci USA 84: 7413–7417

7. Francki RIB, Fauquet CM, Knudson DL, Brown F (1991) Classification and nomenclature of viruses: Fifth Report of the International Committee on Taxonomy of Viruses. Springer, Wien New York (Arch Virol [Suppl] 2)

8. Goldbach R, LeGall O, Wellink J (1991) Alpha-like viruses in plants. Semin Virol 2: 19–25

9. Gorbalenya AE, Koonin EV (1989) Viral proteins containing the purine NTP-binding sequence pattern. Nucleic Acids Res 17: 8413–8440

10. Guinea R, Carrasco L (1990) Phospholipid biosynthesis and poliovirus genome replication, two coupled phenomena. EMBO J 9: 2011–2016

11. Haseloff J, Goelet P, Zimmern D, Ahlquist P, Dasgupta R, Kaesberg P (1984) Striking similarities in amino acid sequence among nonstructural proteins encoded by RNA viruses that have dissimilar genome organization. Proc Natl Acad Sci USA 81: 4358–4362

12. Hayes RJ, Buck KW (1990) Complete replication of a eukaryotic virus RNA by a purified RNA-dependent RNA polymerase. Cell 63: 363–368

13. Hodgman TC (1988) A new superfamily of replicative proteins. Nature 333: 22–23; 578 [Erratum]

14. Kao C, Ahlquist P (1992) Identification of the domains required for direct interaction of the helicase-like and polymerase-like RNA replication proteins of brome mosaic virus. J Virol 66: 7293–7302

15. Kao C, Quadt R, Hershberger R, Ahlquist P (1992) Brome mosaic virus RNA replication proteins 1a and 2a form a complex in vitro. J Virol 66: 6322–6329

16. Kroner PA, Richards D, Traynor P, Ahlquist P (1989) Defined mutations in a small region of the brome mosaic virus 2a gene cause diverse temperature-sensitive RNA replication phenotypes. J Virol 63: 5302–5309

17. Kroner PA, Young B, Ahlquist P (1990) Analysis of the role of brome mosaic virus 1a protein domains in RNA replication, using linker insertion mutagenesis. J Virol 64: 6110–6120

18. Mi S, Stollar V (1991) Expression of Sindbis virus nsP1 and methyltransferase activity in *Escherichia coli*. Virology 184: 423–427

19. Molla A, Paul AV, Wimmer E (1991) Cell-free, de novo synthesis of poliovirus. Science 254: 1647–1651

20. Orr JW, Newton AC (1992) Interaction of protein kinase C with phosphatidylserine. 1. Cooperativity in lipid binding. Biochemistry 31: 4661–4667

21. Perez L, Guinea R, Carrasco L (1991) Synthesis of Semliki Forest virus RNA requires continuous lipid synthesis. Virology 183: 74–82

22. Quadt R, Jaspars EMJ (1990) Purification and characterization of brome mosaic virus RNA-dependent RNA polymerase. Virology 178: 189–194

23. Scheidel LM, Stollar V (1991) Mutations that confer resistance to mycophenolic acid and ribavirin on Sindbis virus map to the nonstructural protein nsP1. Virology 181: 490–499

24. Sekimizu K, Kornberg A (1988) Cardiolipin activation of dnaA protein, the initiation protein of replication in *Escherichia coli*. J Biol Chem 263: 7131–7135

25. Traynor P, Ahlquist P (1990) Use of bromovirus RNA hybrids to map cis- and *trans*-acting functions in a conserved RNA replication gene. J Virol 64: 69–77

26. Traynor P, Young B, Ahlquist P (1991) Deletion analysis of brome mosaic virus 2a protein: effects on RNA replication and systemic spread. J Virol 65: 2807–2815

27. Wu SX, Ahlquist P, Kaesberg P (1992) Active complete in vitro replication of nodavirus RNA requires glycerophospholipid. Proc Natl Acad Sci USA 89: 11136–11140

28. Wu SX, Kaesberg P (1991) Synthesis of template-sense, single-strand Flock house virus RNA in a cell-free replication system. Virology 183: 392–396
29. Dinant M, Janda M, Kroner P, Ahlquist P (1993) Bromovirus RNA replication and transcription require compatibility between the polymerase- and helicase-like viral RNA synthesis proteins. J Virol 67 (in press)

Authors' address: Dr. P. Ahlquist, Institute for Molecular Virology, University of Wisconsin-Madison, 1525 Linden Drive, Madison, WI 53706, U.S.A.

Arch Virol (1994) [Suppl] 9: 147–157

Archives
Virology
© Springer-Verlag 1994
Printed in Austria

Characteristics of the poliovirus replication complex

K. Bienz, D. Egger, and **Th. Pfister**

Institute for Medical Microbiology, University of Basel, Basel, Switzerland

Summary. In the infected cell, the poliovirus replication complex (RC) is found in the center of a rosette formed by many virus-induced vesicles. The RC is attached to the vesicular membranes and contains a compact central part which encloses the replication forks of the replicative intermediate and all proteins necessary for strand elongation. The growing plus strands of the replicative intermediate protrude from the central part of the RC, but are still enclosed by membraneous structures of the rosette. After completion, progeny 36S RNA is set free at the surface of the rosette. In an in vitro transcription system, isolated replication complex-containing rosettes are active in initiation, elongation and maturation (release) of plus strand progeny RNA. Full functionality of the RC depends on an intact structural framework of all membraneous components of the rosette.

Introduction

Polioviral RNA is replicated by primer-dependent viral polymerase $3D^{pol}$ in two steps. First, genomic RNA of the infecting virus is copied into a minus strand, which leads to the formation of the double stranded replicative form (RF) [1, 27]. From the RF a partially double stranded replicative intermediate (RI) [13] is formed and subsequently progeny RNA strands of plus polarity are produced. In vitro, the first step does not depend on any cellular component or structure [33]. The second seems dependent on a specialized cellular membrane-associated structure [28]; for a recent review see [16], called the viral replication complex [7, 14, 28]. Thus, until recently, in vitro plus strand RNA synthesis was only possible in extracts from polio infected cells [7, 11, 27]. Only lately, in vitro systems based on uninfected cell extracts have been reported, which mimic the viral replication cycle and produce infectious virus progeny [2, 21]. Various authors have described the isolation of different replication complexes from infected cells [17], such as "native" or "crude" replication complexes as well as various subunits, which were obtained by treatment with detergents. All replication complexes function in elongating plus strand RNA in the RI but only native, membrane-containing replication complexes, in contrast to detergent-treated subunits, show initiation and maturation (release) of the final genomic (36S) RNA [4].

The work presented here was undertaken to identify the replication complex within the infected cell, to elucidate its architecture and composition, and to define the role of its components in RNA replication and virus morphogenesis.

Visualization and localization of the replication complex in vivo

Polio type I Mahoney and HEp-2 cells in suspension were used throughout. Infection of cells with poliovirus results in the formation of cytoplasmic vesicles, at first (3h p.i.) clustered at different sites, later (4–4.5h p.i.) appearing as a continuous mass of vesicles occupying most of the central cytoplasm (Fig. 1A). The vesicles are derived from the rough endoplasmic reticulum (rER) [3] and are homologous to the intermediate or transport vesicles [3, 19], which carry proteins to the Golgi apparatus. Formation of virus-induced vesicles is probably triggered by the viral protein 2BC [6]. The same vesiculation is found in entero- and rhinovirus infected cultured cells as well as in coxsackievirus infected muscles of newborn mice ([8, 10], Bienz, unpubl.). Formation of these vesicles, leading to profound structural rearrangement of the infected cell, is the ultrastructural equivalent of the cytopathic effect observed in the light microscope [8].

High resolution autoradiography indicated that viral RNA synthesis is associated with these vesicles [5] (Fig. 1B). Subsequently, the replication complex was identified on the vesicular surface [3] (Fig. 1B, inset). Similarly, the replication complex can also be localized by in situ hybridization [31] with a riboprobe recognizing plus strand viral RNA [30]. Localization of viral non-structural proteins with electron microscopic immunocytochemistry using monoclonal antibodies [23], alone [3] or in combination with in situ hybridization (31), shows that the P2 proteins

Fig. 1. **A** Section through a poliovirus-infected HEp-2 cell at 5h p.i. The cell is filled with virus-induced vesicles which are characteristic for a picornavirus infection. Bar: 1 μm. **B** EM-autoradiograph of a poliovirus infected, actinomycin D-treated cell at 4h p.i. ^{3}H-uridine incorporation is observed in the central vesicular mass. Bar: 1 μm. Inset: viral RNA synthesis, i.e. the viral replication complex, is found on the outer surface of the virus-induced vesicles, at the sites where they are budding off from the rough endoplasmic reticulum (rER). Bar: 100 nm

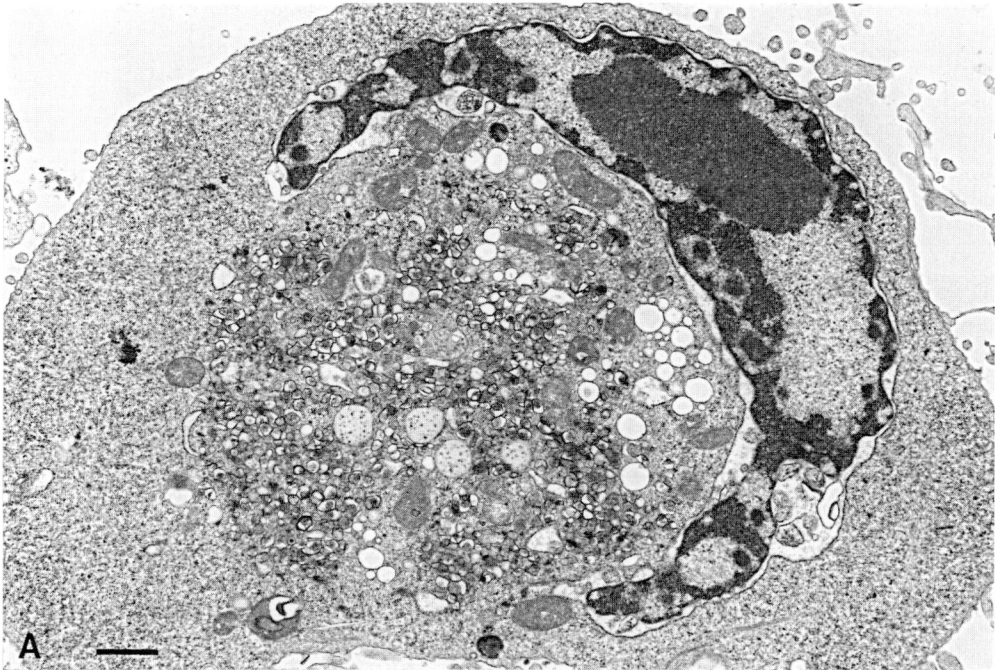

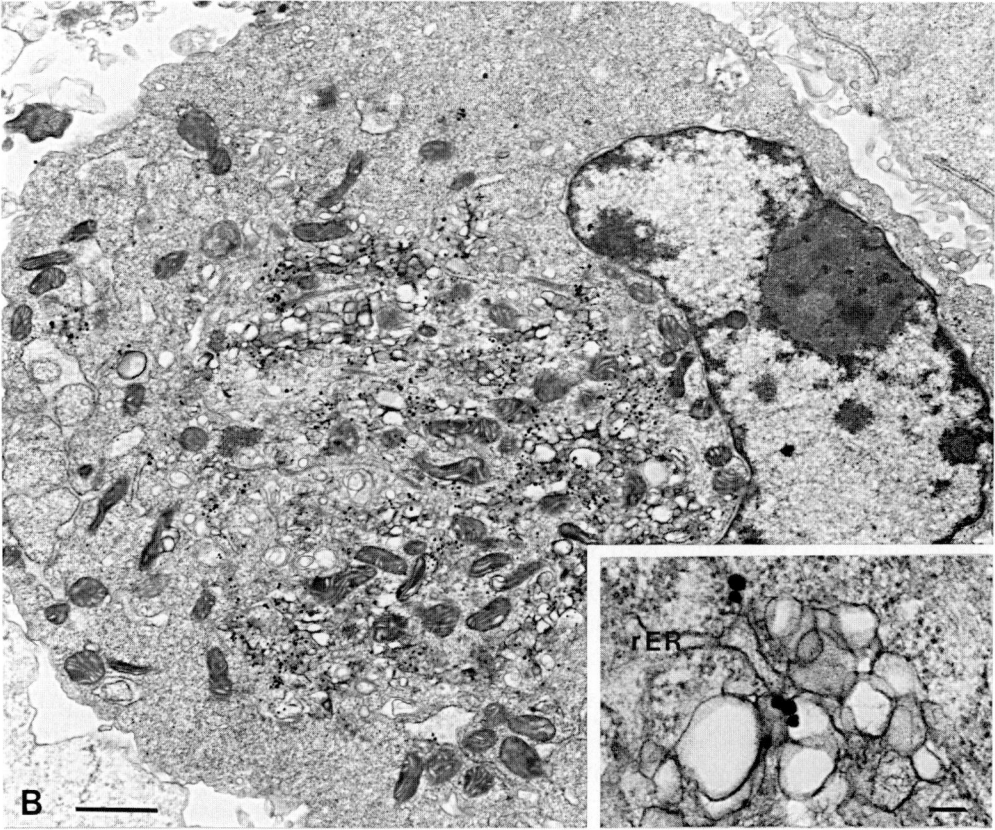

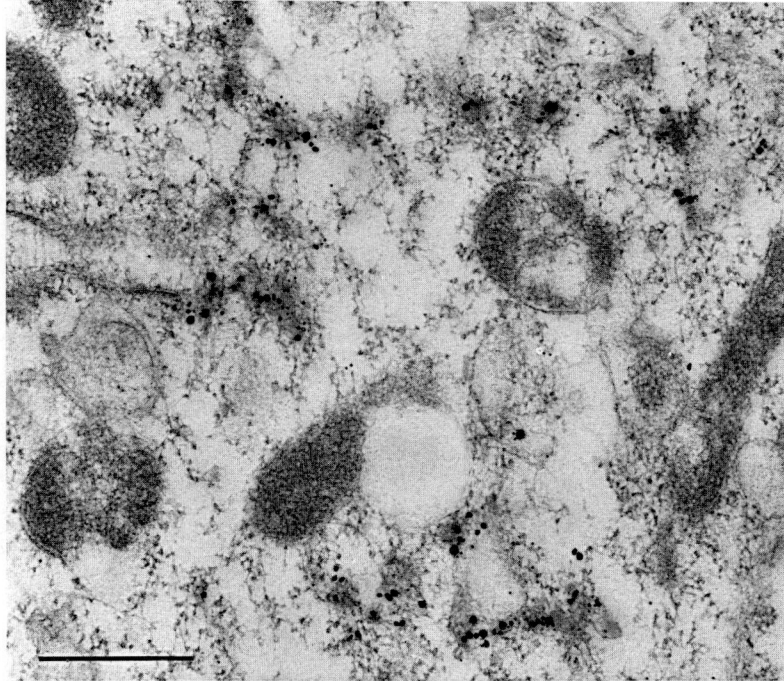

Fig. 2. EM-immunocytochemistry of an infected cell at 4 h p.i. The viral P2-proteins 2C (large grains) and 2B (small grains) are located exclusively over replication complexes. Bar: 0.5 μm

2B, 2C and 2BC are contained exclusively within the replication complex (Fig. 2), whereas the protein 3D^{pol} and its precursors are found, albeit in a lower concentration, also in the peripheral cytoplasm. P2 proteins can thus be used as a histochemical marker for the viral replication complex.

Components of the isolated replication complex

Functionally active replication complexes can be isolated from infected cells, by centrifugation of a cytoplasmic extract onto cushions of sucrose after Dounce homogenization. Two subcellular fractions are obtained: one banding in 30% sucrose (30% fraction) and another fraction banding in 45% sucrose (45% fraction). They differ in content of viral RNA and capsid related particles (see below). Both fractions, however, consist of the same structures of rosette-like arrangements of vesicles surrounding the actual replication complex (Fig. 3) [7], which contains the P2 proteins 2B, 2C and 2BC. Note that several hundreds of such replication complex-containing rosettes make up the whole vesicular mass seen in the sectioned cell (Fig. 1).

As can be seen in Fig. 3 (inset), the rosettes in the vesicular fractions contain two membrane systems: larger, virus induced vesicles surround-

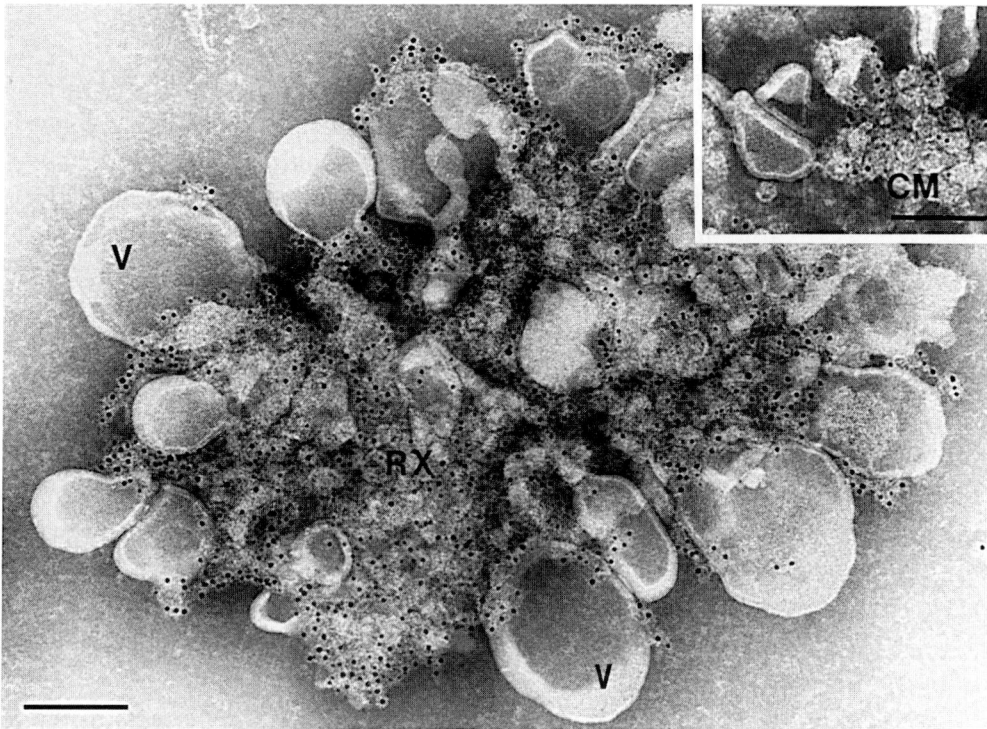

Fig. 3. Isolated, negatively stained 45% vesicular fraction (see text). The replication complex (*RX*) is surrounded by and attached to a rosette of virus-induced vesicles (*V*). Immunocytochemical label: P2-proteins 2B and 2C. Bar: 100 nm. Inset: the replication complex contains a second membrane system ("compact membranes", *CM*) of densely packed vesicles approx. 30 nm in diameter. Label: protein 2C. Bar: 100 nm

ing the replication complex and smaller, densely packed vesicles, "compact membranes", within the complex [4]. P2 proteins are found on membranes of peripheral virus-induced vesicles and on central small vesicles. The immunocytochemical label, however, is not randomly distributed but concentrated at the contact sites of the two types of membraneous structures.

Protein 2C plays an essential role in viral RNA synthesis [18]. To assess its function more closely, we used guanidine-HCl, an inhibitor of viral RNA synthesis. Guanidine-resistant or -dependent poliovirus mutants have been mapped to the genomic region of the viral protein 2C [25] and, thus, any viral function that is guanidine sensitive can be considered dependent on the viral protein 2C. On the ultrastructural level we found that the action of guanidine is a dissociation of the rosettes whereby the protein 2C is detached from the peripheral, virus-induced vesicles [7]. This experiment argues that a function of 2C is in attaching the replication complex to the vesicular membranes, thus maintaining the structure of the rosette as a (functional) entity.

To locate viral RNA and viral proteins with respect to each other and to the rosettes, we used UV-crosslinking of protein and RNA, RNase treatment of the vesicular fraction, and direct visualization of the RNA. It was reported that the protein 2C has two NTP-binding domains [20, 29]. We were able to UV-crosslink viral RNA to all P2 proteins [7], regardless of whether or not protein 2C was blocked by guanidine. Since P2 proteins are associated with membranes (Fig. 3) as well as with RNA, it can be speculated that 2C (or its precursor 2BC) not only attaches the replication complex but the viral RNA to the vesicular membranes as well. Given the fast turning of the RI (unwinding of the RF) during RNA synthesis, attachment of the growing viral plus strand RNA clearly seems necessary from a mechanical point of view [7, 17]. Such a mechanism might prevent back-hybridization of the progeny strand to the template. The attachment of the replication complex to the vesicles, however, seems to be independent of the mechanical fixation of nascent RNA strands, as the 45% fractions, which do not contain RI (see below), still contain intact rosettes.

Sensitivity to RNase was used as a measure of the accessibility of viral RNA in the rosettes. It was found [4] that in the 30% fraction, which contains 36S RNA and RI, the 36S RNA is RNase sensitive, whereas the RI is protected (Fig. 4A). In the 45% fraction, which contains only 36S RNA, only part of this RNA is sensitive and most is protected, the latter being encapsidated. This indicates that 36S RNA, if not encapsidated, is outside or at least freely accessible at the periphery of the rosettes, whereas the RI is tightly enclosed.

These findings were confirmed by directly localizing the viral plus strand RNA by in situ hybridization [4] with a labeled riboprobe [30].

In vitro transcriptional activity of the replication complex: architecture of the rosette and role of the membranes

If introduced into an appropriate in vitro system [7, 11, 27], both subcellular fractions described above are active in synthesizing viral plus strand RNA, which, upon completion, is released from the RI as full length 36S RNA (Fig. 4B) [4].

Similar to the in vivo labeled RNA species (Fig. 4A), the in vitro synthesized 36S RNA is RNase sensitive and the RI is protected. If the membranes of such a rosette are dissolved by Na-deoxycholate (DOC), the RI becomes partially RNase sensitive; nascent plus strands are digested upon addition of RNase and RI is converted to an 18S core (Fig. 4C). This experiment suggests an architecture of the replication

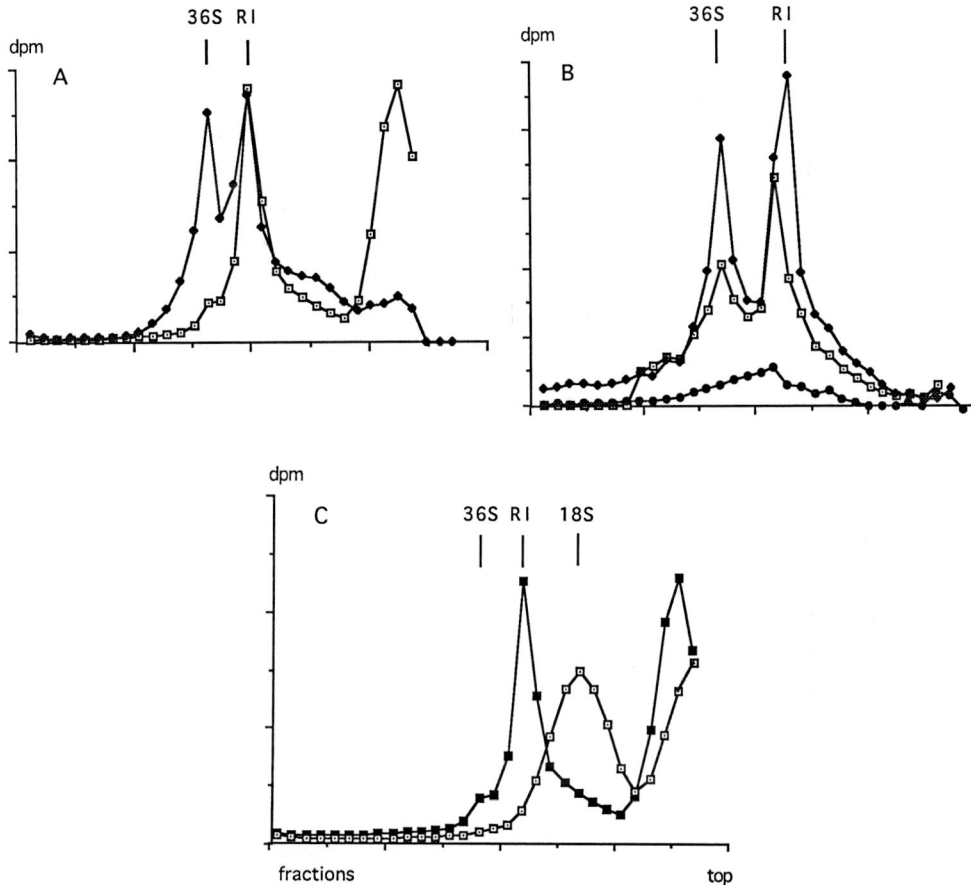

Fig. 4. A In vivo ^{3}H-uridine labeled vesicular fractions before (◆) and after (□) RNase treatment. Analysis of RNA after phenol extraction by rate zonal centrifugation. **B** In vitro RNA synthesis of a 30% vesicular fraction. ^{3}H-UTP labeled RNA analyzed as in Fig. 4A after 15 (●), 30 (□) and 60 (◆) min of in vitro synthesis. **C** A 30% vesicular fraction was RNase treated as in 4A (■). In an aliquot, the membranes were dissolved with DOC and the RNase treatment was repeated (□). The RI is converted into an 18S core, which is RNase sensitive after phenol extraction (not shown)

complex where the central, replication fork-containing part of the RI is confined in additional, presumably proteinaceous, material. By adding DOC to an actively transcribing in vitro transcription system it was found [4] that the replication forks within a membrane-less replication complex are still capable of RNA synthesis, i.e. RNA elongation. Release of mature 36S RNA, however, is dependent on membranes.

Capsid proteins associated with the replication complex

Earlier reports [9, 32] indicate that encapsidation of progeny RNA is spatially and temporally related to plus strand RNA synthesis in the viral

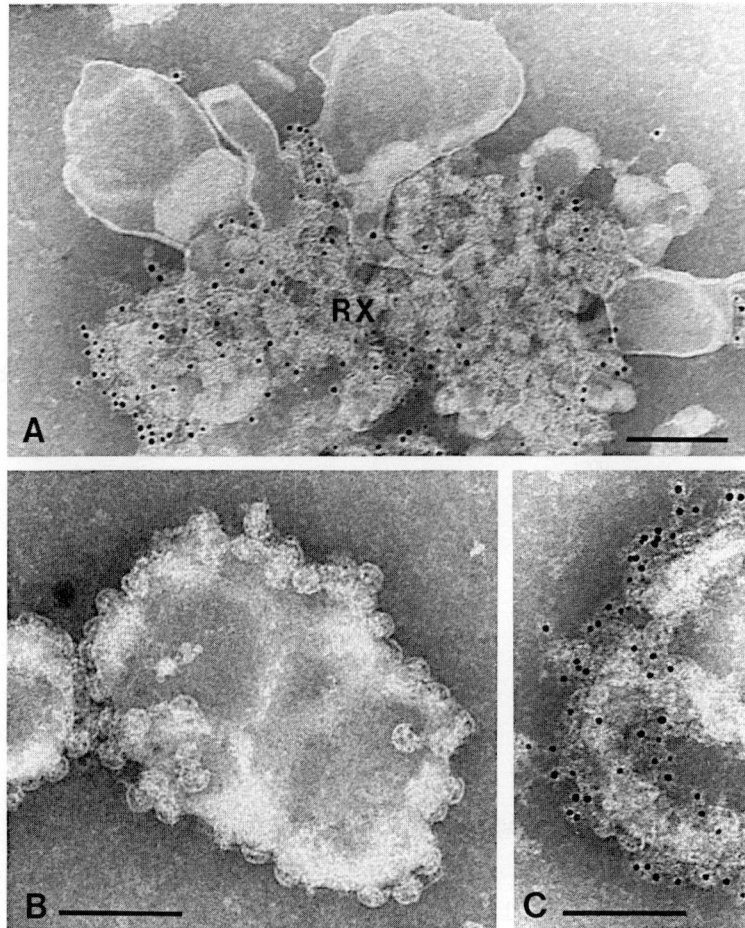

Fig. 5. In 30% vesicular fractions (**A**), capsid protein of 14S (pentamer) antigenicity is found within the replication complex (*RX*). In 45% fractions (**B, C**), 14S pentamers are found as capsid-like structures on the surface of virus-induced vesicles. These structures give the optical impression of convex half shells. **B** without, **A** and **C** with immunolabeling by a 14S-specific monoclonal antibody

replication complex. Immunocytochemistry with monoclonal antibodies [24] recognizing different configurations of capsid proteins (capsid related particles, CRP) indicates an association of 14S pentamers [22] with the vesicular fractions [24]. In the 30% fraction, only 14S pentamers were found within and at the periphery of the replication complex (Fig. 5A). However, in the 45% fraction we observed a novel type of CRP, which we call capsid-like structures, of 14S antigenicity, attached to the outer surface of vesicular membranes and resembling half shells (Figs. 5B and C). In our native electron microscopic preparations, no 65S or 74S empty capsids (nomenclature according to [26]) could be observed. After detergent treatment, 65S and, after detergent treatment and heating to 37°C, 74S empty capsids were found by immuno-electron microscopy, rate zonal centrifugation and immunoprecipitation [24].

These findings indicate that membranes are responsible for maintaining the configuration of 14S pentamers as such and as higher ordered (capsid-like) structures. 65S and 74S empty capsids are observed only after dissolving the membranes. Thus, they are not thought to be naturally occurring in vivo. Whether encapsidation proceeds via 14S pentamers and, if so, whether the 14S pentamers within the replication complex or the capsid-like structures on the vesicular surface are involved, remain to be determined.

Conclusion

The replication complex-containing rosette is the smallest functional entity for plus strand poliovirus RNA production in vivo. The functioning of the replication complex depends on continuing lipid synthesis [15, 19] producing the membraneous components of the replication complex. These membranes then serve as the structural framework necessary for viral plus strand RNA synthesis including initiation, elongation and maturation (release) of plus strand progeny RNA.

The replication complex is contained in the vesicular rosette as a rather compact structure. It contains a central part tightly enclosing the replication forks of the RI and all proteins necessary for strand elongation. The growing plus strands of the RI protrude from this central structure but are still enclosed in DOC-sensitive membraneous structures within the replication complex. The completed 36S RNA is set free at the surface of the rosette.

This architecture of the rosette and the replication complex might also bear on, if not be responsible for, the *cis-* and *trans-*acting properties of viral proteins (see also [12]): *trans-*acting proteins might have their substrate on the surface of the replication complex, whereas *cis-*acting are those that act within the complex and cannot, because of its tightness, be complemented from outside. This reasoning illustrates that methods, which take care to maintain the structural organization of their target, may be helpful in elucidating the mode of action of viral and cellular components involved in RNA replication and virus assembly.

Acknowledgement

This work was supported by grant 31-29910.90 from the Swiss National Science Foundation.

References

1. Baltimore D (1968) Structure of the poliovirus replicative intermediate RNA. J Mol Biol 32: 359–368
2. Barton DJ, Flanegan JB (1993) Coupled translation and replication of poliovirus RNA in vitro: synthesis of functional 3D polymerase and infectious virus. J Virol 67: 822–831
3. Bienz K, Egger D, Pasamontes L (1987) Association of polioviral proteins of the P2 genomic region with the viral replication complex and virus-induced membrane synthesis as visualized by electron microscopic immunocytochemistry and auto-radiography. Virology 160: 220–226
4. Bienz K, Egger D, Pfister Th, Troxler M (1992) Structural and functional charac-terization of the poliovirus replication complex. J Virol 66: 2740–2747
5. Bienz K, Egger D, Rasser Y, Bossart W (1980) Kinetics and location of poliovirus macromolecular synthesis in correlation to virus-induced cytopathology. Virology 100: 390–399
6. Bienz K, Egger D, Rasser Y, Bossart W (1983) Intracellular distribution of polio-virus proteins and the induction of virus-specific cytoplasmic structures. Virology 131: 39–48
7. Bienz K, Egger D, Troxler M, Pasamontes L (1990) Structural organization of poliovirus RNA replication is mediated by viral proteins of the P2 genomic region. J Virol 64: 1156–1163
8. Bienz K, Egger D, Wolff DA (1973) Virus replication, cytopathology, and lysoso-mal enzyme response of mitotic and interphase HEp-2 cells infected with polio-virus. J Virol 11: 565–574
9. Caliguiri LA, Compans RW (1973) The formation of poliovirus particles in associa-tion with the RNA replication complexes. J Gen Virol 21: 99–108
10. Dales S, Eggers HJ, Tamm I, Palade GE (1965) Electron microscopic study of the formation of poliovirus. Virology 26: 379–389
11. Etchison D, Ehrenfeld E (1981) Comparison of replication complexes synthesizing poliovirus RNA. Virology 111: 33–46
12. Giachetti C, Hwang S-S, Semler BL (1992) cis-acting lesions targeted to the hydrophobic domain of a poliovirus membrane protein involved in RNA replica-tion. J Virol 66: 6045–6057
13. Girard M (1969) In vitro synthesis of poliovirus ribonucleic acid: role of the replicative intermediate. J Virol 3: 376–384
14. Girard M, Baltimore D, Darnell JE (1967) The poliovirus replication complex: site for synthesis of poliovirus RNA. J Mol Biol 24: 59–74
15. Guinea R, Carrasco L (1990) Phospholipid biosynthesis and poliovirus genome replication, two coupled phenomena. EMBO J 9: 2011–2016
16. Hambidge SJ, Sarnow P (1992) Early events in poliovirus-infected cells. Semin Virol 3: 501–510
17. Koch F, Koch G (1985) The molecular biology of poliovirus. Springer, Wien New York
18. Li J-P, Baltimore D (1988) Isolation of poliovirus 2C mutants defective in viral RNA synthesis. J Virol 62: 4016–4021
19. Maynell LA, Kirkegaard K, Klymkowsky MW (1992) Inhibition of poliovirus RNA synthesis by brefeldin A. J Virol 66: 1985–1994
20. Mirzayan C, Wimmer E (1992) Genetic analysis of an NTP-binding motif in poliovirus polypeptide 2C. Virology 189: 547–555

21. Molla A, Paul AV, Wimmer E (1991) Cell-free, de novo synthesis of poliovirus. Science 254: 1647–1651

22. Palmenberg AC (1982) In vitro synthesis and assembly of picornaviral capsid intermediate structures. J Virol 44: 900–906

23. Pasamontes L, Egger D, Bienz K (1986) Production of monoclonal and mono-specific antibodies against non-capsid proteins of poliovirus. J Gen Virol 67: 2415–2422

24. Pfister T, Pasamontes L, Troxler M, Egger D, Bienz K (1992) Immunocytochemical localization of capsid-related particles in subcellular fractions of poliovirus-infected cells. Virology 188: 676–684

25. Pincus SE, Diamond DC, Emini EA, Wimmer E (1986) Guanidine-selected mutants of poliovirus: mapping of point mutations to polypeptide 2C. J Virol 57: 638–646

26. Rombaut B, Vrijsen R, Boeyé A (1989) Denaturation of poliovirus procapsids. Arch Virol 106: 213–220

27. Takeda N, Kuhn RJ, Yang C-F, Takegami T, Wimmer E (1986) Initiation of poliovirus plus-strand RNA synthesis in a membrane complex of infected HeLa cells. J Virol 60: 43–53

28. Takegami T, Semler BL, Anderson CW, Wimmer E (1983) Membrane fractions active in poliovirus RNA replication contain VPg precursor polypeptides. Virology 128: 33–47

29. Teterina NL, Kean KM, Gorbalenya AE, Agol VI, Girard M (1992) Analysis of the functional significance of amino acid residues in the putative NTP-binding pattern of the poliovirus 2C protein. J Gen Virol 73: 1977–1986

30. Troxler M, Pasamontes L, Egger D, Bienz K (1990) In situ hybridization for light and electron microscopy. A comparison of methods for the localization of viral RNA using biotinylated DNA and RNA probes. J Virol Methods 30: 1–14

31. Troxler M, Egger D, Pfister Th, Bienz K (1992) Intracellular localization of poliovirus RNA by in situ hybridization at the ultrastructural level using single-stranded riboprobes. Virology 191: 687–697

32. Yin FH (1977) Involvement of viral procapsid in the RNA-synthesis and maturation of poliovirus. Virology 82: 299–307

33. Young DC, Tuschall DM, Flanegan JB (1985) Poliovirus RNA-dependent RNA polymerase and host cell protein synthesize products twice the size of poliovirion RNA in vitro. J Virol 54: 256–264

Authors' address: Dr. K. Bienz, Institute for Medical Microbiology, Petersplatz 10, CH-4003 Basel, Switzerland.

Arch Virol (1994) [Suppl] 9: 159–172

Archives
Virology

© Springer-Verlag 1994
Printed in Austria

Secretory pathway function, but not cytoskeletal integrity, is required in poliovirus infection

J. Doedens[1,2], **L. A. Maynell**[1,2,*], **M. W. Klymkowsky**[1], and **K. Kirkegaard**[1,2]

[1] Department of Molecular, Cellular and Developmental Biology and [2] Howard Hughes Medical Institute, University of Colorado Boulder, Boulder, Colorado, U.S.A.

Summary. We examined the importance of two interactions between poliovirus and its host cell: the putative association between viral proteins and a rearranged intermediate filament (IF) network and the apparent requirement for functional vesicle budding machinery within the host-cell secretory pathway. Poliovirus capsid proteins appeared to associate with reorganized IF proteins during infection. Treatment of cells with cytochalasin D and nocodazole in combination disrupted normal cytoskeletal organization and prevented the poliovirus-induced redistribution of IF proteins to a juxtanuclear location. However, this treatment had no effect on viral yields from single-cycle infections, indicating that neither cytoskeletal integrity nor a specific poliovirus-induced rearrangement of IF proteins is required in the poliovirus life cycle. In contrast, we report that the inhibition of poliovirus replication by brefeldin A (BFA), an inhibitor of secretory membrane traffic, is specific to the host cell. Polioviral yields were not affected by BFA in two BFA-resistant cell lines, demonstrating that BFA targets a host protein or process required by poliovirus. No BFA-resistant virus was detected in these experiments, further supporting the hypothesis that poliovirus replication requires secretory pathway function, perhaps for the generation of vesicles on which viral RNA replication complexes are assembled.

Introduction

A virus relies on its host cell's translational machinery for the synthesis of viral proteins and on the host's metabolism for the production of precursors to be incorporated into viral proteins and nucleic acids. However, there is diversity among viruses in the pathways of genome

*Present address: Gene Expression Laboratory, Salk Institute for Biological Studies, La Jolla, California, U.S.A.

replication and virus assembly and therefore differing requirements for either the integrity or virus-specific alteration of host cell processes. Previous studies have suggested a functional role in poliovirus infection for both a virus-induced rearrangement of intermediate filaments (IF) [12, 23] and a functionally intact secretory pathway [7, 15]. In the experiments reported here, we further examine the potential roles of these host systems in the poliovirus replicative cycle.

Association of poliovirus RNA replication with the host cytoskeleton was suggested by biochemical and morphological analysis of cytoskeletal fractions isolated by extraction of poliovirus-infected cells with non-ionic detergent [12, 23]. In these studies, the localization of intermediate filaments to a vesicle-rich, juxtanuclear region of cytoplasm was observed. This altered region of cytoplasm had been shown by electron microscopy and autoradiography to be the site of poliovirus genome replication and virion assembly [2, 18]. In addition, viral polysomes and newly synthesized viral RNA were found to associate with the cytoskeleton in cell fractionation experiments [12]; viral proteins 2C, 2BC, and 3CD, known to be involved in RNA replication (reviewed in [19]), were also enriched in the cytoskeletal fraction [23]. Together, these data provided a persuasive argument that viral RNA replication complexes are functionally associated with the cytoskeleton during poliovirus infection.

Recently, we found that brefeldin A (BFA) potently inhibits poliovirus replication [15]. BFA is a fungal metabolite that inhibits vesicular transport through the secretory pathway and thus inhibits the replication of many enveloped viruses. Such viruses typically rely on normal secretory pathway function, either for transport out of the cell of enveloped virions that have budded into the endoplasmic reticulum (ER) or Golgi complex, or for transport of viral envelope proteins to the cell surface prior to viral budding at the plasma membrane. Polioviral yields from single-cycle infections were reduced 10^6-fold by BFA; further analysis revealed that viral replication was inhibited at the level of RNA synthesis [15].

Because polioviruses are non-enveloped and do not appear to be transported out of infected cells by the secretory machinery, the mechanism of BFA inhibition of poliovirus replication was not obvious. It is known that poliovirus RNA replication occurs on the surface of membranous vesicles that proliferate throughout the cytoplasm of the host cell and that viral RNA replication complexes are associated with the surface of these vesicles [2–4, 18]. Because BFA has been shown to block ER-to-Golgi transport in vivo [5, 14, 16] and to prevent the formation of transport vesicles from the Golgi in vitro [17], we proposed that BFA inhibited poliovirus RNA synthesis by preventing formation of the virus-induced vesicles on which replication complexes assemble [15]. Others have recently reported the inhibitory effect of BFA on poliovirus replication as well [7]. These authors suggested that poliovirus replication complexes might assemble on the ER and mature as they are transported through the Golgi, with functional membrane-bound replication complexes being released from the trans-Golgi network.

These studies did not address whether the target of BFA's action to inhibit the virus was a host or viral product. It was possible that a viral protein required for RNA synthesis was the target of BFA inhibition, and BFA's effects on membrane traffic were irrelevant to its effect on viral RNA synthesis. Similarly, while an association between poliovirus RNA replication and virus-induced cytoskeletal rearrangements has been observed, there is no direct evidence that this represents a functional association.

In this study, we show that in cells treated with cytochalasin D and nocodazole in combination, poliovirus-induced cytoskeletal rearrangements were disrupted. Despite this disruption, production of virus was unaffected, arguing that poliovirus requires neither an intact host cytoskeleton nor a specific rearrangement thereof. In addition, we present evidence that intact host secretory pathway function is required for poliovirus RNA synthesis by showing that the inhibition of poliovirus by BFA is a property of the host cell and correlates with the sensitivity or resistance of the host secretory pathway to BFA.

Poliovirus replication appears to be associated with redistributed IF proteins

We used immunofluorescence microscopy to examine the organization of cytoskeletal systems and the localization of viral capsid proteins during infection with type I Mahoney poliovirus. The eukaryotic cytoskeleton is composed of three major filament systems: microfilaments, microtubules, and IFs. During poliovirus infection, the organization of actin microfilaments and microtubules was not significantly altered, and no strong co-localization of viral proteins with actin or tubulin was seen (data not shown). Previous studies of this sort also have failed to note any co-localization of poliovirus replication complexes or proteins with microfilaments or microtubules [12, 23].

In contrast, the cellular IF network was extensively altered by poliovirus infection, and viral capsid proteins appeared to be co-localized with both keratin and vimentin-type IF present in HeLa cells (see Fig. 1). In uninfected cells (Figs. 1a,d), the keratin and vimentin-type IFs were distributed throughout the cytoplasm. Infection with poliovirus caused these networks to collapse and concentrate in an intensely stained

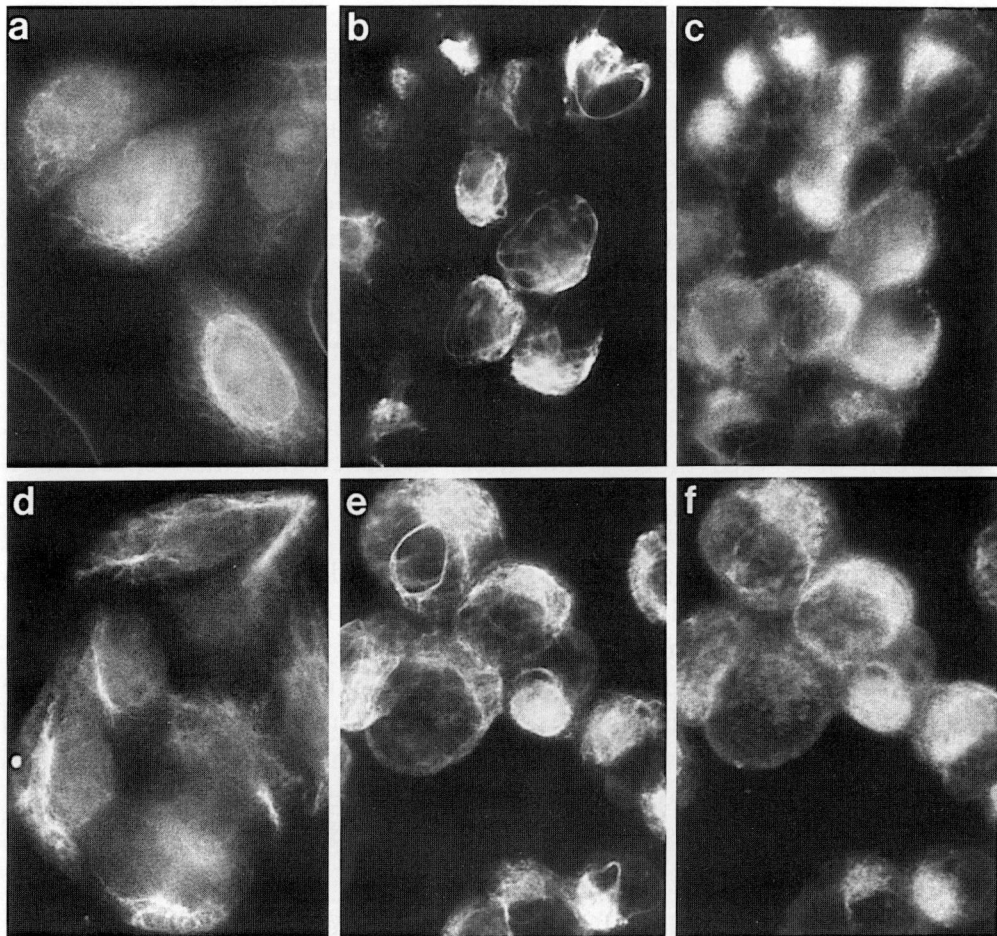

Fig. 1. Immunofluorescence microscopy of IF proteins and poliovirus capsid proteins. **a** Keratin IF system of uninfected HeLa cells; **b,c** double immunolabeling of keratin IF and poliovirus capsid proteins respectively, **d** vimentin-type IF of uninfected cells; **e,f** double immunolabeling of vimentin and poliovirus capsid proteins respectively. HeLa cells grown on coverslips were mock-infected or infected with wild-type poliovirus at an MOI of 10 as described [15]. At 5 h post-infection the cells were fixed for immunofluorescence microscopy as described previously [10] and probed with mouse monoclonal antibodies either to keratin (antibody 1h5 [9]) or to vimentin (clone 3B4, Boehringer-Mannheim Biochemicals) intermediate filament proteins, followed by a FITC-conjugated anti-mouse IgG. Poliovirus capsid proteins were detected with a rabbit anti-capsid antibody (obtained from P. Sarnow, University of Colorado Health Sciences Center) followed by a rhodamine-conjugated anti-rabbit antibody. Immunolabeling was performed as described [10]

juxtanuclear location (Figs. 1b,e). Double immunostaining revealed that this region of the cytoplasm was also highly enriched for viral capsid proteins (Figs. 1c,f), indicating that it was the major site of virus production. Because poliovirus RNA replication and genome packaging are both thought to occur on the surface of poliovirus-induced vesicles [18], this region of cytoplasm is probably the site of viral RNA synthesis as well. These observations are consistent with electron microscopic and cell-fractionation experiments that have been published previously [12, 23].

Effects of cytochalasin D and nocodazole on the subcellular localization of poliovirus antigens

To determine whether virus-induced alterations in cytoskeletal organization are functionally important during poliovirus infection, we attempted to disrupt all three filament systems by treating cells with both cytochalasin D and nocodazole. In addition to disrupting microfilament organization, cytochalasin D has also been shown to disrupt the organization of keratin-type IF [8]. Nocodazole depolymerizes microtubules; because the extended organization of vimentin-type IF is dependent upon microtubules, the vimentin filament system of a cell collapses in response to nocodazole [8].

The microfilament and microtubule networks were both extensively disrupted by treatment with cytochalasin D and nocodazole in both infected and uninfected cells (data not shown). The distribution of polioviral capsid proteins was also drastically altered by drug treatment, with the juxtanuclear staining seen in Fig. 1 giving way to punctate staining throughout the cytoplasm (see Figs. 2b,d). No obvious association of viral antigens with either actin or tubulin was observed in the presence of nocodazole and cytochalasin D (data not shown). Treatment with cytochalasin D and nocodazole also blocked the virus-induced rearrangement of intermediate filaments. In drug-treated cells, both keratin and vimentin-type IF reorganized to form a web of thick cable-like structures in the cytoplasm (Figs. 2a,c). The poliovirus-induced collapse of these filaments to a concentrated juxtanuclear localization (Figs. 1b,e) was not observed in the presence of cytochalasin D and nocodazole. The altered distribution of filaments and poliovirus antigens made it difficult to determine whether any co-localization remained.

Disruption of cytoskeletal organization does not affect viral growth

To determine whether cytoskeletal organization or virus-induced cytoskeletal changes play a functional role in poliovirus infection, we compared viral yields from single-cycle infections in the presence and absence of cytochalasin D and nocodazole (Fig. 3). Although these drugs in combination disrupted normal cytoskeletal organization and pre-vented virus-specific rearrangements of cytoskeletal components, viral yields were not affected by this treatment. Thus, neither cytoskele-

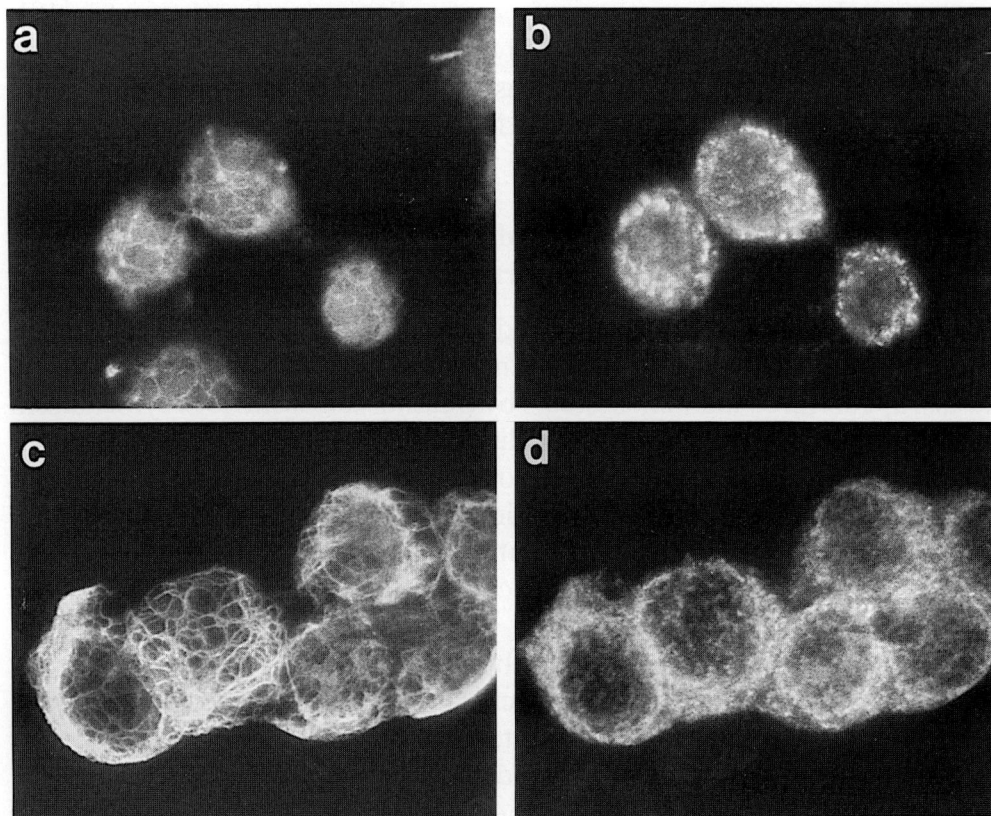

Fig. 2. Immunofluorescence microscopy of the effects of cytochalasin D and noco-dazole on the localization of IF proteins and poliovirus capsid proteins. **a,b** Double immunolabeling of keratin-type IF and poliovirus capsid proteins respectively in polio-virus-infected HeLa cells in the presence of cytochalasin D and nocodazole; **c,d** double immunolabeling of vimentin-type IF and poliovirus capsid proteins in the presence of cytochalasin D and nocodazole. Infection with poliovirus and processing of cells for immunofluorescence was identical to that described in Fig. 1 except cytochalasin D (5 µg/ml) and nocodazole (10 µg/ml) were present throughout the infections

tal integrity nor the virus-specific redistribution of cytoskeletal elements plays a significant role in poliovirus replication.

A model for the inhibition of poliovirus by brefeldin A

In contrast to the apparent irrelevance of the intact cytoskeleton in poliovirus infection, the finding that brefeldin A strongly inhibits viral RNA synthesis [15] implies a requirement by the virus for a functional protein secretory pathway in its host cell. Electron microscopic observa-tions [2, 3] as well as cellular fractionation studies [4, 22] have shown that poliovirus RNA synthesis is a membrane-associated process, occurr-

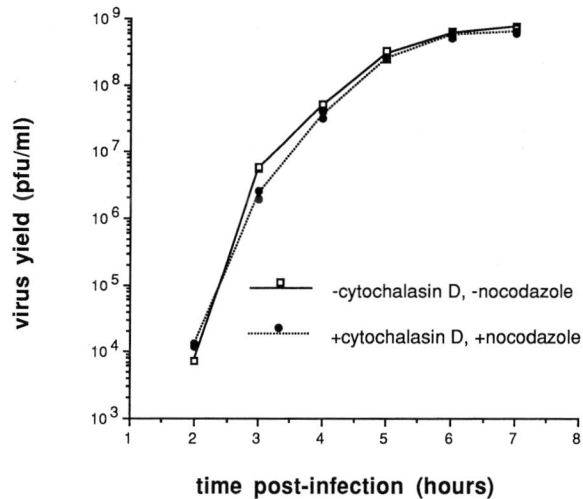

Fig. 3. Single-cycle growth of poliovirus in the presence and absence of cytochalasin D and nocodazole. HeLa cells were infected with poliovirus at an MOI of 0.25 in the presence or absence of cytochalasin D (5 µg/ml) and nocodazole (10 µg/ml). At various times post-infection, intracellular virus was harvested, and viral yields were determined by plaque assay on HeLa cells. Data points shown represent viral yields of duplicate plates assayed at each time point; lines represent the average of these duplicates

ing on the surface of vesicles that proliferate throughout the host cytoplasm. One of the known effects of BFA on the secretory pathway is the inhibition of protein transport from the ER to the Golgi complex [5, 14, 16] by inhibiting the formation of transport vesicles [17]. The model described in Fig. 4 incorporates this information to explain the inhibition of poliovirus RNA synthesis by BFA [15]. In this model, poliovirus causes an accumulation of cytoplasmic vesicles by inhibiting a step in normal vesicular transport subsequent to vesicle budding. As a result, transport vesicles either do not reach or do not fuse with their target organelle. BFA is proposed to inhibit poliovirus RNA synthesis by blocking the formation of the vesicles on which viral RNA replication complexes are assembled.

In Fig. 4, the vesicles are pictured as being derived from the ER, a source consistent both with the BFA studies and with electron microscopy [1]; however, an origin for these vesicles within the Golgi complex is also consistent with the BFA studies, as BFA is known to inhibit budding of vesicles from the Golgi complex in vitro [17]. Also, BFA has been shown to cause mixing of the trans-Golgi-network with early endosomes and to block both endosome to lysosome traffic and transcytosis in some cell types [13]. These processes are thus additional candidates for being the target or targets of BFA in inhibiting poliovirus replication.

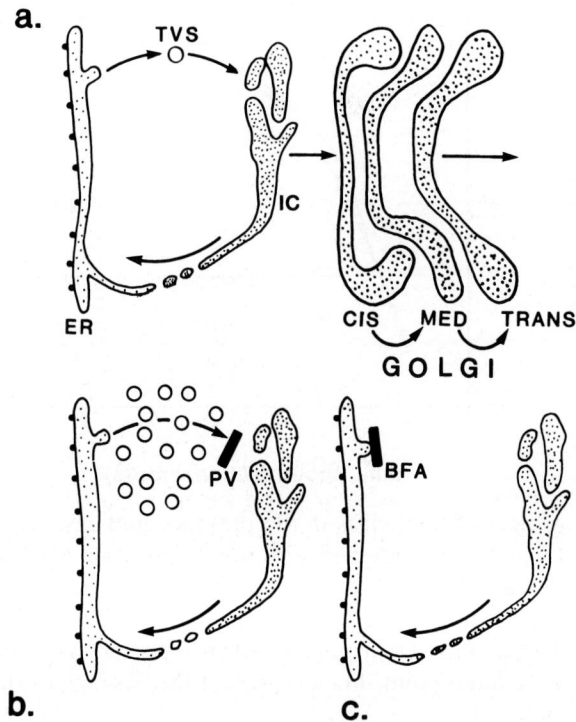

Fig. 4. A model for the inhibition of poliovirus RNA synthesis by BFA. Schematic diagram of the protein secretion apparatus in an uninfected cell (**a**), the proposed mechanism of poliovirus-induced vesicle accumulation (**b**), and the proposed inhibition of poliovirus-induced vesicle accumulation by BFA (**c**). Details of the model are described in the text. For simplicity, we have depicted poliovirus as blocking vesicular transport between the ER (endoplasmic reticulum) and the intermediate compartment, although the blocked step could be between any of the compartments depicted (model taken from [15], reprinted with permission). Abbreviations: *IC* Intermediate compartment; *TVS* transport vesicles; *PV* poliovirus

Inhibition of poliovirus by brefeldin A is host-specific

In the model described above, the inhibition of poliovirus replication by BFA is postulated to be due to action of the compound upstream in the secretory pathway from the putative site of poliovirus action. One would then predict that in a cell possessing a BFA-resistant secretory pathway, that is, a cell in which secretion is unaffected by BFA, poliovirus replication would also be unaffected. One such cell line is the rat-kangaroo kidney line, PtK$_1$. Although endosomal traffic in PtK$_1$ cells is sensitive to BFA [13], the Golgi apparatus of PtK$_1$ cells has been shown to be resistant to the effects of brefeldin A [11]. Mechanisms of BFA-resistance such as exclusion, sequestration, or rapid metabolism of BFA by PtK$_1$ cells were ruled out in these experiments [11].

A.

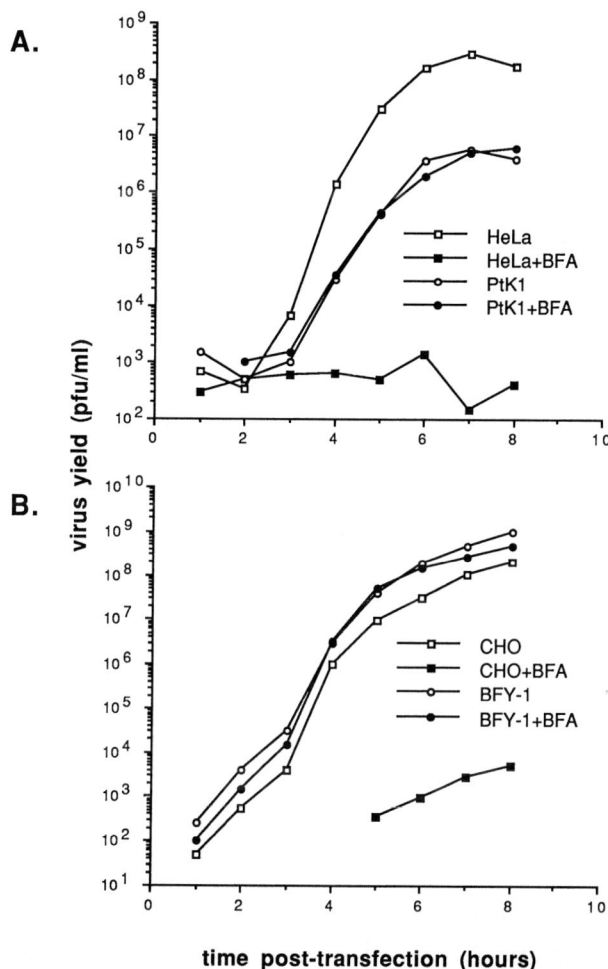

B.

Fig. 5. Single-cycle growth of poliovirus in cell lines possessing BFA-resistant secretory pathways. **A** Growth of poliovirus in PtK$_1$ (BFA-resistant) and HeLa (BFA-sensitive) with and without 5 µg/ml BFA. **B** Growth of poliovirus in BFY-1 (BFA-resistant) and CHO (BFA-sensitive) with and without 5 µg/ml BFA. Cells grown on plates were transfected with approximately 1 µg/plate in vitro-transcribed poliovirus RNA by a liposome procedure [6]. Infections were then allowed to proceed in the presence or absence of BFA (5 µg/ml). At various times post-infection, intracellular virus was harvested, and viral yields were determined by plaque assay on HeLa cells

We found that poliovirus replication was not inhibited by BFA in PtK$_1$ cells (Fig. 5a). PtK$_1$ cells cannot be directly infected with poliovirus because they lack the poliovirus receptor expressed only in human and primate cells. Infections of both HeLa and PtK$_1$ were therefore initiated by transfection with poliovirus RNA transcribed in vitro [20]. Consistent with previous results, the viral yield from HeLa was reduced 6–7 orders of magnitude in the presence of 5 µg/ml BFA. However, viral yield from

PtK_1 was completely unaffected by BFA (Fig. 5a). An identical study was carried out using another BFA-resistant cell line, BFY-1 (Fig. 5b). This cell line was produced by selecting mutagenized CHO cells for growth in the presence of BFA [24]. BFA is not excluded or degraded by the cells, and neither the morphology of the Golgi complex nor transport of proteins through the secretory pathway is affected by BFA in BFY-1 cells [24]. To determine whether poliovirus replication was sensitive to brefeldin A in BFY-1 cells, infection of the parental unmutagenized CHO and BFY-1 cells was initiated as before by transfection with poliovirus RNA produced in vitro. As shown in Fig. 5b, virus yield in CHO cells was diminished 10^5-fold by 5 µg/ml BFA. However, no effect of BFA on virus production was observed in the BFY-1 cells, which possess a BFA-resistant secretory pathway.

These experiments show that brefeldin A does not interfere with poliovirus replication by interacting directly with a viral protein. Instead, it appears that BFA interacts with a host factor or host process required for replication. Any effect of BFA on the formation of poliovirus-induced vesicles has not been directly addressed by these experiments; however, the known effects of BFA make a block in vesicle proliferation a plausible mechanism for its action in inhibiting poliovirus replication.

Poliovirus grown in the presence of brefeldin A remains BFA-sensitive

Prior to this study, we had not detected production of poliovirus in cells that were transfected with poliovirus RNA and subsequently incubated in the presence of BFA [15]. The low level of virus production seen in Fig. 5 in both CHO and HeLa cells treated with BFA might represent a low level of wild-type virus production in the presence of BFA. Alternatively, the low levels of virus produced in both HeLa and CHO in the presence of BFA might represent the spontaneous appearance of a small amount of mutated polioviruses now resistant to BFA. It was therefore of interest to compare the sensitivity to BFA of virus grown in the presence and absence of BFA.

We performed single-cycle growth curves using virus harvested from HeLa cells after 6 h of growth in the presence of 5 µg/ml BFA (Fig. 5a). This virus was grown in HeLa cells in 0, 0.5, and 5 µg/ml brefeldin A in parallel with wild-type virus harvested from untreated cells. The low concentration of brefeldin A was chosen in the hope that it might reveal a decreased BFA-sensitivity of virus that might still be unable to grow in 5 µg/ml BFA. The results of this experiment are shown in Fig. 6. Both wild-type poliovirus and poliovirus passaged once in the presence of BFA were inhibited roughly 40-fold by 0.5 µg/ml BFA. No progeny virus was detected from either sample assayed in 5 µg/ml BFA, a finding

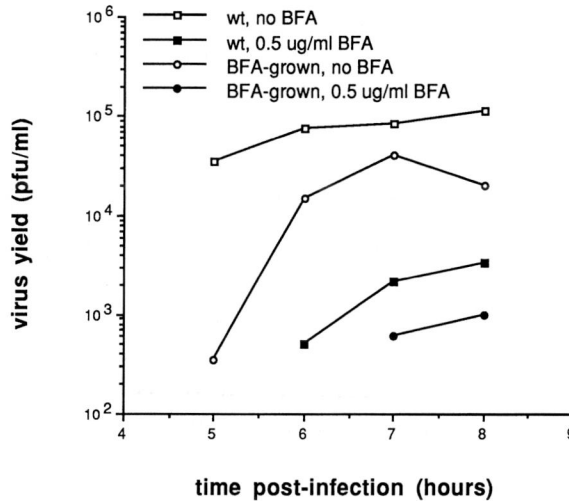

Fig. 6. Sensitivity to BFA of poliovirus passaged once in the presence of 5 μg/ml BFA. Poliovirus grown in HeLa cells in the presence of 5 μg/ml BFA for 6 h (Fig. 5a) was assayed for sensitivity to BFA in parallel with wild type poliovirus. HeLa cells on plates were infected in the presence of 0, 0.5, or 5 μg/ml BFA with either BFA-grown or wild type poliovirus. At the times indicated, intracellular virus was harvested and viral titers were determined by plaque assay on HeLa cells. No virus was detected from infections done in the presence of 5 μg/ml BFA with either wild-type or BFA-grown virus, nor was any virus detected from infections performed in 0.5 μg/ml BFA at the following time points: 5 h from the infection done with wild-type virus, 5 and 6 h from the infection done with BFA-grown virus. The limit of detection in this experiment was 50 pfu/ml

consistent with the 10^6-fold inhibition of wild-type virus by 5 μg/ml BFA shown in Fig. 5a.

Discussion

In this study, we have clarified the importance of two putative host-virus interactions in poliovirus infection. Despite morphological evidence of an association of viral RNA replication with the host cytoskeleton, we could find no evidence that this apparent association was functionally significant. Specifically, treatment of cells with cytochalasin D and nocodazole in combination disrupted the normal organization of microfilaments, microtubules, and IF in infected and uninfected cells and blocked the poliovirus-induced reorganization of IF. In spite of this, no inhibition of viral growth was observed in cells treated with cytochalasin D and nocodazole.

There may still have been some co-localization between poliovirus capsid proteins and the cables of IF present in cells treated with cytochalasin D and nocodazole (see Fig. 2). We note that poliovirus can replicate with apparently equal efficiency in human adenocarcinoma SW13 cells that contain a vimentin-type IF system and SW13 cells that are free of IF [21]. We therefore conclude that the association between IF and poliovirus replication sites is not important in the production of virus, at least in tissue culture. These experiments do not preclude a role for virus-cytoskeleton interactions in cell lysis, since the growth curves in Fig. 3 display the intracellular virus produced during a single cycle of infection.

Because anti-capsid antibodies were used in the immunofluorescence experiments, it is possible that an undetected association of nonstructural proteins with cytochalasin D- and nocodazole-disrupted cytoskeletal systems remained. This possibility seems unlikely, however, given that poliovirus genome replication and packaging appear to occur in association on the surface of the virus-induced vesicles [18].

Our studies extend the evidence that the mechanism of brefeldin A inhibition of poliovirus RNA synthesis is identical to the mechanism by which it inhibits vesicular transport through the secretory pathway, namely by blocking the formation of transport vesicles. That the inhibition of poliovirus by BFA is a property of the host cell was demonstrated in two unrelated cell lines possessing BFA-resistant secretory pathways. The sensitivity of poliovirus to BFA correlates completely with the sensitivity of the host secretory pathway to the compound. In addition, the concentrations of BFA found to inhibit poliovirus are equal to those used to block secretory pathway function. Further, the finding that BFA does not inhibit poliovirus in PtK_1 cells argues that the effects of BFA on endocytic membrane traffic are not relevant to its effects on poliovirus replication because BFA induces altered endosome morphology in PtK_1 cells similar to that seen in other cell types [13]. Given the known association of poliovirus RNA synthesis with cytoplasmic vesicles, it appears that poliovirus infection requires intact function of the vesicle production machinery normally used to generate transport vesicles used in secretory traffic.

An alternative explanation for the action of brefeldin A is that it disrupts a required interaction of host and viral products in a host-specific fashion. Because the BFA-sensitivity of the virus and host correspond, the putative host factor would presumably be identical to the target within the secretory machinery for BFA. If BFA disrupted a simple binding interaction between host and viral products, we would have expected to be able to isolate BFA-resistant virus in our experiments. The lack of detectable BFA-resistant virus further supports the hypothesis that poliovirus interacts with secretory organelles, probably the ER, to divert vesicles from their normal pathway, and BFA inhibits poliovirus by preventing the formation of this required class of vesicles.

Acknowledgements

We wish to thank P. Melançon and J.-P. Yan for providing BFY-1 cells and P. Sarnow and J. Novak for critical reading of the manuscript. J.D. was supported by a fellowship from the Colorado Institute for Research in Biotechnology. M.W.K. was supported by grant DCB 89-0522 from the NSF, and K.K. was supported by NIH grant AI-25166 and by the David and Lucile Packard Foundation. K.K. is an assistant investigator of the Howard Hughes Medical Institute.

References

1. Bienz K, Egger D, Pasamontes L (1987) Association of polioviral proteins of the P2 genomic region with the viral replication complex and virus-induced membrane synthesis as visualized by electron microscopic immunocytochemistry and autoradiography. Virology 160: 220–226
2. Bienz K, Egger D, Rasser Y, Bossart W (1980) Kinetics and location of poliovirus macromolecular synthesis in correlation to virus-induced cytopathology. Virology 100: 390–399
3. Bienz K, Egger D, Troxler M, Pasamontes L (1990) Structural organization of poliovirus RNA replication is mediated by viral proteins of the P2 genomic region. J Virol 64: 1156–1163
4. Caliguiri LA, Tamm I (1970) Characterization of poliovirus-specific structures associated with cytoplasmic membranes. Virology 42: 112–122
5. Doms RW, Russ G, Yewdell JW (1989) Brefeldin A redistributes resident and itinerant Golgi proteins to the endoplasmic reticulum. J Cell Biol 109: 61–72
6. Grakoui A, Levis R, Raju R, Huang HV, Rice CM (1989) A cis-acting mutation in the Sindbis virus junction region which affects subgenomic RNA synthesis. J Virol 63: 5216–5227
7. Irurzun A, Perez L, Carrasco L (1992) Involvement of membrane traffic in the replication of poliovirus genomes: effects of brefeldin A. Virology 191: 166–175
8. Klymkowsky MW, Bachant JB, Domingo A (1989) Functions of intermediate filaments. Cell Motil Cytoskeleton 14: 309–331
9. Klymkowsky MW, Maynell LA, Polson AG (1987) Polar asymmetry in the organization of the cortical cytokeratin system of *Xenopus laevis* oocytes and embryos. Development 100: 543–557
10. Klymkowsky MW, Miller RH, Lane EB (1983) Morphology, behavior, and interaction of cultured epithelial cells after the antibody-induced disruption of keratin filament organization. J Cell Biol 96: 494–509
11. Ktistakis NT, Roth MJ, Bloom GS (1991) PtK1 cells contain a nondiffusible, dominant factor that makes the Golgi apparatus resistant to brefeldin A. J Cell Biol 113: 1009–1023
12. Lenk R, Penman S (1979) The cytoskeletal framework and poliovirus metabolism. Cell 16: 289–301
13. Lippincott-Schwartz J, Yuan L, Tipper C, Amherdt M, Orci L, Klausner RD (1991) Brefeldin A's effects on endosomes, lysosomes, and the TGN suggest a general mechanism for regulating organelle structure and membrane traffic. Cell 67: 601–616
14. Lippincott-Schwartz J, Yuan LC, Bonifacino JS, Klausner RD (1989) Rapid redistribution of Golgi proteins into the ER in cells treated with brefeldin A: evidence for membrane cycling from Golgi to ER. Cell 56: 801–813

15. Maynell LA, Kirkegaard K, Klymkowsky MW (1992) Inhibition of poliovirus replication by brefeldin A. J Virol 66: 1985–1994
16. Misumi Y, Misumi Y, Miki K, Takatsuki A, Tamura G, Ikehara Y (1986) Novel blockade by brefeldin A of intracellular transport of secretory proteins in cultured rat hepatocytes. J Biol Chem 261: 11398–11403
17. Orci L, Tagaya M, Amherdt M, Perrelet A, Donaldson JG, Lippincott-Schwartz J, Klausner RD, Rothman JE (1991) Brefeldin A, a drug that blocks secretion, prevents the assembly of non-clathrin-coated buds on Golgi cisternae. Cell 64: 1183–1195
18. Pfister T, Pasamontes L, Troxler M, Egger D, Bienz K (1992) Immunocytochemical localization of capsid-related particles in subcellular fractions of poliovirus-infected cells. Virology 188: 676–684
19. Rueckert RR (1990) Picornaviridae and their replication. In: Fields BN, Knipe DM, Chanock RM, Hirsch MS, Melnick JL, Monath TP, Roizman B (eds) Virology, vol 1. Raven Press, New York, pp 507–548
20. Sarnow P (1989) Role of 3′-end sequences in infectivity of poliovirus transcripts made in vitro. J Virol 63: 467–470
21. Sarria A, Evans R (1993) University of Colorado Health Sciences Center, Denver, Colorado (pers. comm.)
22. Takeda N, Kuhn RJ, Yang C, Takegami T, Wimmer E (1986) Initiation of poliovirus plus-strand RNA synthesis in a membrane complex of infected HeLa cells. J Virol 60: 43–53
23. Weed HG, Krochmalnic G, Penman S (1985) Poliovirus metabolism and the cytoskeletal framework: detergent extraction and resinless section electron microscopy. J Virol 56: 549–557
24. Yan JP, Beebe L, Melançon P (1992) Isolation and characterization of Chinese hamster ovary cell lines resistant to brefeldin A. Mol Biol Cell [Suppl] 3: 118a

Authors' address: J. Doedens, Department of Molecular, Cellular and Developmental Biology, University of Colorado Boulder, Boulder, CO 80309-0347, U.S.A.

Arch Virol (1994) [Suppl] 9: 173–180

Archives
of
Virology
© Springer-Verlag 1994
Printed in Austria

Role of subgenomic minus-strand RNA
in coronavirus replication

D. A. Brian, R.-Y. Chang, M. A. Hofmann, and **P. B. Sethna**

Department of Microbiology, University of Tennessee, Knoxville, Tennessee, U.S.A.

Summary. Coronavirus subgenomic minus-strand RNAs (negative-strand copies of the 3' coterminal subgenomic mRNAs) probably function in mRNA amplification by serving as templates for transcription from internal (intergenic) promoters, rather than by faithful (full-length) mRNA replication.

Introduction

Coronaviruses replicate by generating a 3' coterminal nested set of 6 to 8 subgenomic mRNA species (reviewed in [1]). When it was first reported that each subgenomic mRNA also had a minus-strand counterpart [2], it was hypothesized that perhaps each mRNA species underwent faithful replication by using the same mechanism(s) as the genome. This hypothesis was based on three sets of observations and an assumption: (i) Coronavirus subgenomic RNA species (both plus and minus strands) underwent amplification early in infection at a rate that was inversely related to their length. (ii) Subgenomic replicative forms existed. (iii) Each subgenomic mRNA species possessed in common with the genome a 3' terminal sequence of approximately 300 nucleotides (excluding the poly(A) tail) and a 5' terminal sequence (a leader) of 65 to 90 nucleotides (the length depends on the species of coronavirus). The assumption was that these termini possessed respectively the promoter and the promoter template for minus- and plus-strand RNA synthesis. More recent reports showing mRNA amplification in other coronavirus species [3, 4], a structural analysis of subgenomic replicative forms [4], and the existence of an antileader on the 3' end of subgenomic minus-strand RNAs [5], in our view supported the notion that mRNAs undergo replication.

To directly test the hypothesis that coronavirus subgenomic mRNAs do undergo faithful replication, clones of mRNA species containing a reporter sequence were used to make transcripts in vitro that were then transfected into infected cells, and evidence of replication was sought.

To date we have found no direct evidence of mRNA replication following transfection of marked transcripts (synthetic mRNAs) into infected cells. These results, along with (i) the successful demonstration that the cloned 5'-terminal 498 nt fragment of a defective-interfering RNA species imparts replicational competence when spliced onto the 5' end of a marked mRNA molecule (described below), and, (ii) the demonstration that a sub-defective-interfering RNA transcript can be generated from a cloned defective-interfering RNA after insertion of a coronavirus intergenic sequence [6], lead us to conclude that perhaps coronavirus subgenomic mRNAs do not amplify by a faithful replication mechanism. Rather, they may amplify by a transcriptional mechanism in which the subgenomic minus-strand templates function by an internal promoter to make new mRNA of smaller size, but not of equal size.

The antileader on subgenomic minus-strand RNA is apparently an insufficient promoter for plus-strand synthesis (at least when the template mRNA enters via transfection)

To determine whether subgenomic mRNA molecules can undergo replication following transfection into infected cells, cloned mRNA for the N proteins of porcine transmissible gastroenteritis virus (TGEV) and bovine coronavirus (BCV) were modified to contain a reporter sequence (either the entire CAT gene or a small portion of it were cloned in-frame within the N ORF) and in vitro generated transcripts were transfected into infected cells and evidence of amplified plus- and minus-strands were sought. In several experiments, no evidence of amplification (i.e., replication) was found (P. Sethna and M. Hofmann, unpubl. data). In one such experiment (Fig. 1), the BCV N mRNA with a 30 nt, in-frame, reporter sequence (a 30 nt fragment of the TGEV N ORF, chosen because it would be unlikely to have a detrimental effect on the stability of the molecule) was used for transfection into BCV-infected cells. Whereas the mRNA species and genome of helper BCV can be seen to increase in amount after infection and transfection (Fig. 1B, lanes 2 through 5), input mRNA transcript carrying a reporter sequence (Fig. 1B, lane 6) did not increase (lanes 7–10). When reporter minus-strand-detecting probe was used, no minus-strand RNA was detected at all after transfection (Fig. 1B, lanes 12–15). Thus, the transfected transcript did not undergo replication.

It is important to note that even though minus-strand RNAs were not sufficiently abundant to be detected in the RNA blotting experiment described in Fig. 1B, we assume they were made in small numbers after

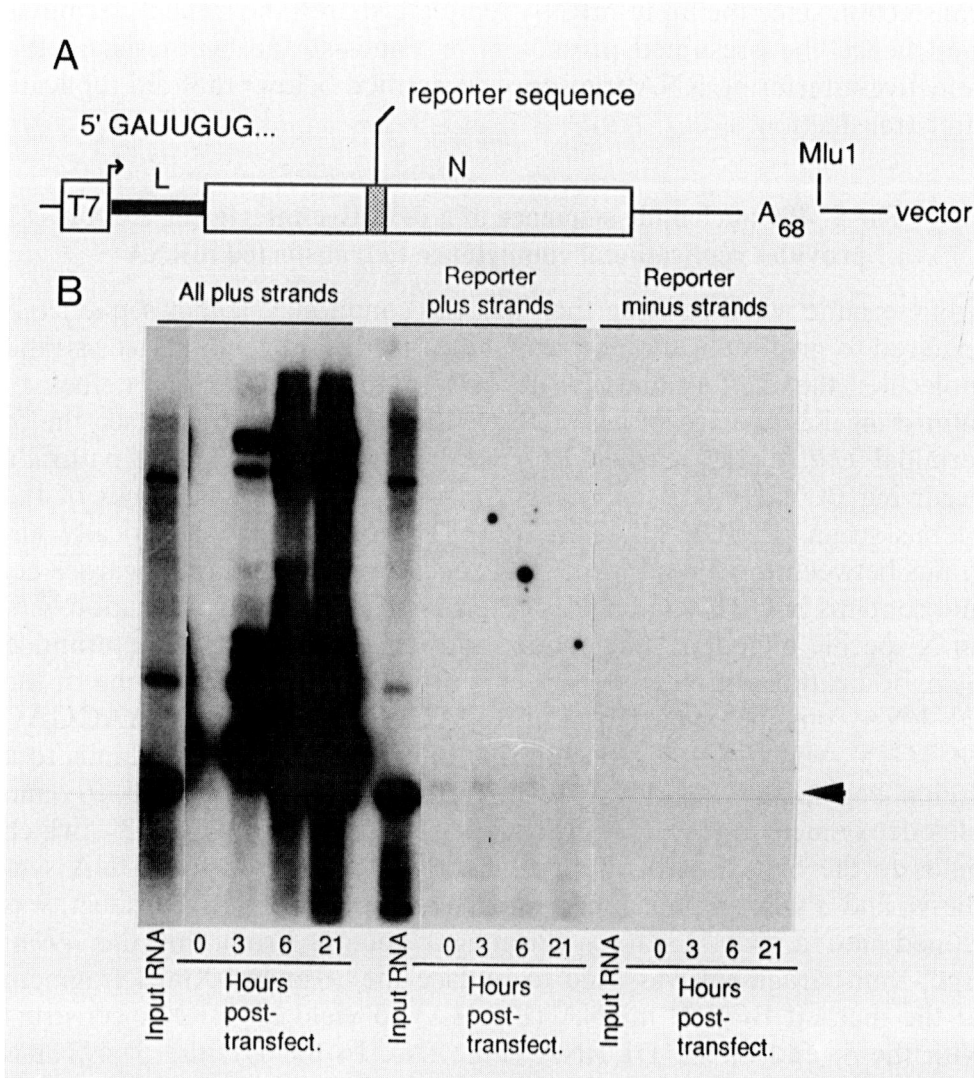

Fig. 1. A Plasmid construct of the cloned bovine coronavirus N mRNA with a 30 nucleotide, in-frame, reporter sequence. Plasmid cut with Mlu1 and transcribed with T7 RNA polymerase generates transcripts beginning with a 5′ GAUUGUG..., (the first bases in the 65 nucleotide leader [L]), and N open reading frame of 478 nucleotides (including the reporter), and a poly A tail of 68 nucleotides. **B** RNA blot (Northern) showing successful transfection of reporter-containing N mRNA, but no replication. HRT-18 cells were infected with the bovine coronavirus (MOI = 5) for 1h, then transfected with transcripts (200 ng/35 mm plate) for 1 h. Cytoplasmic RNA was extracted at the times indicated, electrophoresed in a formaldehyde-agarose gel, transferred to a nylon membrane, and probed with ^{32}P-end-labeled oligonucleotide. An oligonucleotide hybridizing to N plus strands was used to identify all viral RNA species, one hybridizing to reporter plus strands was used to identify reporter-containing plus strands, and one hybridizing to reporter minus strands was used to identify reporter-containing minus strands. Note no increase in the amount of reporter-containing plus strands and the absence of reporter-containing minus strands

transfection since the input mRNA template shared the same 3' terminus (and hence the presumed promoter for minus-strand synthesis) as the defective-interfering RNA transcript (described below) that did replicate after transfection.

The 5' 498 nucleotide sequence of a defective-interfering RNA provides replicational competence to transfected mRNA

To determine whether more than just the common 5' leader sequence is required to give replicational competence to the subgenomic plus-strand molecule, the 5' terminal 498 nt of a naturally occurring defective-interfering RNA of BCV was cDNA cloned and used to replace the 5' terminal 77 nt of the marked BCV N mRNA (Fig. 2A). The naturally occurring BCV DI RNA is present in one of our cloned stocks of the Mebus strain of BCV [3]. The BCV DI RNA electrophoretically migrates between the N and M mRNA species on a denaturing agarose gel and contains N ORF sequence as determined by RNA hybridization with an N-specific oligodeoxynucleotide, and leader sequence as determined by hybridization with a leader-specific probe. To clone the 5' end of the DI RNA, oligodeoxynucleotide 5'CCAGAACGATTTCCAAAGGAC GCTCT5', which contains an internal unique Xmn1 site and binds to a region beginning 34 nt downstream from the start of N ORF [7], and oligodeoxynucleotide 5'GATTGTGAGCGATTTGCGTGCG3', which binds to the first 22 nt of the antileader [8], were used for cDNA synthesis and PCR amplification, and the amplified 557 nt product was cloned into a pGEM3Zf(−) (Promega) vector. From this the 499 nt Bgl2-Xmn1 fragment was used to replace the 78 nt Bgl-Xmn1 fragment of the marked BCV N mRNA (Fig. 1A) to yield an mRNA construct with the 5' end of the DI RNA (Fig. 2A). In the DI, the 5' 498 nt is presumed to represent the 5' end of the BCV genome since it shows extremely high sequence similarity to the 5' end of the MHV genome. In the DI, the 5' DI-specific ORF of 96 amino acids (hatched line in Fig. 2A) is contiguous with the N ORF.

Transcripts of the marked N mRNA containing the 5' DI RNA-specific sequence underwent replication as determined by an increase in plus-strand accumulation (Fig. 2B, lanes 13–14) and the appearance of minus-strand species (Fig. 2B, lane 20).

Possible mechanisms of mRNA amplification from subgenomic minus strands

The rapid appearance of coronavirus subgenomic mRNA species and the abundance of transcriptionally active subgenomic replicative inter-

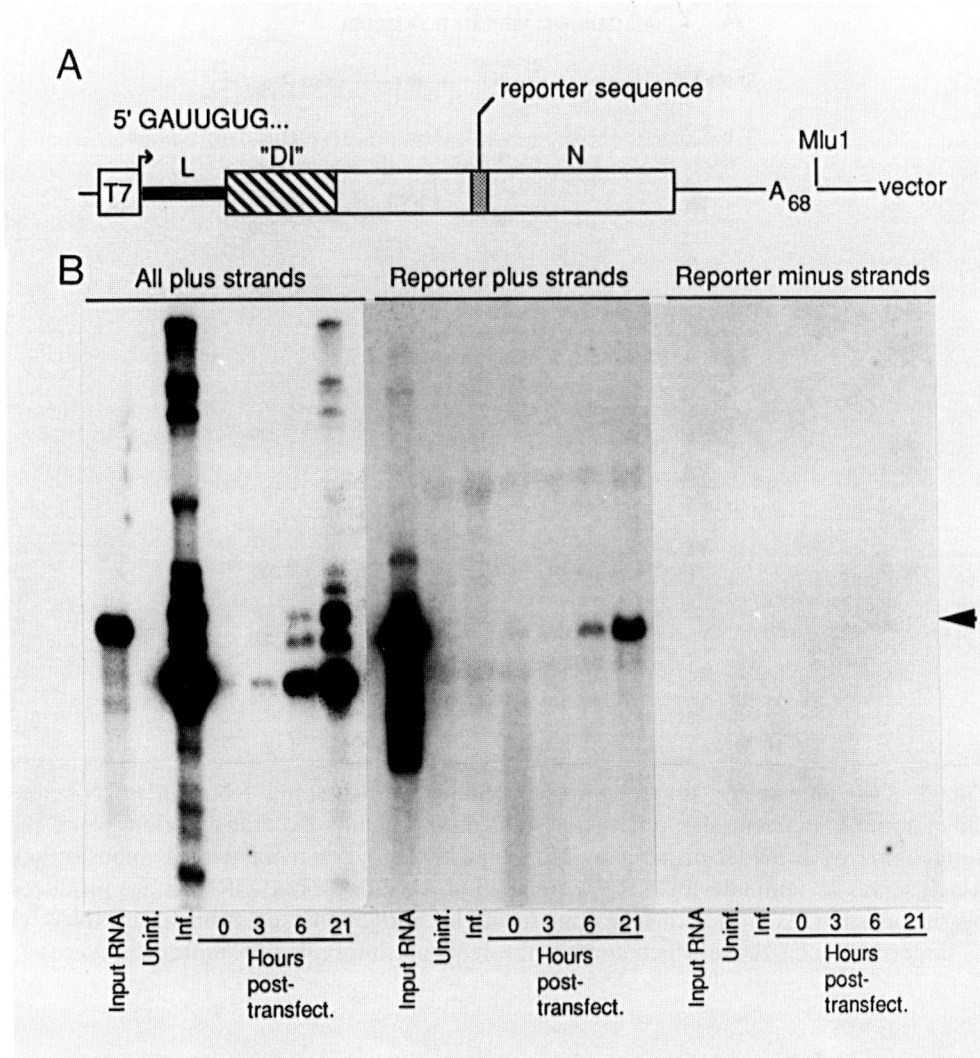

Fig. 2. A Plasmid construct of the cloned bovine coronavirus defective-interfering RNA with a 30 nucleotide, in-frame, reporter sequence. The construct is identical to that described for the N mRNA in Fig. 1A, except for 433 nucleotides of DI-specific sequence inserted between the 65 nt leader and the start codon of N. 288 of the 433 nt make up a DI-specific open reading frame that is in-frame with the N open reading frame. **B** RNA blot (Northern) analysis performed as described in Fig. 1B

mediates suggest that subgenomic minus-strand molecules are important in the process of mRNA amplification.

We envision two possible mechanisms for the amplification of mRNA from subgenomic minus strands. The first is illustrated in Fig. 3A in which the subgenomic mRNA arises, perhaps by a leader priming mechanism, from the antigenomic template (as illustrated) or perhaps by

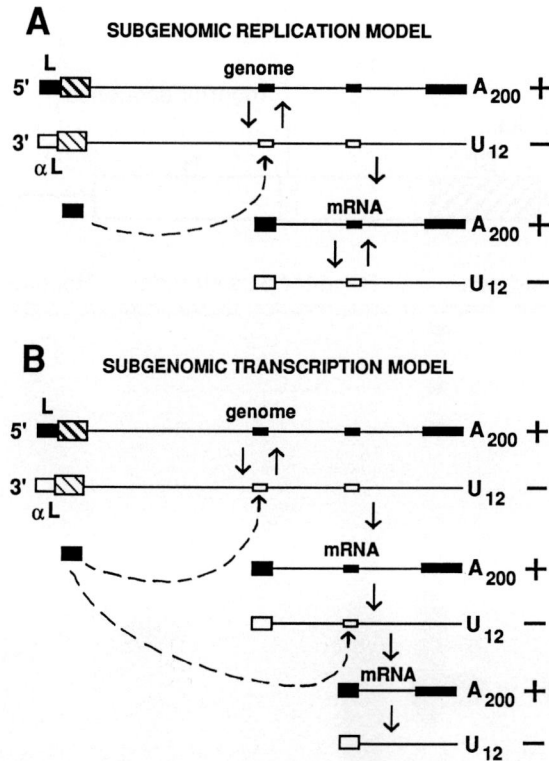

Fig. 3. Two models for the role of subgenomic minus-strand RNA in mRNA amplification. **A** Subgenomic mRNA (produced originally by transcription from the antigenome by a leader-priming or other mechanism) gets copied to a minus strand which serves as template for mRNA replication. **B** Subgenomic mRNAs are produced by the usual transcription mechanism from the antigenome or from antimRNAs of larger size (i.e., those which contain the requisite intergenic promoter sequence)

splicing of the genome or antigenome. Once made, the mRNA (or antimRNA) serves as template for faithful replication using the same mechanism as genome. We interpret the failure of synthetic mRNAs to undergo replication after transfection in the above described experiments to mean either one of two things: (i) mRNAs do not replicate. (ii) mRNAs replicate, but require a special juxtapositional start within the replicational machinery that is gained from the original transcriptional process. That is, mRNA transfected into the cell cannot initiate a cycle of replication without the added cis-acting sequence provided by the genomic (i.e., DI) 5' end. Further experimentation is required to distinguish between these two possibilities.

The second potential mechanism for amplification of mRNA is illustrated in Fig. 3B. In this case, minus-strand RNAs that are synthesized from either genome or subgenomic mRNAs (generated by leader-priming or splicing as described above) can serve as templates for new plus-strand synthesis, but by a leader priming (or splicing) step, not by replication. By this mechanism, subgenomic minus-strand RNA would serve as a template for synthesis of mRNA of smaller size, but not of the same size. Evidence supporting this mechanism comes from the laboratory of Sinji Makino (University of Texas, Austin) in which sub-DI RNA transcripts were generated from a transfected, replicating DI RNA construct into which the conserved UCUAAAC intergenic sequence element (and some flanking sequence) had been internally placed [6]. Presumably this sequence element is functioning on the minus strand to direct priming (or splicing) to generate the subgenomic (sub DI) transcript. With this second mechanism, a subgenomic minus-strand RNA without an antileader could also theoretically function as a template for transcription.

Conclusions

Coronaviruses appear unique among positive strand RNA viruses in that subgenomic minus-strand RNAs are important as templates for the amplification of subgenomic mRNA species. Subgenomic minus-strand RNAs apparently serve as templates for greater than 90% of subgenomic mRNA synthesis (Stan Sawicki, Medical College of Ohio, Toledo, Ohio, personal communication). It remains to be determined, however, whether, subgenomic mRNAs amplify by (i) mRNA replication, (ii) transcription by internal initiation on a subgenomic minus strand template, or (iii) both. At the present time, our data favor the second option.

References

1. Lai MMC (1990) Coronavirus: organization, replication and expression of genome. Annu Rev Microbiol 44: 303–333
2. Sethna PB, Hung SL, Brian DA (1989) Coronavirus subgenomic minus-strand RNAs and the potential for mRNA replicons. Proc Natl Acad Sci USA 86: 5626–5630
3. Hofmann MA, Sethna PB, Brian DA (1990) Bovine coronavirus mRNA replication continues throughout persistent infection in cell culture. J Virol 64: 4108–4114
4. Sawicki SG, Sawicki DL (1990) Coronavirus transcription: subgenomic mouse hepatitis virus replicative intermediates function in mRNA synthesis. J Virol 64: 1050–1056
5. Sethna PB, Hofmann MA, Brian DA (1991) Minus-strand copies of replicating coronavirus mRNAs contain antileaders. J Virol 65: 320–325
6. Makino S, Myungsoo J, Makino JK (1991) A system for study of coronavirus mRNA synthesis: a regulated, expressed subgenomic defective interfering RNA results from intergenic site insertion. J Virol 65: 6031–6041
7. Lapps W, Hogue BG, Brian DA (1987) Sequence analysis of the bovine coronavirus nucleocapsid and matrix protein genes. Virology 157: 47–57

8. Senanayake SD, Hofmann MA, Maki JL, Brian DA (1992) The nucleocapsid protein gene of the bovine coronavirus is bicistronic. J Virol 66: 5277–5283

Authors' address: Dr. D. A. Brian, Department of Microbiology, University of Tennessee, Knoxville, TN 37996-0845, U.S.A.

Arch Virol (1994) [Suppl] 9: 181–194

_Archives_____
Virology
© Springer-Verlag 1994
Printed in Austria

Common replication strategies emerging from the study of diverse groups of positive-strand RNA viruses

G. P. Pogue[2], **C. C. Huntley**[1], and **T. C. Hall**[1]

[1] Institute of Developmental and Molecular Biology, Texas A&M University, College Station, Texas
[2] Division of Hematologic Products, OTRR/CBER/FDA, Bethesda, Maryland, U.S.A.

Summary. Studies using brome mosaic virus (BMV), Sindbis virus and poliovirus have provided evidence that disparate groups of plant and animal positive strand RNA viruses have remarkably similar replication strategies. The conservation of several functional domains within virus-encoded nonstructural proteins implies that, although the precise character of these and interacting host components varies for each virus, they employ similar mechanisms for RNA replication. For (+) strand replication, similarities in *cis*-acting sequence motifs and RNA secondary structures within 5' termini of genomic (+) strands have been identified and have been shown to participate in binding of host factors. The model presented for replication of BMV RNA suggests that binding of these factors to internal control region (ICR) sequence motifs in the double-stranded replication intermediate releases a single-stranded 3' terminus on the (−) strand that may be essential for initiation of genomic (+) strand synthesis. ICR sequences internal to the BMV genome were also found to be required for efficient replication. Asymmetric production of excess genomic (+) over (−) strand RNA, characteristic of all (+) strand viruses, may be accomplished through transition of the replicase from competence for (−) to (+) strand synthesis by the recruitment of additional host factors.

Introduction

The study of brome mosaic virus (BMV), a tripartite RNA virus infecting primarily graminaceous plants, has contributed greatly to our understanding of positive (+)strand RNA virus replication. Each BMV RNA terminates in a multi-functional 3' tRNA-like structure (Fig. 1). Sequence domains within this 3' region directing negative (−)strand synthesis have been extensively characterized using in vitro and in vivo

A. BMV RNA-1

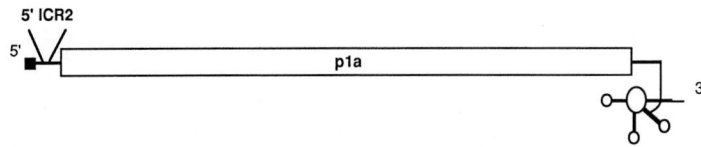

B. BMV RNA-2

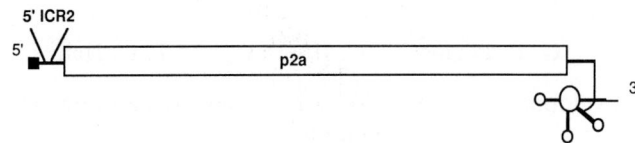

C. BMV RNA-3

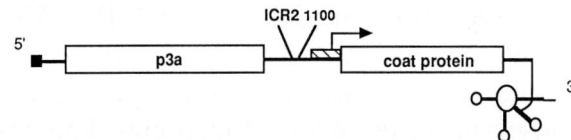

D.

tRNA ICR2 consensus	$G_1 G_2 T_3 T_4 C_5 G_6 A_7 N_8 T_9 C_{10} C_{11}$
RNA-1 5' ICR2	$G_1 G_2 T_3 T_4 C_5 A_6 A_7 \cdot T_8 C_9 C_{10}$
RNA-2 5' ICR2	$G_1 G_2 T_3 T_4 C_5 A_6 A_7 \cdot T_8 C_9 C_{10}$
RNA-3 ICR2$_{1100}$	G G T T C A A T T C C

Fig. 1. Schematic diagrams of BMV genomic RNAs-1 (**A**), -2 (**B**) and -3 (**C**) are shown with the locations of essential ICR2-like sequences indicated by diagonal lines. The 3' tRNA-like structure, conserved among all BMV genomic RNAs, is schematically drawn in cloverleaf form. Open reading frames are designated by open boxes. RNA-1 encodes a putative "helicase-like", NTP binding protein (p1a), while the protein (p2a) encoded by RNA-2 is predicted to be the core RNA-dependent RNA-polymerase [7, 9, 17]. In addition to the virus capsid protein, RNA-3 encodes the putative virus movement protein (p3a; [7, 17]). The arrow (**C**) demarks the initiation site of subgenomic RNA-4, which is produced by a subgenomic promoter (box with diagonal stripes) present in RNA-3. Solid boxes represent 5' RNA cap structures. Comparison (**D**) of ICR2-like sequences found in the BMV genome with the tRNA promoter ICR2 (B-box) consensus [14]. All sequences are given as viral cDNA with spaces (.) inserted to maximize homology. Numbering for the ICR2 motif within RNA-3 denotes location with respect to the position of the 5' nucleotide; the bases within the consensus or individual ICR2-like sequences are numbered 5' to 3' and correspond to the designations used for mutants in the text and references

replication assay systems (reviewed in [17]). In addition, sequences determining the tRNA-like properties of this 3′ structure, repair by nucleotidyl transferase and aminoacylation by host tyrosine tRNA synthetases have been well characterized (reviewed in [17]). The core and modulating elements of the subgenomic promoter (Fig. 1C) responsible for RNA-4 production have been defined both in vitro and in vivo (reviewed in [17]).

BMV is generally regarded as a member of the alphavirus superfamily [7], based primarily on amino acid similarities between virally-encoded nonstructural proteins (reviewed in [7]). Although viruses within this superfamily differ with respect to the nature of their 3′ termini, requirements for proteolytic processing, virion structure and host range, many aspects of virus replication strategy are conserved. Evidence includes the presence of conserved activities putatively encoding methyltransferase, nucleotide binding, RNA-helicase and RNA-dependent RNA-polymerase domains [7, 9]. In addition to conservation of functional protein domains, members of the alphavirus superfamily generally express their structural proteins via subgenomic RNAs. Sequence motifs responsible for promoting subgenomic RNA synthesis are conserved between BMV, Sindbis and other viruses [13, 17]. These similarities suggest that members of the alphavirus superfamily share many aspects of replication strategy. We have recently proposed a model of (+)strand RNA virus replication [22, 23] that is consonant with many aspects of BMV, Sindbis and poliovirus replication. This article provides a concise discussion of this model and details the supportive evidence that led us to our conclusions.

Initiation of (−)strand synthesis

Sequences promoting the synthesis of negative (−)strand RNA, complementary to the infectious messenger (+)sense RNAs, have been rigorously demonstrated to be within the 3′ tRNA-like structure present in all BMV RNAs (reviewed in [17]). It has also been shown (reviewed in [17]) that the single-stranded (ss) 3′-CCA$_{OH}$ terminus of the BMV (+)strand tRNA-like structure is the site of (−)strand RNA initiation (Figs. 2A and B). Although the nascent (−)strand product (Fig. 2B) is likely to be essentially double-stranded (ds), the addition of single-stranded (ss)DNA or ssRNA fragments complementary to the ssRNA 3′ (+)terminus prevents the initiation of (−)strand synthesis in vitro [1]. In contrast, fragments complementary to internal portions of the 3′ tRNA-like region were efficiently read through by the replicase complex [1]. These results suggest that, although the lengthy internal ds regions can be separated by the replicase (presumably by a virally-encoded helicase

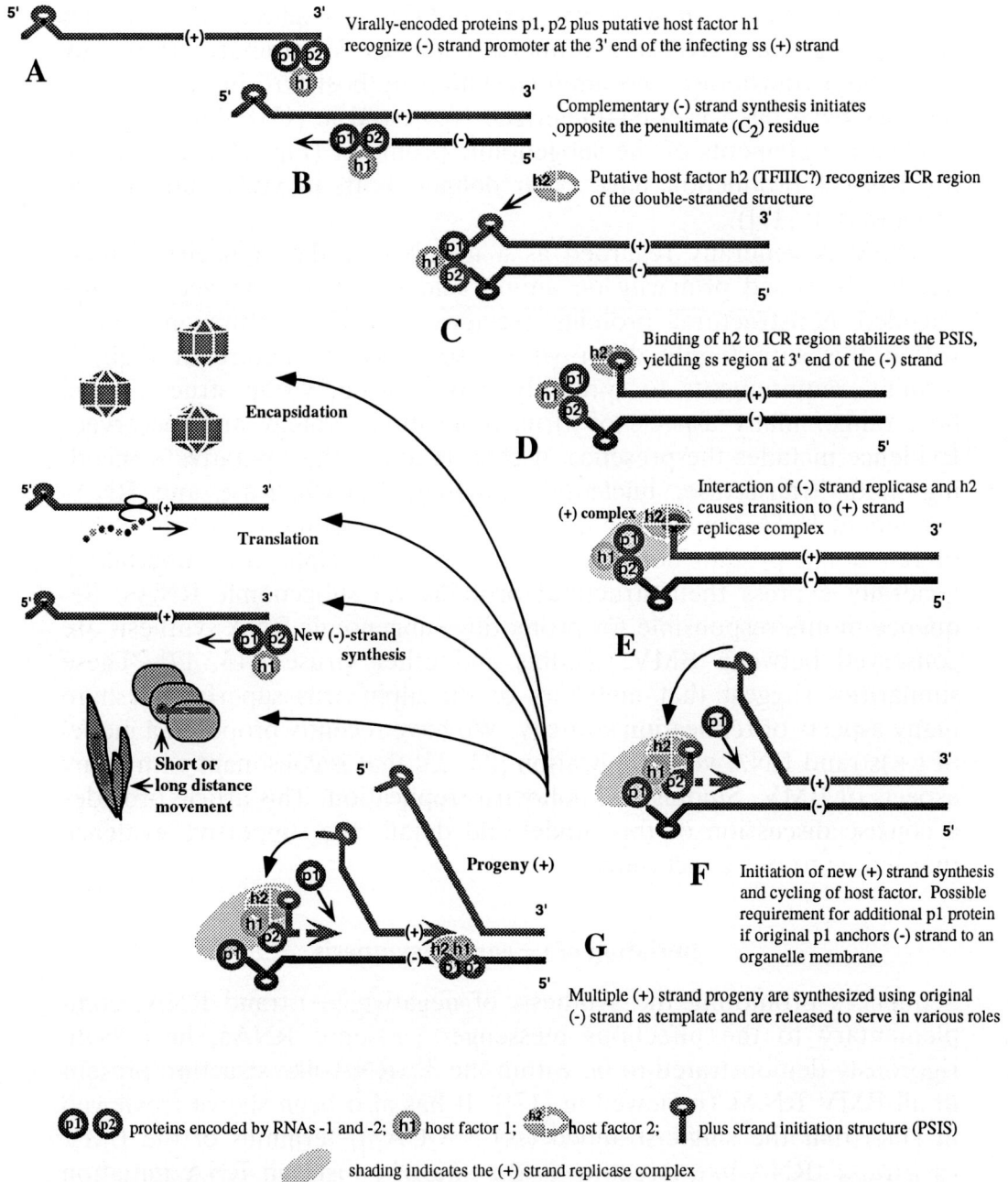

Fig. 2. Model for replication of brome mosaic virus genomic RNA

function; [9]), a ss terminus is mandatory for (−)strand initiation. It is interesting to note that both poliovirus and Sindbis virus (and related (+)strand RNA viruses) initiate genomic replication within a virally-encoded 3′ poly (A) tract. The majority of this 3′ poly (A) sequence is unlikely to participate in complex folding patterns, suggesting a general requirement for a ss region for the initiation of $(-)$strand RNA synthesis. These observations imply a general requirement for a ss $3'$ $(-)$strand terminus in the initiation of genomic $(+)$strand synthesis, so a mechanism must exist to release the $3'$ $(-)$terminus from the dsRNA form in the initial replicative form (Fig. 2C).

The ICR2 motif comprises a sequence-specific cis-acting replication element

The insight guiding our studies into mechanisms of $(+)$strand RNA synthesis was the recognition that sequence motifs at the $5'$ termini of each BMV RNA and other tricornaviruses resemble the internal control regions (ICRs) 1 and 2 (A and B boxes) that promote transcription of tRNA genes ([14]; Figs. 1 and 3). Additional ICR2 motifs were noted within the intercistronic region of BMV RNA-3 ([14]; Fig. 1C), and one, $ICR2_{1100}$, has been shown to be essential for RNA-3 replication [23]. We initially evaluated the role of these sequence motifs in virus replication by constructing a deletion mutant of BMV RNA-2, pRNA-2 M/S (parasitic RNA), whose replication is dependent on the *trans*-acting proteins supplied by BMV RNAs-1 and -2 [21, 22]. Specific single and double base substitutions were introduced into the ICR2 domain of the pRNA replication reporter construct (Fig. 1D) and the replication competence of these derivative mutants was assayed by co-inoculation with BMV RNAs-1 and -2 into barley protoplasts. Results from these studies demonstrated that no pRNA bearing mutations in the ICR2-like sequence replicated at levels >30% of wild type (wt), demonstrating that single base substitutions within this conserved motif greatly reduce the suitability of these RNA templates for replication [21, 22]. Kinetic analysis of mutants with substitutions at positions A_7 and T_8 (Fig. 1D) showed preferential debilitation of $(+)$strand synthesis early in infection, with $(-)$strand synthesis proceeding at wt rates [21]. Such inhibition implies that $(+)$strand replication is primarily debilitated by mutations in the ICR2-like motif.

The functional role of the ICR-like region of BMV RNA-2 was evaluated by co-inoculation of wt RNA-2 and derivative mutants with RNAs-1 and -3 or with RNA-1 alone into barley protoplasts (Fig. 4). With the exception of T_8 (Fig. 1D), alteration of any single nucleotide in the ICR2 sequence of RNA-2 resulted in >80% decrease in RNA replication ([23]; Fig. 4). RNAs bearing double substitutions at positions G_1 and G_2 (Fig. 1D) or deletion of the entire ICR2 motif rendered the RNA-2 template incapable of any replication (Fig. 4). Insertion of a single nucleotide at position 8 (Fig. 1D) of the ICR2 sequence reduced RNA-2 replication to 20% of wt or less [23]. Such intolerance for the

insertion of bases within this motif emphasizes tight sequence conservation and a distinct requirement for the sequential order present in the wt
sequence. There was no observable difference in the stability of wt
RNA-2 and RNA-2 mutants [23], firmly establishing that mutations in
the ICR2 motif dramatically reduced the ability of RNAs to function as a
template for replication. The conservation of sequences between the
ICR-like region of BMV RNAs-1 and -2 strongly suggest a similar
requirement for the ICR2 motif in BMV RNA-1 replication (Fig. 1D).

The ICR2 motif functions within a 5′ (+)strand stem-loop structure

Although these experiments established the crucial role of the ICR2
motif in BMV (+)strand replication, the structural context in which this
sequence acts, or is recognized, required further investigation. The 3′
end of BMV (−)RNAs-1 and -2 can be predicted to fold into a structure
resembling a methionine initiator tRNA (tRNA^{Meti}) with a calculated
thermodynamic stability (ΔG) equalling -12.9 ([17, 22]; Fig. 3). The
complementary bases in the 5′ terminus of the (+)strand can also be
predicted to fold into a stem loop structure of 46 bases in length and
having a ΔG of -13.1 ([22]; Fig. 3). The (+)strand structure consists of
a terminal loop containing the ICR2-like sequence followed by stem (a),
a second loop region (b) and a second stem (c) extending to the 5′
terminal viral base, not including the methylated cap structure (Fig. 3).
Indeed, the 5′ terminal bases of (+)BMV RNAs-1 and -3 and RNAs-1
and -2 of closely related viruses can be predicted to fold into similar
structures, each with an ICR2-like sequence comprising the terminal
stem-loop region [28]. Structural conservation in the midst of sequence
divergence suggests a functional role for the 5′ stem-loop structures in
the virus life cycle.

     To ascertain which stem-loop structure, (+) or (−) strand, is re-
quired for promoter activity, G−U mutations were introduced into the 5′
end of pRNA-2 M/S to selectively disrupt either the (+) or (−)strand
structure. Figure 3 shows the predicted stem-loop regions present at
the 5′ terminal (+)strand and the complementary (−)strand structure
with the bases targeted for mutagenesis boxed in the wt sequence. The
sequence and structural alterations for particularly instructive mutants
are also shown in Fig. 3. Substitutions, such as GG1, disrupting stems
(b) or (c) of the (+)strand stem-loop while maintaining the integrity of
the complementary (−)strand stem resulted in large reductions in RNA
replication ([22]; Fig. 3). When the predicted thermodynamic stability of
the short stem in (b) was improved by substitution of a G for the C at
base 37, the resulting RNA replicated at levels greater than wt [22]. This
result confirmed the importance of maintaining this stem region in the

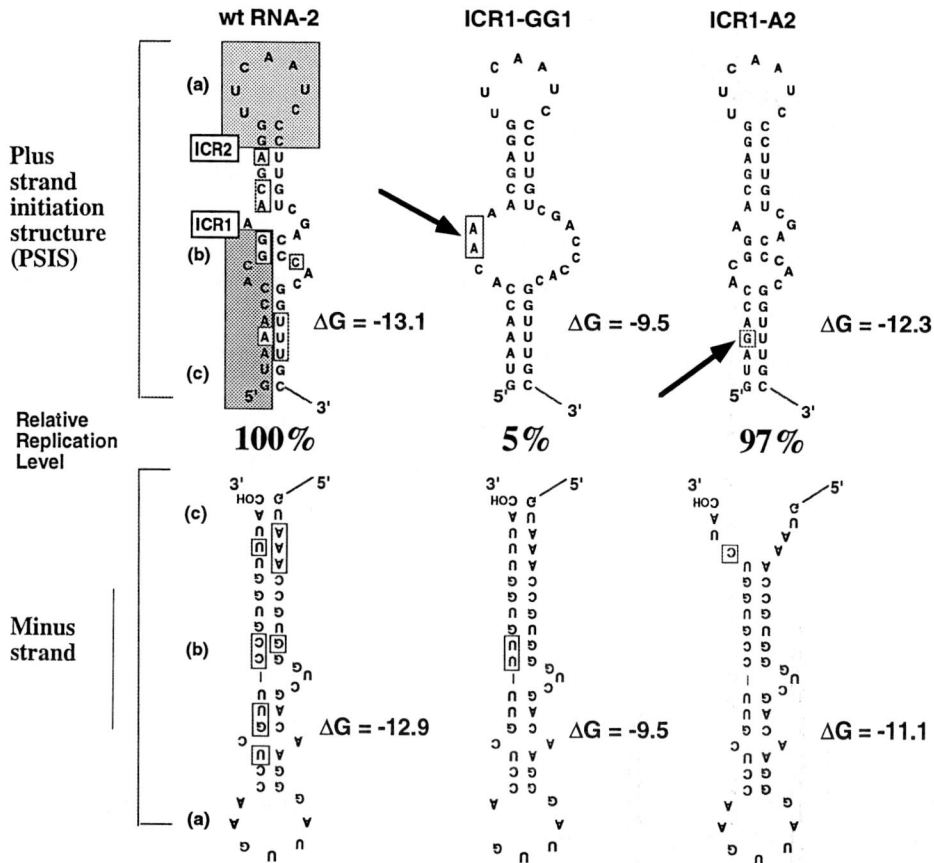

Effect of modifications to the 5' plus-strand initiation structure (PSIS)

Fig. 3. Comparison of predicted stem-loop structures present at the 5' (+)strand terminus and complementary 3' (−)strand terminus of BMV RNA-2 and selected mutant derivatives obtained by site directed mutagenesis [22]. The (+)strand structure is comprised of a terminal loop containing the ICR2-like sequences followed by stem (*a*), a second loop region (*b*) and a second stem (*c*) extending to the 5' terminal viral base. The location of ICR1 and 2 motifs are designated by large boxes. Bases targeted for mutagenesis [22] are boxed (small) on the wt structures, and the sequence changes are also boxed in individual mutant structures and are indicated by arrows. The replication levels of each RNA in transfected barley protoplasts are expressed as a percentage relative to the progeny level (100%) of the wt construct. The predicted structural stabilities (ΔG) of each stem-loop structure are indicated to the right. The name of each mutant is indicated above each pair of structures

(+)strand structure, since much of the complementary (−)strand stem region was reordered by this mutant [22]. Further evidence supporting the participation of a (+)strand stem loop in RNA replication was obtained by testing two mutations (ICR1-A1 and -A2) at the A_4 position of RNA-2 5' sequence (Fig. 3); both result in similar disruptions in the terminal stem of the (−)strand stem-loop structure. ICR1-A1, which has a C substituted for this A, created a bulge in stem (c) of the (+)strand structure and replicated at a level of only 15% of wt [22]. In contrast, pRNA ICR1-A2 (Fig. 3), which has a G substituted at this position that retains the (+)strand structure by G−U base pairing, replicated at nearly wt levels [22].

Results from these experiments [22] demonstrated that maintenance of the stem-loop structure present in the 5′ terminus of the (+)strand RNA is important for virus RNA replication. These data closely resemble the requirement in poliovirus for a 90 nt cloverleaf structure present near the 5′ terminus of (+)strand RNA for viral infectivity [2]. Mutations disrupting this (+)strand structure selectively reduced the accumulation of (+)strand poliovirus RNA in vivo. Following incubation of the poliovirus cloverleaf RNA with extracts from infected cells, two specific ribonucleoprotein complexes were formed comprised of both host and viral proteins [2]. In addition, investigations into Sindbis virus replication have revealed that a stem-loop structure present at the 5′ terminus of the (+)strand or its complement in the 3′ (−)strand terminus are required for virus replication [19]. Deletion of bases within either of these structures differentially reduce virus accumulation, depending on the host cell (mosquito or chicken) used in the assay [19]. Such host-dependent results suggest that host-specific factors interact with one or both of these structures.

Initiation of (+)strand RNA synthesis

Following synthesis of the genomic length (−)strand RNA, the 5′ terminal (+)strand sequences, requisite for BMV replication, are postulated to be recognized by a soluble protein factor (Figs. 2C and D), which induces the formation or stabilization of the 5′ stem loop structure and a ss (−)strand RNA terminus. The well-conserved ICR2-like motif for BMV replication, predicted to comprise the terminal loop of the 5′ (+)stem structure (Fig. 3), is required in a sequence-specific manner, suggesting that this motif may provide a primary binding site for a host or virus-encoded protein [5, 22].

Binding of this protein factor to the ICR2-like motif (Fig. 2D) is postulated to elicit folding of the 5′ terminal stem loop and thus the formation of the *plus strand initiation structure* (PSIS; Figs. 2D and E). The formation of the PSIS will release a single-stranded 3′ (−)terminus for replicase recognition. This model predicts that the 5′ 90 nt cloverleaf structure of poliovirus RNA also functions as a PSIS. Interestingly, the stem-loop "b" (GGUUGUACCC) of the poliovirus RNA cloverleaf structure shares several identical bases with the BMV ICR2 motif and adjacent bases (GGUUCAATCCC). Among these shared bases, a stretch of three cytosine residues, which is also conserved among cloverleaf structures derived for other picornaviruses, is necessary to maintain binding of a host-encoded protein and virus infectivity in cell culture [3]. The specific and high-affinity binding of host factors to rubella [18] and Sindbis virus [20] 3′ (−)strand terminal RNA sequences may reflect the binding of components of the (+)strand replicase complex to the 3′ (−)strand, no longer bound in duplex form. Following replicase recognition, genomic (+)strand synthesis begins, using the (−)sense RNA as its template (Figs. 2E and F). Once synthesis begins, the absence of the replicase complex may elicit the dissociation of the soluble protein factor from the (+)strand stem loop structure and binding to the ICR2-like sequence of the newly-synthesized (+)RNA to form a new PSIS. These events would lead to multiple rounds of (+)strand RNA synthesis from the same (−)strand RNA template (Figs. 2F and G). The newly synthesized (+)strand RNAs will now function in other steps in the replication cycle, such as translation, encapsidation or new (−)strand synthesis (Fig. 2G).

Relevance of the model to (+) and (−)strand replication asymmetry

Transfection of barley protoplasts with brome mosaic virus (BMV) RNAs-1 and -2, in the absence of RNA-3, yields a molar ratio for (+):(−)strand progeny near unity whereas, in the presence of RNA-3, >100 progeny (+)strands are synthesized for every (−)strand progeny RNA [15, 21, 24]. The presence of RNA-3 enhances total (+)strand RNA production over 200-fold and that of RNAs-1 and -2 by nearly 30-fold [15]. Poliovirus replication proceeds in a similarly asymmetric manner, yielding a 20–30:1 (+):(−)strand ratio [2]. This property of asymmetric replication is shared, to varying degrees, by all (+)stranded RNA viruses studied. Consequently a plausible mechanism for asymmetric replication must be included in any model of (+)strand replication. In the replication of BMV RNA, experiments [15] have shown that RNA-3 has the controlling function in the process of strand asymmetry. Inoculations containing mutant RNA-3 constructs incapable of either RNA-4 or coat protein expression still exhibited >50:1 (+):(−)strand progeny ratio. In contrast, transfections containing a RNA-3, mutant lacking a portion of sequences between its two reading frames (intercistronic region, Fig. 1), ΔSGP RNA-3, showed only a <2:1 molar excess of (+)strand progeny [15]. The interference with functions of the subgenomic promoter and upstream intercistronic sequences caused by the ΔSGP deletion abolished asymmetry. This strongly suggests that these regions are responsible for the fundamental switch to asymmetric replication. Further, the presence of these sequences *in trans* appears to redirect (−)strand synthetic activities to the production of (+)strand RNA.

Sequences affected by the ΔSGP mutation include the well-characterized ([17]; Fig. 1C) promoter for subgenomic (+)RNA-4 as well as part of the intercistronic region adjacent to $ICR2_{1100}$ [15]. This ICR2-like sequence, present at nt 1100 (Fig. 1C), is required for genomic (+)strand RNA-3 amplification, and deletion of this motif also dramatically affects the asymmetry of BMV replication [23]. The requirement of an ICR2 motif for BMV RNA-3 replication is consistent with the predictions of our replication model. Virus interference studies further demonstrates the *trans*-acting effect these sequences exert on BMV replication. Negative-sense transcripts corresponding to the intercistronic region inhibited BMV RNA replication by 90% when co-inoculated at a molar ratio of 3:1 into barley protoplasts [10]. Deletion of the $ICR2_{1100}$ motif from these transcripts alleviated 70% of this inhibition [10] suggesting that viral or host factors bind this sequence, or those comprising the subgenomic promoter, and thereby profoundly influence replicase synthetic activities.

Results from recent experiments [23] strongly suggest that the formation of the (−)strand specific replication complex throughout the BMV infection cycle is dependent on the continual translation of RNAs-1 and -2. Any mutation greatly debilitating the replication of RNA-1 leads to a non-productive BMV infection in barley protoplasts [24]. Co-inoculation of wt RNA-2 or RNA-2 mutants incapable of replication (1/ΔICR2, 2/pΔICR 1&2, and 2/G_1G_2-A_1T_2) with wt RNA-1 yields approximately equal levels of (+)strand RNA-1 accumulation (Fig. 4). In contrast to the 1.2:1 (+):(−)strand ratio of wt RNAs-1 and -2, RNA-1 (−)strands accumulate to very low levels, 10-fold less than the complementary (+)strand progeny, when co-inoculated with the non-replicating RNA-2 mutants ([23]; Fig. 4). Under these conditions, translation of p2a, the putative RNA-dependent RNA-polymerase, from these non-replicating RNAs can only occur very early in the infection, since >90% of the inoculum RNA-2 is degraded by 6 h post infection [23]. The p2a protein is known to directly interact with p1a (Fig. 1), a virus helicase-like protein [9, 12], and both proteins co-purify with in vitro replicase activity [17]. Thus the formation of new replicase complexes at later time points during infection would be limited by the reduced levels of p2a. This preferential synthesis of (+)strand RNA-1, when co-inoculated with non-replicating RNA-2 mutants, suggests that sustained translation of p2a is necessary for the continuation of (−)strand RNA synthesis, and that (−)strand replicase complexes are converted irreversibly to a (+)strand specific replicase.

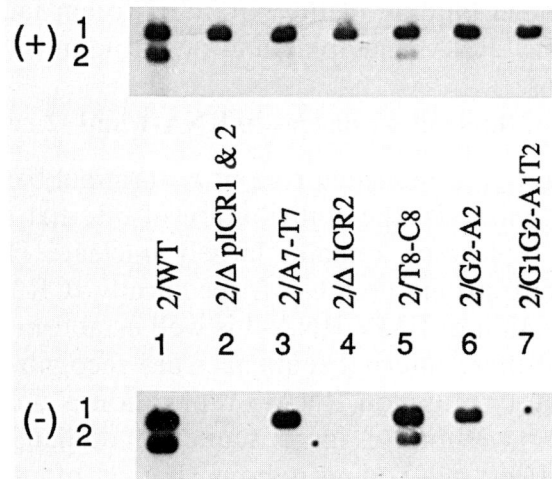

Fig. 4. RNA blot analysis showing replication characteristics in transfected barley protoplasts of selected RNA-2 mutants bearing mutations in the ICR-like sequences. Inocula contained transcripts of genomic RNAs-1 and wt (*1*) or the specified mutant RNA-2. RNA-2 mutants 2/ΔpICR 1&2 (*2*; partial deletion of ICRs 1 and 2), 2/ΔICR2 (*4*; deletion of entire ICR2 motif) as well as the double-substitution mutant, 2/G$_1$G$_2$-A$_1$T$_2$ (*7*), produced no detectable replicative progeny [23]. Mutants 2/A$_7$-T$_7$ (*3*), 2/T$_8$-C$_8$ (*5*) and 2/G$_2$-A$_2$ (*6*) replicated, respectively, at levels of <10%, 15% and 40% of wt RNA-2 [23]. Positions of RNAs-1 and -2 are indicated at the left of each panel. Panels showing (+)strand and (−)strand progeny were exposed for equal lengths of time. Longer exposures of Northern blots allow visualization of RNA-1 (−)strand progeny in *2, 4,* and *7*

A similar dependence of (−)strand synthesis on new protein synthesis has been observed in alphavirus replication. Addition of protein synthesis inhibitors halts the synthesis of alphavirus (−)strand RNA [25], while production of (+)strand RNA is relatively insensitive to translational inhibition. Coincident with the cessation of (−)strand RNA synthesis early (by 3–4 h p.i.) in the alphavirus replication cycle [25] there is a decrease in the level of free nsP4, which encodes the putative virus-encoded polymerase, and the appearance of nsP34 polyprotein [8]. This suggests that temporal regulation of polyprotein cleavage is involved in the transition from (−) to (+)strand synthesis [8]. In addition, studies of tobacco mosaic virus (TMV) have suggested that (−)strand synthesis may be temporally regulated and cease after the initial 6–8 h of infection [11]. These studies support our postulation that newly translated non-structural viral proteins are part of a replicase initially dedicated to the production of (−)strand RNA, and that the replicase complex is later modified to function only in (+)strand RNA synthesis. The nature of this modification event is unclear in BMV replication, although binding of the putative protein factor to the ICR2 motifs within BMV RNAs may play a critical role in this process.

The role of internal sequences in RNA-1 and -2 replication

In contrast to the well established role of 5′ terminal bases in promoting BMV (+)strand synthesis, the contribution of internal sequences to the process of virus replication remains largely unclear. The indispensable nature of the intercistronic region [6], specifically ICR2$_{1100}$ [23], in promoting the replication of BMV RNA-3 is well established. However, the manner by which these internal sequences are recognized and exert the observed stimulatory effect on RNA-3 replication is unknown. In BMV RNAs-1 and -2, deletion of large sequence domains in the central portion of the p1a or p2a coding regions (Fig. 1) often prohibited both (+) and (−)strand synthesis [16, 17]. The crucial question is whether these non-replicating deletion mutants lack essential *cis*-acting sequences or if the removal of these particular regions resulted in nonfunctional or inhibitory RNA folding patterns. Alternatively, translation of a truncated protein with suppression properties from these RNAs could explain their null replication phenotype. However, the considerable increase in RNA stability observed for certain non-replicating mutants [16] argues for the presence of substantial structural alterations. The effect of overall RNA structural characteristics on eukaryotic RNA virus replication is largely unknown. However, studies of Qβ, a (+)stranded RNA coliphage, suggest that the nature of RNA folding patterns are crucial to several steps of replication [4]. Studies to address the role of specific and higher order RNA folding patterns within viral RNAs will be essential before a full understanding replication can be reached.

Similarities between BMV, Sindbis and poliovirus replication suggest that this model may be widely applicable

Although BMV differs in many ways from Sindbis and poliovirus, certain steps in the replication cycle of each virus may be shared. The requirement of 5′ (+)sense structures for (+)strand replication and similarities with respect to asymmetric RNA replication suggest that what has been learned through these studies will benefit those investigating other virus systems. The characterization of sequence elements involved in promoting BMV RNA replication should also augment our understanding of host-viral interactions. The manner in which RNA viruses redirect host factors to express virally-encoded genes and replicate virus RNA should provide new insight into the nature of host enzyme activities and reveal new approaches to inhibit virus infections [16]. Recent advances in identifying host factors interacting with virus sequences from rubella virus and poliovirus are detailed elsewhere in this volume. We hope discussion of this model stimulates further investigations into the mechanisms of virus replication.

Acknowledgements

These studies were supported by NSF Grant DMB-8921023 and fellowships from the Rockefeller Foundation (to C.C.H.), and W.R. Grace & Co. and the ARCS Foundation, Inc. (to G.P.P.).

References

1. Ahlquist P, Bujarski JJ, Kaesberg P, Hall TC (1984) Localization of the replicase recognition site within brome mosaic virus RNA by hybrid-arrested RNA synthesis. Plant Mol Biol 3: 37–44
2. Andino R, Rieckhof GE, Baltimore D (1990) A functional ribonucleoprotein complex forms around the 5′ end of poliovirus RNA. Cell 63: 369–380
3. Andino R personal communication
4. Axelrod VD, Brown E, Priano C, Mills DR (1991) Coliphage Qβ RNA replication: RNA catalytic for single-stranded release. Virology 184: 595–608
5. Blumenthal T, Carmichael GG (1979) RNA replication: function and structure of the Qβ replicase. Annu Rev Biochem 48: 525–548
6. French R, Ahlquist P (1987) Intercistronic as well as terminal sequences are required for efficient amplification of brome mosaic virus RNA3. J Virol 61: 1457–1465
7. Goldbach R (1987) Genomic similarities between plant and animal RNA viruses. Microbiol Sci 4: 197–202
8. deGroot RJ, Hardy WR, Shirako Y, Strauss JH (1990) Cleavage-site preferences of Sindbis virus polyproteins containing the non-structural proteinase. Evidence for temporal regulation of polyprotein processing in vitro. EMBO J 9: 2631–2638
9. Hodgman TC (1988) A new superfamily of replicative proteins. Nature 333: 22–23
10. Huntley CC, Hall TC (1993) Minus sense transcripts of brome mosaic virus RNA-3 intercistronic region interfere with viral replication. Virology 192: 290–297
11. Ishikawa M, Meshi T, Ohno T, Okada Y (1991) Specific cessation of minus-strand RNA accumulation at an early stage of tobacco mosaic virus infection. J Virol 65: 861–868
12. Kao CC, Ahlquist P (1992) Identification of the domains required for direct interaction of the helicase-like and polymerase-like RNA replication proteins of brome mosaic virus. J Virol 66: 7293–7302
13. Levis R, Schlesinger S, Huang HV (1990) Promoter for Sindbis virus RNA-dependent subgenomic RNA transcription. J Virol 64: 1726–1733
14. Marsh LE, Hall TC (1987) Evidence implicating a tRNA heritage for the promoters of positive-strand RNA synthesis in brome mosaic virus and related viruses. In: The evolution of catalytic function. Cold Spring Harbor Symposium on Quantitative Biology, vol 52. Cold Spring Harbor Laboratory Press, Cold Spring Harbor, pp 331–341
15. Marsh LE, Huntley CC, Pogue GP, Connell JP, Hall TC (1991) Regulation of (+):(−) strand asymmetry in replication of brome mosaic virus RNA. Virology 182: 76–83

16. Marsh LE, Pogue GP, Szybiak U, Connell JP, Hall TC (1991) Non-replicating deletion mutants of brome mosaic virus RNA-2 interfere with viral replication. J Gen Virol 72: 2367–2374
17. Marsh LE, Pogue GP, Huntley CC, Hall TC (1991) Insight to replication strategies and evolution of (+) strand RNA viruses provided by brome mosaic virus. In: Miflin BJ (ed) Oxford surveys of plant molecular and cell biology, vol 7. Oxford, University Press, New York, pp 297–334
18. Nakhasi HL, Cao X-Q, Rouault TA, Liu T-Y (1991) Specific binding of host cell proteins to the 3′-terminal stem-loop structure of rubella virus negative-strand RNA. J Virol 65: 5961–5967
19. Nesters HGM, Strauss JH (1990) Defined mutations in the 5′ nontranslated sequence of Sindbis virus RNA. J Virol 64: 4162–4168
20. Pardigon N, Strauss JH (1992) Cellular proteins bind to the 3′ end of Sindbis virus minus-strand RNA. J Virol 66: 1007–1015
21. Pogue GP, Marsh LE, Hall TC (1990) Point mutations in the ICR2 motif of brome mosaic virus RNAs debilitate (+)-strand replication. Virology 178: 152–160
22. Pogue GP, Hall TC (1992) The requirement for a 5′ stem-loop structure in brome mosaic virus replication supports a new model for viral positive-strand RNA initiation. J Virol 66: 674–684
23. Pogue GP, Marsh LE, Connell JP, Hall TC (1992) Requirement for ICR-like sequences in the replication of brome mosaic virus genomic RNA. Virology 188: 742–753
24. Rao ALN, Huntley CC, Marsh LE, Hall TC (1990) Analysis of RNA stability and (−) strand content in viral infections using biotinylated probes. J Virol Methods 30: 239–250
25. Sawicki DL, Sawicki SG (1980) Short-lived minus-strand polymerase for Semliki Forest virus. J Virol 34: 108–118

Authors' address: Dr. T. C. Hall, Institute of Developmental and Molecular Biology, Texas A&M University, College Station, TX 77843-3155, U.S.A.

Arch Virol (1994) [Suppl] 9: 195–204

Archives
of
Virology

© Springer-Verlag 1994
Printed in Austria

Preferential replication of defective turnip yellow mosaic virus RNAs that express the 150-kDa protein *in cis*

T. W. Dreher[1,2] and **J. J. Weiland**[1,*]

[1] Department of Agricultural Chemistry, Program in Genetics and [2] Center for Gene Research and Biotechnology, Oregon State University, Corvallis, Oregon, U.S.A.

Summary. The turnip yellow mosaic virus genome encodes two proteins (the 150-kDa and 70-kDa proteins) that are proteolytically released from a single precursor and which are essential for RNA replication. Genomes with mutations in either of these coding regions were defective for independent replication in turnip protoplasts. The replication *in trans* of genomes with mutations in each region was studied by coinoculation with either a helper genome that carries a deletion in the coat protein gene, or with a second defective RNA that carries a mutation in the region encoding the other essential protein. Inefficient *trans*-replication of the defective RNAs was observed in most cases. In contrast, a defective RNA with a large deletion in the 70-kDa protein coding region could be replicated efficiently *in trans*, demonstrating that the *cis*-preference of replication can be overcome in some cases. Defective RNAs encoding wild type 150-kDa protein and defective 70-kDa protein were more efficiently replicated *in trans* than those encoding defective 150-kDa protein and wild type 70-kDa protein. The results suggest a model in which the 150-kDa and 70-kDa proteins form a relatively stable complex *in cis* on the viral RNA template.

Introduction

Turnip yellow mosaic virus (TYMV) has a 6.3 kb genome that encodes three major open reading frames (ORF) [1] (Fig. 1a). Of these, only ORF-206, which encodes a 206-kDa protein that is proteolytically cleaved to produce an N-terminal 150-kDa protein and a C-terminal 70-kDa protein [2], is essential for RNA replication in protoplasts [3, 4]; Weiland

* Present address: Barley Unit, Northern Crop Sciences Laboratory, USDA-ARS, Fargo, North Dakota, U.S.A.

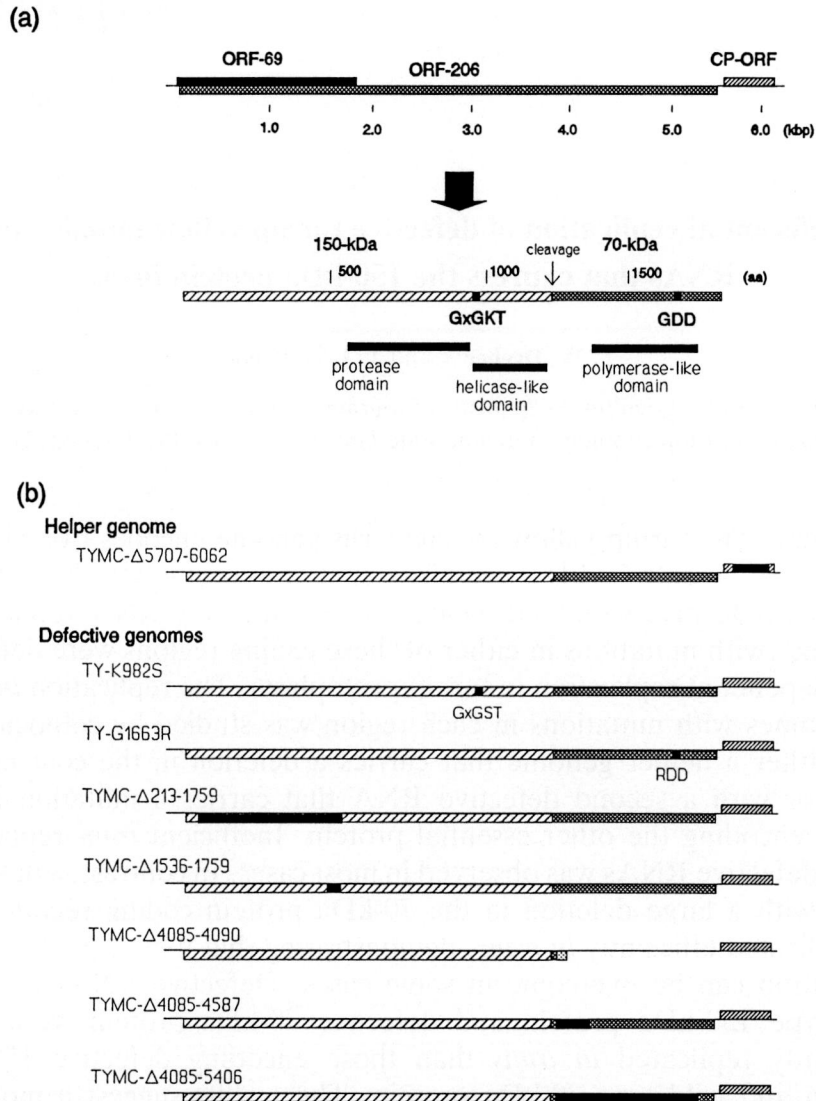

Fig. 1. Diagram of the TYMV genome and various modified genomes studied. **a** The three major ORFs encoded by TYMV RNA are indicated (*CP* coat protein), as is the proteolytic maturation of the ORF-206 product. The conserved sequence elements of the 150-kDa and 70-kDa proteins are shown. **b** Diagram showing the mutations of the TYMV variants studied; all were produced by in vitro transcription after mutagenesis of the full-length cDNA clone pTYMC [3]. ORF-69 has been omitted from the diagrams. Mutant nomenclature reflects the genome nucleotides that are joined in the case of deletion variants, and the ORF-206 codon changes for substitution mutants (TY-K982S and TY-G1331R). Black bars represent deleted sequences and stippled bars are out-of-frame sequences prior to reading frame termination

and Dreher, unpubl. res.). The 150-kDa and 70-kDa proteins possess sequence motifs that are consistent with their involvement in viral RNA replication. A putative nucleotide binding element *Gly-Cys-Gly-Lys-Thr* and other features common to NTP-dependent helicases are present near the C-terminus of the 150-kDa protein (Fig. 1) [5], while the motif *Gly-Asp-Asp* and associated elements that are characteristic of viral RNA-dependent RNA polymerases [5, 6] are present in the 70-kDa protein.

During our studies on the replication of TYMV RNA, we observed that a number of genomes carrying deletions did not replicate efficiently in the presence of helper viral RNAs (Tsai and Dreher, unpubl. res.). Although poor *trans*-replication has been reported for some polioviral defective genomes (reviewed in [7]; see also chapter by B. Flanegan in this book), the literature generally indicates that positive strand RNA viruses are able to support replication *in trans* efficiently. We investigated in more detail the apparently poor replication of defective TYMV RNAs *in trans*, studying the replication in turnip protoplasts of various combinations of the TYMV variants illustrated in Fig. 1b. The substitution, frame-shift and in-frame deletion mutations shown were limited to one of the two essential proteins, such that each defective RNA encoded one functional essential protein. Translation of defective RNAs in reticulocyte lysates yielded the expected products. All the RNAs with mutations in ORF-206 were shown to be defective for independent replication in turnip protoplasts.

The *trans*-replication of defective RNAs was studied by coinoculation with either a helper genome (TYMC-Δ5707-6062 RNA, which lacks much of the coat protein ORF; Fig. 1b) or with another potentially complementing defective RNA. These studies, reported in more detail elsewhere [8], confirmed the existence of a mechanism that results in the *cis*-preferential replication of TYMV RNA.

Replication *in trans* supported by helper genome TYMC-Δ5707-6062 RNA

TYMC-Δ5707-6062 RNA was a convenient helper genome that permitted the detection of low levels of coreplicating RNAs, by using probes corresponding to the sequences deleted from the coat protein ORF to specifically detect defective RNAs in Northern blots. Protoplasts infected with TYMC-Δ5707-6062 accumulated (−)-sense genomic RNA to levels similar to those in TYMC infections, but (+)-sense genomic RNA accumulated to levels 10% those of wild type TYMC infections 30 h post inoculation (Table 1). We assume that the decreased accumulation of (+)-sense genomic RNA is due to increased degradation of the non-encapsidated replication product (cf. [9, 10]).

Table 1. *Trans*-replication of defective RNAs in the presence of helper RNA or complementing defective RNA

	Mock	Defective RNA encoding mut 150K and wt 70K			Defective RNA encoding wt 150K and mut 70K			
		Δ213-1759	Δ1536-1759	K982S	Δ4085-4090	Δ4085-4587	Δ4085-5406	G1663R
Mock	0	0	0	0	0	0	0	0
Helper genome Δ4085-6062	10[a]		<0.1[b]	0.4[b]			9[b]	1.9[b]
Δ4085-5406	0				4[c]	21[c]	49[c]	1[c]
TY-K982S	0	<0.1[c]	3[c]	49[c]				

The values report the accumulations (%) relative to wild type TYMC infections of (+)-sense genomic RNAs in turnip protoplasts 30 h after inoculation (derived from Northern blots; means of at least 3 experiments)

[a] Accumulation of helper RNA
[b] Accumulation of defective RNA
[c] Total accumulation of both defective RNAs

When mutant RNAs TYMC-Δ1536-1759, TY-K982S TY-G1663R, and TYMC-Δ4085-5406 were co-inoculated with helper genome TYMC-Δ5707-6062 RNA, *trans*-replication of the defective RNAs was detected in Northern blots (Table 1). Both genomic and subgenomic RNAs corresponding to the defective inocula were present, verifying that replication of the defective genomes was occurring. The level of *trans*-replication varied widely between the different defective genomes, with RNAs encoding defective 150-kDa protein accumulating to lower levels than did RNAs encoding defective 70-kDa protein. RNAs TYMC-Δ-1536-1759, TY-K982S and TY-G1663R were replicated inefficiently *in trans*, but RNA TYMC-Δ4085-5406 accumulated to levels comparable to the helper RNA. By stripping and reprobing the Northern blots with a probe complementary to the 3′ 259 nucleotides that detects both helper and defective RNAs, it was shown that the varying levels of defective RNA accumulation were not the result of varying helper RNA levels. There was no indication that any of the mutant genomes tested were defective interfering RNAs. These results suggest that the essential proteins encoded by TYMV RNA are usually not freely available for use in replicating another RNA *in trans*, although certain RNAs (such as TYMC-Δ4085-5406 RNA) can be efficiently replicated *in trans*.

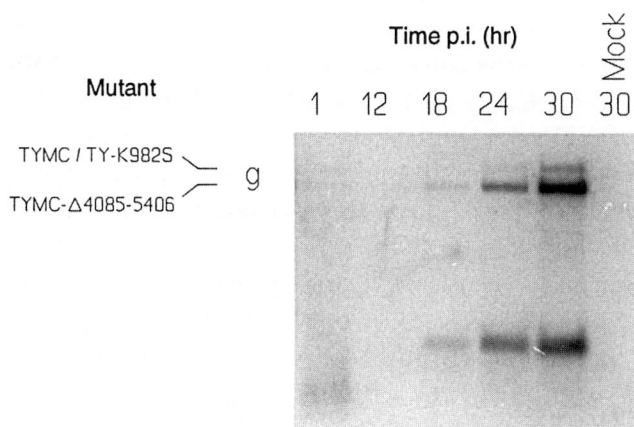

Fig. 2. Mutual complementation between defective RNAs TY-K982S and TYMC-Δ-4085-5406. A Northern blot detecting (+)-sense genomic (*g*) and subgenomic (sg) RNAs in turnip protoplasts at the indicated times after inoculation of 1.4 × 10⁶ protoplasts with 5 μg TY-K982S RNA + 4 μg TYMC-Δ4085-5406 RNA. The migration positions of the genomic RNAs are indicated at left

Complementation between RNAs encoding mutant 150-kDa and 70-kDa proteins

Because the above experiments demonstrated that helper genome TYMC-Δ5707-6062 rescued the 150-kDa mutant RNA TY-K982S and the 70-kDa mutant RNA TYMC-Δ4085-5406 most efficiently within their respective classes, these RNAs were used to test whether RNA accumulation could occur by mutual complementation between two defective RNAs. As shown in the Northern blot of Fig. 2, RNAs corresponding to both defective genomes and the single expected subgenomic species accumulated during the 30 h incubation in protoplasts. TYMC-Δ4085-5406 RNA accumulated to levels 10.5-fold higher than TY-K982S RNA. A similar biased accumulation was observed for genomic (−)-sense RNAs (not shown), suggesting that the skewed accumulations resulted from differential replication rates rather than differential stabilities of progeny RNAs. The apparent absence of rearranged RNAs suggested that significant recombination had not occurred and that replication was the result of mutual complementation between the defective RNAs.

The complementing abilities of further combinations of defective RNAs were tested, and the total accumulations of genomic RNAs relative to wild type infections were quantified (Table 1). The combination represented in Fig. 2 (TY-K982S + TYMC-Δ4085-5406 RNAs) replicated very efficiently, but most of the other combinations studied supported RNA accumulations less than 5% of wild type. As noted in the helper coinoculation experiments and in Fig. 2, the RNA encoding a wild type 150-kDa protein accumulated to higher levels than the coinoculated RNA encoding a wild type 70-kDa protein. This was verified for TY-K982S + TY-G1663R RNA infections by sequencing PCR-amplified cDNA (not shown). This experiment also indicated the retention of the inoculum sequences, confirming the absence of recombination. The existence of recombination for other inoculum combinations was excluded by inoculating the progeny from lysed protoplasts onto turnip plants and observing them for any resultant systemic infections. This is a potent assay for wild type TYMV RNA. For most inoculum combinations no systemic symptoms were observed, but for TY-K982S + TYMC-Δ4085-5406 RNAs we observed a systemic infection characterized by distinct symptoms and by the presence of both input RNA species. Thus, these RNAs are able to constitute a novel bipartite form of the TYMV genome. Because no evidence for recombination was obtained during our studies, we conclude that the observed RNA accumulations resulted from *trans*-replication.

Discussion

Our studies have shown that the TYMV system inherently favors the replication of RNAs encoding functional forms of the two proteins essential for replication, the 150-kDa and 70-kDa proteins; i.e. replication occurs in a *cis*-preferential manner. The general observation that *trans*-replication is inefficient for a variety of RNAs tested in different inoculation combinations makes it highly unlikely that the apparent *cis*-preferential replication can be explained by the loss of *cis*-acting RNA elements (e.g., promoter sequences), to an instability or high turnover of the mutant RNA, or to a gross alteration of ribosomal trafficking that is somehow detrimental to replication [11]. Because the most poorly coreplicating genomes were those encoding a mutant 150-kDa protein – both when coinoculated with helper RNA or with a defective RNA encoding a wild type 150-kDa protein and a mutant 70-kDa protein – the 150-kDa protein appears to be crucially involved in establishing such *cis*-preference. However, most RNAs encoding defective 70-kDa proteins (and wild type 150-kDa proteins) are also poorly replicated *in trans* (e.g., TY-G1663R and TYMC-Δ4085-4090 RNAs), indicating that the 70-kDa protein encoded by a given TYMV RNA variant can also influence the ability of that RNA to be replicated *in trans*.

TYMC-Δ4085-5406 RNA is remarkable for the fact that it is able to be replicated *in trans* with an efficiency that far exceeds that of most other mutants studied. The critical feature of TYMC-Δ4085-5406 RNA is the absence of almost all of the 70-kDa protein coding region, suggesting that the lack of an encoded 70-kDa protein is the key property that permits TYMC-Δ4085-5406 RNA to be replicated *in trans*. Figure 3 outlines a model in which the *cis*-preferential replication of TYMV RNAs is due to the interaction of newly synthesized 150-kDa and 70-kDa proteins preferentially with the RNA genome from which they have been made. This results in the channelled formation of a replication initiation complex *in cis* and the guided transition in the role of an RNA from messenger to replication template. We propose that a 150-kDa/70-kDa complex assembles after the proteolytic cleavage to form these two proteins has occurred, and that this complex interacts with the viral RNA (*in cis*) more strongly than either protein alone. As suggested by the skewed ratio of complementing defective RNAs in favor of the RNA encoding a wild type 150-kDa protein, the 150-kDa protein may itself possess some *cis*-template interaction, and it presumably makes the primary contact with the RNA template (Fig. 3). The dissociation rate of the 150-kDa/70-kDa/RNA template initiation complex is probably relatively slow, limiting the ability to exchange defective subunits with functional subunits supplied by other RNAs *in trans*.

In the context of this model, the poor complementation between defective RNAs such as the substitution mutants TY-K982S + TY-G1663R RNAs (Fig. 3) results from *cis*-inhibition due to the formation of relatively stable, enzymatically inactive 150-kDa/70-kDa complexes. The protein-protein interaction domains implicit in this model presumably are not interrupted by the amino acid substitutions of these mutants. Other defective RNAs with small in-frame deletions (TYMC-Δ1536-1759 and TYMC-Δ4085-4587 RNAs) probably suffered from similar *cis*-inhibition, although it is possible that the intermediate level of *trans*-replication observed for TYMC-Δ4085-4587 RNA resulted from a partially decreased ability of the mutant 70-kDa protein to assemble into stable initiation complexes. The almost complete absence of the 70-kDa protein coding region in TYMC-Δ4085-5406 RNA entirely avoids the formation of the postulated stable complex *in cis*, permitting the involvement of active 70-kDa protein supplied *in trans* in the replication of this defective RNA (Fig. 3).

The presented model appears to be the most plausible mechanism explaining the experimental observations. Nevertheless, the model does not explain the observed accumulations of all the RNAs tested, and it will be necessary to determine the extent to which other factors influence the levels of the various RNA progeny; e.g. the possible loss of promoter elements from the large deletion present in TYMC-Δ213-1759 RNA, or the possibly increased turnover of TYMC-Δ4085-4090 RNA resulting from the presence of a large untranslated region [12].

The formation of the proposed initiation complex on the RNA template *in cis* represents the channelled delivery of the two viral proteins

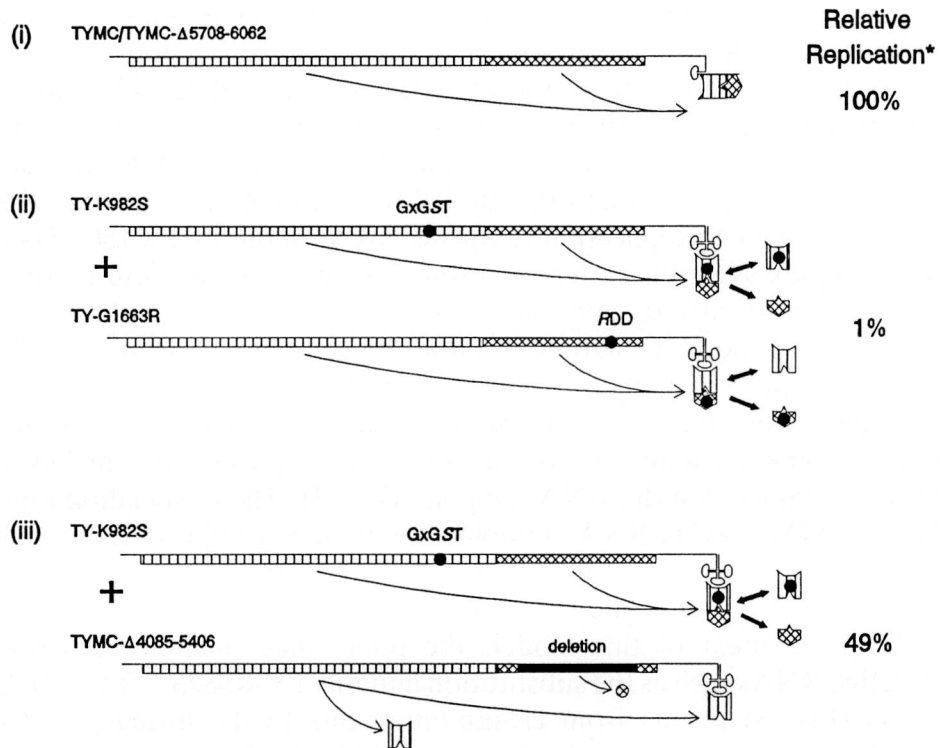

Fig. 3. Model of the proposed molecular interactions between the 150-kDa protein, 70-kDa protein and the RNA template that result in *cis*-preferential replication. The expected fates of the two essential proteins is shown for helper RNA (*i*) and for two combinations of complementable defective RNAs (*ii* and *iii*). The proteins are presumed to interact with the 3'-terminal tRNA-like structure (indicated) that is thought to comprise at least part of the (−) strand promoter; host proteins may participate in replication but are not shown

essential for replication to their site of action on the template. In preventing the free diffusion and resultant dilution of newly made proteins, this channelling is expected to be particularly advantageous early in an infection, when viral RNA and protein concentrations are very low. Other advantages of the proposed form of *cis*-preferential replication are evident: (1) spontaneously derived mutant RNAs that encode defective essential proteins are removed from the replicating pool, and (2) the chance that non-viral RNAs present in the cytoplasm may be used as templates is diminished. The latter point may be especially significant for TYMV, because cytoplasmic transfer RNAs are conformationally very similar to the tRNA-like structure present at the 3' end of the viral RNA [13]. These advantages of *cis*-preferential replication are likely to be generally applicable to positive strand RNA viruses and it is possible that some degree of *cis*-preferential replication will be found to exist for many RNA viruses. Indeed, it was interesting to hear at this meeting indications that mechanisms which *cis*-limit RNA replication may exist in viruses not previously associated in the literature with such a phenomenon: cowpea mosaic virus, Sindbis virus, and nodaviruses (see chapters by J. Wellink, D. Sawicki, A. Ball, respectively, in this book).

Acknowledgement

This work was supported by CSRS-NRICGP of the U.S. Department of Agriculture.

References

1. Morch MD, Boyer JC, Haenni AL (1988) Overlapping reading frames revealed by complete nucleotide sequencing of turnip yellow mosaic virus genomic RNA. Nucleic Acids Res 16: 6157–6173
2. Bransom KL, Weiland JJ, Dreher TW (1991) Proteolytic maturation of the 206-kDa non-structural protein encoded by turnip yellow mosaic virus RNA. Virology 187: 351–358
3. Weiland JJ, Dreher TW (1989) Infectious TYMV RNA from cloned cDNA. Effects in vitro and in vivo of point substitutions in the initiation codons of two extensively overlapping ORFs. Nucleic Acids Res 17: 4675–4687
4. Bozarth CS, Weiland JJ, Dreher TW (1992) Expression of ORF-69 of turnip yellow mosaic virus is necessary for viral spread in plants. Virology 187: 124–130
5. Habili N, Symons RH (1989) Evolutionary relationship between luteoviruses and other RNA plant viruses based on sequence motifs in their putative RNA polymerases and nucleic acid helicases. Nucleic Acids Res 17: 9543–9555
6. Koonin EV (1991) The phylogeny of RNA-dependent RNA polymerases of positive-strand RNA viruses. J Gen Virol 72: 2197–2206
7. Sarnow P, Jacobson SJ, Najita L (1990) Poliovirus genetics. Curr Top Microbiol Immunol 161: 155–188
8. Weiland JJ, Dreher TW (1993) *Cis*-preferential replication of the turnip yellow mosaic virus RNA genome. Proc Natl Acad Sci USA 90: 6095–6099
9. Ogawa T, Watanabe Y, Meshi T, Okada Y (1991) *Trans*-complementation of virus-encoded replicase components of tobacco mosaic virus. Virology 185: 580–584
10. Kaplan G, Racaniello VR (1988) Construction and characterization of poliovirus subgenomic replicons. J Virol 62: 1687–1696
11. White KA, Bancroft JB, Mackie GA (1992) Coding capacity determines in vivo accumulation of a defective RNA of clover yellow mosaic virus. J Virol 66: 3069–3076
12. Atwater JA, Wisdom R, Verma IM (1990) Regulated mRNA stability. Annu Rev Genet 24: 519–541

13. Rietveld K, Pleij CWA, Bosch L (1983) Three-dimensional models of the tRNA-like 3′ termini of some plant viral RNAs. EMBO J 2: 1079–1085

Authors' address: Dr. T. W. Dreher, Department of Agricultural Chemistry, Oregon State University, Agricultural & Life Sciences Room 1007, Corvallis, OR 97331-7301, U.S.A.

Arch Virol (1994) [Suppl] 9: 205–209

Archives
of
Virology
© Springer-Verlag 1994
Printed in Austria

In vivo transfection by hepatitis A virus synthetic RNA

**S. U. Emerson[1], M. Lewis[1], S. Govindarajan[2], M. Shapiro[3],
T. Moskal[3], and R. H. Purcell[1]**

[1] National Institutes of Health, Bethesda, Maryland
[2] Rancho Los Amigos Medical Center, Downey, California
[3] Bioqual, Inc., Rockville, Maryland, U.S.A.

Summary. Marmosets injected intrahepatically with nucleic acids (cDNA and RNA transcripts) representing the full-length genome of the wild-type HM-175 strain of hepatitis A virus experienced acute hepatitis and seroconversion to hepatitis A virus capsid proteins. The hepatitis was comparable in severity to that caused by infection with the wild-type virus. The viral cDNA and the hepatitis A virus recovered from the feces of an injected animal contained the same marker mutation. Therefore, intermediate cell culture steps can be omitted and the virulence of a hepatitis A virus encoded by a cDNA clone can be evaluated by direct transfection of marmosets.

Introduction

Hepatitis A virus (HAV) is a major worldwide cause of enterically-transmitted viral hepatitis. Although HAV is classified as a picornavirus, it differs significantly from more classical picornaviruses and much of the biology of HAV remains a mystery. Most strains of HAV are not cytopathic and it is therefore not clear why HAV infection of the liver results in acute hepatitis nor is it even known how ingested HAV reaches this organ. Virulent strains of HAV that have been successfully propagated in cell culture grow extremely slowly and to low titers. Wild-type HAV requires weeks or months to replicate to detectable levels in cell culture, but mutants that grow more rapidly have been selected by serial passage of infected cells. In the case of the HM-175 strain of virus, one mutant, selected by its ability to grow rapidly in primary African green monkey kidney (AGMK) cells, coincidentally acquired the phenotype of attenuation of virulence for primates [1, 2]. Virus cDNA clones prepared from the liver of a marmoset infected with the parental wild-type virus and from cell cultures infected with this rapidly growing mutant were sequenced [3, 4]. There were only 23 base differences between the two genomes. Genomic length cDNA clones were assembled and inserted into pGEM expression vectors [5]. The viral RNA transcribed from these full-length clones was infectious when transfected into AGMK or FRhK-4 cells. Because only a small number of mutations must account for the phenotypic differences in growth and virulence between the two viruses, we have attempted to map the relevant mutations by constructing chimeric cDNAs and characterizing the virus encoded by each chimera. Determinants of in vitro growth were mapped by analyzing the growth phenotype of each chimeric virus after transfection of cultured cells [6, 7]. Chimeric viruses harvested from these transfected cells were then inoculated into marmosets in order to ascertain their virulence for primates. Although the cell culture-transfection approach worked well for mapping mutations important for HAV growth in cell culture, we were unable to demonstrate virulence using any virus derived by transfection of AGMK cells [7, 8]. However, we were able to demonstrate the virulence of one cDNA-encoded hepatitis A virus when we omitted the cell culture step and transfected the viral nucleic acid directly into the livers of marmosets.

In vivo transfection of liver by synthetic viral RNA

Although the cDNA clone derived from the wild-type virus in the infected liver was assumed to represent the genome of the virulent virus, it was possible that an attenuated mutant arising within the population had inadvertently been cloned. Therefore, before proceeding with mapping studies, it was critical to determine whether this cDNA clone did indeed encode a virulent virus. Previous attempts to recover a virus after transfection of AGMK cells with RNA transcribed from this "wild-type" cDNA clone consistently failed and while transfection of FRhK-4 cells was sporadically successful, the virus resulting from these transfections required over 3 months incubation in cell culture before it was detected [6]. Because replication of wild-type virus in cell culture was so inefficient, a major concern was that any spontaneous, faster growing, mutant would overgrow the culture so that the wild-type virus would quickly constitute a minor fraction of the virus population. Additionally, since the attenuated mutant was actually selected for its ability to grow rapidly in cell culture, it was possible that rapid growth in cell culture and attenuation were co-variable phenotypes and it might be difficult to derive virulent virus through a process as inefficient as transfection of cultured cells. Direct transfection of liver had been previously reported for cDNA genomes of ground squirrel hepatitis virus [9] and for native RNA genomes extracted from virions of rabbit hemorrhagic disease virus [10]. Therefore, we designed an experiment to determine whether in vivo transfection of liver could be accomplished using plus strand RNA transcribed in vitro from a cDNA copy of the HAV genome.

The cDNA clone used in these experiments was identical to that of the "wild-type" virus, except for a single mutation introduced into the 2B gene by oligonucleotide-directed mutagenesis [6]. This mutation served as a convenient marker to identify transfection-derived virus by distinguishing it from circulating wild-type strains. Approximately 10 μg of cDNA was linearized and transcribed into RNA by Sp6 polymerase in a standard transcription reaction [6]. The transcribed sample was then simply diluted with four volumes of phosphate buffered saline and frozen on dry ice. Laparotomy was performed on the marmoset and the diluted transcription mixture containing both the template DNA and transcribed RNA was injected into the liver at three different sites. Serum samples and feces were assayed periodically to monitor the health of the animal.

Four weeks postinoculation, the marmoset developed hepatitis as indicated by a rise in serum levels of liver enzymes and at five weeks, antibody to hepatitis A virus was first detected. The 2B region of genomic RNA of hepatitis A virus in the feces was amplified by polymerase chain reaction and sequenced. The marker mutation in the 2B gene was present, thus confirming that the virus originated from the injected nucleic acid. In all, 3 marmosets were injected with the mixture of template DNA and transcribed RNA and in all 3 cases transfection was successful and virus was produced. In contrast, a single marmoset inoculated with a mock transcription reaction containing untranscribed cDNA was not infected, as evidenced by serum liver enzyme levels remaining at baseline and antibody to hepatitis A virus not produced. Although insufficient numbers of animals were tested to permit a definitive conclusion, it appears that the cDNA itself was not able to initiate infection. The RNA component may be required either because it is more readily recognized by the cell machinery as a functional viral genome or possibly because it was delivered in higher copy number. Regardless of the reason, it appears that transfection of marmoset liver by synthetic HAV genomes can be readily accomplished.

Severity of hepatitis A in transfected animals

Because the transfection of marmosets was successful, it was possible to determine the virulence of the resultant virus. The weekly serum levels of three different liver enzymes (alanine aminotransferase, isocitrate dehydrogenase and gamma glutamyl transferase) were determined for two marmosets. The amplitude and duration of each enzyme elevation were equivalent to that induced by the fully virulent HM-175 strain of

HAV. Additionally, needle biopsies of the liver of transfected marmosets or marmosets infected with wild-type virus were compared under code by a pathologist (S.G.). The pathology reports indicated that the hepatitis A caused by the transfection derived virus was as severe as that caused by the field strain of virulent virus.

Discussion

These experiments demonstrate that in vivo transfection of liver with a synthetic plus-strand RNA was efficient and reproducible. This approach enabled us to demonstrate the virulence of a cDNA-encoded virus that was difficult to propagate in cell culture. We now have the option to proceed in two different ways to map the virulence genes of HAV. In the first approach, we can inject chimeric genomes directly into marmosets as in this study. This approach worked well when the encoded virus was virulent and replicated efficiently in the host. However, it is not yet clear whether this method will succeed as well with attenuated mutants, which grow less efficiently in the liver. Alternatively, because we now know the virus encoded by this cDNA clone is virulent, we can alter the in vitro transfection and cell culture parameters in order to identify a system more amenable to production of wild-type virus. This in vivo transfection protocol may also have a broader application. Because it is often easier to generate a cDNA clone of a virus than it is to establish a cell culture system for its propagation, this in vivo transfection method may prove useful in studying the biology of plus-strand viruses for which no cell culture system currently exists.

Acknowledgement

This study was supported in part by a grant from the World Health Organization Programme for Vaccine Development.

References

1. Daemer RJ, Feinstone SM, Gust ID, Purcell RH (1981) Propagation of human hepatitis A virus in African green monkey kidney cell culture: primary isolation and serial passage. Infect Immun 32: 388–393
2. Karron RA, Daemer RJ, Ticehurst JR, D'Hondt E, Popper H, Mihalik K, Phillips J, Feinstone S, Purcell RH (1988) Studies of prototype live hepatitis A virus vaccines in primate models. J Infect Dis 157: 338–345
3. Cohen JI, Ticehurst JR, Purcell RH, Buckler-White A, Baroudy BM (1987) Complete nucleotide sequence of wild-type hepatitis A virus: comparison with different strains of hepatitis A virus and other picornaviruses. J Virol 61: 50–59
4. Cohen JI, Rosenblum B, Ticehurst JR, Daemer RJ, Feinstone SM, Purcell RH (1987) Complete nucleotide sequence of an attenuated hepatitis A virus: comparison with wild-type virus. Proc Natl Acad Sci USA 84: 2497–2501

5. Cohen JI, Ticehurst JR, Feinstone SM, Rosenblum B, Purcell RH (1987) Hepatitis A virus cDNA and its RNA transcripts are infectious in cell culture. J Virol 61: 3035–3039
6. Emerson SU, Huang YK, McRill C, Lewis M, Purcell RH (1992) Mutations in both the 2B and 2C genes of hepatitis A virus are involved in adaptation to growth in cell culture. J Virol 66: 650–654
7. Cohen JI, Rosenblum B, Feinstone SM, Ticehurst J, Purcell RH (1989) Attenuation and cell culture adaptation of hepatitis A virus (HAV): a genetic analysis with HAV cDNA. J Virol 63: 5364–5370
8. Emerson SU (1993) Unpubl. data
9. Seeger C, Ganem D, Varmus H (1984) The cloned genome of ground squirrel hepatitis virus is infectious in the animal. Proc Natl Acad Sci USA 81: 5849–5852
10. Ohlinger VF, Haas B, Meyers G, Weiland F, Thiel H (1990) Identification and characterization of the virus causing rabbit hemorrhagic disease. J Virol 64: 3331–3336

Authors' address: Dr. S. U. Emerson, Hepatitis Viruses Section, Building 7, Room 203, 9000 Rockville Pike, Bethesda, MD 20892, U.S.A.

RNA recombination

Arch Virol (1994) [Suppl] 9: 213–220

Archives
of
Virology

© Springer-Verlag 1994
Printed in Austria

Recombination between Sindbis virus RNAs

S. Schlesinger and **B. G. Weiss**

Department of Molecular Microbiology, Washington University Medical School,
St. Louis, Missouri, U.S.A.

Summary. The Sindbis virus RNA genome is divided into two modules – one coding for the nonstructural protein genes and the other coding for the structural protein genes. In our studies of recombination, the two parental RNAs were defective in different modules. Analysis of the recombinant RNAs demonstrated that the parental RNAs each contributed its intact module and that the crossovers occurred within the defective modules. The recombinational events giving rise to infectious virion RNAs could create deletions, rearrangements or insertions as long as they occurred outside of the functional module. These crossovers produced RNA genomes that contained two functional subgenomic RNA promoters.

Introduction

The evolution of viruses in the natural world can be an important factor in the appearance of new human and animal diseases. Recombination between viruses is one mechanism by which viruses can evolve and is now a well recognized phenomenon among positive strand RNA viruses. Recombination between RNA viruses which contain nonsegmented genomes was first reported about 30 years ago with mutants of poliovirus [7]. Since then recombination has also been described for apthoviruses [11], coronaviruses [14] and several plant viruses including brome mosaic virus [2] cowpea chlorotic mottle virus [1] and barley stripe mosaic virus [3]. We recently reported that the alphavirus, Sindbis virus, undergoes recombination in cultured cells [22] and there is also evidence that alphaviruses can undergo recombination in nature. Based on a comparison of the sequence of Sindbis virus and that of two other alphaviruses, eastern equine encephalitis virus (EEEV) and western equine encephalitis virus (WEEV), Hahn et al. proposed that WEEV arose by recombination between a Sindbis-like virus and an EEEV-like virus [6].

Mechanisms of recombination between RNA viruses have not been investigated in detail, but template-switching is considered to be the most likely means by which genetic exchange between nonsegmented RNA genomes can occur [10]. Studies with poliovirus have provided evidence that indicates the viral polymerase switches templates during the synthesis of the negative strand [12]. In a recent review, Lai distinguished three types of RNA recombination [13]. The first type, homologous recombination, describes recombinational events in which the crossover is precise. There is, as yet, no evidence for any enzymatic mechanism to achieve this precision. In some crosses a precise crossover may be the only means of maintaining a functional gene and it would be the only way to obtain viable recombinant progeny. The second type is termed aberrant homologous recombination; although the two RNAs undergoing recombination are homologous, the crossover is not precise. This type of recombination can be expected during template switching if there are regions of the genome, either within coding sequences or in noncoding regions, which can tolerate some imprecision at the crossover point. It is the type of recombination we describe here for Sindbis virus. It has also been found with polioviruses and bromoviruses. The third type of recombination is nonhomologous or illegitimate, as the RNAs have no obvious homology. This type of recombination has been described for RNA viruses. We isolated defective interfering (DI) RNAs of Sindbis virus in which the 5′ terminus was replaced by a cellular tRNA^{Asp} [18]. Cellular sequences have been identified in other viral RNAs; a ubiquitin-coding sequence was found in a bovine diarrhea virus [17] and a sequence from 28S rRNA was inserted into the hemagglutinin gene of influenza virus [9]. Recombination also may have occurred between a coronavirus and influenza C virus: the hemagglutinin-esterase gene found in some strains of coronaviruses shares 30% sequence homology with the hemagglutinin gene of influenza C virus [16].

Results

We detected recombination between Sindbis virus RNAs in our studies to develop Sindbis virus as a vector for introducing foreign genes into cells [5, 22]. Sindbis virus is the prototype member of the *Alphavirus* genus of the *Togaviridae* family. The alphavirus genome is a nonsegmented RNA molecule containing approximately 12 kb plus a poly(A) tail [20, 21]. The 5′ two-thirds of the RNA codes for the proteins required for replication and transcription of the RNA. The 3′ one-third codes for the structural proteins – the capsid protein and the proteins that comprise the envelope of the virion. The nonstructural proteins are translated from genomic length mRNAs; the structural proteins are translated from a subgenomic RNA (26S RNA) identical in sequence to the 3' one-third of the genome. This subgenomic RNA is transcribed from the minus strand of genomic RNA by initiation from an internal promoter that spans the junction between the structural and nonstructural genes. The cDNA of the Sindbis virus genome has been cloned downstream of the bacteriophage SP6 DNA dependent RNA polymerase. RNA transcribed in vitro from the cDNA is competent to produce infectious virions when transfected into cultured cells [19]. The availability of cloned cDNAs of both Sindbis virus and the closely related Semliki Forest virus made it possible to create and analyze a wide range of mutations and deletions in their genomes. It also made feasible the use of alphavirus RNAs as vectors for introducing foreign genes into cultured cells [15, 23]. The first self-replicating vector contained the nonstructural protein genes and the cis-acting sequences of the alphavirus genome [23]. The structural protein genes were replaced by the bacterial gene coding for the bacterial enzyme chloramphenicol acetyltransferase. When the vector RNA was transfected into cells, both genomic and subgenomic RNAs were produced, but the genomic RNA was not packaged. The vector RNA was packaged when it was transfected into cells with a defective RNA that produced a subgenomic RNA coding for the viral structural proteins [5]. The defective RNA was also packaged and many particles contained both RNAs. In addition to complementation between the two genomes we sometimes observed recombination – the appearance of an infectious nonsegmented RNA. Both complementation and recombination are illustrated in Fig. 1. The pattern of viral RNAs synthesized in cells transfected with TRCAT and the defective RNA carrying the 26S sequences [DI(26S)] is seen in lane 1. These data show that some cells were transfected with both RNAs since the DI genomic RNA and its subgenomic 26S RNA are dependent for their synthesis on the presence of the nonstructural proteins which would be coded by TRCAT. Virus harvested from the transfected cells was used to infect new cells and the pattern of RNAs produced in those cells is shown in lane 2. A band migrating slower than TRCAT can be seen in this lane. This RNA was shown to be infectious after it was isolated from a gel without prior denaturation and transfected into new cells [22]. Recombinant viruses were then plaque-purified and lanes 3 and 4 represent the RNA patterns obtained from infections with two different plaque isolates.

We have carried out studies on recombination with a variety of Sindbis RNAs (Table 1). In most experiments one of the parental RNAs was the vector RNA TRCAT. The other parental RNA was either the defective-helping RNA DI(26S) or an RNA with a deletion in one of the nonstructural protein genes and an intact structural protein gene

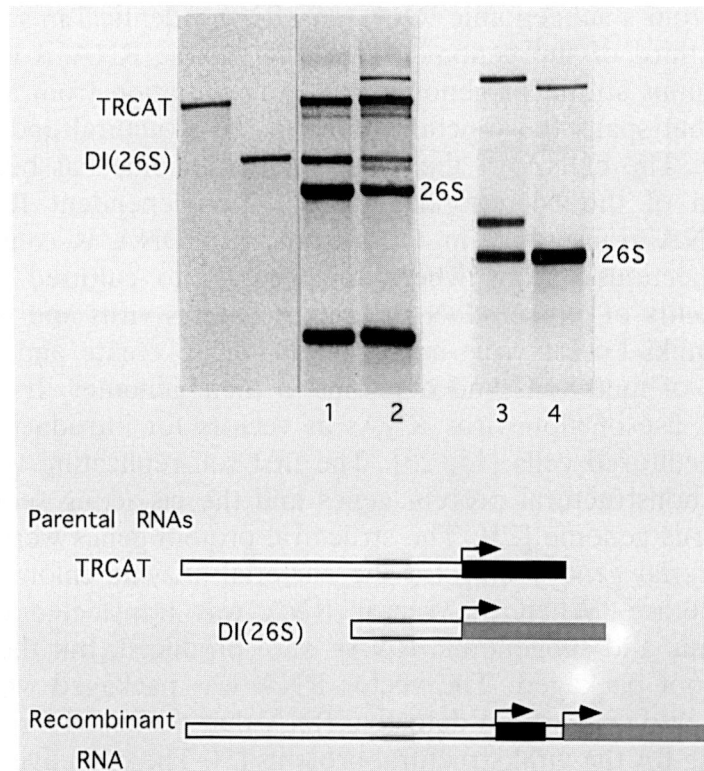

Fig. 1. Complementation and recombination between Sindbis virus RNAs. The upper panel shows the autoradiogram of the ^{3}H-uridine labeled RNA synthesized in cells transfected with Sindbis virus RNAs or infected with Sindbis virus. *1* is the pattern obtained from cells transfected with *TRCAT* and *DI(26S)* RNA; *2* is from cells infected with the virus harvested from the transfected cells; *3* and *4* each represent the pattern obtained from cells infected with a different plaque-purified recombinant virus. The first two unmarked lanes are ^{3}H-uridine labeled transcripts included as markers. The fastest migrating RNAs seen in *1* and *2* are the subgenomic RNAs derived from *TRCAT*. The diagrams below illustrate the parental and recombinant RNAs. The open boxes are either the intact four nonstructural protein genes (*TRCAT* and the recombinant RNA) or a deleted and rearranged version of them *DI(26S)*. The filled in black box represents CAT sequences and Sindbis virus sequences downstream of the start of the subgenomic RNA promoter and at the 3' terminus. The complete *26S* sequence is indicated by the gray box

module. Recombinant viruses from the different crosses listed in Table 1 were isolated by plaque purification. Many of them had two unusual properties: the genomic RNAs were larger than the wild type Sindbis genomic RNA and cells infected with a recombinant virus synthesized an additional RNA species that migrated in an agarose gel between the genomic and subgenomic RNAs. Figure 2 shows the results of a cross between TRCAT and CR3.4, an RNA with a deletion in the nsP3

Table 1. Crosses between Sindbis RNAs that gave rise to infectious recombinants with two subgenomic RNA promoters

Parental RNA with defect in structural protein genes	Parental RNA with defect in nonstructural protein (nsP) gene
TRCAT-structural genes have been replaced by the CAT gene	DI(26S) – 5′ terminal region comes from a DI RNA
TRCAT	CR1.8 – in frame deletion in nsP1 gene
TRCAT	CR3.4 – in frame deletion in nsP3 gene
$E2C_{415}C_{416}$ – two cysteines in the structural gene, PE2, were changed: one to serine; one to alanine [4]	DI(26S)
$E2C_{415}C_{416}$	CR2.4 – in frame deletion in nsP2 gene

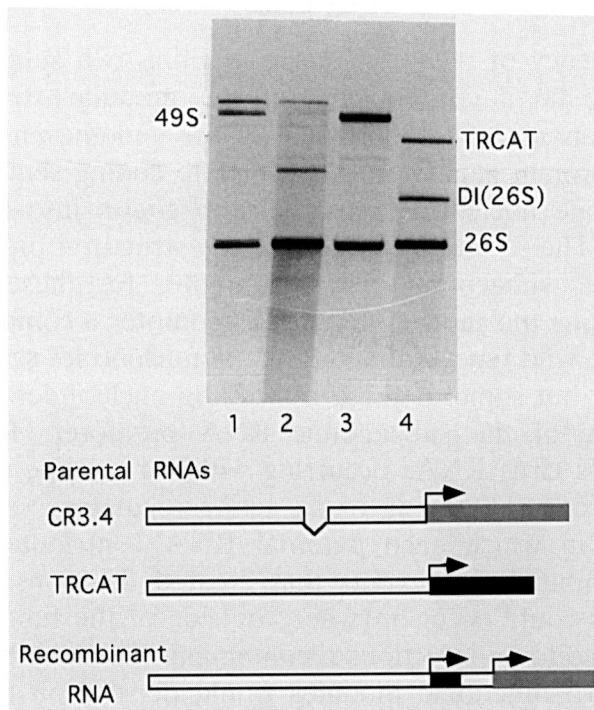

Fig. 2. Recombinant RNA profiles from a cross between TRCAT and CR3.4. The upper panel shows the autoradiogram of the ^{3}H-uridine labeled RNA synthesized in cells infected with Sindbis virus stocks obtained from independent plaques (*1* and *2*). *3* is the pattern of wild type Sindbis virus RNAs and *4* shows the markers. The lower diagrams illustrate the parental and recombinant RNAs. The symbols are identical to those described in Fig. 1

gene. Lanes 1 and 2 show the RNA patterns from cells infected with independent plaque isolates. Both show an RNA species larger than the wild type genomic (49S) RNA (lane 3) as well as an additional RNA species migrating between the genomic and subgenomic RNAs. The new RNA species differed in size from each other and from the new band seen in Fig. 1, lane 3.

Recombinant RNAs from the different crosses were sequenced in the region surrounding the subgenomic RNA promoter revealing that they contained two subgenomic RNA promoters [22]. This finding explains the larger size of the RNAs; they contain additional nucleotides. It explains the new species of RNA; two subgenomic RNAs are synthesized. The diagrams in Figs. 1 and 2 illustrate the structure of two of the recombinant RNAs. Insertions in the different recombinants varied in size ranging from 47 nucleotides to more than 800 nucleotides, indicating that the nucleotides directly surrounding the subgenomic promoter are not a "hotspot" for recombination.

Discussion

The high frequency of recombinants containing two subgenomic RNA promoters may be a consequence of the modular structure of the Sindbis virus genome. The 5′ module of the genome encompasses the nonstructural protein genes but the complete coding sequence of these genes extend one nucleotide plus the stop codon into the 26S RNA sequence [21]. The 3′ module codes for the structural proteins. It must also include the subgenomic RNA promoter for these genes to be expressed, making the subgenomic RNA promoter a component of both modules. In the wild type genomic RNA 48 nucleotides separate the two coding regions, but some of these noncoding nucleotides are important for the activity of the subgenomic RNA promoter. Recombination between Sindbis virus RNAs occurring within a module would have to be precise to conserve coding or cis-acting regulatory sequences. In a recombination in which each parental RNA contributed one of the modules, recombinational events that created deletions, insertions or rearrangements could occur anywhere outside of the functional module and still give rise to an infectious recombinant. The length of the region between the two functional modules would depend on the location at which recombination had occurred and on the stability of the recombinant RNA. The recombinant Sindbis virus genomes containing two subgenomic RNA promoters were larger than the 49S RNA, but they were unstable and eventually evolved to be the same size as wild type RNA. This is an important point; all of the recombinants were analyzed after plaque purification. Several produced the normal pattern of Sindbis virus RNAs (Fig. 1, lane 4) and showed no evidence of any insert in the region of the subgenomic RNA promoter. These recombinants could be ones that had already undergone evolution by the time of analysis.

We have not yet investigated questions about the frequency of recombination between Sindbis virus RNAs or about the influence of sequence homology and structure on the site of crossover. Answers to such questions depend on being able to determine the total number of recombinational events taking place, not just those which lead to the selection of viable progeny. The recent success of Jarvis and Kirkegaard, using polymerase chain reaction to detect recombinant RNA molecules in poliovirus infected cells, makes these studies now seem feasible [8]. A better understanding of the factors involved in RNA recombination may shed further light on the evolution of these viruses.

References

1. Allison R, Thompson C, Ahlquist P (1990) Regeneration of a functional RNA virus genome by recombination between deletion mutants and requirement for cowpea chlorotic mottle virus 3a and coat genes for systemic infection. Proc Natl Acad Sci USA 87: 1820–1824
2. Bujarski JJ, Kaesberg P (1986) Genetic recombination between RNA components of a multipartite plant virus. Nature 321: 528–531
3. Edwards MC, Petty ITD, Jackson AO (1992) RNA recombination in the genome of barley stripe mosaic virus. Virology 189: 389–392
4. Gaedigk-Nitschko K, Schlesinger MJ (1991) Site-directed mutations in Sindbis virus E2 glycoprotein's cytoplasmic domain and the 6K protein lead to similar defects in virus assembly and budding. Virology 183: 206–214
5. Geigenmüller-Gnirke U, Weiss B, Wright R, Schlesinger S (1991) Complementation between Sindbis viral RNAs produces infectious particles with a bipartite genome. Proc Natl Acad Sci USA 88: 3253–3257
6. Hahn CS, Lustig S, Strauss EG, Strauss JH (1988) Western equine encephalitis virus is a recombinant virus. Proc Natl Acad Sci USA 85: 5997–6001
7. Hirst GK (1962) Genetic recombination with Newcastle disease virus, polioviruses, and influenza. Cold Spring Harbor Symp Quant Biol 27: 303–308
8. Jarvis TC, Kirkegaard K (1992) Poliovirus RNA recombination: mechanistic studies in the absence of selection. EMBO J 11: 3135–3145
9. Khatchikian D, Orlich M, Rott R (1989) Increased viral pathogenicity after insertion of a 28S ribosomal RNA sequence into the haemagglutinin gene of an influenza virus. Nature 340: 156–157
10. King AMQ (1988) Genetic recombination in positive strand RNA viruses. In: Domingo E, Holland JJ, Ahlquist P (eds) RNA genetics, vol 2. CRC Press, Boca Raton, pp 150–185
11. King AMQ, McCahon D, Slade WR, Newman JWI (1982) Recombination in RNA. Cell 29: 921–928
12. Kirkegaard K, Baltimore D (1986) The mechanism of RNA recombination in poliovirus. Cell 47: 433–443
13. Lai MMC (1992) RNA recombination in animal and plant viruses. Microbiol Rev 56: 61–79

14. Lai MMC, Baric RS, Makino S, Keck J, Egbert J, Leibowitz JL, Stohlman SS (1985) Recombination between nonsegmented RNA genomes of murine coronaviruses. J Virol 56: 449–456

15. Liljeström P, Garoff H (1991) A new generation of animal cell expression vectors based on the Semliki Forest virus replicon. Bio/Technology 9: 1356–1361

16. Luytjes W, Bredenbeek PJ, Noten AF, Horzinek MC, Spaan WJ (1988) Sequence of mouse hepatitis virus A59 mRNA 2: indications for RNA-recombination between coronavirus and influenza C virus. Virology 166: 415–422

17. Meyers G, Tautz N, Dubovi EJ, Thiel H-J (1991) Viral cytopathogenicity correlated with integration of ubiquitin-coding sequences. Virology 180: 602–616

18. Monroe SS, Schlesinger S (1983) RNAs from two independently isolated defective interfering particles of Sindbis virus contain a cellular tRNA sequence at their 5' ends. Proc Natl Acad Sci USA 80: 3279–3283

19. Rice CM, Levis R, Strauss JH, Huang HV (1987) Production of infectious RNA transcripts from Sindbis virus cDNA clones: mapping of lethal mutations, rescue of a temperature-sensitive marker, and in vitro mutagenesis to generate defined mutants. J Virol 61: 3809–3819

20. Schlesinger S, Schlesinger M (1990) Replication of togaviridae and flaviviridae. In: Fields BN, Knipe DM (eds) Virology. Raven Press, New York, pp 697–711

21. Strauss EG, Strauss JH (1986) Structure and replication of the alphavirus genome. In: Schlesinger S, Schlesinger MJ (eds) The togaviridae and flaviviridae. Plenum Press, New York, pp 35–82

22. Weiss BG, Schlesinger S (1991) Recombination between Sindbis virus RNAs. J Virol 65: 4017–4025

23. Xiong C, Levis R, Shen P, Schlesinger S, Rice CM, Huang HV (1989) Sindbis virus: an efficient, broad host range vector for gene expression in animal cells. Science 243: 1188–1191

Authors' address: Dr. S. Schlesinger, Department of Molecular Microbiology, Washington University Medical School, St. Louis, MO 63110-1093. U.S.A.

Arch Virol (1994) [Suppl] 9: 221–230

Archives
of
Virology

© Springer-Verlag 1994
Printed in Austria

Homologous RNA recombination allows efficient introduction of site-specific mutations into the genome of coronavirus MHV-A59 via synthetic co-replicating RNAs

R. de Groot, L. Heijnen, R. van der Most, and **W. Spaan**

Leiden University, Faculty of Medicine, Institute of Medical Microbiology,
Department of Virology, Leiden, The Netherlands

Summary. We describe a novel strategy to site-specifically mutagenize the genome of an RNA virus by exploiting homologous RNA recombination between synthetic defective interfering (DI) RNA and viral RNA. Marker mutations introduced in the DI RNA were replaced by the wild-type residues during replication. More importantly, however, these genetic markers were introduced into the viral genome; even in the absence of positive selection, MHV recombinants were isolated. This finding provides new prospects for the study of coronavirus replication using recombinant DNA techniques. As a first application, we describe the rescue of the temperature sensitive mutant MHV Albany-4 using DI-directed mutagenesis. Possibilities and limitations of this strategy are discussed.

Introduction

During mixed infection with different MHV strains, RNA recombination occurs at a remarkably high frequency, both in tissue culture and in infected mice [5–7, 11, 12]. Although the mechanism is still unknown, homologous RNA recombination in coronavirus-infected cells presumably occurs via template switching ("copy-choice"). As proposed for picornavirus RNA recombination [4, 8], polymerase complexes containing nascent RNA are thought to dissociate from their original template and anneal to another, after which RNA synthesis proceeds [12].

We are interested in genetic manipulation of coronaviruses. For several other RNA viruses full-length cDNA clones have been constructed from which infectious RNA can be transcribed in vitro [1–10]. However, the construction of a full-length cDNA clone of a coronavirus has been hampered by the extreme length of the coronavirus genome.

Here, we describe an alternative strategy to site-specifically introduce mutations into the MHV genome. This strategy exploits the high-frequency RNA recombination that occurs in MHV-infected cells [11, 12] and involves the use of synthetic defective interfering (DI) RNAs. We provide evidence for homologous RNA recombination between DI RNA and the standard virus genome. Marker mutations introduced into the synthetic DI RNA were replaced by the wild-type residues during replication in MHV-infected cells. More importantly however, these marker mutations were incorporated into the genome of MHV-A59. DI-directed mutagenesis provides exciting new prospects for molecular genetic studies on coronaviruses, as was demonstrated by the rescue of the temperature-sensitive mutant Albany-4. This work was previously published [20].

Results

Construction of pMIDI-C

We previously reported the construction of pMIDI, a full-length cDNA clone of an MHV DI RNA [19]. pMIDI consists of three non-contiguous regions of the MHV genome, namely the most 5′ 3889 nucleotides, the most 3′ 806 nucleotides and, in between, 799 nucleotides derived from ORF1b of the polymerase gene. The three segments are joined in-frame, although in pMIDI the reading frame is interrupted by a UAA termination codon at position 3357 [19]. To study recombination between MHV DI RNAs and the MHV genome, we constructed a pMIDI-derivative, pMIDI-C, in which we introduced three silent point mutations at positions 1778, 2297, 3572 (mutations A, B and C, respectively; Fig. 1). Marker mutations A (G → A) and B (T → C) were introduced in pMIDI by PCR-mutagenesis [18]. The termination codon at position 3357 was replaced by the wild-type CAA codon by exchanging the XhoII-SpeI fragment (nucleotides 3237–3689) of pMIDI for the corresponding fragment derived from an independent, DI-derived cDNA clone, pDI02 [19]. This fragment contained the wild-type CAA codon at position 3357 and an accidental T → C mutation at position 3572 (marker C), presumably acquired during cDNA synthesis.

Point mutations in MIDI-C are replaced by the wild-type sequence during replication in MHV-A59 infected cells

RNA transcribed from pMIDI-C was used for transfection of MHV-infected mouse L-cells as described previously [19]. At 12 h after transfection the tissue-culture supernatant (passage 0 virus) containing virus and DI particles was collected and passaged twice in fresh L-cells,

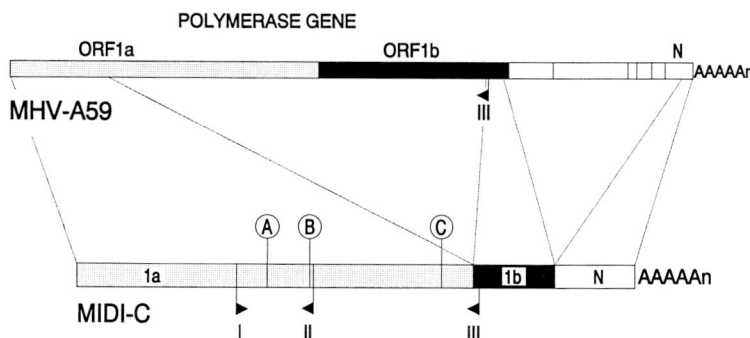

Fig. 1. Schematic representation of the structure of MIDI-C RNA and the MHV-A59 genome. The different parts of MIDI-C, derived from the 5' end (ORF1a), ORF1b and 3' end of the MHV-A59 genome (N) are indicated. The locations are shown of marker mutations *A*, *B* and *C*, and of the oligonucleotides *I*, *II* and *III*. The orientations of these oligonucleotides are indicated by arrowheads

yielding passage 1 and 2 virus. Total intracellular RNA was isolated from passages 0, 1 and 2 and analyzed by hybridization. MIDI-C RNA replicated in MHV-infected cells and strongly interfered with viral mRNA synthesis in cells infected with passage 1 (p1) virus/DI mixture (not shown). To determine whether recombination had occurred between the MHV genome and the synthetic DI RNA, the sequence of p2 MIDI-C RNA was examined. For this purpose, the DI RNA was subjected to cDNA synthesis followed by PCR amplification (RT-PCR) using the oligonucleotide primers I and III (Fig. 1). To prevent in vitro recombination of genomic and DI sequences during RT-PCR, the MIDI-C RNA used as a template was gel-purified. DI-specific PCR-DNA was digested with *Hin*dIII (positions 1985 and 3782 in pMIDI-C) and cloned into pUC20. Sequence analysis of 64 independent clones showed that mutations B and C had been replaced by the wild-type sequence in 14 (22%) and 4 clones (6%) respectively.

Because (i) the mutations were exclusively replaced by the wild-type sequence and (ii) the frequency with which the mutations were replaced depended on their distance from the artificial junction of the 1a and 1b segment, reversion by point mutation appear to be unlikely. In fact, these results can be explained by homologous RNA recombination between MIDI-C RNA and the genome of the standard virus.

Isolation of MHV recombinants

RNA recombination between MIDI-C RNA and the viral genome could, in principle, also yield recombinant viruses carrying the marker mutations. To study this possibility, p2 virus was plaqued and 150 well-isolated plaques were used to inoculate mouse L-cells. Total cellular RNA, isolated from the infected cells, was subjected to RT-PCR using oligonucleotides I and II (Fig. 1). In all cases a DNA fragment with the expected length of 0.6 kb was generated. Evidently, the 0.6 kb fragment would also be produced with MIDI-C RNA as template (Fig. 1). However, MHV DIs are rapidly lost by infecting at a low m.o.i. [13, 19] and the obtained PCR fragments were therefore expected to be genome-specific.

The PCR-DNAs were screened for the presence of marker mutation B by differential hybridization using the 15-mer oligonucleotide probe IV (5′ TGTCAAC<u>G</u>AAATTCT 3′; the guanine residue complementary to the introduced cytidine is underlined). PCR-DNA derived from two plaques, 4 and 138, specifically hybridized with this probe (not shown). The RNA preparations that had been used for RT-PCR were confirmed to be devoid of MIDI-C RNA by Northern blot analysis (not shown).

To determine whether viruses 4 and 138 were true recombinants, four consecutive plaque purifications were performed. From each plaque generation, three to five well-isolated plaques were analyzed by differential hybridization of PCR-amplified cDNA. Wild-type MHV, treated identically, served as a negative control. In all cases, the progeny of viruses 4 and 138 contained mutation B (not shown), confirming that this genetic marker had been introduced into the viral genome and strongly arguing against MIDI-C contamination.

As a final control, we performed direct RNA sequencing. Stocks of viruses 4 and 138, that had been plaque-purified twice, were used to infect L-cells at an m.o.i. of 10. Intracellular RNAs were harvested and used for RNA sequence analysis using the oligonucleotide primers II and III (Figs. 1, 2). In addition to mutation B, the RNAs of viruses 4 and 138 contained mutation A (Fig. 2). Analysis of virus 4 RNA showed that mutation C was absent (not shown).

On the basis of these combined data, we concluded that viruses 4 and 138 are MHV mutants, generated by homologus RNA recombination between the synthetic MIDI-C RNA and the MHV-A59 genome.

Rescue of the MHV Albany-4 ts-*mutant by homologous recombination*

Having demonstrated that sequences of synthetic DI RNAs can be introduced into the viral genome via RNA recombination, we explored "DI-directed mutagenesis" as a means to identify and localize mutations that result in a conditionally lethal phenotype. For this purpose, we used MHV strain Albany-4, a temperature-sensitive (*ts*-) mutant of MHV-A59. The *ts*-phenotype of this virus is the result of an 87 nucleotide deletion in the nucleocapsid (N) gene (nt 1 138–1 224 of the N-ORF). At

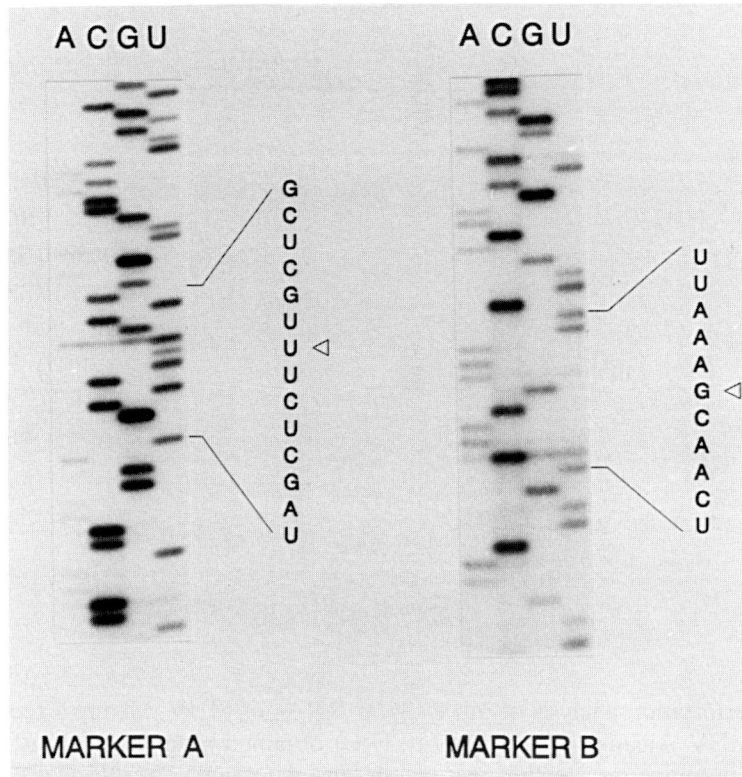

Fig. 2. Sequence analysis of RNA isolated from cells infected with MHV recombinant 4, according to Fichot and Girard [15]. Mutations *A* and *B* were analyzed using oligonucleotides V and II, respectively. Sequences are presented as (−) strand RNA. Arrowheads indicate the introduced mutations. Identical results were obtained for recombinant 138 (not shown)

39°C, virus replication is impaired. Also, incubation of Albany-4 virions at 39°C abolishes infectivity [15].

pMIDI-C contains the 3′ terminal 510 nucleotides of the wild-type N-ORF including the 87 nucleotides that are deleted in Albany-4. Incorporation of the N-sequences of MIDI-C RNA into the Albany-4 genome via RNA recombination should eliminate the *ts*-defect and generate wild-type MHV. To distinguish rescued Albany-4 recombinants from MHV-A59 contaminants, a silent T → A substitution was introduced into the nucleocapsid sequence of pMIDI-C (position 5030, corresponding to position 1200 of the N gene) by PCR-mutagenesis, yielding plasmid pMIDI-C*.

In this set of experiments, we used DNA transfection and in vivo transcription by vaccinia virus-expressed T7 polymerase as an alternative to RNA transfection to introduce MIDI-C* into MHV-infected cells. First, L-cells were infected with recombinant vaccinia virus vTF7-3 [3]

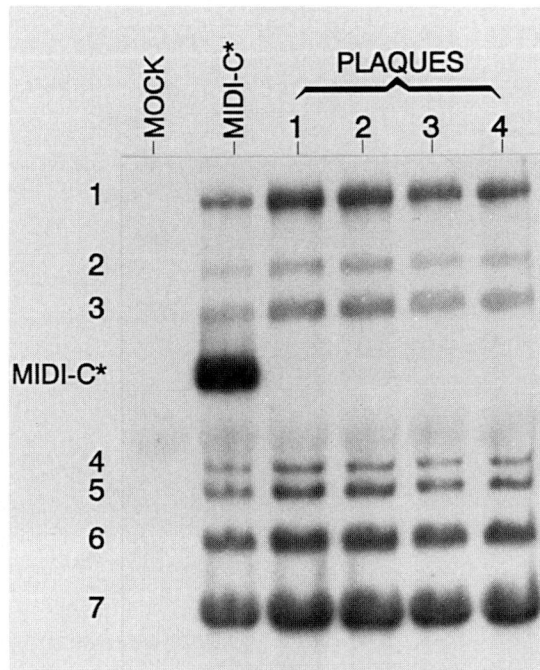

Fig. 3. Hybridization analysis of intracellular RNAs of MHV Albany-4 recombinants. Stocks of MHV Albany-4 (p2, grown at 39°C) obtained after initial transfection with pMIDI-C* or mock-transfection were either used to infect fresh L-cells at 37°C to isolate intracellular RNAs (*Mock* and *MIDI-C**) or plaqued at 39°C. Virus from four randomly-chosen plaques was propagated and intracellular RNAs were isolated (*1–4*)

at an m.o.i. of 5 for 45 min, followed by DNA or mock-transfection at 90 min post infection. Two hours after transfection the cells were superinfected with Albany-4 at an m.o.i. of 10. After a 25 h incubation at 33°C, the tissue culture supernatants were collected and virus passaged once at 33°C in fresh L-cells, and once more at 39°C. Vaccinia virus remains cell-associated and is therefore lost during passage. The 39°C/p2 virus stocks were used to infect monolayers of L-cells at 37°C. Intracelluar RNAs were separated in formaldehyde-agarose gels and hybridized to 5'-end labelled oligonucleotide. This probe binds to the sequence, that has been deleted in Albany-4 (Fig. 3) and is therefore specific for the wild-type N-gene. Strikingly, the probe not only detected MIDI-C* RNA but also the nested set of MHV mRNAs (Fig. 3, lanes "MOCK" and "MIDI-C*"), indicating that during passage, viruses had accumulated that had incorporated MIDI-C* sequences into their genomes.

To obtain direct evidence for this, the p2 virus/DI mixture was plaqued at 39°C and viruses from 4 plaques were passed in L-cells. After hybridization of the intracellular RNAs to oligonucleotide VI, the nested

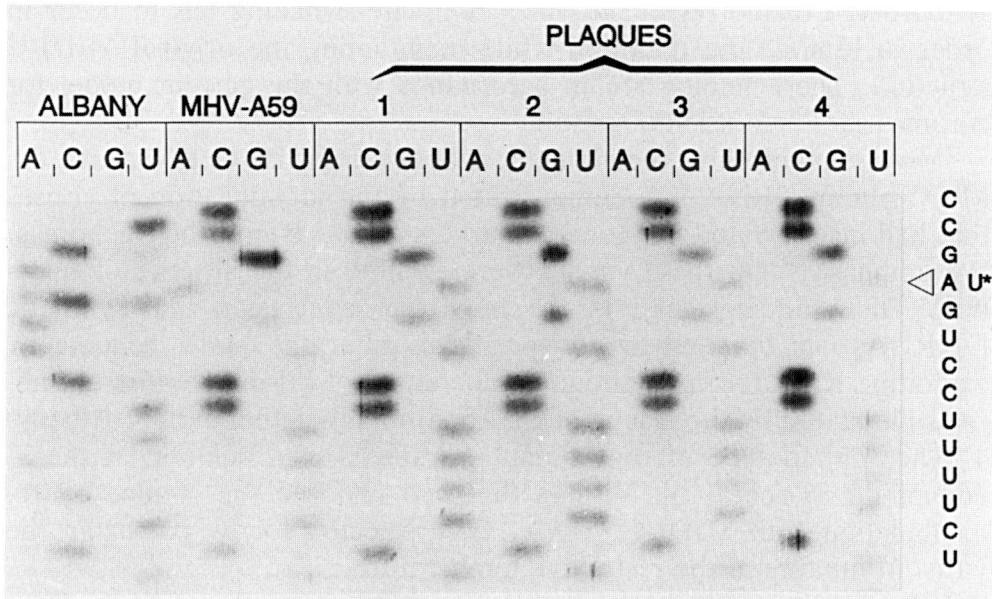

Fig. 4. RNA sequence analysis of recombinant Albany-4 viruses. Intracellular RNA of four plaque-purified recombinant Albany-4 viruses, described in Fig. 3, was subjected to sequence analysis. Intracellular RNA from Albany-4- and MHV-A59-infected cells served as controls. Sequences are presented as (−)strand RNA. An arrowhead indicates the introduced marker mutation

set of MHV RNAs was found; MIDI-C* RNA was not detected (Fig. 3, lanes 1–4). Sequence analysis of these RNA preparations showed that the wild-type N-sequence had been restored and that the U → A marker mutation in the N-ORF was present (Fig. 4), providing further evidence that the isolated viruses were generated by recombination between MIDI-C* RNA and the Albany-4 genomic RNA.

Discussion

Homologous recombination of coronavirus MHV genomic RNAs has been well documented [6, 11, 14]. Here, we extend these observations by providing evidence for recombination between DI RNAs and the MHV-A59 genome. We have shown that silent mutations introduced into the ORF1a sequence of MIDI-C were exclusively replaced by the wild-type residues during DI replication in MHV-infected cells. These marker mutations, however, were not replaced at an equal rate: markers B and C, located 1 500 and 300 nucleotides upstream of the artificial ORF1a/1b border in MIDI-C, were replaced in 22% and 6% of the passage 2 DI RNAs, respectively. Apparently, the frequency of replacement correlated with the distance between the mutation and the ORF1a/

1b border, i.e. the region in which template-switching has to occur in order to remove the mutation while maintaining the original MIDI-C structure. These finding are in accordance with the current model for coronavirus RNA recombination.

Most convincingly, recombination between MIDI-C RNA and the MHV genomic RNA was demonstrated by the identification of viruses that had incorporated the genetic markers A and B into their genomes. We emphasize that these viruses were isolated in the absence of selection. These findings lead to the important conclusion that synthetic DI RNAs can be used to site-specifically alter the MHV genome by exploiting RNA recombination. The potential of DI-directed mutagenesis is illustrated by the rescue of the conditionally lethal mutant Albany-4. The *ts*-phenotype of this mutant is caused by a deletion at the 3' end of the N-gene [15]. Viruses that had restored the N-gene by recombination with MIDI-C* RNA, were selected for by passaging the virus/DI mixture at the restrictive temperature.

Clearly, there will be limitations in this system: thus far, we have introduced mutations only in the 5' and 3' terminal regions of the MHV genome. It remains to be determined whether mutations can be introduced efficiently into more internal regions, which this would require double recombination events. Also, in the case of mutations causing reduced replication in vitro, the screening for recombinant viruses will be difficult. Presumably, such problems can be solved by improving screening procedures and by applying selection, e.g. via rescue of *ts*-markers (this paper) or by using neutralizing antibodies [14]. Studies to address these issues are currently in progress.

A recent publication by Koetzner et al. [9] described the rescue of MHV Albany-4 by targeted RNA recombination using a synthetic mRNA 7 homologue. However in this case, the observed recombination frequency was considered too low to allow direct identification of recombinants without selection. In fact, it was anticipated that a more general applicability of targeted RNA recombination would require finding conditions that favour higher rates of recombination of exogenous RNA. The results described here show that by using synthetic co-replicating DI RNAs site-specific mutations can be introduced efficiently into the MHV genome.

Acknowledgements

We thank Dr. P. Masters for generously providing MHV Albany-4 and Dr. W. Luytjes for stimulating discussions. R.G.M. was supported by grant 331-020 from the Dutch Foundation for Chemical Research (SON). We gratefully acknowledge Oxford University Press for permission for the use of copyright material [20].

References

1. Ahlquist P, French R, Janda M, Loesch-Fries LS (1984) Multicomponent RNA plant virus infection derived from cloned viral cDNA. Proc Natl Acad Sci USA 81: 7066–7070
2. Fichot O, Girard M (1990) An improved method for sequencing of RNA templates. Nucleic Acids Res 18: 6162
3. Fuerst TR, Niles EG, Studier FW, Moss B (1986) Eukaryotic transient-expression system based on recombinant vaccinia virus that synthesizes bacteriophage T7 RNA polymerase. Proc Natl Acad Sci USA 83: 8122–8126
4. Jarvis TC, Kirkegaard K (1991) The polymerase in its labyrinth: mechanisms and implications of RNA recombination. Trends Genet 7: 186–191
5. Keck JG, Stohlman SA, Soe LH, Makino S, Lai MMC (1987) Multiple recombination sites at the 5′-end of murine coronavirus RNA. Virology 156: 331–341
6. Keck JG, Soe LH, Makino S, Stohlman SA, Lai MMC (1988) RNA recombination of murine coronaviruses: recombination between fusion-positive mouse hepatitis virus A59 and fusion-negative mouse hepatitis virus 2. J Virol 62: 1989–1998
7. Keck JG, Matsushima GK, Makino S, Fleming JO, Vannier DM, Stohlman SA, Lai MMC (1988) In vivo RNA-RNA recombination of coronavirus in mouse brain. J Virol 62: 1810–1813
8. Kirkegaard K, Baltimore D (1986) The mechanism of RNA recombination in poliovirus. Cell 47: 433–443
9. Koetzner CA, Parker MM, Ricard CS, Sturman LS, Masters PS (1992) J Virol 66: 1841–1848
10. Lai C-J, Zhao B, Hori H, Bray M (1991) Infectious RNA transcribed from stably cloned full-length cDNA of dengue type 4 virus. Proc Natl Acad Sci USA 88: 5139–5143
11. Lai MMC, Baric RS, Makino S, Keck JG, Egbert J, Leibowitz JL, Stohlman SA (1985) Recombination between nonsegmented RNA genomes of murine coronaviruses. J Virol 56: 449–456
12. Makino S, Keck JG, Stohlman SA, Lai MMC (1986) High-frequency RNA recombination of murine coronaviruses. J Virol 57: 729–737
13. Makino S, Taguchi F, Fujiwara K (1984) Defective interfering particles of mouse hepatitis virus. Virology 133: 9–17
14. Makino S, Fleming JO, Keck JG, Stohlman SA, Lai MMC (1987) RNA recombination of coronaviruses: localization of neutralizing epitopes and neuropathogenic determinants on the carboxyl terminus of peplomers. Proc Natl Acad Sci USA 84: 6567–6571
15. Masters PS, Sturman LS (1990) Background paper: Functions of the coronavirus nucleocapsid protein. Adv Exp Med Biol 276: 235–238
16. Rice CM, Levis R, Strauss JH, Huang HV (1987) Production of infectious RNA transcripts from Sindbis virus cDNA clones: mapping of lethal mutations, rescue of a temperature-sensitive marker, and in vitro mutagenesis to generate defined mutants. J Virol 61: 3809–3819
17. Rice CM, Grakoui A, Galler R, Chambers TJ (1989) Transcription of infectious yellow fever virus RNA from full-length cDNA templates produced by in vitro ligation. New Biol 1: 285–296
18. Sambrook J, Fritsch EF, Maniatis T (1989) Molecular cloning: a laboratory manual. Cold Spring Harbor Laboratory Press, Cold Spring Harbor

19. Van der Most RG, Bredenbeek PJ, Spaan WJM (1991) A domain at the 3′ end of the polymerase gene is essential for encapsidation of coronavirus defective interfering RNAs. J Virol 65: 3219–3226
20. Van der Most RG, Heijnen L, Spaan WJM, De Groot RJ (1992) Homologous RNA recombination allows efficient introduction of site-specific mutations into the genome of coronavirus MHV-A59 via synthetic co-replicating RNAs. Nucleic Acids Res 20: 3375–3381

Authors' address: Dr. W. Spaan, Leiden University, Faculty of Medicine, Institute of Medical Microbiology, Postbus 320, 2300 AH Leiden, The Netherlands.

Arch Virol (1994) [Suppl] 9: 231–238

Archives
of
Virology
© Springer-Verlag 1994
Printed in Austria

Targeting of the site of nonhomologous genetic recombination in brome mosaic virus

J. J. Bujarski and **P. D. Nagy**

Plant Molecular Biology Center and the Department of Biological Sciences,
Northern Illinois University, DeKaib, Illinois U.S.A.

Summary. The genome of brome mosaic virus (BMV) consists of three positive strand RNA segments that share a high degree of sequence homology in the 3' noncoding region. The phenomenon of both homologous and nonhomologous intersegment RNA-RNA recombination has been demonstrated within the 3' noncoding region of BMV RNAs. It has been postulated that nonhomologous crossovers occur at local heteroduplexes formed between the recombining BMV RNA substrates of the same polarity and that the formation of double-stranded regions facilitates strand switching by the replicase. To test the hypothesis of hybridization-mediated recombination in BMV, RNA-3 constructs carrying short antisense RNA1-derived sequences have been used to induce nonhomologous recombination events between RNA-1 and RNA-3 at or near the site of hybridization. We find that both the incidence of recombination and the location of recombinant junctions depends on the structure and the stability of heteroduplexes. Furthermore, our preliminary results demonstrate that mutations in the helicase-like domain of BMV protein 1a affect the location of recombinant junctions. This provides experimental evidence that BMV replicase protein 1a participates in recombination.

Introduction

The genome of brome mosaic virus (BMV) includes three RNA segments that share a high degree of sequence homology within portions of their 3' noncoding regions [1]. The phenomenon of RNA-RNA recombination in BMV was first demonstrated within the 3' noncoding region when a systemic BMV host was inoculated with a mixture of wild type (wt) RNA1, wt RNA2 and mutated RNA3 (designated M4) [2]. Both homologous (legitimate) and nonhomologous (illegitimate) recombinants were identified. Characterization of a large number of nonhomologous crossovers suggested that recombining BMV RNA substrates of the same polarity can form local heteroduplexes at the crossover sites [3, 4]. We postulated that the creation of double-stranded regions facilitated strand switching by the replicase [3, 4].

As in BMV, the formation of local heteroduplex structures at crossover sites was proposed for poliovirus [5, 6]. In coronaviruses, some form of discontinuous mechanism has been suggested to explain high frequency recombination in mouse hepatitis virus (MHV) [7]. Based on the observation that sequences at crossover sites were similar to the recognition sequences of turnip crinkle virus (TCV) replicase, reassociation of polymerase complex has been proposed to be responsible for recombination in TCV [8]. Among alphaviruses, a mutationally altered Sindbis virus RNAs induced illegitimate recombinants that contained both virally-derived and nonviral insertions [9].

A major goal of this research was to test the hypothesis of hybridization-mediated recombination in BMV. RNA-3 constructs carrying short antisense RNA-1 sequences have been used to induce recombination events between RNA-1 and RNA-3 at or near the site of hybridization. We find that both the incidence of recombination and the location of recombinant junctions depends on the structure and stability of heteroduplexes formed between the recombining RNAs. In addition, we provide preliminary results demonstrating that certain mutations in helicase domain of BMV protein 1a affected the location of recombinant junctions. This supported the involvement of BMV replicase in recombination.

Effect of sequence modifications in the RNA3 3' noncoding region on recombination

The M4 RNA-3 deletion mutant generated recombinant molecules with a low frequency, when tested on *Chenopodium hybridum* local lesion host [4]. However, another RNA-3 mutant (designated DM4), which contained duplication of the M4 3' noncoding region (regions A and B in Fig. 1 and [3, 4]), recombined at high levels. The DM4-generated recombinant molecules had the crossovers (indicated by arrows in Fig. 1) distributed along the length of region B. In order to make possible studies on sequences required for recombination, the DM4 construct was modified by insertion of a 197 bp CCMV RNA-3 3' noncoding fragment (region C in PN0 construct). Our previous data revealed the role of selection in BMV recombination [4]. We hypothesized that, due to the extended length and the possibility of competition for BMV replicase proteins [10], progeny recombinants, which contain the CCMV insert, will not be viable. Consequently, the crossover events should be directed

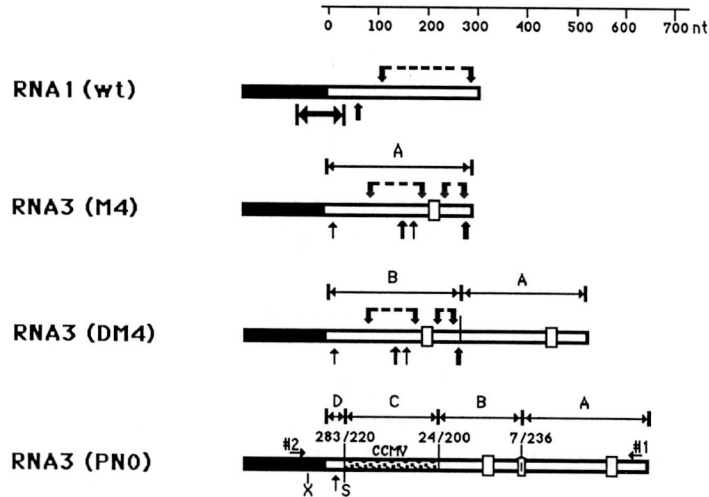

Fig. 1. Schematic representation of the 3′ terminal noncoding regions in the wt BMV RNA1 and in three mutated BMV RNA3 (*M4*, *DM4*, and *PN0*). The DM4 RNA-3 mutant has a duplicated 3′ noncoding region (open boxes designated as regions A and B), each containing a 20 nt deletion (designated as *M4*, see [29]) between nucleotides 81–100 (shown by a small vertical rectangular open box). The PN0 has an additional 197 nt 3′ noncoding region of CCMV RNA-3 (region C; cross-hatched box). Regions B and C lack the last 6 and 23 essential nucleotides (marked by the smallest rectangular box), respectively. The numbers above PN0 RNA-3 show the positions of the first and the last nucleotides of each region, counted from the 3′ end. The sites of crossovers are marked by either above (legitimate recombination) or below (illegitimate recombination) arrows. *S* and *X* mark the position of Spe I and Xba I restriction sites. Coding regions (not to scale) are represented by solid boxes. The location of the RNA-1 sequence that was used as antisense insert in PN constructs (see Fig. 2) is marked by a double horizontal arrow below the RNA-1 molecule

upstream to the CCMV insert. That the upstream region of BMV RNA3 had been inefficient in recombination in M4 and DM4 infections suggested that PN0 might be inefficient in recombination, as well. As predicted, none of the PN0-induced local lesions on *C. quinoa* contained recombinants (Fig. 1). In *C. hybridum*, 5% local lesions accumulated recombinants, all having crossovers at predicted upstream locations (marked by arrows in Fig. 1). These results demonstrated that the crossovers could occur only inefficiently at upstream locations and that the elimination of the 3′ half of the CCMV insert was important for recombinant accumulation.

Antisense sequences direct the site of crossovers

We speculated that the recombination activity of PN0 would be restored by insertion of sequences that fulfill the structural requirements of recombination. To test our hypothesis that local hybridization facilitates crossover events, a 66 nucleotide sequence derived from RNA-1 was inserted at the SpeI restriction site of region D, in both the sense or antisense orientation. Figure 2 demonstrates that the corresponding antisense construct PN2(−) accumulated recombinants and that the crossovers were localized within or near the targeted sequence (shown by arrows). Northern blot analysis revealed that the recombinant RNA-3 components accumulated to high levels within the lesions (data not shown).

The antisense inserts have been further modified in order to test the effect of their length and sequence composition on recombination. Constructs PN1(−) and PN3(−) through PN5(−) contained 140, 40, 30 and 20 bp antisense RNA1 cDNA fragments nested at a common 243 RNA-1 position. PN1(−), PN3(−), and PN4(−), but not PN5(−), generated recombinant RNA3 progeny in *C. quinoa* (Fig. 2). This demonstrated that an antisense sequence longer than 20 nt is necessary to promote efficient recombination. The incidence of recombination depended on the length of the antisense sequence, whereas the recombination crossovers clustered at several hot spots within or near the left side of the putative heteroduplex (using the convention shown in Fig. 2). The latter suggested that the left side of the heteroduplex is associated with recombination.

In order to test the involvement of left side of the heteroduplex as well as the role of selection in recombination, sequence alterations were introduced outside the hot spot sequences observed for PN1(−) through PN4(−) recombinants. The PN6(−) construct was derived from PN1(−) by substitution of four C residues with four U residues. A shift of the crossover sites towards the central part of the duplex was observed. The C to U modifications in PN6(−) were located downstream of the PN1(−) through PN4(−) recombination hot spots. Therefore, the region where the recombination hot spots for PN1(−) through PN4(−) occurred, was not mutated in PN6(−). Consequently, the fact that none of the PN6(−)-induced crossovers occurred at the same positions as those generated by PN1(−) through PN4(−) could not be explained by selection. This conclusion was supported by results obtained with PN7(−) through PN10(−) constructs. Plasmid PN7(−) had a 5-base deletion in the upstream part of the antisense region and a 6-base heterologous sequence that disrupted the left side of heteroduplex (Fig. 2). The crossover sites in PN7(−)-generated recombinants were shifted further towards the central part of the heteroduplex. The PN8(−) construct, which contained three downstream mismatches (Fig. 2), generated recombinants with crossovers shifted even more into the central part of heteroduplex. That crossovers generated by some PN constructs were not observed for the other PN constructs, although all PN constructs

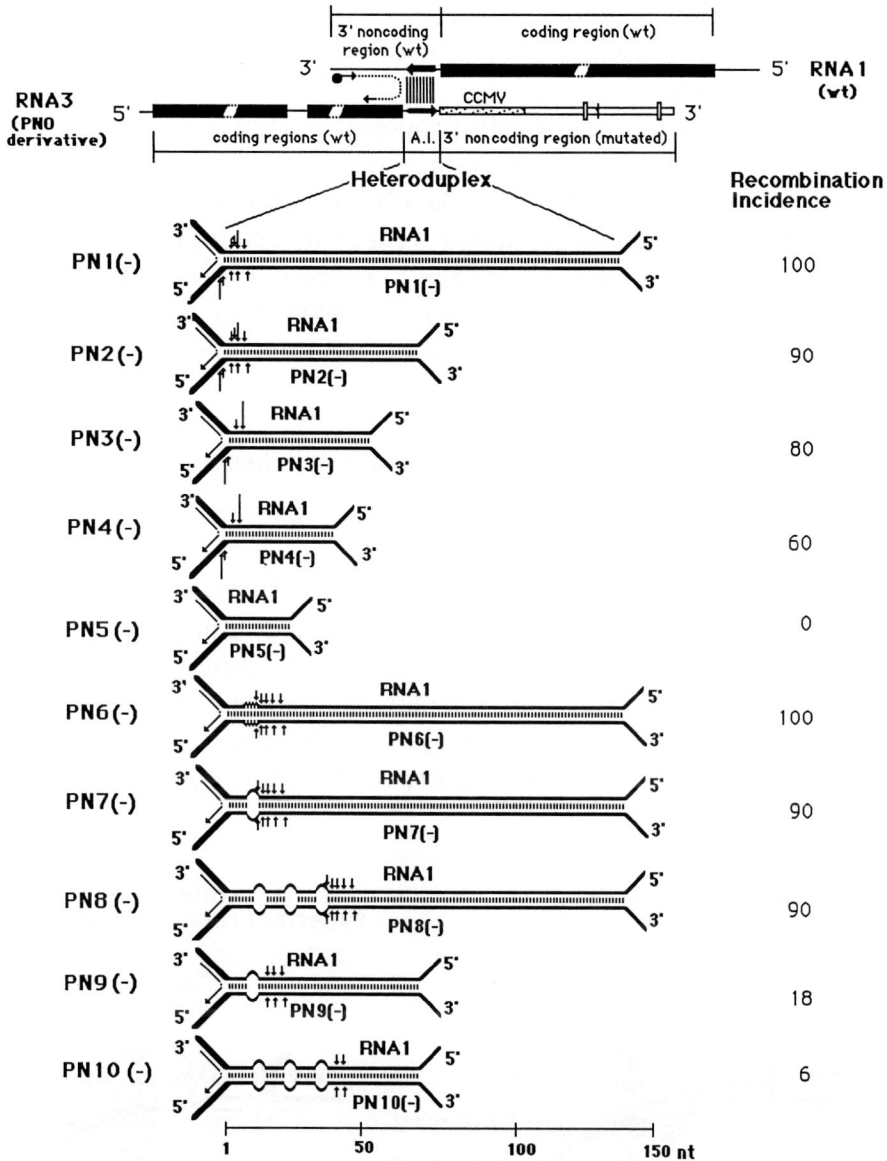

Fig. 2. Top: Schematic representation of the wt RNA-1 molecules as hybridized to antisense inserts in PN1(−) through PN10(−) RNA-3 constructs (marked by a central cluster of short vertical lines). The viral replicase (represented by a small solid circle) initiates at the 3′ end of the wt RNA-1 and switches to the RNA-3 sequences at the left side of the heteroduplex (following the dotted line). Bottom: Magnification of the heteroduplex region (represented by double lines) and location of crossover sites (represented by vertical arrows). The curbed region on the left side of the PN6(−) double line depicts the area of C to U mutations. One or three loops in PN7(−) through PN10(−) double lines represent mismatch mutations. The recombination incidence is defined as the percentage of local lesions on *C. quinoa* that accumulated the recombinants

had recombinationally active sequences, suggested that the crossovers were primarily determined by the structure of the antisense region at the recombination sites.

In order to imitate previously proposed partial hybridizations naturally occurring between recombining BMV RNAs at the sites of crossovers [3, 4], constructs PN9(−) and PN10(−) were designed (Fig. 2). They could form energetically much weaker heteroduplexes with RNA1 than did the PN1(−) through PN8(−). Indeed, PN9(−) and PN10(−) demonstrated a significantly reduced recombination incidence. As our model predicted, recombinant crossovers were located within sequences capable of heteroduplex formation.

Model of heteroduplex-mediated template switches by BMV replicase

Based on the results described above we speculate that the replicase can operate in two modes: regular RNA synthesis mode (A) and aberrant (recombinant) RNA synthesis mode (B) (Fig. 3). During regular RNA synthesis, the helicase domain of the replicase complex (in protein 1a) precedes that of viral RNA polymerase (in 2a protein) while moving across the RNA template. This enables the helicase portion to unwind double-stranded regions. Such factors as temperature, substrate or RNA template concentration, among other, might contribute to the mainte-

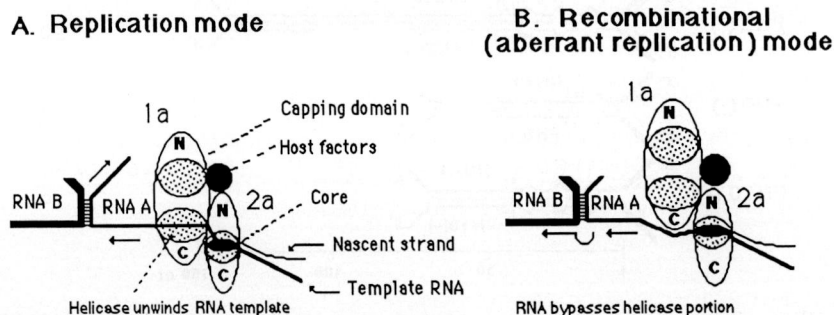

Fig. 3. The model of strand switching by BMV replicase. Proteins 1a and 2a (represented by ellipses) contain functional domains (represented by smaller internal ellipses) of nucleotidyl transferase, helicase and core RNA polymerase. The proteins interact with each other and with host factors (represented by a smaller solid circle). During replication mode (**A**), the RNA template (represented by a solid line) interacts with both helicase and core RNA polymerase domains. Helicase unwinds double-stranded (heteroduplex) regions and the original RNA template is copied through. During recombinational (aberrant replication) mode (**B**), the conformation of the replicase complex is different so that the RNA can interact only with the 2a core domain. In this case, the RNA polymerase passes through undissociated double-stranded regions synthesizing a recombinant nascent strand. The arrows indicate the direction of replicase migration

nance of the proper conformation of the replicase complex. The non-covalent nature of 1a–2a binding over long regions has been demonstrated recently [11, 12]. Conformational fluctuations may occasionally release the RNA template from the helicase portion, thus promoting aberrant RNA replication. In such a case, double-stranded RNA structures bypass the helicase and "slide" directly through the active site of the RNA polymerase component. The removal of double-stranded structures would result in generation of recombinants with the crossovers clustered at one side of the heteroduplex.

Discussion

The results reported in this communication provide the first direct evidence that local complementarity between viral RNA molecules can induce recombination events. The crossovers occurred between nonconserved, upstream parts of the 3′ noncoding regions in wt RNA1 and in mutated RNA3. This generated different-than-wt-size RNA3 recombinants, with most of the 3′ noncoding region taken from the RNA1 segment. Hence, these recombinants can be defined as nonhomologous (illegitimate) ones. Our system does not explain homologous recombination events, as defined by Lai [13]. One can envision the involvement of heteroduplexes in homologous recombination if double-stranded regions were formed by palindromic RNA sequences. It remains to be demonstrated whether the heteroduplex-driven mechanism could operate in other regions on the BMV genome or within the genomes of other RNA viruses.

Clustering of crossovers at the left side of the putative heteroduplex supports the hypothesis of replicase strand switching mechanism. Such a mechanism is also supported by preliminary data indicating that mutations within the helicase-like domain of BMV replicase 1a protein affected the recombination incidence, the incorporation of nontemplated nucleotides, and the number of asymmetric crosses (P. D. Nagy et al., unpubl. res.). BMV RNA replication involves multiple functions, including template recognition and binding for positive- and negative-strand synthesis, subgenomic promoter recognition and binding, initiation and elongation of RNA synthesis, strand separation, and capping. If the model shown in Fig. 3 is correct, template binding, strand separation and elongation of RNA synthesis are the steps most likely involved in recombination. The molecular details of the operation of RNA polymerase around double-stranded regions are not known. One possibility is that the enzyme detaches just before and reattaches right after double-stranded structure. However, our data did not reveal conserved sequences at crossover sites, neither sequences resembling BMV RNA genomic or subgenomic promoters. We speculate that BMV replicase may not be able to reattach de novo at random internal RNA positions. Therefore, the enzyme relocations may occur inside the replicase complex, so that the flanking RNA template sequences are still being held by the replicase portions. Although the model predicts that the 2a protein component switches the RNA templates that have been released from the 1a protein portion, both proteins might participate in these events. Kinetically, the model presumes that strand switching represents a unimolecular RNA copying reaction.

References

1. Ahlquist P, Dasgupta R, Kaesberg P (1981) Near identity of 3′ RNA secondary structure in bromoviruses and cucumber mosaic virus. Cell 23: 183–189
2. Bujarski JJ, Kaesberg P (1986) Genetic recombination in a multipartite plant virus. Nature 321: 528–531
3. Bujarski JJ, Dzianott AM (1991) Generation and analysis of nonhomologous RNA-RNA recombinants in brome mosaic virus: sequence complementarities at crossover sites. J Virol 65: 4153–4159
4. Nagy PD, Bujarski JJ (1992) Genetic recombination in brome mosaic virus: effect of sequence and replication of RNA on accumulation of recombinants. J Virol 66: 6824–6828
5. Romanova LI, Blinov VM, Tolskaya EA, Viktorova EG, Kolesnikova MS, Guseva EA, Agol VI (1986) The primary structure of crossover regions of intertypic poliovirus recombinants: a model of recombination between RNA genomes. Virology 155: 202–213
6. Tolskaya EA, Romanova LA, Blinov VM, Viktorova EG, Sinyakov AN, Kolesnikova MS, Agol VI (1988) Studies on the recombination between RNA genomes of poliovirus: the primary structure and nonrandom distribution of crossover regions in the genomes of intertypic poliovirus recombinants. Virology 161: 54–61
7. Makino S, Keck JG, Stohlman SA, Lai MMC (1986) High frequency RNA recombination of murine coronaviruses. J Virol 57: 729–737
8. Cascone PJ, Carpenter CD, Li XH, Simon A (1990) Recombination between satellite RNAs of turnip crinkle virus. EMBO J 9: 1709–1715
9. Weiss BG, Schlesinger S (1991) Recombination between sindbis virus RNAs. J Virol 65: 4017–4025
10. Pacha RF, Ahlquist P (1991) Use of bromovirus RNA3 hybrids to study template specificity in viral RNA amplification. J Virol 65: 3693–3703
11. Kao CC, Quadt R, Hershberger RP, Ahlquist P (1992) Brome mosaic virus RNA replication proteins 1a and 2a form a complex in vitro. J Virol 66: 6322–6329
12. Kao CC, Ahlquist P (1992) Identification of the domains required for direct interaction of the helicase-like and polymerase-like RNA replication proteins of brome mosaic virus. J Virol 66: 7293–7302
13. Lai MMC (1992) RNA recombination in animal and plant viruses. Microbiol Rev 56: 61–79

Authors' address: Dr. J. J. Bujarski, Plant Molecular Biology Center Northern Illinois University, Dekalb, IL 60115, U.S.A.

Arch Virol (1994) [Suppl] 9: 239–244

_Archives___
Virology
of
© Springer-Verlag 1994
Printed in Austria

Natural recombination in bovine viral diarrhea viruses

J. F. Ridpath[1], **F. Qi**[2], **S. R. Bolin**[1], and **E. S. Berry**[2]

[1] Virology Cattle Research Unit, National Animal Disease Center, USDA,
Agricultural Research Service, Ames, Iowa
[2] Department of Veterinary Science, North Dakota State University, Fargo,
North Dakota, U.S.A.

Summary. BVDV isolates exist as two biotypes differentiated at the molecular level by production of a p80 polypeptide. Insertions consisting of host cell sequences and/or duplicated and rearranged viral sequences have been observed in the portion of the genome coding for the p80 polypeptide in some, but not all, cytopathic BVDV. The significance of these insertions to biotypic expression has yet to be demonstrated. It has been hypothesized that recombination results in the production of the p80 polypeptide by introduction of a cleavage site into a precursor polypeptide or the introduction of a second copy of the p80 gene. Because inserts have not been identified in all cytopathic BVDV examined, it appears that recombination may not be the only mechanism involved in biotypic determination.

Introduction

Bovine viral diarrhea virus (BVDV) is an ubiquitous pathogen of cattle and the type species of the genus *Pestivirus*. The other members of this genus are hog cholera virus (HCV) and border disease virus (BDV) of sheep. Elucidation of the genomic organization of pestiviruses [1] has resulted in their reclassification into the family *Flaviviridae* [2]. While not yet assigned to this genus, it has been noted that human hepatitis C virus has some pestivirus-like features [3].

While only one serotype of BVDV is recognized, isolates of these viruses vary genomically [4,5], antigenically [6–8], and biotypically [9]. Two biotypes of BVDV, cytopathic and noncytopathic, were first differentiated by their activity in vitro [9]. Noncytopathic and cytopathic BVDV isolates interact in an undefined manner to cause a highly fatal condition termed mucosal disease [10, 11]. Mucosal disease results when an animal, with a congenital persistent noncytopathic BVDV infection, becomes superinfected with cytopathic BVDV. Not all combinations of noncytopathic and cytopathic BVDV isolates cause mucosal disease [12]. Initial studies, characterizing certain pairs of cytopathic and noncytopathic BVDV isolated from cattle that had mucosal disease, reported antigenic similarity between the viruses [8, 13]. This observation led to the suggestion that cytopathic BVDV arises by mutation from noncytopathic BVDV.

Biotypic differences in viral proteins

Of the eleven BVDV polypeptides described by Collett et al. [14], six are readily observed via immunoprecipitation using sera from convalescent cattle [15, 16]. Five of these polypeptides, p125, gp53, gp48, gp25, and p20 are produced by both cytopathic and noncytopathic BVDV isolates [16]. The sixth polypeptide, p80, is produced only by cytopathic BVDV isolates [16]. Peptide mapping suggests that the p80 is colinear with the carboxyterminus of the p125 [17]. The size of the p80 polypeptide is highly conserved among cytopathic BVDV isolates while the size of the p125 polypeptide varies among both cytopathic and noncytopathic BVDV [15, 16, 18].

Recombinations in the p125 coding region

Insertions into the portion of the genome coding for the p125 polypeptide have been reported for several, but not all, cytopathic BVDV [19–23] (Fig. 1). All reported insertions have been in frame with respect to the large open reading frame of BVDV. The viral sequences flanking the carboxyterminal end of all sequenced insertions correspond to either amino acid position 1 626 or position 1 662 (numbering based on the published BVDV-NADL sequence [24]). However, insertion at amino acid position 1 626 has only been observed in the BVDV-NADL sequence (Fig. 1B). The insertions consist of host cell-derived sequences and/or duplicated and rearranged viral genomic sequences (Figs. 1B, C–F). While the organization of sequences within the insert may differ, the viral genomic sequences downstream from the carboxyterminus of the insertion are conserved (Table 1). The sequences flanking the corresponding putative recombination positions in viral and cellular RNAs exhibit little homology [22, 23]. This observation led to the proposal that insertions in cytopathic BVDV arise by random RNA recombination, with viable virus being generated only when recombination is in the correct orientation [22]. The conserved 5′ flanking sequences, observed at the C-terminus of the insertions, may result from a requirement for a specific p80 amino terminus.

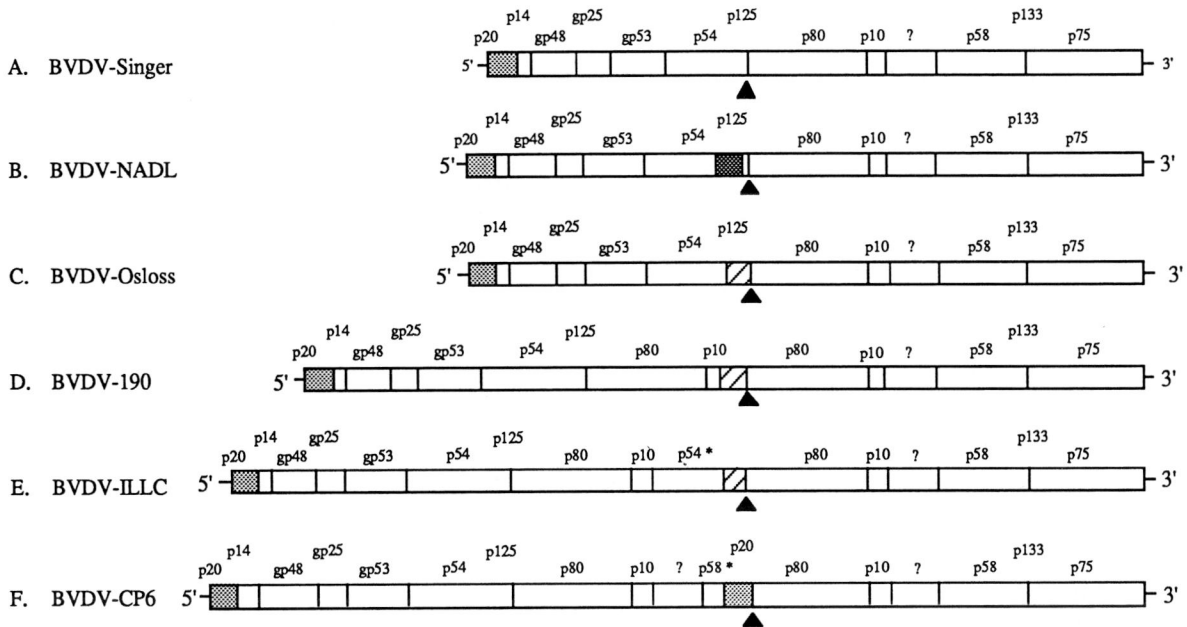

Fig. 1. Genomic organization of cytopathic BVDV isolates. Examples of genomic organization in cytopathic BVDV are shown. The simplest, in which there are no insertions or duplications in the p125 encoding regions, is shown in *A* (BVDV-Singer). The genomic organization of BVDV-NADL, shown in *B*, involves a host cell-derived insertion of unknown function (▨) into the region encoding the p125 polypeptide. Other cytopathic BVDV have been reported to contain inserts consisting of ubiquitin encoding sequences (▧) and/or duplicated rearranged viral genomic sequences (*C* BVDV-Osloss, *D* BVDV-190, *E* BVDV-Ill-c, and *F* BVDV-CP6). Triangles point to sequences homologous to position 5 423 of the published BVDV-NADL genomic sequence [24]. Asterisks indicate a truncated polypeptide encoding sequence

Table 1. Conservation of BVDV sequences immediately downstream from the carboxyterminus of genomic insertions into the p125 region

Cytopathic BVDV	3′ insert	5′ BVDV (p80 amino terminus)
BVDV-Osloss	CGT CTG AGG GGT AGT	GGG CCT GCC GTG TGC
BVDV-190	CGT CTG AGG GGT GGC	GGG CCT GCC GTG TGT
BVDV-Ill-C	CGC CTC AGA GGT GGG	GGA CCT GCC GTG TGC
BVDV-CP6	TGG GTT ACA AGC TGC	GGG CCT GCC GTG TGT

BVDV genomic sequences flanking the carboxyterminus of insertions into the genomic sequences of cytopathic BVDV. For all viruses shown, flanking BVDV sequence is homologous to a sequence starting with nucleotide 5423 of the BVDV-NADL genome

Correlation between recombination and synthesis of the p80 polypeptide

While the amino acid sequence of the p80 polypeptide is colinear with the carboxyterminal portion of the p125 polypeptide [14, 15, 17, 18], its origin is a matter of speculation. Current data indicates the p80 polypeptide may arise by either of two different mechanisms; cleavage from the p125 polypeptide [14] or expression of a second copy of the gene encoding the p80 polypeptide [22, 23]. It has been hypothesized that recombination results in the production of the p80 polypeptide by introduction of a cleavage site into the p125 polypeptide or the introduction of a second copy of the sequence coding for the p80 poly-peptide [22]. However, because inserts have not been identified in all cytopathic BVDV examined [21–23, 25], it is possible that recom-bination may not be the only mechanism involved in biotypic deter-mination. In fact, surveys of BVDV field isolates indicate that the majority of cytopathic BVDV do not contain insertions ([25], J. Ridpath, unpubl. data).

Possible recombination in a pestivirus isolated from sheep

Within the genus *Pestivirus*, the presence of insertions may not be restricted to BVDV genomes. Recently a cytopathic border disease virus (BDV) was isolated from a lamb in North America [26]. Host cell tropism, MAB binding, and PCR amplification suggest that this virus is a true BDV isolate and not a BVDV isolated from sheep. Viral proteins produced by this cytopathic BDV were similar to viral proteins generated by noncytopathic BDV, with two exceptions. The cytopathic BDV produced a 80 kilodalton polypeptide while the noncytopathic BDV did not and the p125 polypeptide of the cytopathic BDV had a greater molecular weight than the corresponding polypeptide of the noncytopathic BDV. Studies are underway to determine if the cytopathic BDV con-tains a genomic insertion similar to those identified in some cytopathic BVDV.

References

1. Collett MS, Larson R, Belzer SK, Retzel E (1988) Proteins encoded by bovine viral diarrhea virus: the genomic organization of a pestivirus. Virology 165: 200–208
2. Wengler G (1991) Family Flaviviridae. In: Francki RIB, Fauquet CM, Knudson DL, Brown F (eds) Classification and Nomenclature of Viruses: Fifth Report of the International Committee on Taxonomy of Viruses. Springer, Wien New York, pp 232–233 (Arch Virol [Suppl] 2)
3. Choo Q-L, Richman KH, Han JH, Berger K, Lee D, Dong D, Gallegos C, Coit D, Medina-Selby A, Barr PJ, Weiner AJ, Bradley DW, Kuo G, Houghton M (1991) Genetic organization and diversity of the hepatitis C virus. Proc Natl Acad Sci USA 88: 2451–455

4. Ridpath JF, Bolin SR (1991) Hybridization analysis of genomic variability among isolates of bovine viral diarrhea virus using cNDA probes. Mol Cell Probes 5: 291–298

5. Dubovi EJ (1992) Genetic diversity and BVD virus. Comp Immunol Microbiol Infect Dis 15: 155–162

6. Bolin SR, Moennig V, Kelso Gourley NE, Ridpath JF (1988) Monoclonal antibodies with neutralizing activity segregate isolates of bovine viral diarrhea virus into groups. Arch Virol 99: 117–124

7. Bolin SR, Littledike ET, Ridpath JF (1991) Serologic detection and practical consequences of antigenic diversity among bovine viral diarrhea viruses in a vaccinated herd. Am J Vet Res 52: 1033–1037

8. Corapi WV, Conis RO, Dubovi EF (1988) Monoclonal antibody analysis of cytopathic and noncytopathic viruses from fatal bovine viral diarrhea virus infections. J Virol 62: 2823–2827

9. Gillespie JH, Baker JA, McEntee K (1960) A cytopathogenic strain of virus diarrhea virus. Cornell Vet 50: 73–79

10. Bolin SR, McClurkin AW, Cutlip RC, Coria MF (1985) Severe clinical disease induced in cattle persistently infected with noncytopathic bovine viral diarrhea virus by superinfection with cytopathic bovine viral diarrhea virus. Am J Vet Res 46: 573–576

11. Brownlie J, Clarke MC, Howard CJ (1984) Experimental production of fatal mucosal disease in cattle. Vet Rec 114: 535–536

12. Bolin SR, McClurkin AW, Cutlip RC, Coria MF (1985) Response of cattle persistently infected with noncytopathic bovine viral diarrhea virus to vaccination of bovine viral diarrhea and to subsequent challenge exposure with cytopathic bovine viral diarrhea virus. Am J Vet Res 46: 2467–2470

13. Howard CJ, Brownlie J, Clarke MC (1987) Comparison by the neutralisation assay of pairs of non-cytopathogenic and cytopathogenic strains of bovine virus diarrhoea virus isolated from cases of mucosal disease. Vet Microbiol 13: 361–369

14. Collett MS, Larson R, Belzer SK, Retzel E (1988) Proteins coded by bovine viral diarrhoea virus: the genomic organization of a pestivirus. Virology 165: 200–208

15. Donis RO, Dubovi EJ (1987) Characterization of bovine viral diarrhoea-mucosal disease virus-specific proteins in bovine cells. J Gen Virol 68: 1597–1605

16. Donis RO, Dubovi EJ (1987) Differences in virus-induced polypeptides in cells infected by cytopathic and noncytopathic biotypes of bovine virus diarrhea-mucosal disease virus. Virology 158: 168–173

17. Purchio AF, Larson R, Collett MS (1984) Characterization of bovine viral diarrhea virus proteins. J Virol 50: 666–669

18. Greiser-Wilke I, Dittmar KE, Liess B, Moennig V (1992) Heterogeneous expression of the non-structural protein p80/p125 in cells infected with different pestiviruses. J Gen Virol 73: 47–52

19. Meyers G, Rumenapf T, Thiel H-L (1989) Ubiquitin in a togavirus. Nature 341: 491

20. Meyers G, Rumenapf T, Thiel H-J (1990) Insertion of ubiquitin-coding sequence identified in the RNA genome of a togavirus. In: Brinton MA, Heinz FX (eds) New aspects of positive strand RNA viruses. American Society for Microbiology, Washington, pp 25–29

21. De Moerlooze L, Desport M, Renard A, Lecomte C, Brownlie J, Martial JA (1990) The coding region for the 54-kDa protein of several pestiviruses lacks host insertions but reveals a "zinc finger-like" domain. Virology 177: 812–815

22. Meyers G, Tautz N, Stark R, Brownlie J, Dubovi EJ, Collett MS, Thiel H-J (1992) Rearrangement of viral sequences in cytopathogenic pestiviruses. Virology 191: 368–386
23. Qi F, Ridpath JF, Lewis T, Bolin SR, Berry ES (1992) Analysis of the bovine viral diarrhea virus genome for possible cellular insertions. Virology 189: 285–292
24. Collett MS, Larson R, Gold C, Strinck D, Anderson DK, Purchio AF (1988) Molecular cloning and nucleotide sequencing of the pestivirus bovine viral diarrhea virus. Virology 165: 191–199
25. Greiser-Wilke I, Haas L, Dittmar K, Liess B, Moennig V (1992) Bovine viral diarrhoea (BVD) virus: How frequent are insertions in the coding region for the nonstructural proteins p125 from cytopathogenic BVD virus strains? Proc Second Symp Ruminant Pestiviruses. Fondation Marcel Mérieux, Lyon, pp 69–70
26. Ridpath JF, Bolin SR, Sawyer M, Osburn BI (1992) Molecular characterization of a North American isolate of cytopathic border disease virus. Proc Second Symp Ruminant Pestiviruses. Fondation Marcel Mérieux, Lyon, pp 35–40

Authors' address: Dr. J. Ridpath, USDA, ARS, National Animal Disease Center, P.O. Box 70, Ames, IA 50010, U.S.A.

Arch Virol (1994) [Suppl] 9: 245–251

Archives
of
Virology
© Springer-Verlag 1994
Printed in Austria

Sequences at the ends of RNA-2 of I6, a recombinant tobravirus

D. J. Robinson

Scottish Crop Research Institute, Invergowrie, Dundee, U.K.

Summary. Tobravirus isolate I6 is an anomalous strain of tobacco rattle virus (TRV), which has biological properties typical of TRV, but which is serologically related to pea early-browning virus (PEBV). Its RNA-2 species is a natural recombinant molecule which contains internal sequences, including the particle protein gene, resembling those of PEBV, but with some TRV-like sequences at each end. Sequencing of RNA-2 of isolate I6 showed that, at the 3′ end, it is almost identical to RNA-2 of PEBV strain SP5, and TRV-like sequences are limited to the 25 residues that are common to both viruses. However, unlike PEBV RNA-2, I6 RNA-2 can be replicated by TRV RNA-1-coded enzymes. Thus, specific recognition of RNA-2 by TRV replicase seems not to involve sequences at the 3′ end. About 275 residues at the 5′ end of I6 RNA-2 are similar to those in RNA-2 of TRV strain TCM, and are joined to the PEBV-SP5-like subgenomic promoter and coding sequences by about 100 residues of uncertain origin. These results support the idea that I6 RNA-2 originated by recombination between TRV and PEBV RNA-2 molecules, although the exact position of the junction and therefore the mechanism involved are unclear.

Introduction

The tobravirus group comprises tobacco rattle virus (TRV), pea early-browning virus (PEBV) and pepper ringspot virus (PRV). Their genomes consist of two RNA segments, contained in separate rod-shaped particles. RNA-1 contains genes for putative RNA replicase components and can replicate in infected plants that do not contain RNA-2 [8]. RNA-2, in contrast, replicates only in plants that also contain RNA-1 of the same virus, and must therefore be dependent in a virus specific way on RNA-1-coded enzymes [9].

Many distinct strains of TRV and, to a lesser extent, of PEBV have been isolated. The RNA-1 species of all TRV strains have highly homologous nucleotide sequences [13, 15], but have only low homology with those of PEBV and PRV strains. In contrast, the nucleotide sequence of RNA-2 can differ greatly among isolates of TRV and of PEBV [13, 14].

At both 3′ and 5′ ends, RNA-1 and RNA-2 of naturally-occurring TRV isolates have identical sequences [4, 15]. However, the extent of this homology is different in different strains. Moreover, although these terminal regions include conserved motifs, they are not identical in different strains. Therefore, there seems to be a mechanism that maintains the identity of the ends of the two genome segments of natural isolates [2]; one involving RNA recombination seems likely. Direct homology between the ends of TRV and PEBV RNA species is limited to six residues at the 5′ terminus and 25 at the 3′ terminus, although internal homologous sequences can also be discerned, particularly in the 3′ non-coding region [12].

Natural recombinant isolates of TRV

Several naturally-occurring tobravirus isolates, including I6, N5 [15] and TCM [1, 4], are serologically similar to PEBV, but have biological properties typical of TRV. Moreover, pseudorecombinant isolates, in which RNA-1 is derived from a typical TRV strain (SYM) and RNA-2 from isolate I6 or N5, are readily obtained [15], but when RNA-2 is derived from PEBV, viable pseudorecombinants are not produced [11, 14]. Thus, the gene product(s) of TRV RNA-1 that are involved in RNA replication recognize I6 or N5 RNA-2 but not PEBV RNA-2.

Nucleic acid hybridization experiments showed that the RNA-1 sequences of I6, N5 and TCM were similar to those of typical TRV strains. I6 RNA-2 contained sequences resembling those of the British serotype of PEBV, but with some TRV-like sequences at the 3′ and 5′ ends, whereas RNA-2 of N5 and TCM contained more extensive TRV-like 3′ and 5′ ends flanking sequences that were related to those of the Dutch serotype of PEBV [1, 4, 15]. These molecules have presumably arisen by a mechanism involving RNA recombination.

Sequence at the 3′ end of I6 RNA-2

A primer complementary to the 20 residues found at the 3′ terminus of all TRV and PEBV RNA species was used to synthesize cDNA from the 3′ end of I6 RNA-2. Four clones were sequenced, and no differences were found between them. The longest clone contained 369 residues, and its sequence was 95% homologous to the 3′ terminal sequence of PEBV strain SP5 RNA-2 (Fig. 1). RNA-1 and RNA-2 of PEBV-SP5 are 97% homologous to one another in the 3′-terminal 266 residues, but further upsteam their sequences diverge [6], and here I6 RNA-2

```
                    10              30              50              70
I6  RNA-2  5'-- GGGUUCUUUU UCUGUUCUUC UAAAUAUUGU UUUAAUUGUU AUUUUAAUUU UAUUUUUAGA AGUGUAUUGC GAUGUUUGCC
                |||||||||  ||||||||||  |||||||| |  ||| ||||||  |||||||            |||||||  ||||| | |  ||||||||||
PEBV RNA-2  5'-- GGGUUCUUUC UCUGUUCUUC UAAAUAUCGU UUUUAUUGUU AUUUUAA...  ...UUUUAGA AGUGUGUAGU·GAUGUUUGCC
                |          |  ||    | | | ||  ||      | ||   |  | | |             ||      |    ||||||          |||| |
PEBV RNA-1  5'-- UUAGUAAAGA UGGAUUUGAA AAUUGAUUGG AGGCAAUUUU AAAUUGAUGA ACGUUCACUA CAUUAGUAGU UUGUUUUUCA
                |             ||| |||  | |  |  | | | |||   ||  |||   | |      |               ||||
TRV  RNA-1  5'-- AAAAAGUCUC AUAAUUCGAA GACCUCUAAG AAGAAAUUCA AAGAGGACAG AGAAUUUGGG ACACCAAAAA GAUUUUUAAG


                    90             110             130             150
I6  RNA-2      CGUAGUGGGA CAUCCGUUGC AAUUA..UUG UGUCGUGUAA ACGACAUUUG UGUUAUGC.U AUUUGUUUGU GUAUGAACUA
                ||||||||||  ||||||||||  ||||||||   ||||||||||  ||||||||||  |||||||||  ||||||||||  |||||| |||
PEBV RNA-2     CGUAGUGGGG CAUCCGUUGC AAUUA..UUG UGUCGUGUAA ACGACAUUUG UGUUAUGC.U AUUUGUUUGU GUAUGAGCUA
                |||          |   | |    |||| |||   ||||||||    ||||||||||  ||||||||||  |||||||||||
PEBV RNA-1     AUUAGCAUUA GUUUAAGUAC UAUUAUUUUG UGUCGUGUGA GCGACAUUUG UGUUAUGCAU AUUUGUUUGU GUAUGAGCUA
                |  |         ||   |    |   ||   | | |    || |      | ||  |||  || ||||  | | |
TRV  RNA-1     AGAUGAUGUU CCUUUCGGGA UUGAUCGUUU GUUUGCUUUU UGAUUUUAUU UUAUAUUGUU AUCGUUUUCU GUGUAUAGAC


                   170             190             210             230
I6  RNA-2      GGUAUUGGUU AAUUCCAU.. ..GCUUAAGU GGUUCUACCA UAGGAACGAU AGUUUGUUAU UAUUAUUAUU UGAAGUUUGU
                ||||||||||  ||||||||   ||||||||||  ||||||||||  ||||||||||  |||||||| |  ||||||||||  ||||||||||
PEBV RNA-2     GGUAUUGGUU AAUUCCAU.. ..GCUUAAGU GGUUCUACCA UAGGAACGAU AGUUUGUUGU UAUUAUUAUU UGAAGUUUGA
                ||||||||||  ||| |||||   ||||||||||  ||||||||||  |||| ||||  || |||  |  ||||||||||  |||  ||
PEBV RNA-1     GGUAUUGGUU AAUCCCAU.. ..GCUUAAGU GGUUCUACCG UAGGGACGAU AGUUUGUUAU UAUUAUUAUU UUAAGUUUGA
                ||  |        || |||   |  | | | |    | |  |   ||||  |  |||||  | |  |||||  ||| | |
TRV  RNA-1     UGUUUGAGAU UGGCGCUUGG CCGACUCAUU GUCUUACCAU AGGGGAACGG ACUUUGUUUG UGUUGUUAUU UUAUUUGUAU


                   250             270             290             310
I6  RNA-2      UUUACGAAAA UCUUCUGCUU UCUGAAAA.G UC.GUUGACC AAGAGAAAGU GGGAGGGUGA GUAAGUACUC UUGAAGUGGU
                ||||  |||||  ||||||||||  ||||||||   ||||||||||  ||||||||||  ||||||||||  ||||||||||  ||||||||||
PEBV RNA-2     UUUAUGAAAA UCUUCUGCUU UCUGAAAAGG UC.GUUGACC AAGAGAAAGU GGGAGGGUGA GUAAGUACUC UUGAAGUGGU
                |||  ||||||  ||||||||||  ||||||||    ||||||||||  ||||||||||  ||||||||||  ||||||||||  ||||||||||
PEBV RNA-1     UUUGUGAAAA UCUUCUGCUU UCUGAAAAGG UC.GUUGACC AAGAGAAAGU GGGAGGGUGA GUAAGUACUC UUGAAGUGGU
                |||  ||||||        |||||||||||  |   |     | |      |           |||||||||||  || ||||| |
TRV  RNA-1     UUUAUUAAAA UUCUCAAUGA UCUGAAAAGG CCUCGAGGCU AAGAGAUUAU UGGGGGGUGA GUAAGUACUU UUAAAGUGAU


                   330             350             370
I6  RNA-2      GAUGGUUACA AGGCAUAGAA GGG.AAAACC CUUCCGCCUA CGUAAGCGUU AUUACGCCC 3'
                ||||||||||  ||||||||||  ||| |||||  ||||||||||  ||||||||||  |||||||||
PEBV RNA-2     GAUGGUUACA AGGCAUAGAA GGGUUAAACC CUUCCGCCUA CGUAAGCGUU AUUACGCCC 3'
                ||||||||||  ||||||||||  ||||||||||  ||||||||||  ||||||||||  |||||||||
PEBV RNA-1     GAUGGUUACA AGGCAUAGAA GGGUUAAACC CUUCCGCCUA CGUAAGCGUU AUUACGCCC 3'
                ||||||||||  |||| |||||   |||| || ||  |||||||||  ||||||||||  |||||||||
TRV  RNA-1     GAUGGUUACA AAGGCAAAAG GGGUAAAACC CCU.CGCCUA CGUAAGCGUU AUUACGCCC 3'
```

Fig. 1. Nucleotide sequence of the 3' end of RNA-2 of isolate I6 (top line), compared with the 3'-terminal regions of PEBV (strain SP5) RNA-2 [6], PEBV (strain SP5) RNA-1 [13] and TRV (strain SYM) RNA-1 [8]. Hyphens represent residues not determined, dots represent spaces inserted for alignment

resembled PEBV-SP5 RNA-2 rather than RNA-1. Similarities between these three sequences and that of TRV strain SYM RNA-1 [8] are less exact and limited to the terminal 170 residues. Only the terminal 25 residues are identical in all four sequences (Fig. 1).

Robinson et al. [15] suggested that terminal sequences of RNA-2 were involved in its recognition by RNA-1 coded enzymes. However, the sequence at the 3' end of I6 RNA-2 is almost identical to that of PEBV-SP5 RNA-2, and contains no feature that explains the ability of TRV RNA replicase to distinguish between them. The obvious conclusion is that specific recognition of tobravirus RNA by replicase depends on sequences only at the 5' end of the virus RNA, although the recognition event may actually involve the 3' end of the complementary strand. If this conclusion is correct, it is not clear why the 3' ends of RNA-1 and RNA-2 of naturally-occurring tobravirus isolates have identical sequences. However, a pseudorecombinant isolate comprising RNA species with non-identical 3′ ends was viable and stable [2], showing that 3′ terminal identity is not essential, and suggesting that the process by which it is achieved is not rapid.

Sequence at the 5′ end of I6 RNA-2

A primer complementary to a region near the 5′ end of the particle protein gene of PEBV-SP5 was used to synthesize cDNA to the 5′ non-coding region of I6 RNA-2. At its 5′ terminus, I6 RNA-2 showed strong homologies with RNA-2 sequences from other TRV strains, which are all quite similar to one another [8], and in particular was identical to that of strain TCM (Fig. 2). All these RNA-2 sequences, including that of I6, contain a segment which is lacking in TRV RNA-1 [7, 8] (Fig. 2). This supports the suggestion [7] that anomalous isolates, such as I6, have acquired their 5′ termini by recombination from TRV RNA-2, rather than RNA-1. Homology with PEBV-SP5 RNA-2 in this region is limited, apart from a sequence which is conserved in the RNA-2 molecules of all tobraviruses [7] (boxed in Fig. 2).

The identity of the sequences of I6 RNA-2 and TCM RNA-2 ceases at residue 275. At residue 376, the homology between I6 RNA-2 and PEBV-SP5 RNA-2 increases markedly, and continues to be high into the beginning of the coat protein gene (Fig. 2). The beginning of this PEBV-like region is about 60 residues upstream of the start of subgenomic RNA-2a, the mRNA for particle protein synthesis (Fig. 2).

Possible recombination sites and mechanisms

The structure of I6 RNA-2, in comparison with those of PEBV-SP5 RNA-2 and TCM RNA-2, is summarized in Fig. 3.

The 5′ end of I6 RNA-2 most closely resembles TCM RNA-2 among the tobravirus RNA species for which sequence information is available. However, TCM is itself a recombinant isolate, whose RNA-2 comprises internal sequences similar to those of PEBV together with TRV-like ends [1]. The homology between I6 RNA-2 and TCM RNA-2 at their 5′ ends is so striking as to suggest that these sequences are derived from the same TRV strain. Goulden et al. [7] identified a conserved region in the RNA-2 of isolates of all three tobraviruses, and suggested that the generation of TCM RNA-2 might have involved recombination within this region. However, the identity of the I6 and TCM RNA-2 sequences continues for about 130 residues beyond this region (Fig. 3). It seems more likely, therefore, that the whole of this 5′ terminal section, comprising the first 275 residues of RNA-2 in both I6 and TCM, is derived from the TRV parent, and that the sequence identified by Goulden et al. [7] is probably not involved in the recombination process.

```
                  10           30              50              70
TCM RNA-2 5' AUAAAACAUU GCACCAAUGG UGCUGCCCUG GCUGGGGUAU GUCUUUGAAC GCAGUAGAAU GUGCUAAUUG ACAAGUUG.G
                 ||||| |||||||||| |||||||||| |||||||||| |||||||||| |||||||||| |||||||||||
I6  RNA-2 5' ---------- ---CAAUGG UGCUGCCCUG GCUGGGGUAU GUCUUUGAAC GCAGUAGAAU GUGCUAAUUG ACAAGUUG.G
                 ||| |  |    |     |   |  |   |   | ||   |  |   | | |   |  |   |     |
SP5 RNA-2 5' AUAAAAUUUG UGAAGCUUGG ...GGGGACC CCCUUGCUGA ACUACAAGAU G.UGUGAUAG AUGACUUUGA GGUUAACUAC
                                   └_____┘

                  90          110             130             150
TCM RNA-2  AGAACGCGGU AGAACGUACU UAUCCGACAC AGCCUUUAUC CCUUUGUUGA GAGGUUUUUC UCAACUGCAC CGAAAUUCUG
           |||||||||| |||||||||| |||||||||| ||||||||||| |||||||||| |||||||||| |||||||||| ||||||||||
I6  RNA-2  AGAACGCGGU AGAACGUACU UAUCCGACAC AGCCUUUAUC CCUUUGUUGA GAGGUUUUUC UCAACUGCAC CGAAAUUCUG
           |  |  |                         ||||| |||| |||| ||||| |||||||||| |||||| |  |  |||| |||
SP5 RNA-2  AUCAUGAC.. .......... .......... ....UUUAUC CCUUAGUUGA GAGGUUUUUC UCAACCGUAA CAAAAUCCUG

                 170          190             210             230
TCM RNA-2  GAUUUAGUUG GUCGCUAGAA GACUGAAGGA AUGUAGGUGU AUUUUUAGAU AG.UUAAUAG GUAUCUUCUC AAAUACCAUA
           |||||||||| |||||||||| |||||||||| |||||||||| |||||||||| ||.||||||| |||||||||| |||||||||
I6  RNA-2  GAUUUAGUUG GUCGCUAGAA GACUGAAGGA AUGUAGGUGU AUUUUUAGAU AG.UUAAUAA GUAUCUUCUC AAAUACCAUA
           |||||||||| |||||||||| |||||||| | |||| || || |||||||| | || |||| ||||  | ||||||| | | ||||
SP5 RNA-2  GAUUUAGUUG GUCGCUAGAA GACUGAAAGG AUGGAGUUGC GUUUUUAGGU AGAAGUAUAA GUGUCUUCUC AUACA.CAUA

                 250          270             290             310
TCM RNA-2  AA.CAUGAGA CGCAUCGCUU GCGAAAGUAG CA.UUAAAAG CUAAACCUUA CUAAAUGACA AAUUUUAGUU UGGUGGUUCA
           |||| ||||| |||||||||| |||| ||||| || || ||| || ||| |  |  || || ||| ||||| |
I6  RNA-2  AA.CCUGAGA CGCAUCGCUU GCGAGAGUAG CAUUUUAAAA ACAAGCGUGA CGGUGUUGCA UGUUGUCGUC ACGUGGUU.A
           || ||||||| || ||  |   ||||  |    || |||| ||  || ||||| | |||| |   | | ||| | ||||  |
SP5 RNA-2  AAACCUGAGA CGUUGGGAGU ACGAAAACUC UUUUACAAAA AGAAUUGUGA GGCAUUGCUC UUUUGUGCAC ACAAGGUUAA

                 330          350             370             390
TCM RNA-2  AAAACGUCAC UCAAAGUCAA UUUGAGUGUG CGCAGAACAG AACGAUGUCA CGUUCCUGUU CGGGGUUUCA CACUAUCUUA
           |||||||||| |   |   | |    ||||    ||||     | |   | |||| | ||||| ||||||||| ||||||||||
I6  RNA-2  AAAACGUCAG GUUCGGU..U CUCCAAUCCA CGCAAGUUCA AGAACAGCAA CGUUCUGAGC UGGGGUUUCA CACUAUCUUU
           ||| |   ||  || | | | |     |   | |                    | |   ||  | |||||||||| ||||||||||
SP5 RNA-2  AAACGCUGUA GUAAUACAUG CGCAAGAACA GGCUG..... ........AG CAUCCUGUUC UGGGGUUUCA CACUAUCUUU

                 410          430             450             470
TCM RNA-2  AGAGAAAGUG UUAAAAUAGU GAGAAUAUUC UCACUAUGAG CAUAAUUAUA CUGGUUAUCC UCUCGCUGAU AGAGACUAUC
           |||||||||| ||||   | | |   |||||| || ||||||  |||||||||| |||||||||  |||||||||| ||||||||||
I6  RNA-2  AGAGAAAGUG UUAAGCUAGU GAGAAUACUC UCACUAAGAG CAUAAUUAUA CUGAUUUUGC UCUCGUUGAU AGAGACUAUC
           |||||||||| ||||| |  | ||   | || | | |||||| |||||||||| ||||||||  | |||||||||| |||| |||||
SP5 RNA-2  AGAGAAAGUG UUAAGUUAAU UAAGUUAUCU UAAUUAGAG CAUAAUUAUA CUGAUUUGUC UCUCGUUGAU AGAGUCUAUC

                 490          510             530             550
TCM RNA-2  ...AUUCUUA AAAUCACUUU GAAAGGUAGU CUUGUUGAUA CACAGGUUGC UUUAUCACUU GCACUACUUA CUUAUGGCAG
              ||| |||| | ||| || | ||| |||||| ||||  | ||  || ||| |||| ||||| ||||||||
I6  RNA-2  GAAAUUGUUA CUGAGAAUUU GACAACUCGG UUUGCUGAUC UACUGGUUAC UGUAUCACUC ACCCGAGUUA ..........
           |  |||||   | |  |||| ||||||||| |||||||||| ||||||| | |||||||||| |||||||||| ||||||||||
SP5 RNA-2  .AUUCUGUUA UUAAAAAUUU GACAACUCGG UUUGCUGACC UACUGGUUAC UGUAUCACUU ACCCGAGUUA ..........

                 570          590
TCM RNA-2  GUAGUUAUGG GGAAACGUUC GAUGGAAAAA UUCUCG--- 3'
             |  |||| |||| |    | ||| || ||
I6  RNA-2  .ACGAAAUGG UGAAAGGAAA GUAUGAAGGG UUUUCU--- 3'
           ||||| |||| ||||||||| ||| |||||| |||||
SP5 RNA-2  .ACGAGAUGG UGAAAGGAAA GUACGAAGGG UUUUCG--- 3'
```

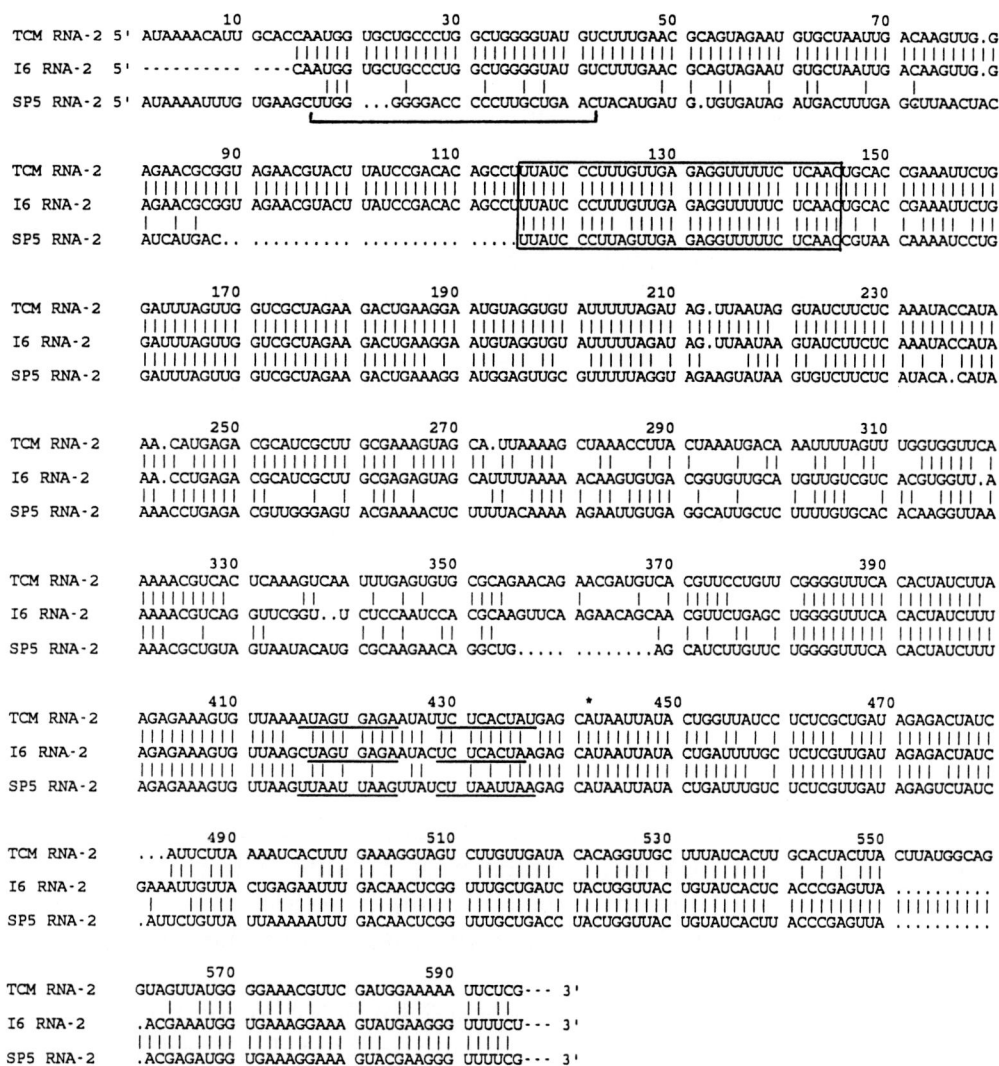

Fig. 2. Nucleotide sequence of the 5′ end of RNA-2 of isolate I6 (middle line), compared with the 5′-terminal regions of RNA-2 of TRV strain TCM [1] and PEBV strain SP5 [6]. Hyphens represent residues not determined, dots represent spaces inserted for alignment. The conserved region identified by Goulden et al. [7] is boxed. Asterisk indicates the probable 5′ terminus of the subgenomic mRNA for particle protein, and inverted repeats that may be part of the promoter for synthesis of this RNA are underlined. The brace at positions 17–42 indicates the segment which is lacking in RNA-1 of TRV

The first 100 residues of the PEBV-like part of I6 RNA-2 (B in Fig. 3) contain the start site for subgenomic messenger RNA-2a, and probably include elements with promoter activity [6]. The equivalent segments in RNA-2 of I6, PEBV-SP5 and TRV-TCM are 80–90% homologous to one another, and 54–61% homologous to that in RNA-2

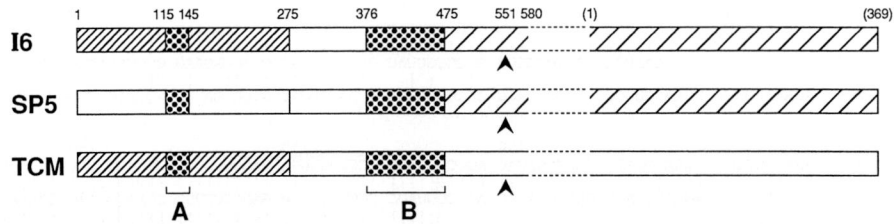

Fig. 3. Diagrammatic representation of the structure of *I6* RNA-2 (top), compared with RNA-2 of PEBV strain *SP5* (middle) and of TRV strain *TCM* (bottom). Similarly hatched segments are at least 90% homologous in nucleotide sequence. Numbers refer to nucleotide residues counted from the 5′ end of the I6 sequence; those in brackets at the 3′ end are arbitrary. Segment *A* is the conserved sequence identified by Goulden et al. [7]. Segment *B* contains the start site for subgenomic messenger RNA-2a and probably has promoter activity. The position of the initiation codon of the particle protein gene is marked (▲). The central section, indicated by the dashed outline, has not been sequenced and is about 2.5 kb in length

of TRV strain PSG. These homologies correlate with the origins of the particle protein genes to which the sequences are attached, and suggest that in both I6 and TCM this segment is derived from PEBV RNA-2. If so, this segment is unlikely itself to have a role in the recombination process, and the recombination junction must be further upstream.

Thus, the recombination event in I6 RNA-2 probably took place between the end of the TCM-like sequence at residue 275 and the beginning of the PEBV-like sequence at residue 376 (Fig. 3). The sequence in this region is not sufficiently similar to any of the previously sequenced tobravirus RNA species to allow its origin to be deduced with certainty.

Mechanisms that have been proposed for RNA recombination are variants of template switching (copy choice) or of breakage and religation, and the results described here do not support either in particular. Whatever the mechanism, some homology between the parental sequences may be required, yet none of the conserved sequences in tobravirus RNA-2 seem likely candidates. However, the homologous sequence can be as short as eight bases [5], or it may not be precisely at the crossover site [10]. Alternatively, the sequence requirements may be for local complementarity between the parents [3]. Moreover, recombination may not be site-specific [10], and different recombinant tobravirus RNA-2 species may have arisen as a result of one or more events at different sites.

Acknowledgements

I thank Mrs. A. Grant for technical assistance. This work was supported by the Scottish Office Agriculture and Fisheries Department.

References

1. Angenent GC, Linthorst HJM, Van Belkum AF, Cornelissen BJC, Bol JF (1986) RNA 2 of tobacco rattle virus strain TCM encodes an unexpected gene. Nucleic Acids Res 14: 4673–4682
2. Angenent GC, Posthumus E, Brederode FT, Bol JF (1989) Genome structure of tobacco rattle virus strain PLB: further evidence on the occurrence of RNA recombination among tobraviruses. Virology 171: 271–274
3. Bujarski JJ, Dzianott AM (1991) Generation and analysis of nonhomologous RNA-RNA recombinants in brome mosaic virus: sequence complementarities at crossover sites. J Virol 65: 4153–4159
4. Cornelissen BJC, Linthorst HJM, Brederode FT, Bol JF (1986) Analysis of the genome structure of tobacco rattle virus strain PSG. Nucleic Acids Res 14: 2157–2169
5. Edwards MC, Petty ITD, Jackson AO (1992) RNA recombination in the genome of barley stripe mosaic virus. Virology 189: 389–392
6. Goulden MG, Lomonossoff GP, Davies JW, Wood KR (1990) The complete nucleotide sequence of PEBV RNA2 reveals the presence of a novel open reading frame and provides insights into the structure of tobraviral subgenomic promoters. Nucleic Acids Res 18: 4507–4512
7. Goulden MG, Lomonossoff GP, Wood KR, Davies JW (1991) A model for the generation of tobacco rattle virus (TRV) anomalous isolates: pea early browning virus RNA-2 acquires TRV sequences from both RNA-1 and RNA-2. J Gen Virol 72: 1751–1754
8. Hamilton WDO, Boccara M, Robinson DJ, Baulcombe DC (1987) The complete nucleotide sequence of tobacco rattle virus RNA-1. J Gen Virol 68: 2563–2575
9. Harrison BD, Robinson DJ (1978) The tobraviruses. Adv Virus Res 23: 25–77
10. Kirkegaard K, Baltimore D (1986) The mechanism of RNA recombination in poliovirus. Cell 47: 433–443
11. Lister RM (1968) Functional relationships between virus-specific products of infection by viruses of the tobacco rattle type. J Gen Virol 2: 43–58
12. MacFarlane SA, Taylor SC, King DI, Hughes G, Davies JW (1989) Pea early browning virus RNA1 encodes four polypeptides including a putative zinc-finger protein. Nucleic Acids Res 17: 2245–2260
13. Robinson DJ, Harrison BD (1985) Unequal variation in the two genome parts of tobraviruses and evidence for the existence of three separate viruses. J Gen Virol 66: 171–176
14. Robinson DJ, Harrison BD (1985) Evidence that broad bean yellow band virus is a new serotype of pea early-browning virus. J Gen Virol 66: 2003–2009
15. Robinson DJ, Hamilton WDO, Harrison BD, Baulcombe DC (1987) Two anomalous tobravirus isolates: evidence for RNA recombination in nature. J Gen Virol 68: 2551–2561

Author's address: Dr. D. J. Robinson, Scottish Crop Research Institute, Invergowrie, Dundee, DD2 5DA, U.K.

RNA-protein interactions and
host-virus interactions

Archives
of
Virology

© Springer-Verlag 1994
Printed in Austria

Identification and characterization of host factor interactions with *cis*-acting elements of rubella virus RNA

H. L. Nakhasi[1], **N. K. Singh**[1], **G. P. Pogue**[1], **X.-Q. Cao**[1,*], and **T. A. Rouault**[2]

[1] Division of Hematologic Products, CBER, Food and Drug Administration, and [2] Cell Biology and Metabolism Branch, National Institute of Child Health and Human Development, National Institutes of Health, Bethesda, Maryland, U.S.A.

Summary. We have analyzed the function of *cis*-acting elements of rubella virus RNA and the components which interact with these elements in viral RNA replication. We demonstrated that the 5′- and 3′-terminal sequences from RV RNA promote translation and negative-strand RNA synthesis of chimeric chloroamphenicol acetyltransferase (CAT) RNAs. These sequences have a potential to form stem-loop (SL) structures and bind cellular proteins specifically in RNA gel-shift and UV cross-linking assays. The 5′ end binding proteins were identified to be Ro/SSA-associated antigens by virtue of being recognized in an RNA complex by an autoimmune patient serum with Ro antigen type specificity. Purification and sequence analysis of the 3′ end binding protein revealed that it is a homologue of human calreticulin. The role of host proteins in RV replication is discussed.

Introduction

Rubella virus (RV; family *Togaviridae*, genus *Rubivirus*) is the causative agent of German measles, a disease characterized by rash and mild fever in children and young adults. If contracted during the first trimester of pregnancy, RV can cause fetal death or multisystem birth defects including deafness, cataracts, mental retardation, and congenital heart disease [1]. Rubella virus consists of a 40S single-stranded polyadenylated genomic RNA of positive polarity, encapsidated by a capsid protein and

* Present address: Laboratory of Pulmonary and Molecular Immunology, NHLBI, NIH, Bethesda, Maryland, U.S.A.

contained within a lipid bilayer envelope in which the two virus-specific glycoproteins, E1 and E2, are embedded [11]. The genomic RNA is 9 757 nucleotides in length and contains two long open reading frames (ORFs): a 5′-proximal ORF of 6 615 nucleotides, which encodes the non-structural proteins and a 3′-proximal ORF of 3 189 nucleotides, which encodes the structural proteins [3]. In infected cells, a subgenomic RNA is synthesized from which the 3′-ORF is expressed [11].

Little is known about the replication of RV. However, the organization of RV genome is similar to that of the alphaviruses, implying a shared replication strategy [3]. Sequence comparison of RV genomic RNA with alphaviruses does not reveal significant overall homology; though some of the sequence elements conserved among alphaviruses also are found within the RV genome [3]. Functional analysis of conserved sequences in Sindbis virus RNA showed that the elements are important for virus replication [15]. While analyzing conserved sequences of both Sindbis and RV RNA, it became apparent that they can fold into stem-loop (SL) structures, implicated into translation and RNA replication (Fig. 1; [3, 15]). Because of the possible role of SL structures as protein recognition sites, specific interaction of cellular proteins with these elements may be crucial for viral replication.

The significance of understanding the mechanism of RNA virus replication lies in its role in virus pathogenicity. Despite its importance, little is known about the details of replication processes in higher eukaryotic cells. Often, host factors are required for RNA replication and may play a variety of roles in this process, including enzymatic, structural, regulatory and RNA-binding. In most cases, host factors are poorly characterized, and their functions are not well defined. Recently it has been shown that conserved sequence elements of viral RNAs are recognized by specific host proteins and that these interactions may be necessary for viral replication [4, 6, 8–10, 12]. As part of an on-going program to study RV replication, we have analyzed the function of conserved sequence elements in viral replication and identified host proteins that interact with these elements.

Functional role of *cis*-acting elements

Functions of SL structures in the RV RNA can be postulated based on their location within the RV genome: the 5′(+) SL in translational initiation; 3′(+) SL in negative strand initiation; 3′(−) SL in positive strand initiation, and the subgenomic stem-loop in initiation of subgenomic RNA synthesis (Fig. 1A). In order to test the functional role of the terminal sequence elements in various steps of the virus replication cycle, the RV RNA 5′(+) SL and 3′(+) SL sequences were fused to the

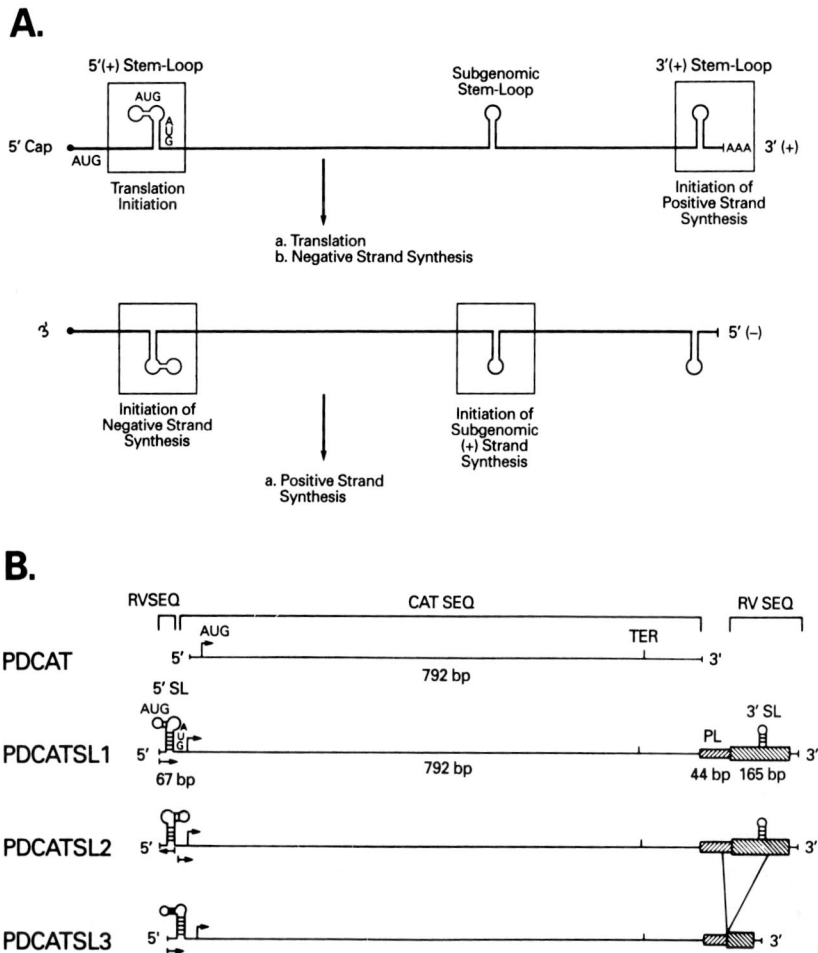

Fig. 1. A Schematic diagram of RV (+) and (−) strand RNAs. The boxed areas represent the *cis*-acting conserved sequence elements. Under each boxed area the possible function of the conserved sequence is denoted. The sequences are drawn as stem-loop structures as has been predicted by the RNA secondary structure analysis [3]. **B** Schematic diagram of the RV and CAT RNA chimeric constructs. Representative constructs have been designated PDCAT through PDCATSL3 and in each construct regions of rubella (RVSEQ) and CAT sequences (CATSEQ) are shown. Arrow near the 5′ end of each construct shows the direction of the RV AUGs with respect to CAT AUG. PL represents the 44 base poly linker region of the PD5 vector. The two lines coming out of the 3′ end of PDCATSL2 show the region of 3′ SL sequence which has been deleted in PDCATSL3

open reading frame of the chloroamphenicol acetyl transferase (CAT) gene (Fig. 1B). The 3′(+) SL sequence used in these constructs includes the authentic poly A tract from RV RNA. The chimeric constructs were generated so that protein synthesis could be directed from either the second AUG present in the rubella 5′(+) SL structure or from the

CAT-encoded AUG, both being in the same reading frame. Various mutations were introduced in the 5'(+) and 3'(+) SL structures of the chimeric RV/CAT constructs in order to ascertain their importance in virus translation or RNA replication. Representative constructs designated PDCAT through PDCATSL3 are diagrammed schematically in Fig. 1B. PDCAT and the derivative constructs were cloned either into a PGEM7Zf(+) vector, allowing in vitro transcription with SP6 RNA polymerase, or a PD5 vector, which allows in vivo expression of RNAs from the adenovirus major late promoter.

The ability of the RV SL structures to direct translation was first tested in vitro (Fig. 2). Equal quantities of transcribed RNAs corresponding to PDCAT and derivative constructs, containing rubella sequence elements, were incubated in rabbit reticulocyte lysate and products were labeled with [^{35}S]-methionine. PDCAT RNA, which lacks both 5' and 3' SL sequences from RV, including the poly A tail, functioned very poorly as a translation template (Fig. 2A, lane 2). The addition of the RV RNA 5' and 3' SL sequences to the PDCAT RNA, resulted in the synthesis of two translation products (Fig. 2A, lane 3). The predominant product is a chimeric protein (RV/CAT hybrid) produced by initiation from the second AUG codon present in the 5'(+) SL sequence, which allowed translation to continue through the entire CAT reading frame. Initiation from the CAT AUG codon produced the less abundant protein, of lower molecular weight (Fig. 2A, lane 3). Reversing the orientation of the 5'(+) SL (PDCATSL2), resulted in no detectable translation product of both RV/CAT and CAT proteins (Fig. 2A, lane 4). The crucial role of the 5'(+) SL structure in translation was further demonstrated by significant decrease in RV/CAT protein synthesis from RNA templates bearing deletions in this SL region [13].

Deletion of the 3'(+) SL sequence, (PDCATSL3), without altering the 3' poly A tail or the 5'(+) SL sequence, led to >10-fold reduction in the translation of the RV/CAT product (Fig. 2A, lane 5). The ratio of protein products synthesized from PDCATSL1 RNA was 3.2:1 (RV/CAT: CAT) (Fig. 2B). A 16-fold decrease in this ratio (0.2:1) was observed in reactions containing PDCATSL3 RNA (Fig. 2B).

To confirm the requirement of these structures for translation in vivo, the chimeric RV/CAT constructs under the control of the adenovirus major late promoter in a PD5 vector were transiently expressed in Vero 76 cells. Twelve hours after transfection, cells expression PDCATSL1 RNA displayed 3-fold greater CAT activity than cells expressing PDCATSL3 RNA [13]. Upon deletion of the entire 5'(+) SL structure from PDCATSL1, CAT activity decreased 6-fold when compared with cells expressing PDCATSL1 RNA [13]. The differences in activity observed in these experiments are in agreement with the profile of in

A.

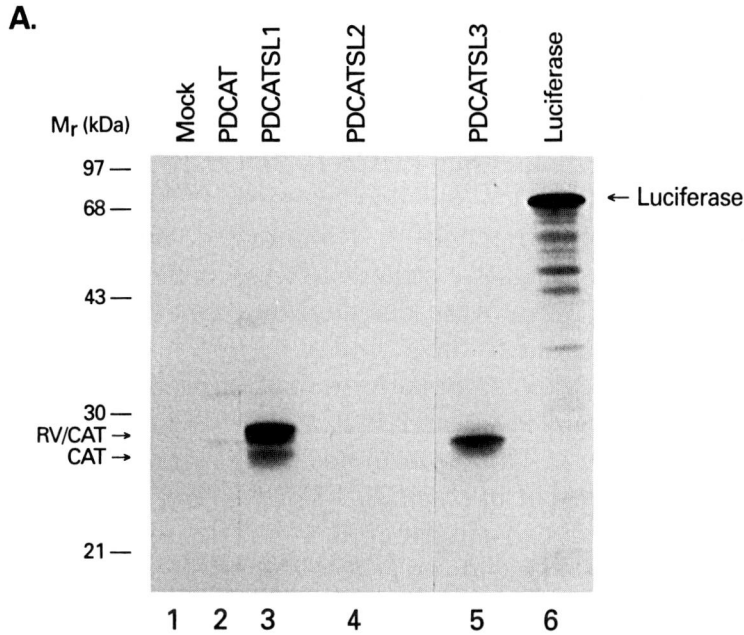

B.

Ratios of *In Vitro* Synthesized Protein Products

RNA Construct	PDCAT	PDCATSL1	PDCATSL2	PDCATSL3
Level of translation (O.D. Area) RV[1]	—	6.3	N.D.	0.58
CAT[2]	0.16	1.2	N.D.	3.9
Ratio of translation products RV:CAT	0:1	3.2:1	N.D.	0.2:1

[1]Translation products initiated from rubella encoded AUGs.
[2]Translation products initiated from CAT encoded AUG.
N.D.   Not detectable.

Fig. 2. A Autoradiogram of SDS-PAGE of in vitro translated, [35S]-methionine labeled proteins from transcribed RNAs of several chimeric constructs. No RNA, (*1*); RNA transcribed from PDCAT vector (*2*); from PDCATSL1 vector (*3*); from PDCATSL2 vector (*4*); from PDCATSL3 vector (*5*); from luciferase RNA (*6*). RV and CAT AUG initiated proteins are shown by arrows. **B** The bands in the autoradiogram were quantitated by densitometry and the area under each band was calculated. Ratios of RV/CAT and CAT products synthesized from various RNAs were determined

vitro synthesized protein products from various RV/CAT chimeric RNAs and are not reflected in the level of CAT RNA present in the transfected cells (Fig. 2; [13]). However, it is not possible to distinguish which proteins, initiated from either the RV or CAT AUG, are responsible for the differences in the CAT activity observed in vivo. Results from both in vitro and in vivo assays demonstrate that the 5′ and 3′(+)

SL sequences are necessary, and appear to function synergistically, in directing translational initiation from RV encoded AUGs [13].

In order to investigate the role of the RV RNA SL sequences in negative-strand synthesis, Vero 76 cells were stably transformed with PD5 constructs containing either PDCAT, PDCATSL1, PDCATSL2 or PDCATSL3 (Fig. 1B). Each cell line expressing PDCAT or PV/CAT chimeric RNAs was either mock- or RV-infected, and total RNA was harvested 48h later. Negative-strand RNAs, complementary to the expressed RV/CAT RNAs, were detected by primer extension analysis using a primer homologous to the 3′ end of the CAT reading frame (Fig. 3). Following RV infection, a novel termination product is seen in cells transformed with PDCATSL1 and PDCATSL2 (Fig. 3, lanes 3 and 4 indicated by an arrow). The 5′ terminus of this RNA maps to the 3′ terminal sequences present in the PDCATSL1 RNA. This primer extension product is notably absent from RNA derived from RV-infected cells expressing either PDCAT or PDCATSL3 (Fig. 3, lanes 1 and 5). Such results suggest that the 3′(+) SL, which is absent from PDCATSL3 and PDCAT, comprises an essential cis-acting element involved in promotion of RV/CAT negative-strand RNA synthesis. In addition, these results demonstrate that the negative-strand initiation events from the chimeric RNAs are dependent on RV infection, which strongly implicates the involvement of RV replicase in this process. Interestingly, the replicase complex may be unable to recognize the RV sequence present in reverse orientation at the 3′ end of the PDCATSL2 negative-strand RNA, thereby reducing the amount of newly synthesized positive-strand RNA (Fig. 3, lane 4). Differences in the levels of positive-strand templates would be reflected in the lower intensity of the novel primer extension product observed with PDCATSL2 compared to PDCATSL1 (Fig. 3, lanes 3 and 4). Consistently, additional background primer extension products were seen in all reactions irrespective of the virus infection (Fig. 3, lanes 1–5).

Characterization of host protein interaction with the cis-acting elements of rubella virus RNA

Previous studies demonstrated that translation and negative-strand synthesis of the RV/CAT RNAs is dependent upon the presence of both the 5′ and 3′ SL structures. Such a requirement suggests that these RNA sequences are recognized by components of the replicase complex. To look for proteins in such complexes that may interact with the SL RNAs, we synthesized RNA molecules encompassing the nucleotides of the possible SL structures. Nucleotide sequence of these potential SL structure is shown in Figs. 4A, B and C, and have been designated as RV 5′(+)

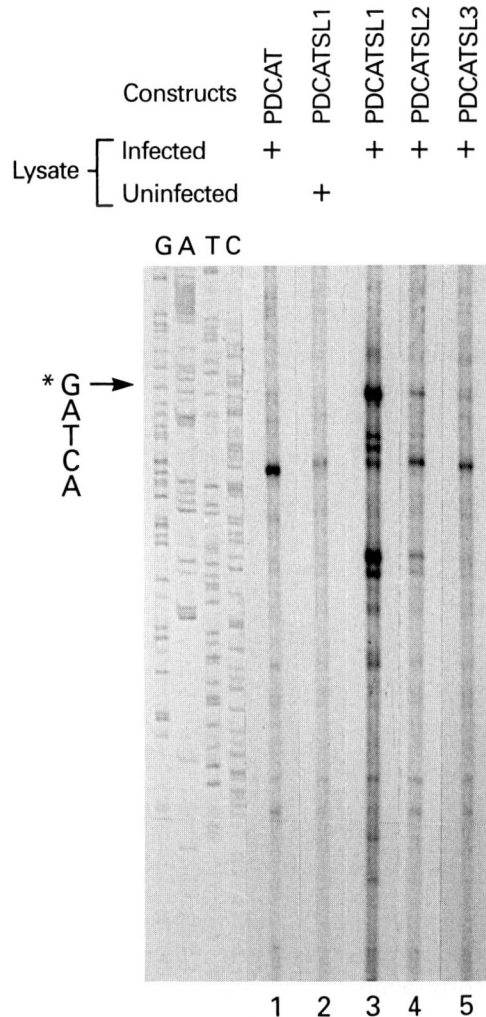

Fig. 3. Primer extension analysis of negative-strand CAT RNA from cells expressing chimeric CAT RNA infected with RV. RNA isolated from infected cells expressing PDCAT (1); RNA isolated from uninfected cells (2) and infected cells (3) expressing PDCATSL1; RNA isolated from infected cells expressing PDCATSL2 (4); and RNA isolated from infected cells expressing PDCATSL3 (5). Lanes labeled as G A T C represent the nucleotide sequence of the 3′ end region of RV genomic RNA. The arrow points to the nucleotide corresponding to the novel primer extension product

SL RNA; 3′(−) SL RNA and 3′(+) SL RNA, respectively. The radio-labeled RNA was used as a probe to search for specific RNA-binding proteins present in both uninfected and infected cytosols by the use of the gel mobility shift and UV cross-linking techniques [8].

Products from UV crosslinking reactions containing cell lysates and 5′(+) SL RNA probe were resolved by SDS-PAGE. Two major protein bands of 52 and 59 kDa and minor protein band of ~110 kDa were

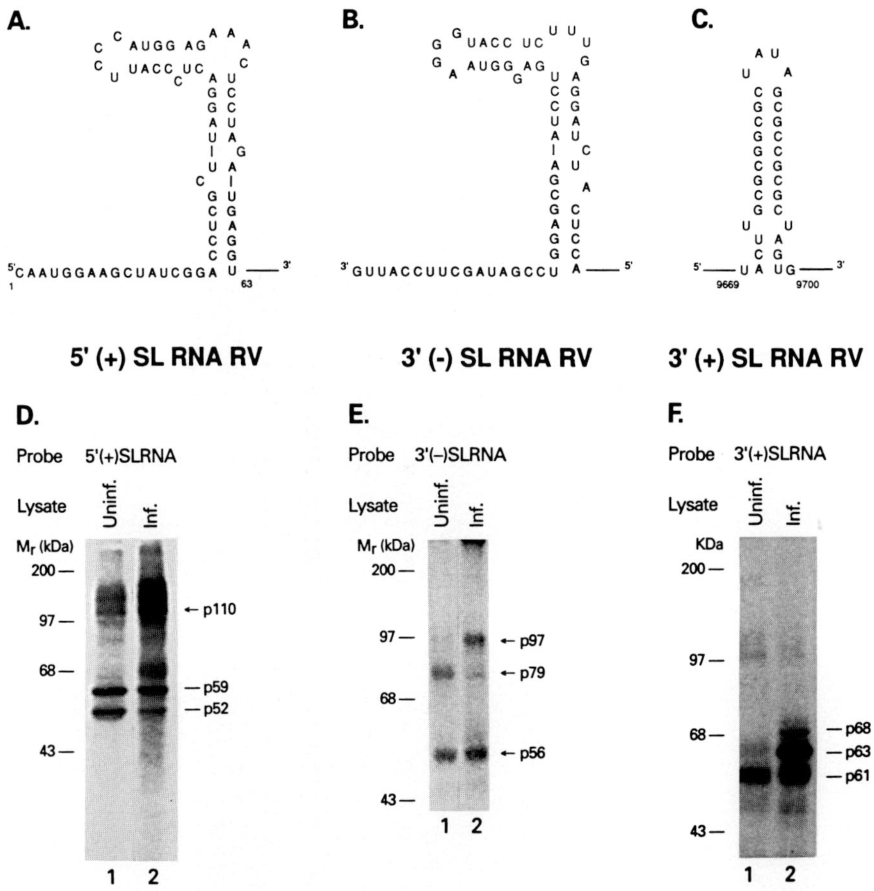

Fig. 4. Nucleotide sequence of the RV RNA stem-loop structures and SDS-PAGE analysis of UV cross-linked proteins. The nucleotide numbers are given at the 5′ and 3′ ends and the numbers correspond to sequence numbers in [4]. **A** 5′(+) SL RNA; **B** 3′(−) SL RNA and **C** 3′(+) SL RNA. Cell lysates from uninfected and rubella virus infected Vero 76 cells were incubated, cross-linked with labeled RNA probes from 5′(+) SL **D**; 3′(−) SL **E**; and 3′(+) SL **F** and analyzed on SDS-PAGE. Arrows denote the sizes of proteins interacting with the RNAs

observed from both uninfected and infected lysates (Fig. 4D, lanes 1, 2). An excess unlabeled RNA of identical sequence and polarity could specifically block the interaction of only 52 and 59 kDa proteins, whereas unrelated RNAs did not block this activity (data not shown).

Three cytosolic proteins with molecular weights of 56, 79 and 97 kDa were observed to specifically interact with the 3′(−) SL RNA (Fig. 4E, lanes, 1, 2 and [9]). Altering the SL structure by deleting sequences in either of the two potential loops (Fig. 4B) abolished the binding interaction [9]. Of the three proteins, the 56 kDa protein also appeared to bind specifically to 3′(+) SL RNA [9]. Further, the 3′(−) SL RNA of

RV seems to share protein-binding activities with similar structures in alphaviruses [9].

Similar analysis using the $3'(+)$ SL RNA as a probe revealed that three specific high-affinity binding proteins, with relative molecular masses (M_r) of 61, 63 and 68 kDa, interacted with the RNA (Fig. 4F, lanes 1, 2). An increase in the binding activity of 63 and 68 kDa proteins after infection coincided temporally with the appearance of the negative-strand RNA synthesis and this increase was sensitive to protein synthesis inhibitors [8]. Both alteration in the stem structure by removal of specific bases or treatment of cell lysates with alkaline phosphatase abrogated the binding interaction [8].

In summary, we have shown that the possible SL structures at the ends of RV RNA, which are necessary for translation and negative-strand synthesis, bind specifically to cellular proteins. Further, one of the cellular proteins has recognition sites for both the $3'$ positive- and $3'$ negative-strand SL RNAs, suggesting that this protein may be involved in more than one process in viral replication.

Identification of host proteins

Having established that various host proteins interact with the cis-acting elements of RV RNA, we identified the individual RNA-binding proteins. Initially, the cellular protein (56 kDa) which interacts with both the $3'(+)$ and $3'(-)$ SL RNAs was isolated and characterized. The 56 kDa protein was purified to homogeneity from Vero 76 cytosolic extracts using various chromatography procedures [14]. Products from each step of the purification were monitored by RNA gel retardation assay using the $3'(+)$ SL RNA as a probe. The purified protein appears to be homogeneous on silver stained SDS-PAGE (Fig. 5B, lane 1; [14]). The homogeneous protein interacts specifically with the $3'(+)$ SL RNA in gel mobility shift assays (Fig. 5A, lanes 1–5) and on SDS-PAGE after UV cross-linking [14]. An excess of unlabeled RNA of identical sequence and polarity could specifically block the binding activity (Fig. 5A, lanes 2 and 3), whereas unrelated RNAs, such as globin or poly (I: C), did not interfere in the activity (Fig. 5A, lanes 4 and 5).

Further analysis was necessary to confirm that the 56 kDa protein is the only purified component interacting with the $3'(+)$ SL RNA. We, therefore, separated the purified 56 kDa protein on SDS-PAGE, and renatured. The renatured protein bound specifically to the RV $3'(+)$ SL RNA in subsequent gel mobility shift assays [14]. From such an analysis it was evident that the 56 kDa protein was a single polypeptide and we therefore proceeded to further identify this protein. The protein was digested with trypsin, and tryptic preptides were purified

A.

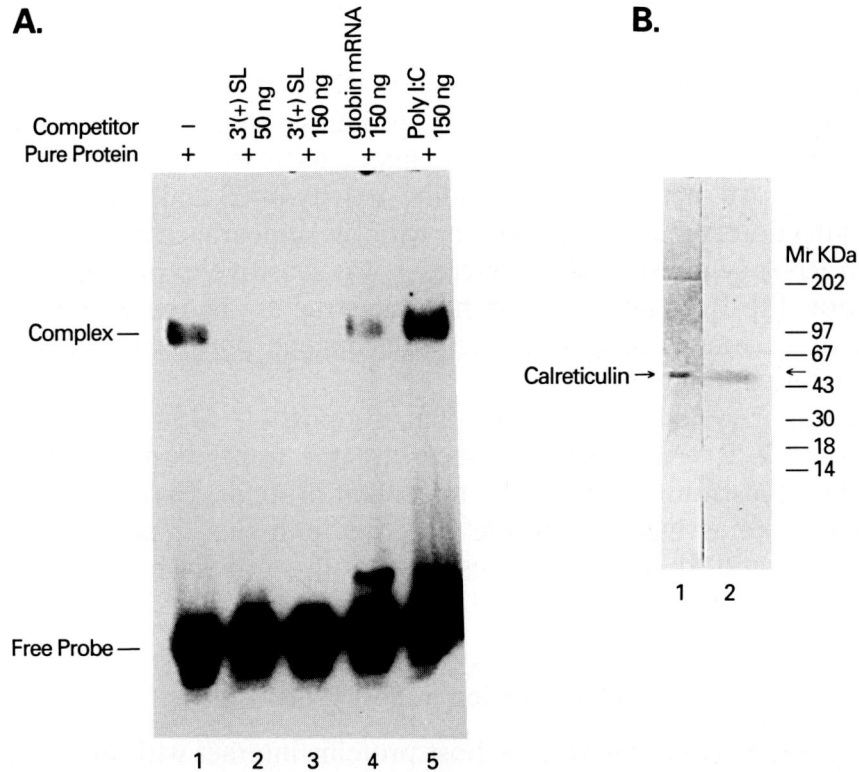

B.

Fig. 5. **A** Gel-retardation assay of 3′(+) SL RNA probe with purified 56 kDa protein. Purified protein was incubated with labeled RNA probe alone (*1*); in the presence of increasing amounts of specific RNA (*2* and *3*); and in the presence of non-specific RNAs such as globin (*4*) and poly (I: C) (*5*). **B** Silver stained SDS-PAGE of the purified 56 kDa protein (*1*). *2* is Western blot analysis of the protein in (*1*) with the human calreticulin antibody

for amino acid sequence analysis. Comparison of peptide amino acid sequence of the 56 kDa protein with the protein data bank revealed that the 3′(+) SL RNA binding protein is 100% homologous to human calreticulin [5]. The similarity of the protein with human calreticulin was further confirmed by Western blot analysis, using either antisera against the entire human calreticulin or two individual peptides derived from it (Fig. 5B, lane 2; [14]).

Calreticulin is a Ca^{2+} binding protein primarily localized in the lumen of the endoplasmic reticulum in a variety of tissues [7]. However, it has been suggested that calreticulin can be associated with a group of small molecular weight cytoplasmic RNAs known as hYRNAs [5]. These RNAs are usually known to bind Ro/SS-A antigen, which is the target of autoantibodies present in the serum of many patients with systemic lupus erythematosus and Sjogren's syndrome [2, 16]. To determine the

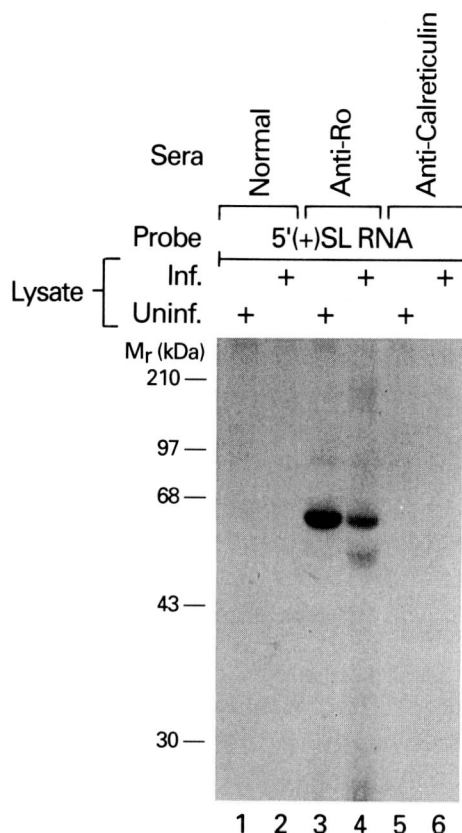

Fig. 6. SDS-PAGE analysis of imunoprecipitated RNA-protein complexes with the 5'(+) SL RNA probe. *1* (uninfected) and *2* (infected) cell lysates cross-linked with labeled probe and immunoprecipitated with non-immune sera. *3* (uninfected) and *4* (infected) cell lysates cross-linked with labeled probe and immunoprecipitated with human anti-Ro antibody. *5* (uninfected) and *6* (infected) cell lysates cross-linked with labeled probe and immunoprecipitated with anti-calreticulin antibody

specificity of the interaction of calreticulin with the 3'(+) SL RNA, UV cross-linked RNA-protein complexes from Vero 76 cell cytosolic extracts were immunoprecipitated with human anti-calreticulin sera [14]. Two RNA-protein complexes containing labeled 3'(+) SL RNA were immunoprecipitated using anti-calreticulin sera, whereas no complexes were observed from reactions incubated with normal sera [14]. In a control experiment, we failed to immunoprecipitate a protein-3'(+) SL RNA complex by either mono- or polyclonal anti-Ro/SS-A sera [14]. We also performed similar analyses with the 5'(+) SL RNA binding proteins using calreticulin and Ro/SS-A antibodies. Surprisingly, we observed that the Ro/SS-A antibody immunoprecipitated RNA-protein complexes of approximately 59 and 52 kDa with the 5'(+) SL RNA (Fig. 6, lanes 3, 4) but not with the 3'(+) SL RNA [13, 14]. In contrast, calreticulin antibody did not immunoprecipitate a RNA-protein complex when the 5'(+) SL RNA was used as a probe (Fig. 6, lanes 5, 6). Non-immune sera did not immunoprecipitate the RNA-protein complexes (Fig. 6, lanes 1, 2). Because 5'(+) SL RNA of RV was shown to be essential for RV/CAT protein synthesis (Fig. 2, lanes 3, 4), the interaction of Ro/SS-A-associated antigens with the 5'(+) SL RNA raises the possibility that these proteins may be involved in RV RNA translation [13]. However, at present, the role of Ro/SS-A antigen in RV/CAT translation is speculative.

In summary we have demonstrated that 5'(+) and 3'(+) SL structures provide important *cis*-acting sequences for RV RNA translation and replication. We have also identified two host cellular proteins, Ro/SS-A-associated antigens and calreticulin, which can interact in vitro with the 5'(+) SL and 3'(+) SL RNAs, respectively, and are currently characterizing their possible role in RV replication and translation.

Acknowledgements

We thank Drs. J. D. Capra and R. Sontheimer (Univ. Texas), Dr. M. Michalak (Univ. Alberta) and Dr. L. Rokeach (Univ. Montreal) for providing calreticulin antibodies. We would like to thank Dr. E. Chan (Scripps Institute, La Jolla) for providing antibodies to Ro/SS-A antigen.

References

1. Cooper LZ, Buimovici-Klein E (1985) Rubella. In: Fields BN, Knipe DM, Chanock RM, Melnick JL, Roizman B, Shope RE (eds) Virology. Raven Press, New York, pp 1005–1020

2. Deutscher SL, Harley JB, Keene JD (1988) Molecular analysis of the 60-kDa human Ro bibonucleoprotein. Proc Natl Acad Sci USA 85: 9479–9483

3. Dominguez G, Wang C-Y, Frey TK (1990) Sequence of the genome RNA of rubella virus: evidence for genetic rearrangement during togavirus evolution. Virology 177: 225–238

4. Jang SK, Wimmer E (1990) Cap-independent translation of encephalomyocarditis virus RNA: structural elements of the internal ribosomal entry site and involvement of a cellular 57-kD RNA-binding protein. Genes Dev 4: 1560–1572

5. McCauliffe DP, Lux FA, Lieu T-S, Sanz I, Hanke J, Newkirk MM, Bachinski LL, Itoh Y, Siciliano MJ, Reichlin M, Sontheimer RD, Capra JD (1990) Molecular cloning, expression, chromosome 19 localization of a human Ro/SS-A autoantigen. J Clin Invest 85: 1379–1391

6. Meerovitch K, Pelletier J, Sonenberg N (1989) A cellular protein that binds to the 5'-noncoding region of poliovirus RNA: implications for internal translation initiation. Genes Dev 3: 1026–1034

7. Michalak M, Milner RE, Burns K, Opas M (1992) Calreticulin. Biochem J 285: 681–692

8. Nakhasi HL, Rouault TA, Haile D, Liu T-Y, Klausner RD (1990) Specific high-affinity binding of host cell proteins to the 3' region of rubella virus RNA. New Biol 2: 255–264

9. Nakhasi HL, Cao X-Q, Rouault TA, Liu T-Y (1991) Specific binding of host cell proteins to the 3′-terminal stem-loop structure of rubella virus negative-strand RNA. J Virol 65: 5961–5967

10. Najita L, Sarnow P (1990) Oxidation-reduction sensitive interaction of a cellular 50 kDa protein with an RNA hairpin in the 5′ noncoding region of the poliovirus genome. Proc Natl Acad Sci USA 87: 5846–5850

11. Oker-Blom C, Ulmanen I, Kaariainen L, Pettersson R (1984) Rubella virus 40S genomic RNA specifies a 24S subgenomic mRNA that codes for a precursor to structural proteins. J Virol 49: 403–408

12. Pardigon N, Strauss JH (1992) Cellular proteins bind to the 3′ end of Sindbis virus minus-strand RNA. J Virol 66: 1007–1015

13. Pogue GP, Cao X-Q, Singh NK, Nakhasi HL (1993) 5′ sequences of Rubella virus RNA stimulate translation of chimeric RNAs and specifically interact with host-encoded proteins. J Virol 67 (in press)

14. Singh NK, Rouault TA, Liu T-Y, Nakhasi HL (1993) Purification and characterization of rubella virus 3′(+) SL RNA-binding host protein (in preparation)

15. Strauss JH, Kuhn RJ, Niesters HGM, Strauss E (1990) Functions of the 5′-terminal and 3′-terminal sequences of the Sindbis virus genome in replication. In: Brinton MA, Heinz FX (eds) New aspects of positive-strand RNA viruses. Am Soc Micro, Washington, pp 61–66

16. Tan EM (1982) Autoantibodies to nuclear antigens (ANA): their immunobiology and medicine. Adv Immunol 33: 167–177

Authors' address: Dr. H. L. Nakhasi, DHP/CBER/FDA, Bldg. 29, Rm. 107, 8800 Rockville Pike, Bethesda, MD 20892, U.S.A.

Arch Virol (1994) [Suppl] 9: 269–277

Archives
of
Virology
© Springer-Verlag 1994
Printed in Austria

Interaction of cellular proteins with the poliovirus 5′noncoding region

E. Ehrenfeld[1] and **J. G. Gebhard**[2]

[1] Department of Molecular Biology and Biochemistry, University of California, Irvine, California, and [2] Department of Cellular Viral and Molecular Biology, University of Utah School of Medicine, Utah, U.S.A.

Summary. The 5′ noncoding region (NCR) of poliovirus RNA is folded into a complex structure comprised of multiple, critically spaced, stem-loop domains. Mutations in at least one of these domains markedly affects the neurovirulence of the virus. Two proteins have been identified recently which bind and apparently mediate functions of the 5′ NCR in translation. We have demonstrated specific binding of three additional proteins in a Hela cell ribosomal salt wash that can be crosslinked to specific stem-loop segments of the 5′ NCR. These same RNA segments inhibit translation of polio RNA in vitro, presumably by competing for protein binding. The Sabin vaccine strain of polio RNA exhibits a reduced affinity of binding for specific proteins. The determinant for this reduction appears to be a single nucleotide difference at position 480 between the neurovirulent and attenuated viral strains.

Introduction

Viral versus host cell translation

From the earliest studies of the molecular events associated with the growth and replication of polioviruses in cultured cells, it was recognized that the virus induces a specific inhibition of host cell protein synthesis (for review see [1]). The virus-induced block occurs at the initiation step of protein synthesis such that cellular mRNAs fail to bind to ribosomal subunits. Under these conditions of host cell protein synthesis inhibition, viral protein synthesis nevertheless proceeds efficiently. These observations implied that initiation of viral translation occurred by a mechanism that differed in some fundamental way from host cell protein synthesis.

The proposal by Kozak [2] of the scanning mechanism of translation of capped, cellular mRNAs, and the identification of protein factors required to catalyze the various steps in translation initiation [3], confirmed the need to postulate another mechanism for translation of poliovirus and other picornaviral RNAs: Their uncapped, long 5′noncoding regions (NCRs) with multiple unused AUG codons and extensive secondary and tertiary structure, precluded application of the Kozak rules for translation and prohibited interactions with cap-binding proteins thought to direct the scanning machinery. Indeed, numerous experiments were conducted that collectively suggested that ribosome binding leading to initiation of translation occurred by recognition of specific RNA structural or sequence elements located not at the 5′end, but rather internally, within the 5′NCR [4]. The region of the 5′NCR that functions as the ribosome binding site is called the IRES, for Internal Ribosome Entry Site [5] or the RLP, for Ribosome Landing Pad [6]. It has been demonstrated for poliovirus, encephalomyocarditis virus, rhinovirus, foot and mouth disease virus and hepatitis A virus (for review see [7]).

Structure and function of the 5′noncoding region

In recent years, intensive efforts have been directed toward defining the features of the poliovirus 5′NCR responsible for its unique, cap-independent binding to ribosomal subunits. Secondary structure maps have been predicted by several computer-assisted RNA folding procedures [8–11] and, fortunately, all predict similar overall structures, with some differences in the details of particular stem-loop domains (Fig. 1). Biochemical probings for nucleotides involved in base pairing were consistent with many of the structural predictions [9, 10]. In addition, an analysis of sequence variation found in independent poliovirus isolates supported the existence of most of the proposed stem-loop structures by revealing extensive structure-conserving substitutions in the stems [12].

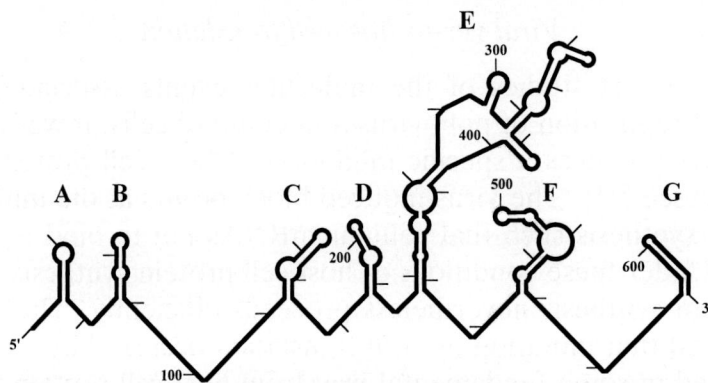

Fig. 1. Schematic representation of the predicted secondary structure of the poliovirus 5′NCR. The figure has been adapted from [12]

In order to dissect which regions of the poliovirus 5′NCR control translation initiation, numerous mutations were introduced into cDNAs representing the 5′NCR so as to generate RNA transcripts whose translation could be evaluated, both in vitro and in vivo. The results showed that a large segment of approximately 450 nucleotides between positions ≈140 and ≈600, comprising several stem-loop domains (C–F, Fig. 1) is required for IRES function (reviewed in [4, 7]). Within this region, most deletions and point mutations eliminate IRES function, although deletion of the entire domain D can be tolerated. Often, spacing, rather than sequence, between stem-loops is important.

Neurovirulence and the 5′ noncoding region

Single nucleotide substitutions in the lower stem of domain F (approximate position 480) of the poliovirus 5′NCR drastically modify the neurovirulence of poliovirus [13]. Although a weakening or disruption of this stem might affect any of several functions of the viral RNA, the increased attenuation of RNAs carrying these mutations has been linked to a reduction in their translation efficiencies relative to their neurovirulent counterparts [14, 15]. It has been suggested that this differential translation was more pronounced in cells of neural origin than in other cells [16, 17]. This, in turn, suggested that there might be cell specificity in the expression of the translation phenotype. The idea of cell-specific modulation of picornavirus translation efficiencies fit well with a number of previous observations that poliovirus RNA behaved as a poor messenger RNA in lysates of rabbit reticulocytes, but that both stimulation of translation initiation as well as correction of the fidelity of initiation site utilization could be achieved by addition to the reticulocyte translation reaction of extracts from Hela cells. Thus, some factors present (in sufficient amounts) in Hela cells but not in rabbit reticulocyte lysates, or factors perhaps limiting in cells of the nervous system, appeared to be required for efficient and correct translation of poliovirus RNA.

Interaction of cellular proteins with the 5′ noncoding region

The postulated existence of cell-specific factors that mediate IRES utilization prompted numerous investigators to search for cellular proteins that could specifically interact with the poliovirus 5′NCR. Meerovitch et al. [18] identified a 52 kD protein in Hela cell extracts that was shown to bind to nts 559–624 in domain G. This protein has recently been identified as the La antigen (N. Sonenberg, pers. comm.). A 57 kD protein was isolated from the ribosomal salt wash of Hela cells and was shown to bind to the poliovirus 5′NCR upstream of the oligopyrimidine tract [19]. This protein is identical to the protein that binds to a critical stem-loop in the 5′NCR of EMCV RNA essential for IRES function [20], and it has been identified as the polypyrimidine tract-binding protein (Wimmer and Jackson, pers. comm.). In addition, other regions of the 5′NCR have been shown to specifically bind to a number of cellular proteins [21–23], although the identities of these proteins and their possible functions in translation have not been ascertained.

Results

Rationale

Recently, our laboratory also became interested in identifying cellular proteins that interact with the poliovirus 5′NCR and that may function to facilitate its internal ribosome binding and translation initiation. Concern about two types of issues influenced our approach. The first was the use of small RNA fragments representing particular domains of the 5′NCR as binding probes to search for proteins. Preliminary experiments with some small RNA fragments showed specific crosslinking (by irradiation with ultraviolet light) to cellular proteins that did not crosslink to the intact 5′NCR [24]. This is likely because small RNA fragments may sometimes adopt folded conformations different from these adopted by the same sequence within the context of a larger RNA. Our second concern involved demonstrating that a given RNA-protein interaction was significant for translation. The mere demonstration that a protein binds to an RNA segment important for translation can not automatically be interpreted to mean that the protein functions in the translation reaction.

Competition of 5′ noncoding region RNA fragments for translation

We attempted to address the latter problem by first trying to identify regions of the poliovirus (Mahoney, type 1) 5′NCR that could compete for translation of poliovirus RNA, presumably by binding to factors needed for translation [24]. We constructed a nested set of RNAs from the 5′end of poliovirus RNA, and we added increasing concentrations of these RNAs (up to a 25-fold molar excess over mRNA) to a rabbit reticulocyte lysate translating poliovirus RNA, supplemented with Hela cell factors. Table 1A shows the RNAs tested and their effects on the translation of poliovirus RNA [24]. RNAs containing poliovirus sequences 1–626 and 1–456 both competed efficiently for translation; shorter RNA of 1–286 or shorter failed to compete. Control RNAs of similar lengths and nonsense sequences showed no competition.

When the same RNAs were tested for their abilities to compete for translation of other (uncapped) mRNAs such as chloramphenicol acetyl

Table 1. Competition of poliovirus 5′NCR RNA fragments for translation

Competitor RNA	Messenger RNA		
	Poliovirus RNA	CAT mRNA	VSV mRNA
A Translation in rabbit reticulocyte lysates supplemented with Hela cell extracts			
PV 1–626	+	+	+
1–456	+	–	–
1–286	–	–	–
1–66	–	–	–
nonsense RNA (~500 nts)	–	–	–
B Translation in rabbit reticulocyte lysates without supplementation			
1–626	+	+	
1–456	–	–	
1–286	–	–	
1–66	–	–	

transferase (CAT) or vesicular stomatitis virus (Indiana) NS protein mRNAs, only 1–626 competed; 1–456 did not (Table 1A). These data suggested that the 3′end of the IRES region (nts 456–626) binds some general translation factors required by all mRNAs, whereas the upstream end of the IRES (nts 286–456) binds factors specifically required for the translation of poliovirus RNA but not utilized by other mRNAs.

To determine whether the factors bound by the various RNA fragments were located in the reticulocyte lysate or were supplied by the Hela cell extract, the competition experiments were performed again during translation of the truncated poliovirus RNA in rabbit reticulocyte lysate alone, without supplementation with Hela cell extract (Table 1B). The RNA fragment that contained sequences that bound general translation factors (1–626) competed effectively, as expected. The RNA fragment that bound factors specific for poliovirus RNA sequences (1–456) did not compete for translation of poliovirus RNA in the absence of Hela cell extract. Thus, the poliovirus-specific factors were supplied by the Hela cell extract, and were not present (or only in small amounts) in rabbit reticulocyte lysates.

UV crosslinking of proteins to the poliovirus 5′noncoding region

Efforts to determine what proteins in Hela cell extracts were binding to the poliovirus 5′NCR involved UV crosslinking of Hela cell extracts to ^{32}P-labelled 5′NCR RNA (nts 1–865) followed by ribonuclease digestion and analysis of the tagged proteins by SDS-PAGE. Four major poly-

Table 2. Competition of poliovirus 5′NCR RNA fragments for crosslinking

Competitor RNA	p52	p48	p38	p35
1–865 (self)	+	+	+	−
1–626	+	+	+	−
1–456	−	+	+	−
1–286	−	−	−	−
1–66	−	−	−	−
VSV mRNA	−	−	−	−

peptides were identified, with molecular masses of approximately 52, 48, 38 and 35 kDa. Some additional higher molecular weight proteins ($\approx$100 kD and $\approx$70 kD) are also variably crosslinked. Competition for crosslinking by the same RNA fragments utilized in the translation experiments mapped the binding sites of three of the four major proteins (Table 2). (The fourth, 35 kDa protein, rarely showed competition by any RNA, perhaps due to its presence in very high concentration, and thus was not mapped). Applying these data to the secondary structure map of the 5′NCR (Fig. 1) places the 52 kDa protein on stem-loop G, and both the 48 and 38 kD proteins on stem-loop E. Both the molecular weight and the binding location of p52 are consistent with its being the La protein already identified by Meerovitch et al. [7, 18]. The 48 and 38 kD proteins appear to interact with the core of the IRES. They are specific for poliovirus RNA and are present in Hela cell extracts but not rabbit reticulocyte lysates, and thus are good candidates for mediating internal ribosome binding. At least one of these proteins is likely responsible for causing a gel mobility shift of isolated E loop RNA ([23]; Blyn et al., unpubl. obs.), and experiments are currently underway to identify these proteins.

Sabin versus Mahoney

As mentioned above, a major determinant of the attenuated phenotype of the Sabin strains of poliovirus is a single nucleotide change that weakens or disrupts the stem of the F domain. The attenuated strains are translated less well than their neurovirulent counterparts. We have constructed a series of RNAs to use for both translation and crosslinking studies to examine the effects on protein binding of the single nucleotide substitutions at position 480 in type 1 poliovirus. A diagram of these RNAs is shown in Fig. 2. There are a total of five nucleotide differences in the 5′NCR of Mahoney (M) and Sabin (S) RNA. At position 480,

Translation

X-linking

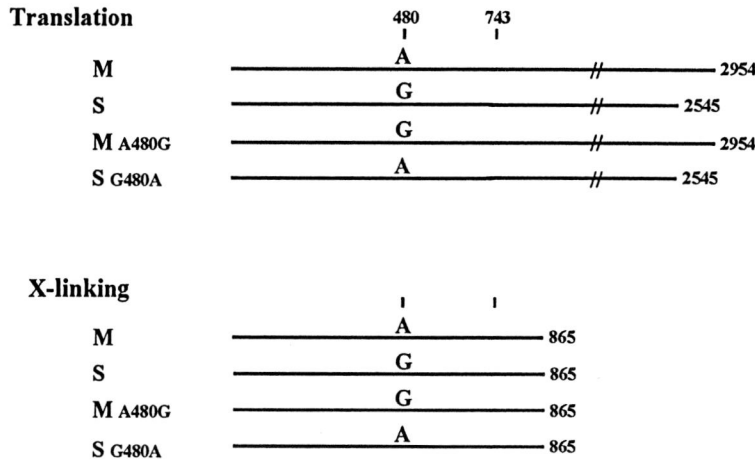

Fig. 2. Diagram of poliovirus RNAs transcribed from engineered cDNAs of Mahoney and Sabin strain sequence. The RNAs were used for translation studies or crosslinking, as indicated

Mahoney RNA contains an A residue, whereas Sabin RNA has a G. Reciprocal mutations were introduced so as to generate RNAs identical to the parents except at position 480. Translation experiments in Hela cell extracts show that Sabin RNA translates less well than Mahoney RNA at any RNA concentration. Competition translation experiments and translation data for the engineered mutant RNAs are not yet completed, so we do not know whether changing the G residue in Sabin RNA to an A improves its translational efficiency.

Similar sets of RNAs representing nucleotides 1–865 were prepared for crosslinking studies. Preliminary analysis of crosslinking experiments with M and S RNAs and the 480 mutants indicate that Sabin RNA crosslinks to the same set of four major (35–52 kDa) proteins and several higher molecular weight proteins as Mahoney. The affinity of crosslinking, however, is significantly reduced. Substitution of a G residue at position 480 in Mahony RNA did not detectably change its binding affinity. Substitution of A_{480} in Sabin RNA, however, improved its crosslinking affinity significantly, restoring it to the same level as seen with Mahoney.

Conclusions

The observations described above represent a preliminary approach to dissecting the interactions between the poliovirus 5'NCR and cellular proteins that mediate translation initiation from the poliovirus IRES. A full understanding of this complex reaction will require identification and characterization of all of the trans-acting factors involved, as well as the cis-acting elements within the RNA. We may anticipate a likely scenario in which the spectrum of mRNA-specific cell-specific and general translation factors utilized by viral mRNAs will match the complexity of the transcription factor pattern required to regulate gene expression from viral DNAs.

Acknowledgement

This work was supported by grant AI-12387 from the National Institutes of Health.

References

1. Sonenberg N (1990) Poliovirus translation. Curr Top Microbiol Immunol 161: 23–47
2. Kozak M (1988) The scanning model for translation: an update. J Cell Biol 108: 229–241
3. Hershey JWB (1991) Translational control in mammalian cells. Annu Rev Biochem 60: 717–755
4. Agol VI (1991) The 5′-untranslated region of picornaviral genomes. Adv Virus Res 40: 103–180
5. Jang SK, Davies MV, Kaufman RJ, Wimmer E (1989) Initiation of protein synthesis by internal entry of ribosomes into the 5′non-translated region of encephalomyocarditis virus RNA in vivo. J Virol 63: 1651–1660
6. Pelletier J, Sonenberg N (1988) Internal initiation of translation of mRNA. Nature 334: 320–325
7. Meerovitch K, Sonenberg, N (1993) Internal initiation of picornavirus RNA translation. Semin Virol (in press)
8. Rivera VM, Welsh JD, Maizel JV (1988) Comparative sequence analysis of the 5′noncoding region of the enteroviruses and rhinoviruses. Virology 165: 42–50
9. Pilipenko EV, Blinov VM, Chernov BK, Dmitrieva TM, Agol VI (1989) Conserved structural domains in the 5′-untranslated region of picornaviral genomes: an analysis of the segment controlling translation and neurovirulence. Virology 168: 201–209
10. Skinner MA, Racaniello VR, Dunn G, Cooper J, Minor PD, Almond JW (1989) New model for the secondary structure of the 5′non-coding RNA of poliovirus is supported by biochemical and genetic data that also show that RNA secondary structure is important in neurovirulence. J Mol Biol 207: 379–392
11. Le S-Y, Zucker M (1990) Common structures of the 5′non-coding RNA in enteroviruses and rhinoviruses: thermodynamic stability and statistical significance. J Mol Biol 216: 729–741
12. Pöyry T, Kinnunen L, Hovi T (1992) Genetic variation in vivo and proposed functional domains of the 5′noncoding region of poliovirus RNA. J Virol 66: 5313–5319
13. Racaniello V (1988) Poliovirus neurovirulence. Adv Virus Res 34: 217–246
14. Svitkin YV, Maslova SV, Agol VI (1985) The genomes of attenuated and virulent poliovirus strains differ in their in vitro translation. Virology 147: 243–252
15. Svitkin YV, Cammock N, Minor PD, Almond JW (1990) Translation deficiency of the Sabin type 3 poliovirus genome: association with an attenuating mutation C472-U. Virology 175: 103–109

16. Agol VI, Drozdov SG, Ivannikova TA, Kolesnikova MS, Korolev MB, Tolskaya EA (1989) Restricted growth of attenuated poliovirus strains in cultured cells of a human neuroblastoma. J Virol 63: 4034–4038
17. La Monica N, Racaniello VR (1989) Differences in replication of attenuated and neurovirulent polioviruses in human neuroblastoma cell line SH-S4S4. J Virol 63: 2357–2360
18. Meerovitch K, Pelletier J, Sonenberg N (1989) A cellular protein that binds to the 5′-noncoding region of poliovirus RNA: implication for internal translation initiation. Genes Dev 3: 1026–1034
19. Pestova TV, Hellen CUT, Wimmer E (1991) Translation of poliovirus RNA: role of an essential cis-acting oligopyrimidine element within the 5′nontranslated region and involvement of a cellular 57-kilodalton protein. J Virol 65: 6194–6204
20. Jang SK, Wimmer E (1990) Cap-independent translation of encephalomyocarditis virus RNA. Structural elements of the internal ribosomal entry site and involvment of a cellular 57-kd RNA binding protein. Genes Dev 4: 1560–1572
21. delAngel RM, Papavassiliou AG, Fernandez-Tomas C, Silverstein SJ, Racaniello V (1989) Cell proteins bind to multiple sites within the 5′untranslated region of poliovirus RNA. Proc Natl Acad Sci USA 86: 8299–8303
22. Najita L, Sarnow P (1990) Oxidation-reduction sensitive interaction of a cellular 50-kDa protein with an RNA hairpin in the 5′noncoding region of the poliovirus genome. Proc Natl Acad Sci USA 87: 5846–5850
23. Dildine SL, Semler BL (1992) Conservation of RNA-protein interactions among picornaviruses. J Virol 66: 4364–4376
24. Gebhard JR, Ehrenfeld E (1992) Specific interactions of Hela cell proteins with proposed translation domains of the poliovirus 5′noncoding region. J Virol 66: 3101–3109

Authors' address: Dr. E. Ehrenfeld, Department of Molecular Biology and Biochemistry, 3205 Bio Sci II, University of California, Irvine, CA 92717, U.S.A.

Arch Virol (1994) [Suppl] 9: 279–289

Archives
of
Virology
© Springer-Verlag 1994
Printed in Austria

IRES-controlled protein synthesis and genome replication of poliovirus

M. Schmid and **E. Wimmer**

Department of Microbiology, School of Medicine, State University
of New York at Stony Brook, Stony Brook, New York, U.S.A.

Summary. Initiation of translation of the single-stranded genomic RNAs of picornaviruses such as poliovirus (PV) and encephalomyocarditis virus (EMCV) is cap-independent and controlled by a long segment within the 5′non-translated region (5′NTR), termed internal ribosomal entry site (IRES). Cellular RNA-binding proteins have been identified that are involved in IRES function in trans. One of these proteins (p57) has been found to be identical to the polypyrimidine tract binding protein (pPTB), a nuclear protein implicated in various processes involving pre-mRNA. Anti-pPTB antibodies inhibit picornavirus mRNA, but not globin mRNA translation, in vitro. Proof for the 5′-independent initiation of translation in vivo was obtained by inserting the EMCV IRES into the ORF of PV thereby constructing a dicistronic, viable poliovirus with the genotype [PV] 5′NTR-P1-[EMCV] IRES-[PV] P2-P3-3′NTR. Dicistronic polioviruses were also constructed that served as novel expression vectors where a foreign gene has been inserted into the PV genome. Incubation of poliovirus RNA in a HeLa cell-free extract leads to the synthesis and processing of viral proteins, viral RNA replication followed by formation of infectious virions. Cell-free synthesis of PV has nullified the dictum that no virus can multiply in a cell-free medium. The genome replication of poliovirus and the mechanism of recombination in poliovirus replication is still not fully understood. Biochemical evidence has been obtained that the conserved NTP-binding motif in PV protein 2C is essential for RNA replication and virus propagation. Finally by using genetic studies we found that during viral RNA synthesis a poliovirus containing two tandemly arranged VPgs (3A-VPg1-VPg2-3C^{pro}) led to the removal of the 3C-proximal VPg copy.

Introduction

Although much information about poliovirus (genus *Enterovirus*, family *Picornaviridae*) with respect to structure, gene organization, gene expression, replication and pathogenicity has been obtained, multiple aspects of the biology of this virus remain to be solved. This review summarizes some recent results of picornavirus protein synthesis, poliovirus replication and recombination.

Cap-independent translation of picornaviruses

Poliovirus, the prototype picornavirus, is a nonenveloped icosahedral virus with a single-stranded RNA genome of plus strand polarity. On entry into the cell, the genomic RNA, prior to its function as mRNA, is thought to be cleaved from its 5′-terminal protein VPg to yield a pUp 5′ end. Indeed, all poliovirus-specific mRNA isolated from polysomes has this pUp 5′ end.

Poliovirus mRNA has several characteristics, shared by mRNAs of other members of the family *Picornaviridae*, that are distinct from eukaryotic mRNA and are difficult to reconcile with the scanning model for translation initiation [1]. These unique features include: (i) a long 5′ nontranslated region (5′NTR) ranging from 650–1 300 nt., (ii) multiple non-initiating AUGs upstream of the translation initiation site, and (iii) absence of a cap structue at the 5′ end [2]. Experimental evidence has shown that cap-independent initiation of translation of picornavirus mRNA occurs by binding of ribosomes, or their subunits, to a specific segment of RNA downstream of the 5′ end but within the 5′NTR [3–6]. We have termed this approximately 400 nt. long segment the internal ribosomal entry site (IRES) [3, 4].

Based on differences in RNA sequences and proposed secondary structures, these cis-acting elements can be divided into two types: (i) IRES elements of the genera *Enterovirus* and *Rhinovirus* (type 1) and (ii) of the genera *Cardiovirus*, *Aphthovirus* and *Hepatovirus* (type 2) [2, 7–9]. Despite their dissimilarities in sequence and higher order structures, the two IRES types share common features highly conserved among all picornaviruses. An example is the YnXmAUG motif that consists of a stretch of pyrimidine residues (Yn; n = 8–10) separated from the downstream AUG triplet by a non-conserved nucleotide sequence (Xm; m = 18–22) [2, 10]. Generally, this motif resides at the 3′ border of the IRES. In type 2 IRES elements the AUG of the YnXmAUG motif serves as the initiating codon for polyprotein synthesis (Fig. 1A). In type 1 IRES elements, the AUG of the motif is "silent" for translation. Instead, an AUG codon downstream from the YnXmAUG serves as the polyprotein initiating codon. In poliovirus mRNA, the "spacer

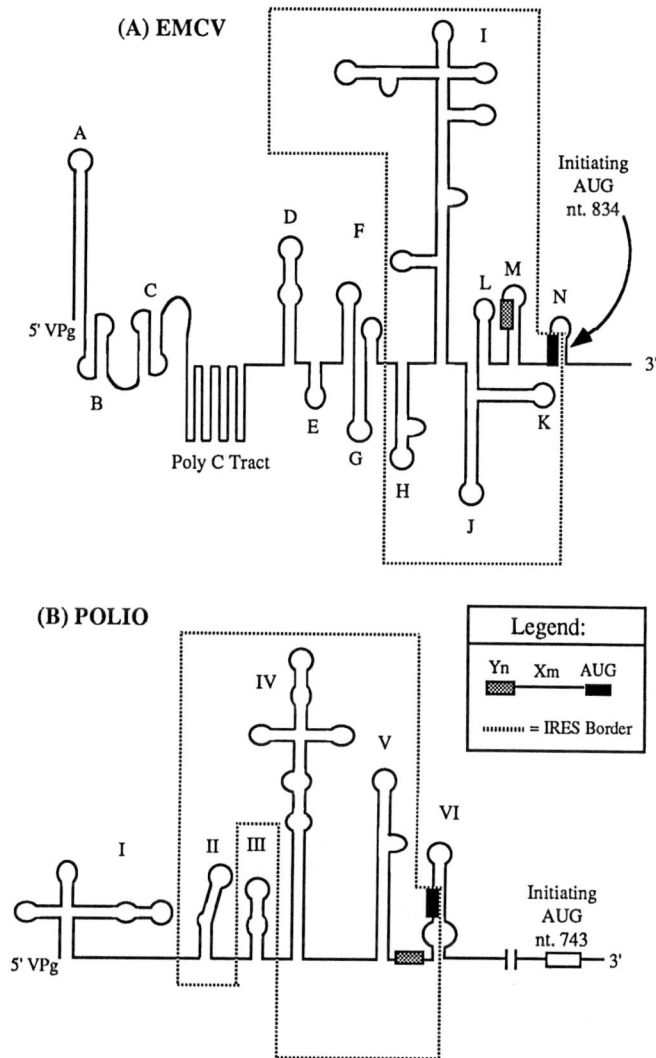

Fig. 1. The structure of the IRES. Schematic representations of the predicted secondary structures for the EMCV (**A**) and poliovirus (**B**) IRES elements. Yn-Xm AUG motifs are indicated by shaded boxes (where the light shaded box = Yn, the dark shaded box = AUG and the space between these boxes = Xm). Stem loop structures of the EMCV IRES and poliovirus IRES elements are labeled with letters and roman numerals, respectively. The poliovirus AUG start codon at nt 743 is labeled with an open box and the spacer distance between the YnXmAUG motif and the initiating AUG is 154 nucleotides. The borders of the IRES are indicated by dashed boxes (text and figure from [40])

sequence" between motif and initiation codon is 154 nt. long (Fig. 1B) whereas in rhinovirus mRNA, it is 31 nt. long [2, 10]. Figure 1 illustrates the borders of the IRES elements and the proposed structures used for ribosomal binding to initiate translation in the encephalomyocarditis virus (EMCV) and poliovirus 5′ NTR regions [7, 8, 11]. Recently, internal ribosome binding has also been shown for several cellular mRNAs, including Bip [12], antennapedia [13] and bFGF mRNAs [14]. These RNAs appear to contain structures that function in 5′-independent translation as is the case with the IRES elements of picornaviruses.

Internal ribosomal entry depends on the interaction of the IRES element with specific trans-acting factors. Some of these factors are derived from the host cell. A ribosome associated protein (p57) has been found to interact specifically with the IRES of EMCV [11]. Mutations that abolish p57 binding to a specific segment of the EMCV IRES (the H-loop; formerly designated as the E-loop) correlate with a decrease in cap-independent translation [11]. The binding of p57 to IRES elements has also been observed for EMCV by Borovjagin et al. [15], for foot- and mouth disease virus (FMDV) by Luz and Beck [16], and for poliovirus by Pestova et al. [17].

Using UV cross-linking to [^{32}-P]-labeled EMCV IRES followed by nuclease digestion, p57 was resolved by SDS PAGE into doublet or triplet of bands. On the basis of this property as well as similarities in its purification characteristics, we have explored the possibility that p57 and a nuclear RNA binding protein (pPTB) are related [18]. The polypyrimidine tract binding protein (pPTB), also known as heterogenous nuclear ribonucleoprotein (hnRNP) I [19], is a 57 kDa nuclear protein that may be involved in 3′ splice site selection as well as in events leading to alternate splicing [20–22]. We used antisera directed against purified recombinant pPTB to study the possible relationship between p57 and pPTB. P57 from a HeLa S10 cytoplasmic fraction bound to an EMCV IRES specific RNA probe (nt. 260–488) was immunoprecipitated with antibodies prepared against both purified recombinant pPTB and a peptide corresponding to amino acid residues 219–238 of pPTB. UV cross- linking of recombinant pPTB which contained 12 vector-derived amino acid residues to the same probe yielded a single band with slightly lower mobility than that of p57. This single band was precipitated with anti- pPTB serum, as was p57 that had been purified from Krebs 2 ascites carcinoma cells. These results suggest that human and murine cytoplasmic proteins that bind to picornavirus IRES elements share antigenic deter- minants with pPTB. Because pPTB was identified as a nuclear protein, we examined the binding of the EMCV IRES RNA probe to HeLa nuclear extract. In this experiment we detected co-migrating 57 kDa species in the cytoplasmic and nuclear extract. Conclusive evidence for the identity of p57 and pPTB was provided by the similarities in the patterns of [^{32}-P]-labeled peptides which resulted from cleavages by cyanogen bromide and V8 protease. This experiment was carried out by binding recombinant pPTB or cytoplasmic proteins of a HeLa extract to

[^{32}-P]-labeled H-loop, followed by UV cross-linking, RNAse A digestion and PAGE. The [^{32}-P]-labeled bands were then eluted from the gel and subjected to analysis. New data, obtained by nitrocellulose filter binding and UV cross-linking assays indicated that purified pPTB binds with lower specificity to the EMCV IRES than does endogenous pPTB in crude cellular extracts. Binding specificity of purified pPTB can be restored by adding back cellular extract, an observation suggesting that factors in the extract interact with pPTB, thereby altering its binding parameters [23]. Additional support for the involvement of p57 in the IRES dependent internal initiation was based on the observation that anti-pPTB antibodies inhibit cap-independent translation of mRNAs containing either type 1 (polio) or type 2 (EMCV) IRES elements. In contrast, translation of uncapped and capped β-globin mRNA was unaffected by addition of anti-pPTB, anti-peptide or pre-immune sera [18]. However, IRES-dependent translation was not restored by adding back purified pPTB to the immunodepleted extract. The inability to restore IRES function after immunodepletion demonstrates that p57/pPTB acts together with other factors, probably forming a ribonucleo-protein complex and that these factors were depleted together with p57/pPTB. The observation that a nuclear protein is an essential transacting factor in IRES-mediated translation is surprising. Provided that pPTB is indeed involved in the metabolism of pre-mRNA, this implies interesting similarities between eukaryotic pre-mRNA splicing and the initiation of translation by internal ribosomal entry. Both processes require the precise assembly of the RNA into RNP complexes. P57 could play a key role in translation by virtue of its ribosome-association and its affinity for a specific structural element within the IRES [18].

Direct evidence for the function of the IRES in vivo has been obtained by constructing a dicistronic poliovirus in which the EMCV IRES was inserted into the open reading frame of poliovirus polyprotein between the regions encoding P1 and P2. This virus (W1-P1/E/P2,3-1) (Fig. 2B) was found to express a small plaque phenotype. In HeLa cells, the dicistronic virus synthesized all virus-encoded proteins in yields comparable to those of wild type virus [24]. Significantly, an analysis of virus concentration and plaque number showed that each virus particle of W1-P1/E/P2,3-1 had the propensity to initiate an infectious cycle [24]. This observation proofs that the IRES element functions in vivo without degradation to yield monocistronic RNAs. Insertion of the EMCV IRES between 2A and 2B also yielded a viable poliovirus (W1-P1,2A/E/2BC, P3-1) (Fig. 2C) expressing a small plaque phenotype [25]. However, insertion of the IRES elements into all remaining cleavage sites of P2 or P3 (Figs. 2D–G) abolished viral proliferation. Available evidence

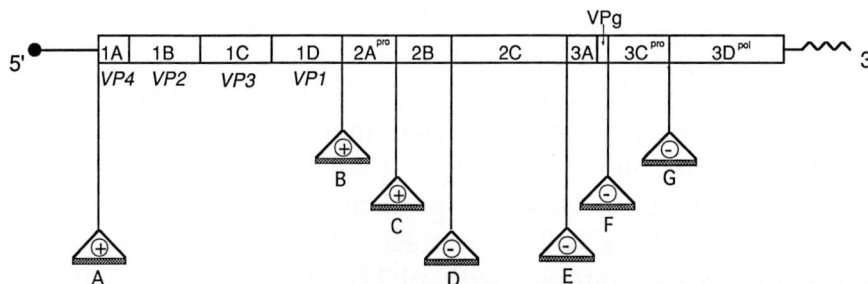

Fig. 2. Schematic representation of dicistronic poliovirus constructs containing the EMVC IRES at various sites. Insertions of the EMCV IRES are indicated by *A* to *G*. Viability and nonviability are indicated by (+) and (−) respectively

suggests that this inactivation is due to the disruption of cis cleavages of the P2–P3 polyprotein (Paul et al., unpubl. res.).

In another study, a viable poliovirus was constructed by inserting the EMCV IRES between the 5′NTR and the open reading frame of poliovirus (W1-PNENPO) (Fig. 2A). After transfection into HeLa cells, the hybrid poliovirus that contained two heterologous IRES elements in tandem, replicated and expressed a small plaque phenotype. Insertion of the chloramphenicol acetyl transferase (CAT) gene between the two IRES-elements of W1-PNENPO yielded a viable dicistronic poliovirus (W1-DICAT). This remarkable virus, which contains a 17% larger genome than wild type poliovirus RNA, has similar growth characteristics to W1-PNENPO. Moreover, CAT assays and Western blots illustrate that the foreign CAT gene is expressed in vivo [26].

Dicistronic polioviruses provide a powerful tool to study functional aspects of the poliovirus polyprotein as well as the function of viral polypeptides and the poliovirus 5′NTR. In addition, these viruses may prove useful in the construction of novel expression vectors and vaccines.

Cell-free synthesis of poliovirus and RNA replication

An important advance in our studies of picornaviral replication and genetics has been the development of a HeLa cell-free extract that, if programmed with poliovirus RNA, directs the synthesis of viral proteins, followed by RNA replication and RNA packaging to yield infectious virus. After incubation for 10 or more h, this cell-free system releases synthesized infectious virions [27]. The time course of protein synthesis and processing is shown in Fig. 3. We have used this system to explore the role of membranes, or membrane components, in RNA replication or study the effect of specific inhibitors such as brefeldin A. For example, we have determined that oleic acid inhibits viral protein synthesis at high concentrations (100 μM) and virus forma-

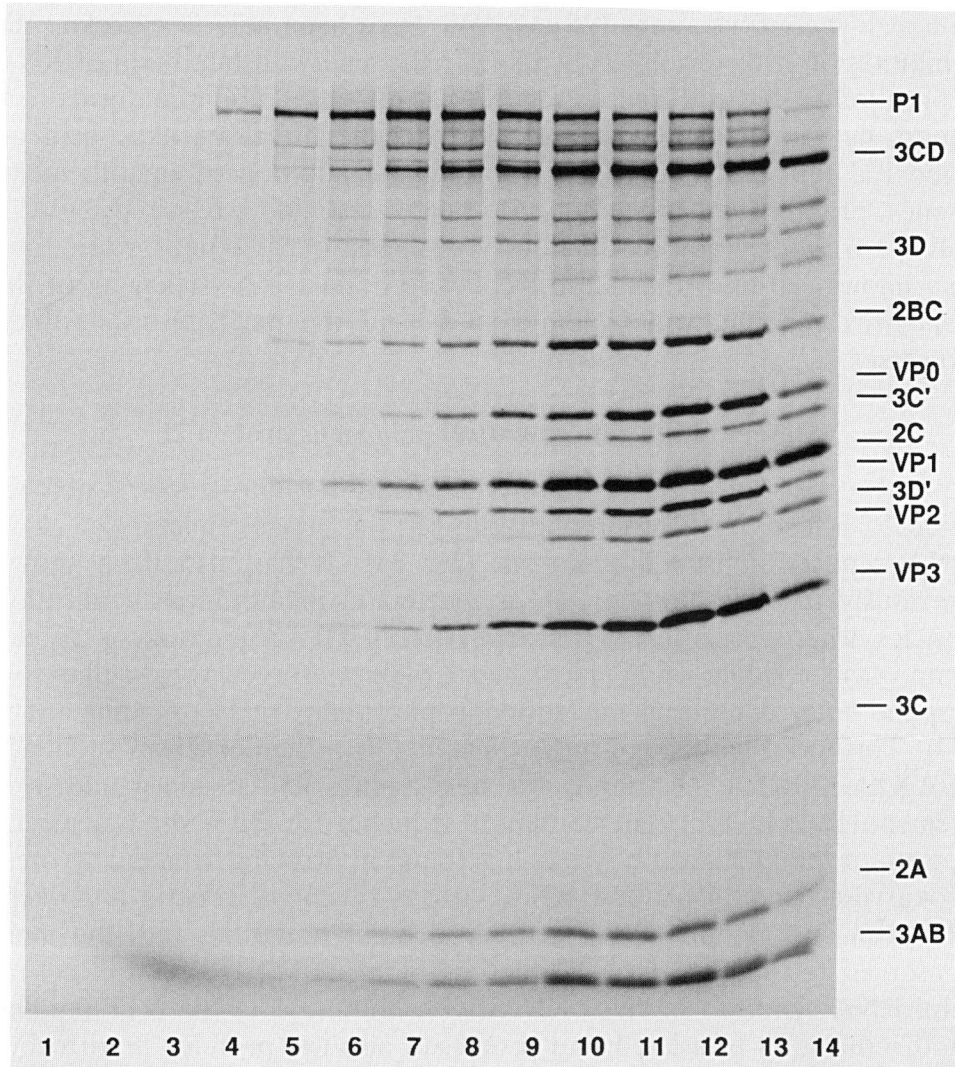

Fig. 3. Time course of poliovirus polypeptide synthesis in vivo. The in vitro incubation mixtures (250 µl) contained 93 µl of S10 extract with the following additions: 1 mM ATP, 50 µM GDP, 10 mM creatine phosphate, 6 µg creatine phosphokinase, 2 mM DTT, 6 µM of calf liver tRNA, 12 µM each of all amino acids except methionine, 300 µCi of [^{35}S] methionine 18 mM Hepes (pH 7.4), 240 µM spermidine, 0.1 M K acetate, 0.35 mM Mg acetate, 0.75 mM MgCl$_2$ and 2 µg of PV1(M) RNA. Incubation was for 18 h at 30°C. At each time point samples (20 µl) were removed, mixed with RNAse A (to 20 µg/ml), RNAse Ti (to 100 U/ml), and cyclohexamide (to 5 µg/ml), incubated for 5 min at 30°C, and frozen immediately, Portions (6 µl) were analysed on 12.5% SDS-polyacrylamide gels. Gels were fixed, treated with EN3HANCE, dried, and exposed for 15 to 24 h to x-ray film at 70°C. *1* no RNA; *2–13* incubation at 0, 0.5, 1, 2, 3, 4, 5, 6, 9, 12, 15, and 18 h, respectively: *14* cytoplasmic extract of [^{35}S] methionine-labeled PVI(M) infected HeLa cells as marker (figure and text from [27])

tion at low concentrations (25 μM). We have been able to correlate the inhibition of virus formation at 25 μM oleic acid with inhibition of RNA replication in the cell-free extract [28]. This observation conforms to a report by Guinea and Carrasco [29] on the effect of oleic acid on polioviral replication in vivo. Following the addition of smooth membranes, protein synthesis at 100 μM oleic acid could be restored, but this was not true for virion production. We suggest that the new system, one that allows cell-free translation, replication and RNA packaging of the poliovirus genome, will facilitate studies of the mechanisms of these processes.

Studies of the non-structural poliovirus protein 2C

Several reports have shown that mutations in the non-structural polypeptide 2C affect viral RNA replication, but the precise function of 2C in RNA replication is still unknown [30, 31]. It has also been shown previously that viral protein 2C is located in membranous replication vesicles where viral RNA synthesis occurs [32, 33]. Analysis of the amino acid residues of 2C identified a well conserved region that correspond to a consensus nucleoside triphosphate (NTP)-binding motif [34]. This motif consists of an A element, with the consensus sequence G/AXXGXGKS/T (X stands for any amino acid residue) and a B element (D or DD/E). The consensus sequence for the B site is a special version of the D-E-A-D box which is found in proteins from eukaryotes, prokaryotes and DNA and RNA viruses. Using a genetic approach, Mirzayan and Wimmer [35] introduced point mutations into the most conserved residues of the NTP-binding motif and tested their effect on viral RNA synthesis in vivo. RNAs with mutations in the NTP-binding motif efficiently translated and processed all viral proteins in vitro but did not produce infectious viruses. After transfection into HeLa cells these RNAs displayed defects in RNA replication, an observation suggesting that the conserved sites of the NTP-binding motif of 2C are important in RNA replication and virus proliferation [35]. In addition Mirzayan and Wimmer (submitted) using a baculovirus vector, have expressed 2C in insect cells. The subsequent purification of 2C has led to the demonstration that this polypeptide expresses an ATPase activity that is stimulated by RNA.

Genetic studies involving VPg

The 5′ end of all newly synthesized viral genomic RNAs is covalently linked to a small virus encoded protein, VPg [36]. Available evidence has led to the hypothesis that a uridylylated form of VPg or 3AB serves as a primer for both positive- and negative- strand synthesis [37]. The genome of aphthoviruses, such as FMDV, contain three copies of tandemly arranged VPg coding regions. All three VPgs encoded by the aphthoviruses RNAs are used in viral genome replication. In contrast, the genomes of all other picornaviruses, poliovirus included, encode only one VPg. Using cartridge mutagenesis, a set of poliovirus cDNA clones were constructed [pT7-2VPg(L6M1)] to contain two tandemly arranged VPgs. Transcript RNAs were then transfected into HeLa cells to determine whether poliovirus could tolerate two or more VPgs and, if so, which copy would be attached to the viral genome. The results revealed that after transfection of HeLa cells with pT7-2VPg(L6M1) RNA, virus could be isolated but it contained only a single VPg-encoding sequence. By using a genetic marker, we determined that all recovered viruses contained only the 3A-linked VPg copy, on observation suggesting that a very specific event, either homologous recombination or loop out deletion, led to the removal of the 3C-proximal VPg copy [38]. Analysis by RT-PCR of viral RNAs extracted from transfected cells at various times post-transfection indicated the presence of two RNA species: one containing a single VPg sequence, the other contains two VPg sequences. However, as the infection progressed the quantity of the 2xVPg-genome decreased concomitant with an increase of the quantity of the 1xVPg-genome. These results suggest that two homologous VPgs are not stable during polioviral RNA replication. The genetic pressure leading to the selection of the 1xVPg-genome appears to reside in the process of proteolytic cleavages [39].

Acknowledgements

We thank all members of our laboratory for comments and contributions from their work. This work was supported by the National Institute of Allergy and Infectious Diseases and the National Cancer Institute of the NIH. M. Schmid is a recipient of postdoctoral fellowship from the Deutsche Forschungsgemeinschaft (DFG).

References

1. Kozak M (1978) How do eukaryotic ribosomes select initiation regions in messenger RNA? Cell 15: 1109–1123
2. Jang SK, Pestova TV, Hellen CUT, Witherell GW, Wimmer E (1990) Cap-independent translation of picornavirus RNAs: structure and function of the internal ribosomal entry site. Enzyme 44: 292–309
3. Jang SK, Kräusslich HG, Nicklin MJH, Duke GM, Palmenberg AC, Wimmer E (1988) A segment of the 5′ nontranslated region of encephalomyocarditis virus RNA directs internal entry of ribosomes during in vitro translation. J Virol 62: 2636–2643
4. Jang SK, Davis MV, Kaufman RJ, Wimmer E (1989) Initiation of protein synthesis by internal entry of ribosomes into the 5′ nontranslatd region of encephalomyocarditis virus RNA in vivo. J Virol 63: 1651–1660
5. Pelletier J, Sonenberg N (1988) Internal initiation of translation of eukaryotic mRNA directed by a sequence derived from poliovirus RNA. Nature 334: 320–325

6. Pelletier J, Sonenberg N (1989) Internal binding of eukaryotic ribosomes on poliovirus RNA: translation in HeLa cell extracts. J Virol 63: 441–444

7. Pilipenko EV, Blinow VM, Romanova LI, Sinyakov AN, Maslova SV, Agol VI (1989a) Conserved structural domains in the 5'-untranslated region of picornaviral genomes: an analysis of the segment controlling translation and neurovirulence. Virology 168: 201–209

8. Pilipenko EV, Blinow VM, Chernov BK, Dmitrieva TM, Agol VI (1989b) Conservation of the secondary structure elements of the 5'-untranslated region of cardio- and aphthovirus RNAs. Nucleic Acids Res 17: 5701–5711

9. Brown EA, Day SP, Jansen RW, Lemon SM (1991) The 5' nontranslated region of hepatitis A virus RNA: secondary structure and elements required for translation in vitro. J Virol 65: 5828–5838

10. Pilipenko EV, Gmyl AP, Maslova SV, Svitkin YV, Sinyakov AN, Agol VI (1992) Prokaryotic-like cis elements in the cap-independent internal initiation of translation on picornavirus RNA. Cell 68: 119–131

11. Jang SK, Wimmer E (1990) Cap-independent translation of encephalomyocarditis virus RNA: structural elements of the internal ribosomal entry site and involvement of a cellular 57-kDA RNA-binding protein. Genes Dev 4: 1560–1572

12. Macejak DG, Sarnow P (1991) Internal initiation of translation mediated by the 5' leader of a cellular mRNA. Nature 353: 90–94

13. Oh SK, Scott MP, Sarnow P (1992) Homeotic gene Antennapedia mRNA contains 5' noncoding sequences that confer translational initiation by internal ribosome binding. Genes Dev 6: 1643–1653

14. Prats AC, Vagner S, Prats H, Amalric F (1992) Cis-acting elements involved in the alternative translation initiation process of human basic fibroblast growth factor mRNA. Mol Cell Biol 12: 4796–4805

15. Borovjagin AV, Evstafieva AG, Ugarova TY, Shatsky IN (1990) A factor that specifically binds to the 5'-untranslated region of the encephalomyocarditis virus RNA. FEBS Lett 261: 237–240

16. Luz N, Beck E (1990) A cellular 57 kDa protein binds to two regions of the internal translation initiation site of foot-and mouth disease virus. FEBS Lett 269: 311–314

17. Pestova TV, Hellen CUT, Wimmer E (1991) Translation of poliovirus RNA: role of an essential cis-acting oligopyrimidine element within the 5' nontranslated region and involvement of a cellular 57-kilodalton protein. J Virol 65: 6194–6204

18. Hellen CUT, Witherell GW, Schmid M, Shin SH, Pestova TV, Gil A, Wimmer E (1993) A cytoplasmic 57 kDa protein (p57) that is required for translation of picornavirus RNA by internal ribosomal entry is identical to the nuclear polypyrimidine tract-binding protein. Proc Natl Acad Sci USA 90: 7642–7646

19. Ghetti A, Pinol-Roma S, Michael WM, Morandi C, Dreyfuss G (1992) HnRNP I, the polypyrimidine tract-binding protein: distinct nuclear localization and association with hnRNAs. Nucleic Acids Res 20: 3671–3678

20. Patton JG, Mayer SA, Tempst P, Nadal-Ginard B (1991) Characterization and molecular cloning of polypyrimidine tract-binding protein: a component of a complex necessary for pre-mRNA splicing. Genes Dev 5: 1237–1251

21. Gil A, Sharp PA, Jamison SF, Garcia-Blanco MA (1991) Characterization of cDNAs encoding the polypyrimidine tract-binding protein. Genes Dev 5: 1224–1236

22. Mulligan GJ, Guo W, Wormsley S, Helfman DM (1992) Polypyrimidine tract binding protein interacts with sequences involved in alternative splicing of β-Tropomyosin pre-mRNA. J Biol Chem 267: 25480–25487

23. Witherell GW, Gil A, Wimmer E (1993) Interaction of polypyrimidine tract binding protein and the encephalomyocarditis virus internal ribosomal entry site. Biochemistry 32: 8268–8275

24. Molla A, Jang SK, Paul AV, Reuer Q, Wimmer E (1992) Cardioviral internal ribosomal entry site is functional in a genetically engineered dicistronic poliovirus. Nature 356: 255–257

25. Molla A, Paul AV, Schmid M, Jang SK, Wimmer E (1993b) Studies on dicistronic polioviruses implicate viral proteinase 2A^{pro} in RNA replication. Virology 196: 739–747

26. Alexander L, Lu HH, Wimmer E (1993) Studies of poliovirus containing type 1 and/or type 2 elements: genetic hybrids and the expression of a foreign gene (submitted)

27. Molla A, Paul AV, Wimmer E (1991) Cell-free, de novo synthesis of poliovirus. Science 254: 1647–1651

28. Molla A, Paul AV, Wimmer E (1993a) Effects of temperature and lipophilic agents on poliovirus formation and RNA synthesis in a cell free system. J Virol 67: 5932–5938

29. Guinea R, Carrasco L (1991) Effect of fatty acids on lipid synthesis and viral RNA replication in poliovirus infected cells. Virology 185: 473–476

30. Pincus SE, Diamond DC, Emini EA, Wimmer E (1986) Guanidine-selected mutants of poliovirus: mapping of point mutations of polypeptide 2C. J Virol 57: 638–646

31. Li JP, Baltimore D (1988) Isolation of poliovirus 2C mutants defective in viral RNA synthesis. J Virol 62: 4016–4021

32. Takegami T, Semler BL, Anderson CW, Wimmer E (1983) Membrane fractions active in poliovirus RNA replication contain VPg precursor polypeptides. Virology 128: 33–47

33. Bienz K, Egger D, Troxler M, Pasamontes I (1990) Structural organization of poliovirus RNA replication is mediated by viral proteins of the P2 genomic region. J Virol 64: 1156–1163

34. Gorbalenya AE, Koonin EV, Donchenko AP, Blinow VM (1988) A conserved NTP-motif in putative helicases. Nature 333: 22

35. Mirzayan C, Wimmer E (1992) Genetic analysis of an NTP-binding motif in poliovirus polypeptide 2C. Virology 189: 547–555

36. Lee YF, Nomoto A, Detjen BM, Wimmer E (1977) A protein covalently linked to poliovirus genome RNA. Proc Natl Acad Sci USA 74: 59–63

37. Wimmer E (1979) The genome-linked protein of picornaviruses: discovery, properties and possible functions. In: Perez-Bercoff R (ed) The molecular biology of picornaviruses. Plenum Press, New York, pp 175–189

38. Wimmer E, Hellen CUT, Cao X (1993) Genetics of poliovirus. Annu Rev Genet 27: 353–435

39. Cao X, Kuhn RJ, Wimmer E (1993) Replication of poliovirus RNA containing two VPg genes leads to a specific deletion event. J Virol 67: 5572–5578

40. Harber J, Wimmer E (1993) Aspects of the molecular biology of picornaviruses. In: Carrasco L, Sonenberg N, Wimmer E (eds) Proceedings of the NATO ASI on regulation of gene expression in animal viruses. Plenum Press, New York, pp 189–224

Authors' address: Dr. M. Schmid, Department of Microbiology, School of Medicine, State University of New York at Stony Brook, Stony Brook, NY 11794-8621, U.S.A.

Arch Virol (1994) [Suppl] 9: 291–298

Archives
of
Virology
© Springer-Verlag 1994
Printed in Austria

Analysis of hepatitis A virus translation in a T7 polymerase-expressing cell line

L. E. Whetter[1], **S. P. Day**[1,*], **E. A. Brown**[1], **O. Elroy-Stein**[2], and **S. M. Lemon**[1]

[1] Department of Medicine, The University of North Carolina at Chapel Hill, Chapel Hill, North Carolina, U.S.A.
[2] Department of Cell Research and Immunology, Tel Aviv University, Ramat Aviv, Tel Aviv, Israel

Summary. Hepatitis A virus (HAV) exhibits several characteristics which distinguish it from other picornaviruses, including slow growth in cell culture even after adaptation, and lack of host-cell protein synthesis shut-down. Like other picornaviruses, HAV contains a long 5′ non-translated region (NTR) incorporating an internal ribosomal entry site (IRES), which directs cap-independent translation. We compared HAV IRES-initiated translation with translation initiated by the structurally similar encephalomyocarditis virus (EMCV) IRES, using plasmids in which each of the 5′NTRs is linked inframe with the chloramphenicol acetyltransferase (CAT) gene. Translation was assessed in an HAV-permissive cell line which constitutively expresses T7 RNA polymerase and transcribes high levels of uncapped RNA from these plasmids following transfection. RNAs containing the EMCV IRES were efficiently translated in these cells, while those containing the HAV IRES were translated very poorly. Analysis of translation of these RNAs in the presence of poliovirus protein 2A, which shuts down cap-dependent translation, demonstrated that their translation was cap independent. Our results suggest that the HAV IRES may function poorly in these cells, and that inefficient translation may contribute to the exceptionally slow replication cycle characteristic of cell culture-adapted HAV.

Introduction

Picornavirus RNAs, which lack the 5′ 7-methyl-guanosine cap of cellular mRNAs, have a long 5′ nontranslated region (NTR) which facilitates initiation of translation by promoting internal binding of the 40 S ribo-

* Present address: State Laboratory of Hygiene, Madison, Wisconsin, U.S.A.

some subunit to the RNA [8, 15]. This appears to be true as well for hepatitis A virus (HAV), a unique picornavirus which has a strikingly slow and noncytolytic replication cycle in cultured cells. The predicted secondary structure of the HAV 5′NTR [2] shares a number of features with that of encephalomyocarditis virus (EMCV), a rapidly replicating, cytolytic picornavirus. Similarities in secondary structure extend to the 5′NTR sequence forming the internal ribosomal entry site (IRES) of EMCV, which is highly efficient at promoting internal translation initiation. An IRES element has also been identified between bases 151 and 735 of the HAV 5′NTR. This IRES mediates internal initiation of translation in rabbit reticulocyte lysates (RRL) programmed with bicistronic RNAs in which segments of the HAV 5′NTR were placed within the intercistronic space to control translation of a downstream cistron [3]. However, like poliovirus, HAV translation is inefficient in RRL [2, 10, 15]. In order to gain an appreciation of the extent to which low translational efficiency of the HAV 5′NTR might contribute to its slow and noncytolytic replication cycle, we considered it important to gain an understanding of HAV IRES function in a biologically relevant in vivo system.

Toward this end, we constructed an HAV-permissive cell line, BT7-H, which constitutively expresses bacteriophage T7 RNA polymerase. These cells are derived from continuous African green monkey kidney (B-SC-1) cells. The T7 polymerase supports transcription from transfected plasmid DNA which contains the T7 promoter. In a murine cell line which constitutively expresses T7 polymerase, translation of T7 transcripts was not detected unless the coding region was placed under control of the EMCV IRES [6]. These data suggested that the T7 transcripts were not capped in this cell line, and that T7-producing cells might be useful for studying IRES function in vivo.

HAV 5′NTR-initiated translation is inefficient in BT7-H cells

HAV translation was evaluated in BT7-H cells transfected with plasmids containing the reporter gene chloramphenicol acetyltransferase (CAT), fused in-frame with the HAV 5′NTR and placed within a transcriptional unit with flanking T7 promoter and T7 terminator sequences. To facilitate cloning, it was necessary to create minor changes in the nucleotide sequence surrounding the two possible initiator AUG codons of HAV [16] or in the aminoterminal codons of the CAT gene. Thus four different constructs were evaluated, each of which differs slightly in this region (Fig. 1). pHAV-CAT1 and pHAV-CAT2 contain the complete HAV 5′NTR, while pHAV-CAT3 and pHAV-CAT4 lack 45 bases at the 5′ end of the 5′NTR. Deletion of these bases has little effect on

A.

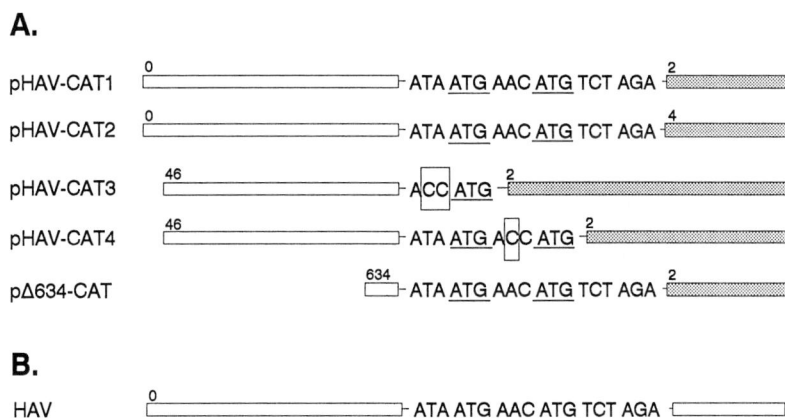

B.

HAV

Fig. 1. A Plasmids containing HAV 5′NTR sequences fused in-frame with CAT. The nucleotide sequence of each is shown at the junction between the HAV 5′NTR (open boxes) and CAT coding region (stippled boxes). Initiator AUG codons [16] in the HAV sequence are underlined and nucleotides deviating from the HAV sequence are boxed. Numbers over the CAT coding region indicate the CAT codon at which the authentic CAT sequence begins. pHAV-CAT1 adds 2 amino acids derived from HAV sequence to the N-terminus of the CAT protein, while pHAV-CAT2 changes the first 2 CAT codons to HAV sequence. Numbers placed over the HAV coding region represent the 5′ limit of the HAV 5′NTR sequence. **B** Nucleotide sequence surrounding the initiator AUG codons in the authentic HAV sequence (open boxes)

translation in vitro [2]. pHAV-CAT1 and pHAV-CAT2 contain the 5′NTR sequence of HM175/P16 HAV [9], while the 5′NTR sequence of pHAV-CAT3 and pHAV-CAT4 are derived from HM175/P35 virus (pHAV/7) [4]. Both of these HAV variants replicate relatively well in BS-C-1 cells.

BT7-H cells were transfected with plasmid DNA using a liposome-mediated transfection system (Lipofectin, BRL), and CAT expression was assessed 48 h later by a phase-extraction, liquid scintillation assay (Promega; pCAT Reporter Gene Systems). As shown in Fig. 2, the HAV IRES was far less active than the EMCV IRES in promoting expression of CAT, with average values for CAT activity nearly one thousand-fold less in cells transfected with pHAV-CAT1 than in those transfected with pEMCV-CAT (an analogous construct containing the EMCV IRES fused in-frame to CAT). This result was not due to differences in RNA transcription or stability, based on evaluation of CAT RNA levels by northern blot (data not shown). Similar low levels of CAT expression were noted following transfection of BT7-H cells with pHAV-CAT2, pHAV-CAT3 and pHAV-CAT4 (data not shown).

Removal of much of the HAV IRES from pHAV-CAT1 increased expression of CAT about 3-fold (compare expression from pHAV-CAT1

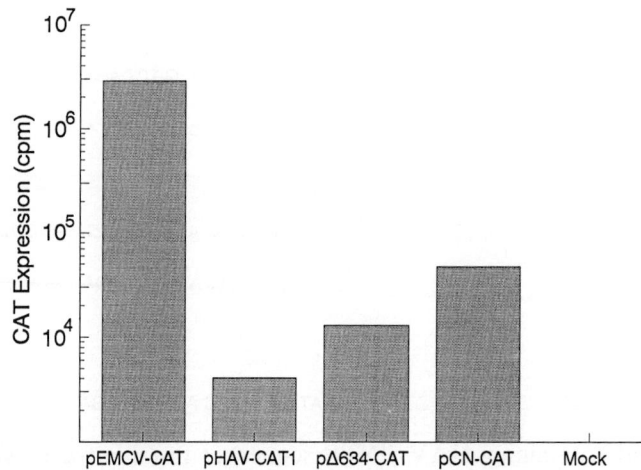

Fig. 2. CAT activity expressed in BT7-H cells transfected with pEMCV-CAT, pHAV-
CAT1, pΔ634-CAT or pCN-CAT. See text for details

with expression from pΔ634-CAT, in which all of the HAV sequence 5′
of base 634 has been removed) (Fig. 2). Since a similar deletion in
bicistronic constructs eliminated HAV IRES activity in RRL [3], it
appeared likely that the translation of pΔ634-CAT transcripts did not
depend on internal ribosomal entry, and perhaps occurred by ribosome
scanning from the 5′ end of the uncapped RNA. To determine whether
this translation was dependent upon specific sequences in the residual
HAV 5′NTR segment, a construct (pCN-CAT) was made in which the
HAV IRES sequence was entirely replaced by a 53 base leader sequence
derived from the pCAT-Control vector (Promega). This plasmid contains an AUG codon in excellent context (AAATGG) for initiation
of CAT by scanning. As shown in Fig. 2, CAT expression in cells
transfected with pCN-CAT exceeded expression in cells transfected with
pΔ634-CAT.

Cap-independent translation is enhanced in the presence of poliovirus protein 2A^{pro}

To verify that CAT translation in the BT7-H cells was cap-independent
and to evaluate the effect of host-cell shutdown on HAV translation,
expression from each of the CAT plasmids was assessed in the presence
of the poliovirus protease, 2A^{pro}. Poliovirus 2A^{pro} induces a proteolytic
cleavage of the p220 subunit of the cap-binding complex eIF-4f, which is
accompanied by shut-down of cap-dependent translation of cellular
mRNAs (for review see [1, 17]). BT7-H cells were co-transfected with
CAT plasmids and p2A-WT, which contains the T7 promoter and the
EMCV IRES fused in-frame with the coding region for poliovirus 2A^{pro}

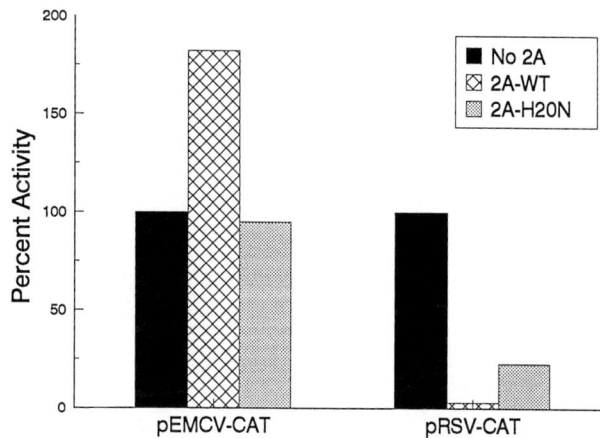

Fig. 3. CAT expression in cells transfected with pEMCV-CAT or pRSV-CAT, with or without co-transfection with p2A-WT (expressing active poliovirus $2A^{pro}$) or p2A-H20N (expressing protease-defective poliovirus $2A^{pro}$). Results were normalized so that CAT activity in the absence of co-transfection with either plasmid represents 100% activity

(a gift from R. Lloyd), and cell lysates were assayed for CAT activity as before. Since the EMCV IRES itself may interfere with cap-dependent host-cell translation [5], we also assessed expression from CAT plasmids following co-transfection with p2A-H20N (also provided by R. Lloyd), containing a poliovirus $2A^{pro}$ coding sequence which has been altered to allow expression of a protease-inactive $2A^{pro}$ [19]. As shown in Fig. 3, expression of CAT from pEMCV-CAT was significantly enhanced in the presence of the active, but not inactive, poliovirus $2A^{pro}$. As an additional control, we evaluated the expression of CAT from pRSV-CAT (provided by S. Kenney), in which capped CAT transcripts are produced by nuclear (rather than cytoplasmic) transcription under the direction of the Rous sarcoma virus promoter. CAT expression from pRSV-CAT was virtually eliminated when cells were co-transfected with p2A-WT (Fig. 3), and reduced to a lesser extent in cells co-transfected with p2A-H20N. These results are consistent with the notion that expression of poliovirus $2A^{pro}$ abolishes cap-dependent translation, but also support the hypothesis that the EMCV IRES inhibits translation of capped transcripts by competing for necessary translation factors.

Like pEMCV-CAT, pΔ634-CAT and pCN-CAT expressed higher levels of CAT in the presence of poliovirus $2A^{pro}$ (Fig. 4). These results confirm that translation of transcripts from these plasmids is cap-independent. Moreover, these data indicate that an intact IRES is not essential for translation of certain uncapped transcripts in vivo, particularly in the absence of competing cap-dependent translation of cellular

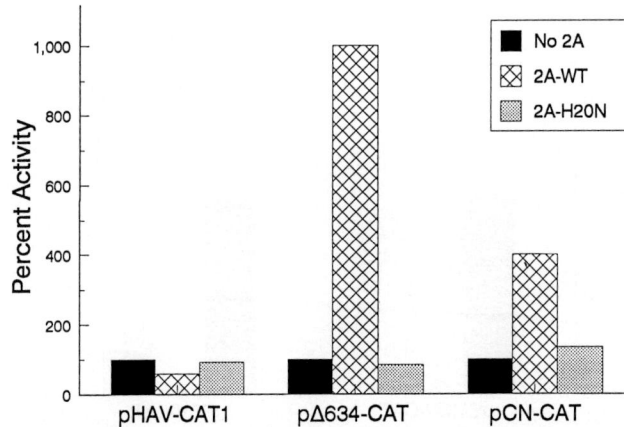

Fig. 4. CAT expression in cells transfected with pHAV-CAT1, pΔ634-CAT or pCN-CAT, with or without co-transfection with p2A-WT or p2A-H20N. Results have been normalized so that CAT activity in the absence of co-transfection with either plasmid represents 100% activity

mRNAs. Surprisingly, translation of CAT from pHAV-1 was not enhanced in the presence of poliovirus 2A^{pro} (Fig. 4).

Discussion

Conventional expression systems which produce mRNA by nuclear transcription complicate in vivo studies of picornavirus IRES function, since the transcripts are capped by cellular enzymes. The use of bicistronic constructs [13], or direct transfection of uncapped RNAs [11] are two alternative strategies which have been used to circumvent this problem and to analyze cap-independent initiation of picornavirus translation. We have chosen to use an HAV-permissive cell line in which there is constitutive cytoplasmic expression of bacteriophage T7 polymerase to study HAV IRES function in vivo. This system has several advantages. First, there is no possibility that the secondary structure or function of an IRES element might be altered by the extensive upstream sequences present by necessity in all bicistronic constructs. In addition, greater quantities of cytoplasmic RNA would be expected in DNA-transfected T7-producing cells than in normal cells transfected directly with RNA. Finally, an advantage of this system over the use of recombinant vaccinia virus expressing T7 polymerase [7, 14] is the lack of capping of T7 transcripts, which may occur in up to 20% of transcripts with the vaccinia-T7 system. Our results confirm that cells which constitutively express T7 polymerase are useful for characterizing the functions of picornaviral IRES elements in vivo.

These experiments have demonstrated that the HAV IRES is far less active in promoting translation initiation in vivo than the structurally similar EMCV IRES. Transcripts in which most of the HAV 5'NTR was deleted, or replaced with a 5' leader which contained no HAV sequence, expressed significantly greater quantities of CAT than transcripts containing the entire HAV 5'NTR. The data suggest that translation of uncapped transcripts having reduced secondary structure at the 5' end (e.g., pΔ634-CAT and pCN-CAT) occurs by scanning of ribosomes from the 5' end and does not require an intact IRES. However, it appears that stable secondary structures present in the HAV sequence upstream of base 634 are inhibitory to translation by this mechanism [2], requiring that translation initiation occur by internal ribosomal entry. Further studies will be needed to better understand the function of the HAV IRES in BT7-H cells. However, the results presented here support the hypothesis that poor HAV IRES function may contribute to the slow growth and low virus yields characteristic of HAV infection in cell culture [12, 18].

References

1. Anthony DD, Merrick WC (1991) Eukaryotic initiation factor (eIF)-4f: implications for a role in internal initiation of translation. J Biol Chem 266: 10218–10226
2. Brown EA, Day SP, Jansen RW, Lemon SL (1991) The 5' nontranslated region of hepatitis A virus RNA: secondary structure and elements required for translation in vitro. J Virol 65: 5828–5838
3. Brown EA (1993) Manuscript in prep.
4. Cohen JI, Ticehurst JR, Feinstone SM, Rosenblum B, Purcell RH (1987) Hepatitis A virus cDNA and its RNA transcripts are infectious in cell culture. J Virol 61: 3035–3039
5. Duke GM, Hoffman MA, Palmenberg AC (1992) Sequence and structural elements that contribute to efficient encephalomyocarditis virus RNA translation. J Virol 66: 1602–1609
6. Elroy-Stein O, Moss B (1990) Cytoplasmic expression system based on constitutive synthesis of bacteriophage T7 RNA polymerase in mammalian cells. Proc Natl Acad Sci USA 87: 6743–6747
7. Elroy-Stein O, Fuerst TR, Moss B (1989) Cap-independent translation of mRNA conferred by encephalomyocarditis virus 5' sequence improves the performance of the vaccinia virus/bacteriophage T7 hybrid expression system. Proc Natl Acad Sci USA 86: 6126–6131
8. Jang SK, Pestova TV, Hellen CUT, Witherell GW, Wimmer E (1990) Cap-independent translation of picornavirus RNAs: structure and function of the internal ribosomal entry site. Enzyme 44: 292–309
9. Jansen RW, Newbold JE, Lemon SM (1988) Complete nucleotide sequence of a cell culture-adapted variant of hepatitis A virus: comparison with wild-type virus with restricted capacity for in vitro replication. J Virol 163: 299–307
10. Jia X-Y, Scheper G, Brown D, Updike W, Harmon S, Richards O, Summers D, Ehrenfeld E (1991) Translation of hepatitis A virus RNA in vitro: aberrant internal initiations influenced by 5' noncoding region. Virology 182: 712–722

11. Hambridge SJ, Sarnow P (1991) Terminal 7-methyl-guanosine cap structure on the normally uncapped 5' noncoding region of poliovirus mRNA inhibits its translation in mammalian cells. J Virol 65: 6312–6315

12. Lemon SM, Murphy PC, Shields PA, Ping LH, Feinstone SM, Cromeans T, Jansen RW (1991) Antigenic and genetic variation in cytopathic hepatitis A virus variants arising during persistent infection: evidence for genetic recombination. J Virol 65: 2056–2065

13. Pelletier J, Sonenberg N (1988) Internal initiation of translation of eukaryotic mRNA directed by a sequence derived from poliovirus RNA. Nature 334: 320–325

14. Percy N, Belsham GJ, Brangwyn JK, Sullivan M, Stone DM, Almond JW (1992) Intracellular modifications induced by poliovirus reduce the requirement for structural motifs in the 5' noncoding region of the genome involved in internal initiation of protein synthesis. J Virol 66: 1695–1701

15. Sonenberg N, Meerovitch K (1990) Translation of poliovirus mRNA. Enzyme 44: 278–291

16. Tesar M, Harmon SA, Summers DF, Ehrenfeld E (1992) Hepatitis A virus polyprotein synthesis initiates from two alternative AUG codons. Virology 186: 609–618

17. Thach RE (1992) Cap recap: the involvement of eIF-4f in regulating gene expression. Cell 68: 177–180

18. Ticehurst J, Cohen JI, Feinstone SM, Purcell RH, Jansen RW, Lemon SM (1988) Replication of hepatitis A virus: new ideas from studies with cloned cDNA. In: Semler BL, Ehrenfeld E (eds) Molecular aspects of picornavirus infection and detection. American Society for Microbiology Press, Washington, pp 27–50

19. Yu SF, Lloyd RE (1991) Identification of essential amino acid residues in the functional activity of poliovirus 2A protease. Virology 12: 615–625

Authors' address: Dr. S. M. Lemon, Department of Medicine, 547 Burnett-Womack, CB No. 7030, The University of North Carolina at Chapel Hill, Chapel Hill, NC 27599-7030, U.S.A.

Arch Virol (1994) [Suppl] 9: 299–306

_Archives_____
V̈irology
ⓒ Springer-Verlag 1994
Printed in Austria

Purification and characterization of the U-particle, a cellular constituent whose synthesis is stimulated by Mengovirus infection

M. R. Mulvey, H. Fang, and **D. G. Scraba**

Department of Biochemistry, University of Alberta, Edmonton, Alberta, Canada

Summary. We have isolated a cellular protein particle whose synthesis is induced by infection with Mengovirus or TMEV. The U-particle inhibits translation in vitro and binds to both capped and uncapped mRNA's. It is spherical, 12 nm in diameter, and is composed of multiple copies of two polypeptide subunits having molecular weights of 23 000 and 25 000 which do not appear to be glycosylated or phosphorylated. U-particles are capable of inhibiting mRNA translation in vitro.

Introduction

Several years ago our laboratory developed a protocol for the purification of the 14S capsid protein pentamers of Mengovirus in order to study virion assembly. During these experiments we discovered that the 12–18S fractions from sucrose gradient centrifugation of infected L-cell lysates contained cellular particles as well as the viral capsid subunits. Moreover, a particular population of such cellular particles was labeled efficiently with radioactive amino acids at a time when total cellular protein synthesis was reduced to 15–20% of normal [3]. Further examination of these particles revealed that their synthesis was actually stimulated approximately 3-fold during the first 6 h of virus infection. Electrophoresis of the particles in denaturing urea-phosphate gels revealed a single polypeptide component of apparent molecular weight 20 000; this polypeptide was designated "U" for protein of unknown function. Electron microscopy of the U-particles showed them to be spherical entities of ≈12 nm diameter, large enough to accommodate 20–25 U-protein molecules. The synthesis of U-particles was induced by Mengovirus in both mouse (L) and human (HeLa) cell lines by Mengovirus; it was not induced by heat shock, or by reovirus infection [2]. Recently, we have resumed our investigations of the U-particle, and in this communication present a progress report on our experiments.

Purification of the U-particle

Since purification of the U-particles by sucrose density gradient centrifugation was limited in terms of both capacity and efficiency, we have turned to conventional chromatographic techniques for this purpose. Cells were infected, labeled (when necessary), harvested 5 h post-infection, and lysed in hypotonic buffer as previously described [2]. Since the U-particle had no known activity to monitor, two assays for its localization were used at each step of the purification procedure. First, because the synthesis of U-particles is induced by Mengovirus infection, U-proteins can be labeled in vivo with [3H]-leucine. Fractions from the various purification steps were subjected to SDS-polyacrylamide gel electrophoresis (SDS-PAGE; [5]) and autoradiographed to determine the location of the U-proteins. Second, samples from the various fractions were negatively stained with sodium phosphotungstate and examined in a Philips EM 420 electron microscope in order to visualize the U-particles.

To begin purification, ammonium sulfate was added to clarified lysates and U-particles were precipitated between 40 and 80% saturation. The precipitated protein was resuspended in reticulocyte standard buffer (RSB; 10 mM Tris-HCl, 10 mM NaCl, 1.5 mM $MgCl_2$, pH 8.5), loaded onto a column (60 × 3 cm) of Sephacryl S-300 (Pharmacia), and eluted with the same buffer. Fractions containing the U-particle were pooled, loaded onto a 16 × 1 cm column of DEAE Sephacel (Pharmacia), and washed with 0.1 M ammonium bicarbonate until the absorbance at 280 nm had returned to baseline. The proteins were then eluted with a continuous ammonium bicarbonate gradient (0.1–1.0 M) at pH 8.5. Fractions containing the U-particle were pooled, concentrated with a Centricon 30 (Amicon), and loaded onto a 20 × 0.8 cm column of Affigel Blue (Biorad) equilibrated with RSB. The U-particles eluted in the first protein (A280 nm) peak. These fractions were pooled, and the U-particles concentrated by ethanol precipitation and resuspended in RSB. A summary of the purification procedure is given in Table 1. Approximately 10 mg of U-particles could be purified from 100 mg of total cellular protein (four large roller bottles; $\approx 10^9$ Mengovirus-infected L-cells).

Characterization of the U-particle

Purified U-particles electrophoresed as a single band in non-denaturing gels (Fig. 1A, lane 1). When the band was excised, eluted from the gel, and subjected to SDS-PAGE [5], two distinct protein bands were visualized; these have molecular masses of 23 and 25 kDa, respectively (Fig. 1A, lane 2). Densitometric scans of the silver-stained U-protein bands showed that about 70% of the mass of the particle is contributed

Table 1. Summary of the purification of the U-particle

	Protein concentration (mg/ml)	Volume (ml)	Total protein (mg)
Lysate	5.12	18.5	95
40–80% $(NH_4)_2SO_4$	1.21	2.65	32
Sephacryl S-300	0.41	25.0	10.3
DEAE Sephacel	3.64	0.3	1.1
Affigel Blue	0.50	0.02	0.01

by the 23 kDa protein. The discrepancy between the previously reported single protein subunit of 20 kDa [2] and the two protein subunits reported here can be attributed to the type of gel system used for electrophoresis: in SDS-urea-phosphate gels the two U-proteins migrate as a single 20 kDa species. Electron microscopic examination showed that the morphology of U-particles purified chromatographically is identical to that of the particles purified by sucrose density gradient centrifugation (Fig. 1B; [2]). Some of the 12 nm spherical U-particles show a stain-filled indentation or hole which could be due to a loss of some protein molecules from the structure. Alternatively, all of the particles may have such an indentation at one end; those appearing to be uniform spheres may simply be oriented such that the hole is not visible.

U-particles are resistant to a variety of proteases. Incubation with trypsin, chymotrypsin, papain or V8 protease (20:1 protease molecules: U particle) at 37° for 4 h had no effect on the morphology of the particle or its subunit size in SDS-PAGE. Digestion with pronase E or proteinase K under the same conditions did result in proteolytic cleavage of the subunits, but electron microscopic examination did not reveal any changes in the morphology of the U-particle.

There were no N- or O-linked sugars detected (by digestion with specific glycosidases followed by SDS-PAGE) on either of the two U-protein subunits, supporting the idea that the particle is localized in the cytoplasmic compartment. Neither was phosphate detected in the U-particle by adding [^{32}P]-orthophosphate to Mengovirus-infected cells in the presence of okadaic acid (a phosphatase inhibitor); this makes it unlikely that any activity the particle may have is regulated by phosphorylation/dephosphorylation. Also, our inability to label the U-particle with phosphate or with [^{3}H]-uridine under conditions where it is being synthesized suggests that it contains little, if any, RNA.

The two U-protein subunits were separated by SDS-PAGE (12% polyacrylamide) and blotted onto a PVDF membrane (Biorad). Attempts to determine their N-terminal sequences by automated Edman degrada-

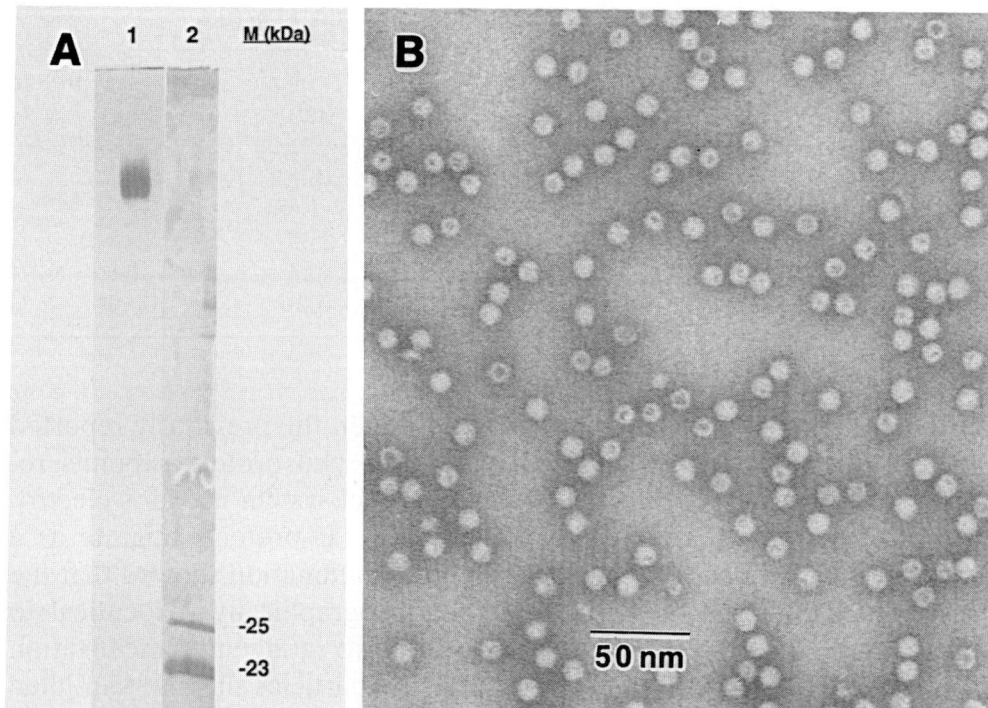

Fig. 1. A Electrophoretic mobility of the U-particle in non-denaturing (*1*) and denaturing (*2*) polyacrylamide gels. Following electrophoresis in a 7% polyacrylamide gel, the Coomassie blue stained U-particle band shown in *1* was excised and eluted into 1.0 ml of 2.3% N-methylmorpholine (pH 7.6) at 37° overnight. The eluted protein was lyophilized, resuspended in electrophoresis sample buffer [5], and boiled for 4 min. The sample was electrophoresed by conventional SDS-PAGE in a 12% gel, then dried and silver stained. Two protein bands (subunits) with molecular weights of 23 000 and 25 000 are evident (*2*). **B** Electron micrograph of purified U-particles. The sample was negatively stained with 2% sodium phosphotungstate (pH 7) and photographed in a Philips EM420 electron microscope operated at 100 kV

tion were unsuccessful, probably because of a blocking group on the N-terminus of each of the protein subunits.

Induction of U-particle synthesis by Theiler's murine encephalitis virus

Since Theiler's virus (TMEV) is now generally recognized as a cardiovirus [6], we were interested to examine its ability to induce the synthesis of U-particles in infected cells. Cytoplasmic extracts of L-cells which had been mock-infected or infected with the GDVII strain of TMEV (kindly provided by Dr. R. Grant at the Harvard Medical School) were centrifuged in sucrose gradients under conditions where U-particles would migrate to near the bottom of the tube [3]. Fractions were collected, bovine plasma albumin was added as a carrier, proteins

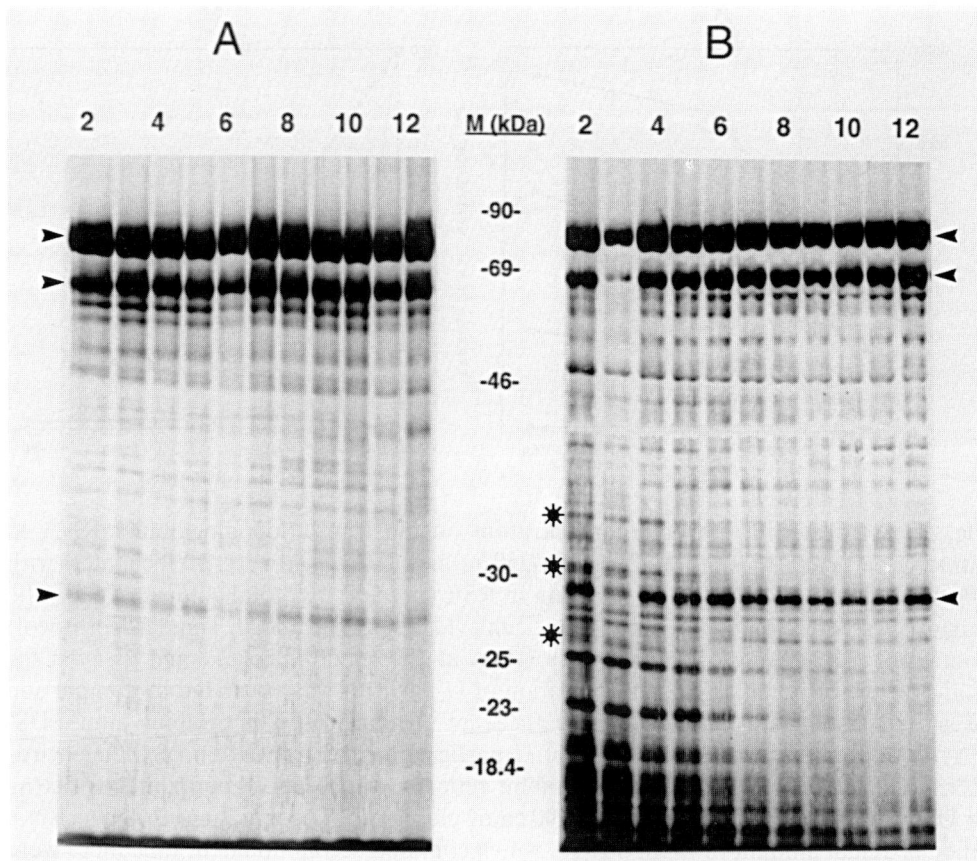

Fig. 2. Induction of the U-particle in TMEV infected L-cells. Cytoplasmic lysates were collected 25 h post-infection and centrifuged in 5–20% sucrose gradients as described previously [3]. Gradient fractions (0.2 ml) were collected, bovine albumin (Fraction V, Sigma; 0.01% final concentration) was added as carrier, and proteins were precipitated with trichloroacetic acid (15% final concentration). The precipitates were collected by centrifugation, washed with acetone and resuspended in electrophoresis sample buffer [5]. Proteins were separated by SDS-PAGE (12% acrylamide) and visualized by silver staining. Electropherograms from mock infected (**A**) and TMEV strain GDVII infected (**B**) L-cells are shown. The prominant bands of MWs 78 000, 68 000, and 28 000 (arrowheads) which appear in all fractions are derived from the bovine plasma albumin. Bands corresponding to the two U-protein subunits having molecular weights of 23 000 and 25 000 are apparent in gel B in fractions 2 to 6, as are bands which probably represent the TMEV proteins VP0 (MW ≈ 37 000), VP1 (MW ≈ 31 000) and VP3 (MW ≈ 26 000) – these are marked by asterisks

were precipitated with TCA and subjected to SDS-PAGE. After electrophoresis the gels were sliver-stained and photographed. The results are shown in Fig. 2. Only very small amounts of the U subunit proteins were detected in the 12–18S region of the sucrose gradient (Fig. 2A, lanes 2–6). In contrast, cells that had been infected with TMEV pro-

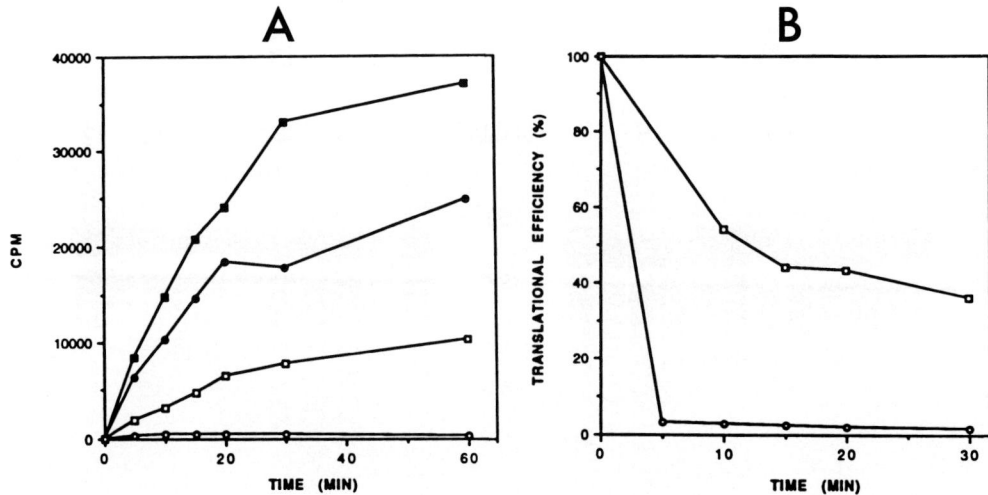

Fig. 3. A Effect of U-particle concentration on the translation of globin mRNA in vitro. U-particles (■-0 ng, ●-125 ng, □-250 ng and ○-500 ng) were preincubated with 100 ng of globin mRNA at 30° for 30 min in RSB (pH 8.5) in a total volume of 5 μl. The mixture was then added to a rabbit reticulocyte lysate (Promega; standard protocol) containing [³H]-leucine. Samples were taken at the times indicated and assayed for acid-precipitable counts. The concentration of U-particles was estimated by comparison of silver-stained band intensities with those of standard marker proteins after SDS-PAGE. **B** Comparison of the effect of U-particles on the translation of Mengo virus RNA (□) and globin mRNA (○) in rabbit reticulocyte lysates. Approximately 500 ng of U-particles was mixed with either 400 ng of globin mRNA or 1.2 μg of Mengo RNA (the maximum amount of Mengo RNA which could be accommodated; thus there were four times as many molecules of globin mRNA as Mengo RNA in the assay mixture) in a total reaction volume of 5 μl. The preincubation and translation reaction conditions were identical to those for **A**

duced easily visible bands, 23 kDa and 25 kDa in size, in the same region of the gradient (Fig. 2B, lanes 2–6). This pattern is similar to that which was obtained with Mengovirus-infected L-cells (data not shown), strongly suggesting that the GDVII strain of TMEV induces the synthesis of U-particles in infected cells.

The ability of Sabin type 3 poliovirus to induce the synthesis of U-particles in HeLa cells was also examined. Cells were infected and harvested 5 h post-infection and processed as described above. Examination of the denaturing gels revealed no evidence that this strain of poliovirus induces U-particle synthesis (data not shown).

Possible function of the U-particle

In 1987, Akhayat et al. [1] isolated a cytoplasmic particle from duck erythroblasts which they referred to as a "prosome-like particle": it was reported to be composed of a small cytoplasmic RNA and multimers of a 21 kDa protein. Most interesting was the presentation of evidence showing that this particle was able to inhibit the translation of globin mRNA in a rabbit reticulocyte lysate. Since in electron micrographs the erythroblast particles appear similar to U-particles, we were encouraged to examine the effect of U-particles on the translation of globin mRNA in vitro.

U-particles were preincubated with globin mRNA (BRL) before being added to a leucine-deficient rabbit reticulocyte lysate. Samples were removed at various times and assayed for acid-precipitable incorporation of added [^{3}H]-leucine. The results are shown in Fig. 3A. As the concentration of U-particles was increased there was a corresponding decrease in the efficiency of translation of the globin mRNA. These results indicate a stoichiometric rather than an enzymatic mechanism of inhibition. Heating the U-particles at 100° for 10 min prior to incubation abolished their inhibitory ability (data not shown).

The effect of the U-particles on the translation of Mengo RNA was also examined, and the results are shown in Fig. 3B. Assuming that all mRNA molecules are intact, approximately 2–5 U-particles per RNA strand can inactivate globin mRNA by >95% in this assay. For the Mengo RNA, at 1/4 the molecular concentration, the binding of 8–20 U-particle per RNA strand only inhibits translation by about 60%. Since the synthesis of the U-particle is induced approximately 3-fold during first 6 h of Mengovirus infection [2], it is tempting to speculate that the initial decrease of cellular mRNA translation in infected cells [4] may be related to the increase in the number of U-particles, and that this situation may also provide an advantage for viral RNA translation. These observations do not, however, throw any light on the mechanism whereby the synthesis of the U-proteins is induced.

Note added in proof

The 23- and 25 kDa subunits of the U-particle have now been identified as the H- and L-chains, respectively, of murine apoferritin (M. Mulvey, H. Fang, C. Holmes, D. Scraba [Virology; in press]).

Acknowledgements

We thank P. Carpenter and R. Bradley for excellent technical assistance, and D. Kobasa for helpful discussions. M.R.M is a Postdoctoral Fellow of the Alberta Heritage Foundation for Medical Research, and these studies were supported by a grant from the Medical Research Council of Canada to D.G.S.

References

1. Akhayat O, Infante AA, Infante D, Martins de Sa C, Grossi de Sa MF, Scherrer K (1987) A new type of prosome-like particle, composed of small cytoplasmic RNA

and multimers of a 21 kDa protein, inhibits protein synthesis in vitro. Eur J Biochem 170: 23–33

2. Boege U, Hancharyk R, Scraba DG (1987) The synthesis of a particle-forming cellular protein is enhanced by Mengo virus infection. Virology 159: 358–367

3. Boege U, Ko DSW, Scraba DG (1986) Toward an in vitro system for picornavirus assembly: purification of Mengovirus 14S capsid precursor particles. J Virol 57: 275–284

4. DeStefano J, Olmsted E, Panniers R, Lucas-Lenard J (1990) The a subunit of eukaryotic initiation factor 2 is phosphorylated in Mengovirus-infected mouse L cells. J Virol 64: 4445–4453

5. Laemmli UK (1970) Cleavage of structural proteins during the assembly of the head of bacteriophage T4. Nature 227: 680–685

6. Palmenberg AC (1989) Sequence alignments of picornaviral capsid proteins. In: Semler BL, Ehrenfeld E (eds) Molecular aspects of picornavirus and detection. American Society for Microbiology, Washington, pp 211–241

Authors' address: Dr. D. G. Scraba, Department of Biochemistry, University of Alberta, Edmonton, Alberta T6G 2H7, Canada.

Arch Virol (1994) [Suppl] 9: 307–316

Archives
of
Virology
© Springer-Verlag 1994
Printed in Austria

B-lymphocytes are predominantely involved in viral propagation of hepatitis C virus (HCV)

H. M. Müller[1], **B. Kallinowski**[1], **C. Solbach**[1], **L. Theilmann**[1], **T. Goeser**[1], and **E. Pfaff**[2]

[1] Department of Internal Medicine, University of Heidelberg, Heidelberg
[2] Federal Research Centre for Virus Diseases of Animals, Tübingen,
Federal Republic of Germany

Summary. Recent reports have shown that HCV infection is not only restricted to hepatocytes. Like hepatitis B virus (HBV), which also was thought to be strictly hepatotropic in early molecular and cellular investigations, infection of lymphoid cells by HCV in vivo has been demonstrated. We showed that total peripheral blood leukocytes of chronically HCV-infected patients are infected by detection of plus- and minus-stranded HCV RNA using strand-specific oligonucleotide primers in the RT-PCR. These cells also represent extrahepatic sites for the viral replication, as demonstrated by incorporation of [^{3}H]-uridine into nascent RNA after stimulation of the cells with a mitogen. Furthermore, total PBML from an uninfected person could be infected in vitro using an HCV-positive serum. It could be shown that replication of HCV RNA takes place in these cells. Examination of different subsets of PBML showed predominant infection of B-lymphocytes during HCV disease. Additionally, infection of T-lymphocytes was detected in about 50% of all chronically HCV-infected patients.

Introduction

Hepatitis C virus (HCV) is responsible for most cases of posttransfusional hepatitis [1, 2], with a chronic course in up to 60% of individuals, leading to cirrhosis and eventually hepatocellular carcinoma in 10–15% of all individuals [6, 9].

Based on sequence analyses [3, 11], the genome of the virus is a single-stranded RNA of about 9 416 nucleotides with positive polarity. A single open reading frame encodes a precursor polyprotein of 3 010/3 011 amino acids. These characteristics, as well as the genomic organization and sequence homologies suggest that HCV is related to flavi- and pestiviruses [7].

HCV can be detected in serum, liver, and peripheral blood leuko-cytes (PBML) using reverse transcription and subsequent amplification of the cDNA by PCR. In this study we used strand-specific oligonucleotide primers in the reverse transcriptase (RT) reaction to investigate the presence of minus-stranded RNA as a marker for viral replication of HCV in serum, liver tissue, total PBML, and different lymphocyte subpopulations from patients chronically infected with HCV. Although

a)

	Uridine incorporation (cpm)			
	HCV +		HCV −	
	HCV	HBV	HCV	HBV
day 2:	191.5	5	33.2	17.2
day 4:	69.0	15	23.7	18.7
day 6:	115.5	16	24.0	20.0

b)

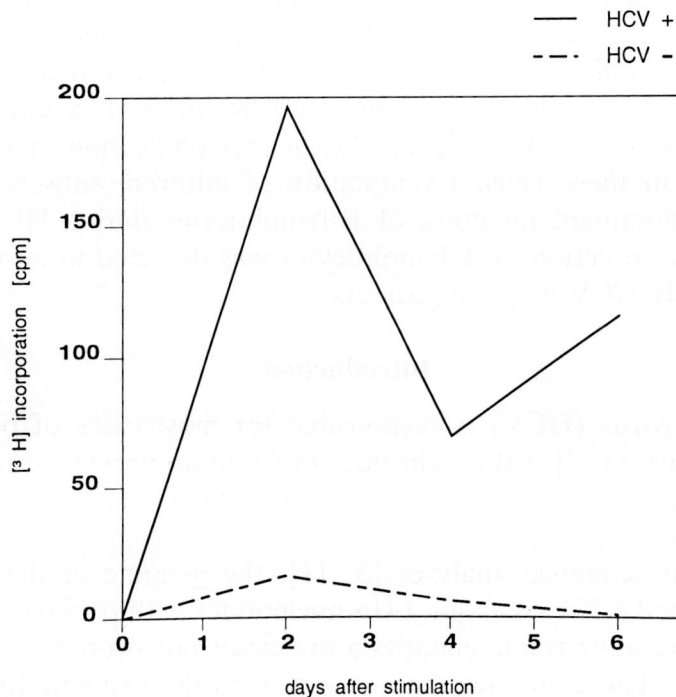

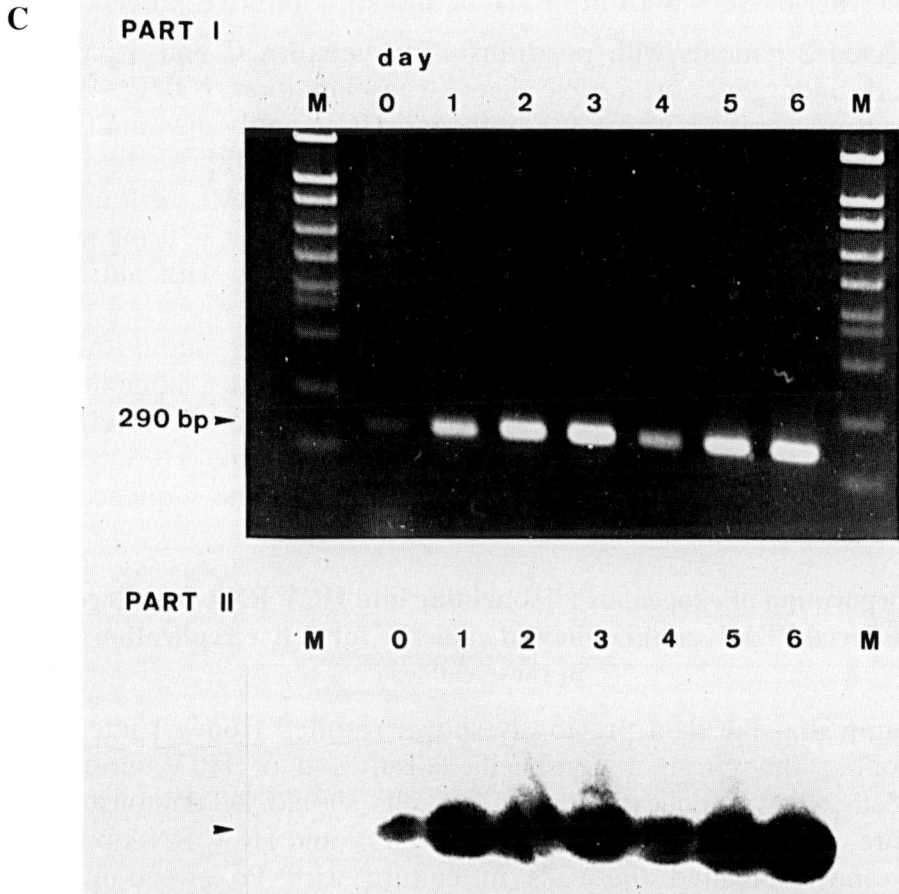

Fig. 1. Pokeweed-mitogen (PWM) stimulation of HCV-infected PBML in culture. **a** Absolute cpm-values of the hybrid release assay from HCV-negative and HCV-positive PBML RNA hybridized to HCV and HBV fragments. **b** Background corrected cpm-values (minus nonspecific HBV binding) shown as a diagram. **c** PCR products (arrowheads) from PBML stimulated with Pokeweed-mitogen (PWM), separated by agarose gel electrophoresis, stained with ethidium bromide (*part I*), and the corresponding Southern blot hybridization (*part II*). The numbers represent the day of cell harvesting after initiation of stimulation, the PCR product of day 0 refers to nonstimulated fresh PBML

PBML were found to be infected with HCV [12, 13], as shown by amplification of minus-stranded HCV RNA in these cells, viral replication has not yet been confirmed functionally. In practice, HCV-specific replication was demonstrated in in vivo and in vitro infected total PBML employing a modified "hybrid release assay" detecting radiolabeled RNA. The aim of the study was to identify possible sites of extrahepatic replication which may serve as a source for infectious virions causing reinfection of liver cells.

Presence of HCV RNA in PBML of anti-HCV positive patients

We selected 2 patients with posttransfusion hepatitis C and 3 patients who had undergone orthotopic liver transplantation (OLT). These patients were serum positive for both anti-HCV antibodies and HCV RNA before OLT and got graft-reinfection with HCV [8].

The amplification of HCV RNA extracted from PBML of four anti-HCV negative and five anti-HCV positive patients by PCR using nested primers gave a positive result in 3 of the 5 patients with anti-HCV antibodies.

Amplification products of RNA from serum, liver and PBML of these patients were cloned and sequenced. We found a homology of about 82% at the nucleotide level and 90% at the amino acid level when compared to the sequence of the American prototype HCV [3, 5], and 91% and 97%, respectively, when compared to the sequence of a Japanese isolate [11].

Incorporation of exogenous [^{3}H]-uridine into HCV RNA in mitogen induced PBML culture showed evidence for active replication in these cells

To confirm that the data previously shown resulted from an infectious event rather than from a nonspecific adsorption of HCV virions to blood cells, active replication in these cells should be demonstrated. Therefore, we prepared PBML of an anti-HCV and HCV RNA positive patient and stimulated the cells in culture with Pokeweed-mitogen (PWM) over a period of seven days. Incorporation of [^{3}H]-uridine into nascent RNA was determined by hybridization to a HCV-specific fragment fixed on nitrocellulose filters. Specificity was demonstrated by hybridization to an unrelated fragment (hepatitis B virus, HBV). Stimulation of normal PBML served as control. Figures 1a and 1b show the levels of absolute uridine incorporation (in cpm), and the background corrected values (in cpm) as a diagram, respectively. Figure 1c represents the stained agarose gel of PCR products derived from viral genomic RNA (part I) and the corresponding Southern blot hybridization (part II). Maximal uridine incorporation into HCV RNA was observed on day 2, corresponding to the highest level of HCV replication. This is consistent with the data obtained by PCR (Fig. 1c). We believe that the decrease of uridine incorporation on the following days was due to the loss of viable cells observed after day 4 to 5.

Detection of minus-stranded RNA in mononuclear leukocytes

Hepatitis C virus contains a single-stranded RNA molecule with positive polarity. In infected cells, RNA molecules with negative polarity repre-

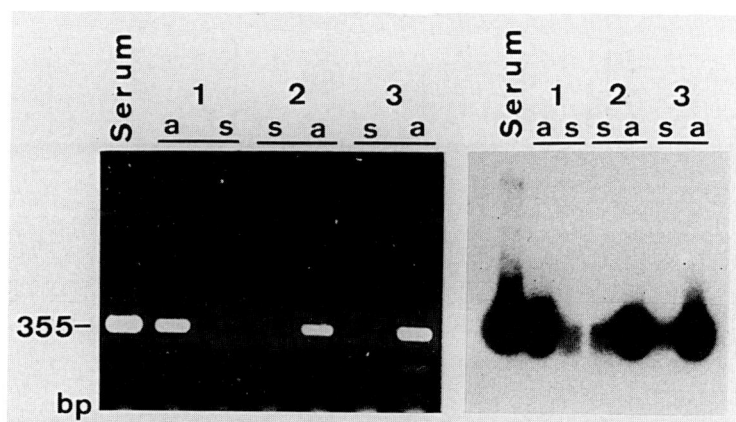

Fig. 2. Agarose gel and Southern blot hybridization after nested PCR of HCV cDNA of RNA prepared from PBML of three patients after OLT. HCV RNA from serum served as a positive control (*serum*). Synthesis of cDNA was done either with an antisense primer (*a*) or with a sense primer (*s*). Hybridization was carried out with a ^{32}P-labeled HCV probe lacking the PCR primer sequences (map position 83–465)

senting replicative intermediates indicate active replication of the viral genome. To test for the presence of plus-stranded and minus-stranded RNA molecules, RNA extracted from PBML was transcribed into cDNA with antisense or sense primers. The reverse primer was then added for PCR amplification. Subsequently the DNA was reamplified by PCR with nested primers.

To exclude any contamination with plus-stranded RNA or residual activity of the reverse transcriptase in the assay for the detection of minus-stranded RNA, the reaction was stopped by heating and by addition of RNases (A, H, and T1). Phenol/chloroform extraction was done prior to DNA amplification. In Fig. 2, an HCV specific amplification product (lanes 1–3, s), using sense primers in the RT reaction, was clearly detected by Southern blot hybridization analysis. Lanes 1–3 (Fig. 2a) show amplification products derived from PBML RNA transcribed with the antisense primer. In this case detection was possible with both the ethidium bromide stained gels and the Southern blot hybridization. No minus-stranded RNA was detected in the serum samples. No amplification product was found in the final wash supernatant of the PBML preparation, thereby ruling out the possibility of contamination with free, serum-associated viral particles (data not shown).

HCV susceptibility of B- and T-lymphocyte subsets of chronically HCV-infected patients

To further clarify which subsets of peripheral leukocytes are infected in vivo, total PBML of four chronically HCV-infected patients were

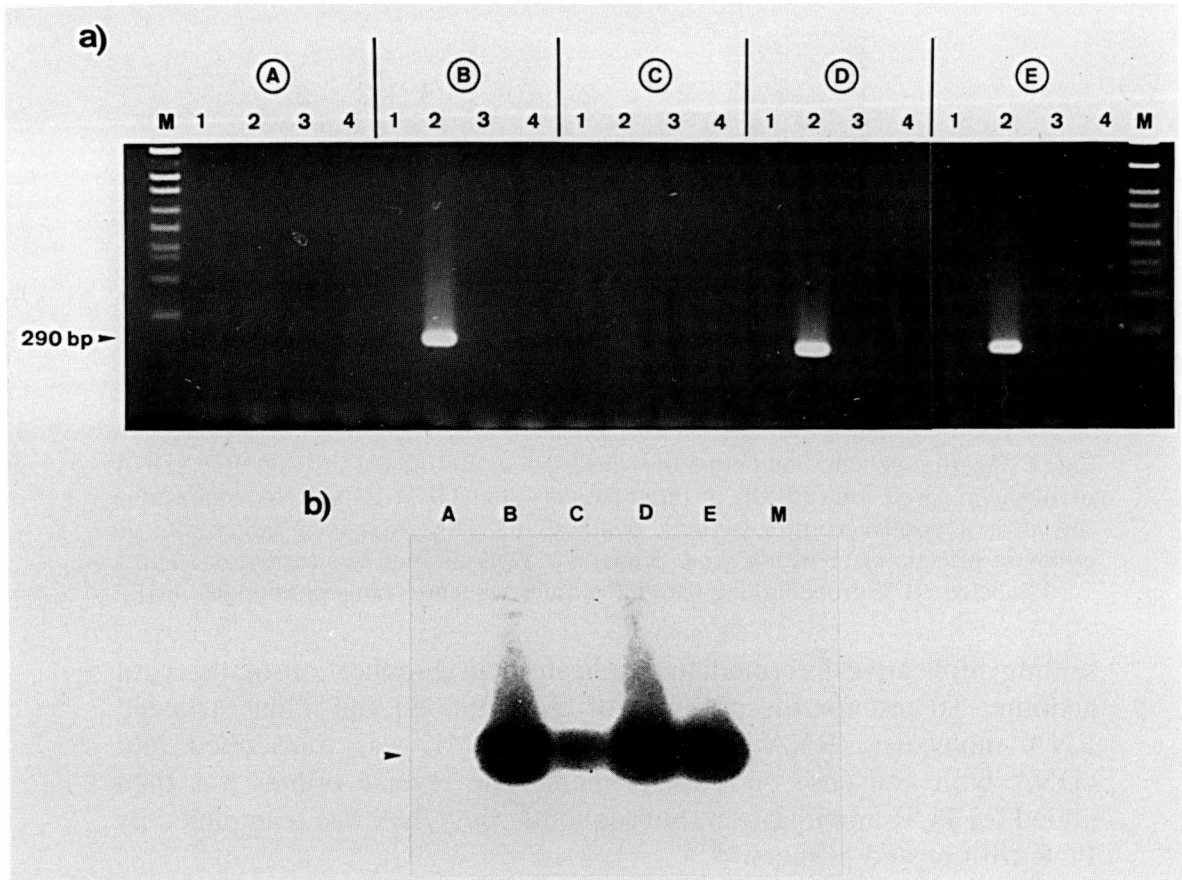

Fig. 3. a Agarose gel of HCV-specific products after nested PCR. Natural killer cells (NK; *1*), B-lymphocytes (*2*), CD4- (*3*), and CD8-T-lymphocytes (*4*) of HCV-positive patients (*B–E*) and one HCV-negative control person (*A*) were examined to detect viral RNA of HCV. **b** Southern blot analysis to detect minus-stranded HCV RNA from B-lymphocytes of patients *A–E*

separated. RNA extracted from each subset, including natural killer cells (NK), B- and T-lymphocytes, and CD4- and CD8-subsets, was transcribed into cDNA using strand-specific primers to detect plus-stranded (Fig. 3a) and minus-stranded RNA (Fig. 3b). HCV RNA was detected in B-lymphocytes (Fig. 3a; A–E, 2), but not seen in the T-lymphocyte subsets (A–E, 3 and 4) nor in NK cells (A–E, 1).

Further studies using 10 patients with a histologically and serologically proved hepatitis C demonstrated that not only B-lymphocytes, but also T-lymphocytes (6 of 10 specimens) were positive for HCV RNA by PCR (Table 1). None of the control patients (No. 11–13) showed HCV sequences amplifiable with primer sets in serum or in B- or T-lymphocytes.

Table 1. HCV susceptibility of B- or T-lymphocyte subsets of chronically HCV infected patients

Patient	Clinical status[a]	Transaminases		Anti-HCV ELISA 2nd	HCV RNA	Lymphocytes	
		AST[b]	ALT[c]			B-cells	T-cells
1	CAH	>100	>100	+	+	+	−
2	CAH	28	50	+	+	+	+
3	CAH	61	121	+	+	+	−
4	CAH	50	120	+	+	+	−
5	CAH	40	91	+	+	+	+
6	CAH	25	60	+	+	+	+
7	Cirrh	30	25	+	+	+	−
8	Cirrh	40	25	+	+	+	+
9	Cirrh	45	30	+	+	+	+
10	LTX-CAH	45	153	+	+	−	+
11	−	normal		−	−	−	−
12	−	normal		−	−	−	−
13	−	normal		−	−	−	−

[a] *CAH* Chronic active hepatitis; *Cirrh* cirrhosis; *LTX-CAH* chronic active hepatitis after liver transplantation

[b] *AST* Aspartate-aminotransferase

[c] *ALT* Alanine-aminotransferase

In vitro infection of PBML

The data presented in this study show that PBML were infected in vivo during the natural course of chronic hepatitis C. To establish a cell culture system for HCV, it must be shown that PBML can be infected in vitro. Therefore, PBML from a healthy donor were isolated, cultured in 24-well plates, and incubated with an anti-HCV serum for 12 h. After medium was changed, [^{3}H]-uridine was added as a 24 h pulse for each timepoint. RNA synthesis and HCV specificity were confirmed by the modified "hybrid release assay", as described above. A maximal incorporation of radioisotopes was observed on day 3 (Fig. 4). This was equated with PWM stimulation of PBML infected in vivo (see Fig. 1b). Uninfected, but stimulated PBML were used as controls.

Discussion

There have been several reports on the detection by PCR of HCV genomic and antigenomic RNA in liver tissues [5] and peripheral blood mononuclear cells [12, 13]. Detection of minus-stranded HCV RNA in liver tissue indicates an active viral replication in these cells. Although

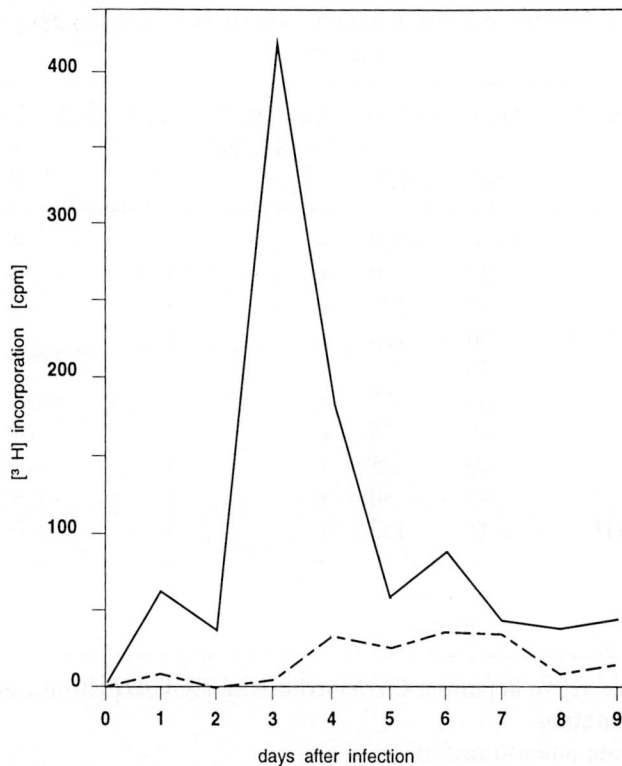

Fig. 4. [^{3}H]-uridine incorporation into HCV RNA of normal PBML infected in vitro (solid line) and noninfected control PBML (dashed line), as measured by the hybrid release assay. The cpm-values represent the new synthesized HCV RNA, minus non-specific HBV binding

the presence of HCV RNA in lymphocytes has been detected by PCR, a mitogenic activation of HCV replication in circulating PBML has not yet been reported. In this study, we present evidence that HCV RNA can be found in PBML and that the virus replicates in these cells.

The detection of HCV RNA in PBML appears to be specific. It seems unlikely that detection of viral HCV RNA is an artefact of serum-associated viral particles or simple adhesion of virions to the cells. Several authors have described the detection of minus-stranded RNA in serum [4, 13], suggesting that this RNA can also be packed into viral particles. This, however, would not explain the de novo synthesis of viral nucleic acids in PBML. The most conclusive argument for susceptibility and active viral replication was the incorporation of exogenous [^{3}H]-uridine in nascent RNA with a proven HCV specificity in in vivo and in vitro infected PBML.

The presence of minus-stranded HCV RNA as a replicative form in PBML demonstrates active viral replication in these cells. In a first assay, only the B-lymphocyte subpopulation of PBML were shown to be susceptible for HCV. Further studies using a larger number of patients demonstrated that HCV RNA could be found not only in B-lymphocytes but also in T-lymphocytes. However, it is not yet clear whether T-lymphocytes are infected in vivo. This lack of evidence might be due to contamination of the T-lymphocyte subset with B-lymphocytes. In this context, Shimizu [10] recently reported a productive in vitro infection of a human T-cell line (MOLT-4) with HCV and described the detection of minus-stranded viral RNA as a replication marker.

No amplification product can be detected in PBML of patients with acute hepatitis C infection; it may be that the appearance of viral RNA in PBML is related to the chronic carrier status. It remains unclear at which time during infection extrahepatic cells, such as PBML, become infected and whether they remain infected and for how long. We suggest that PBML of chronically infected patients could serve as a source of virus for infection of new hepatocytes. It has not yet been shown, however, whether PMBL can serve as a source of formation and release of infectious particles. No amplifiable HCV RNA was detected in medium from cultured PBML infected in vivo or in vitro, suggesting that viral particles are not being released.

All current data are consistent with our suggestion that PBML may be a potential replicative site of ongoing chronic viral infection. Nevertheless, HCV does not appear to be present in PBML of all chronically HCV-infected patients. We speculate that replication of HCV in PBML takes place at low levels or at infrequent periods. More specific information is needed about the role of the PBML in the etiology of the chronic stage of hepatitis C and about the kinetics of infection.

References

1. Alter HJ, Purcell RH, Shih JW, Melpolder JC, Houghton M, Choo Q-L, Kuo G (1989) Detection of antibody to hepatitis C virus in prospectively followed transfusion recipients with acute and chronic non-A, non-B hepatitis. N Engl J Med 321: 1494–1500
2. Choo Q-L, Weiner AJ, Overby LR, Kuo G, Houghton M, Bradley DW (1990) Hepatitis C virus: the major causative agent of viral non-A, non-B hepatitis. Br Med Bull 46: 423–441
3. Choo Q-L, Richman KH, Han JH, Berger K, Lee C, Dong C, Gallegos C, Coit D, Medina-Selby A, Barr PJ, Weiner AJ, Bradley DW, Kuo G, Houghton M (1991) Genetic organization and diversity of the hepatitis C virus. Proc Natl Acad Sci USA 88: 2451–2455
4. Fong T-L, Shindo M, Feinstone SM, Hoofnagle JH, Di Bisceglie AM (1991) Detection of replicative intermediates of hepatitis C viral RNA in liver and serum of patients with chronic hepatitis C. J Clin Invest 88: 1058–1060
5. Han JH, Shyamala V, Richman KH, Brauer MJ, Irvine B, Urdea MS, Tekamp-Olson P, Kuo G, Choo Q-L, Houghton M (1991) Characterization of the terminal

regions of hepatitis C viral RNA: identification of conserved sequences in the 5′ untranslated region and poly(A) tails at the 3′ end. Proc Natl Acad Sci USA 88: 1711–1715

6. Kiyosawa K, Sodeyama T, Tanaka E, Gibo Y, Yoshizawa K, Nakamo Y, Furuta S (1990) Interrelationship of blood transfusion, non-A, non-B hepatitis and hepatocellular carcinoma: analysis by detection of antibody to hepatitis C virus. Hepatology 12: 671–675

7. Miller R, Purcell RH (1990) Hepatitis C virus shares amino acid sequence similarity with pestiviruses and flaviviruses as well as members of two plant virus subgroups. Proc Natl Acad Sci USA 87: 2059–2061

8. Müller HM, Otto G, Goeser T, Arnold J, Pfaff E, Theilmann L (1992) Recurrence of hepatitis C virus infection after orthotopic liver transplantation. Transplantation 54: 743–745

9. Sakamoto M, Hirohashi S, Tsuda H, Ino Y, Shimosato Y, Yamasaki S, Makuuchi M, Hasegawa H, Terada M, Hosoda Y (1988) Increasing incidence of hepatocellular carcinoma possibly associated with non-A, non-B hepatitis in Japan, disclosed by hepatitis B virus DNA analysis of surgically resected cases. Cancer Res 48: 7294–7297

10. Shimizu YK, Iwamoto A, Hijikata M, Purcell RH, Yoshikura H (1992) Evidence for in vitro replication of hepatitis C virus genome in a human T-cell line. Proc Natl Acad Sci USA 89: 5477–5481

11. Takamizawa A, Mori C, Fuke I, Manabe S, Murakami S, Fujita J, Onishi E, Andoh T, Yoshida I, Okayama H (1991) Structure and organization of the hepatitis C virus genome isolated from human carriers. J Virol 65: 1105–1113

12. Wang J-T, Sheu J-C, Lin J-T, Wang T-H, Chen DS (1992) Detection of replicative form of hepatitis C virus RNA in peripheral blood mononuclear cells. J Infect Dis 166: 1167–1169

13. Zignego AL, Macchia D, Monti M, Thiers V, Mazzetti M, Foschi M, Maggi E, Romagnani S, Gentilini P, Brechot C (1992) Infection of peripheral mononuclear blood cells by hepatitis C virus. J Hepatol 15: 382–386

Authors' address: Dr. H. M. Müller, Department of Internal Medicine, University of Heidelberg, Bergheimer Strasse 58, D-69115 Heidelberg, Federal Republic of Germany.

Protein expression and virion maturation

Arch Virol (1994) [Suppl] 9: 319–328

_Archives_____

V̈irology

© Springer-Verlag 1994
Printed in Austria

Folding of the mouse hepatitis virus spike protein and its association with the membrane protein

D.-J. E. Opstelten, P. de Groote, M. C. Horzinek, and **P. J. M. Rottier**

Institute of Virology, Department of Infectious Diseases and Immunology,
Veterinary Faculty, Utrecht University, The Netherlands

Summary. Coronaviruses are assembled by budding into pre-Golgi membranes. Using different approaches we have demonstrated that the spike (S) protein and the membrane (M) protein of mouse hepatitis virus (MHV) associate to form large complexes. Newly synthesized M was found in these complexes almost immediately after its synthesis, whereas the S protein started to appear in heterocomplexes after 10–20 min. This is consistent with the slow rate of folding of S and with the observation that folding of S preceeds its association with M. While the folding of S involves the formation of multiple disulfide bonds, folding of M is disulfide-independent. This contrast was reflected by the differential sensitivity of the two proteins to reduction with dithiothreitol (DTT). Addition of DTT to the culture medium of MHV-infected cells drastically impaired the folding of S, but not of M. Consequently, the S protein was unable to interact with M. Under these conditions, S stayed in the ER while M was transported efficiently beyond the site of budding to the Golgi complex. We conclude that the association of S with M is an essential step in the formation of the viral envelope and in the accumulation of both proteins at the site of virus assembly.

Introduction

Budding through cellular membranes is the last step in the assembly of enveloped viruses. The assembly process is driven by specific interactions between the nucleocapsid and the viral envelope proteins [12]. Depending on the virus, budding takes place at the plasma membrane or at intracellular membranes. The site of budding appears to be determined by the envelope proteins because virus assembly occurs where these proteins accumulate. Accordingly, viruses that assemble at the plasma membrane have envelope proteins that are rapidly transported to the cell surface after synthesis. In contrast, membrane proteins from in-tracellularly budding viruses are retained in the budding compartment [4, 8].

We study the assembly of the coronavirus mouse hepatitis virus (MHV). MHV particles are composed of three structural proteins. The nucleocapsid (N) protein is complexed with the genome, thereby form-ing the helical nucleocapsid. The spike (S) glycoprotein constitutes the large peplomers and functions in cell attachment and fusion during virus entry. The membrane (M) protein is a small glycoprotein, which is largely embedded in the lipid bilayer. MHV matures by budding into intracellular smooth membranes located between the endoplasmic reticulum (ER) and the Golgi complex [13], and it has been concluded that the M protein determines the site of budding. The correlation between this site and the intracellular accumulation of M protein strongly argues for such a role [13]. In addition, tunicamycin treatment of MHV-infected cells resulted in the secretion of spikeless virions suggesting that only the M protein is required for budding [5, 9]. When expressed independently, however, the M protein is transported beyond the site of budding to the trans side of the Golgi complex [7, 11]. The same holds true for the S protein which, when not incorporated into virions in infected cells or when expressed independently, is transported to the plasma membrane (Vennema and Rottier, unpubl. res.). Clearly, neither envelope proteins localize to the budding compartment by themselves, implying that in MHV-infected cells they have to interact in order to be retained and to co-accumulate at the site of budding. Such an interaction is probably specific because cellular membrane proteins are virtually absent in virions.

Complex formation of the viral envelope proteins

Until now no experimental data in support of the proposed interaction between the two coronaviral envelope proteins have been reported. We reasoned that any complexes between S and M might simply have escaped detection due to the analytical conditions used, e.g. by disrup-tion of the complexes during solubilization of the infected cells. There-fore, we studied the effects of different detergents with the aim of finding conditions that might preserve the interaction between the two proteins. A large panel of buffer compositions was tested and the conclu-sion was reached that the nature of the detergent(s) used for cell lysis and during further analysis indeed had profound effects and that M/S complexes do exist. Optimal preservation of the complexes was achieved when we used a combination of the non-ionic detergent Nonidet-P40 (NP-40) and the ionic detergent sodium deoxycholate (DOC), both at

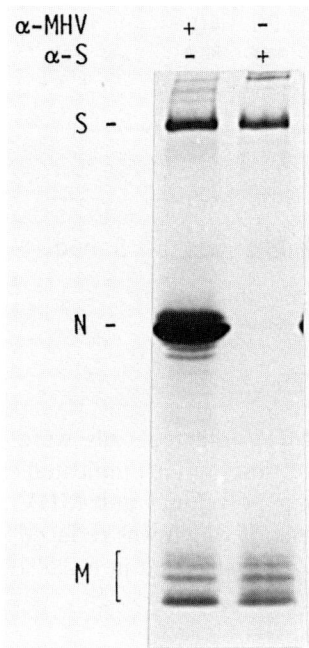

Fig. 1. Co-immunoprecipitation of the MHV M and S protein. MHV-infected cells were labeled with ^{35}S-methionine for 1 h and lysed in a buffer containing 0.5% Nonidet-P40 and 0.5% Na-deoxycholate. Viral proteins were precipitated from half of the lysate with a polyclonal anti-MHV serum (α-MHV); for the other half a monoclonal antibody to S was used (α-S). MHV structural proteins are indicated (S, N, M)

0.5%. We have characterized the specificity and nature of the interaction in several ways.

(i) We detected heterocomplexes of S and M by co-precipitation of the S protein with monospecific antibodies to the M protein, and vice versa. Similar amounts of M protein can be precipitated with a monoclonal antibody to S as with a polyclonal anti-virion serum (Fig. 1). Our experiment also illustrates the specificity of the interaction because only the viral envelope proteins were co-precipitated by the monoclonal antibodies: scarcely any nucleocapsid protein or cellular protein was observed in the immunoprecipitates.

(ii) Pulse-chase analysis demonstrated that the complexes are formed post-translationally. Interestingly, we found that M and S engage in complex formation with different kinetics. Our data indicate that M associates with S very soon after its synthesis while newly synthesized S protein starts to appear in complexes only after a lagtime of 10–20 min. This implies that immediately after synthesis M molecules associate with S molecules synthesized some time previous.

(iii) Sucrose gradient analysis under the detergent conditions described above demonstrated that S and M occur in huge multimeric complexes. These have been observed after detergent treatment of virions as well as in lysates of infected cells, being more heterogeneous in the latter.

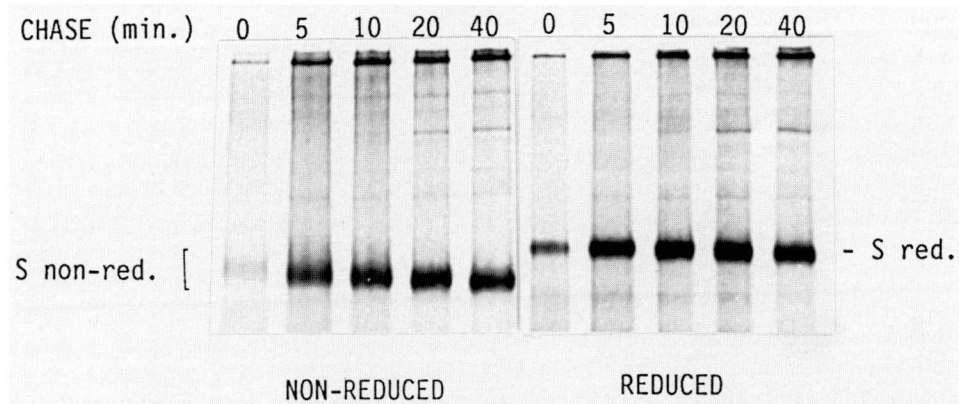

Fig. 2. Disulfide bond formation in the MHV S protein. MHV-infected cells were pulse-labeled for 5 min with ^{35}S-methionine and chased for the time periods indicated. Viral proteins were precipitated from the cell lysates with a polyclonal anti-MHV serum. The immunoprecipitates were split into two portions one of which was reduced with 20 mM DTT. The samples were heated for 5 min at 95°C and analyzed in a 7.5% SDS-polyacrylamide gel

On the basis of these results we hypothesize that S and M congregate at the site of budding to form a matrix into which viral nucleocapsids can bud.

Folding of the spike protein

The typical surface projections of coronavirions are formed solely by the S protein. In previous work we studied their biogenesis by analyzing the oligomerization process [15]. It was found that S forms oligomers with a half-time of 40–60 min, rather slow as compared to most other viral spike oligomers [6]. Since the conditions used in our earlier experiments did not preserve the M/S interactions we were not able to link the oligomerization of the S protein to the complex formation. It is now obvious that both processes take place slowly. This suggested to us that the folding of the S protein is the rate-limiting step.

We studied the folding of S by following the formation of disulfide bonds. These play an important role in the folding and stability of secretory and membrane proteins and are usually crucial for the generation of functional structures. We used the approach that has recently been described for the hemagglutinin protein (HA) of influenza virus [2]. As illustrated in Fig. 2, different folding intermediates of the S protein could be visualized in non-reducing gels on the basis of their differences in electrophoretic mobility. The large mobility difference between the fully reduced form and the unreduced folding intermediates demonstrates that the formation of the disulfide bonds in the S molecules has a pronounced effect on the protein's conformation. Disulfide bond formation apparently starts co-translationally; even after very short pulse labelings the S intermediates always migrated faster than the fully reduced species. In contrast to influenza virus HA no distinct intermediates were detected. Instead, the S protein appeared to undergo many conformational changes that did not resolve. This probably reflects the high number of cysteines present in the luminal domain of the S protein, giving rise to a wide spectrum of forms as a result of the formation or redistribution of disulfide bonds.

After synthesis the S protein undergoes its major folding transitions during the first 10–20 min, as judged from electrophoretic analyses. The more compact, faster migrating conformations occur after some 20–30 min. This time-course corresponds well with the lag-time, after which newly synthesized S protein starts to appear in complexes with the M protein. It also suggests that the S molecule must have reached a certain conformational maturity before it can engage in complex formation. To verify this point we analyzed the S protein present in M/S complexes in non-reducing gels. Indeed, only the faster migrating forms of S were detected in the heterocomplexes, which indicates that the molecule acquires its competence to associate with M as a result of folding.

Manipulation of disulfide bond formation in the spike protein

By adding dithiothreitol (DTT) to the cell culture medium the oxidizing state in the lumen of the ER can be drastically affected. As Braakman et al. [3] have demonstrated, this treatment prevents disulfide bond formation in newly synthesized HA and even leads to the reduction of oxidized HA present in the ER. We wondered whether in vivo reduction of the MHV S protein would also affect its folding and what implications this would have for its association with M.

When 5 mM DTT was added to the culture medium of MHV-infected cells we observed reduction of partially as well as fully oxidized S protein. The effect was monitored in non-reducing gels; the oxidized forms were converted to the slower migrating reduced form. Interestingly, this reduction was accompanied by changes in epitopes, as judged from the loss of recognition by several S-specific monoclonal antibodies. We used some of these antibodies to localize the S protein in MHV-infected cells by indirect immunofluorescence. In untreated cells the antibodies predominantly stained the ER, with additional intense fluorescence in a distinct perinuclear region. Double immunofluorescence identified the latter region as the major site of M protein in MHV-infected cells, presumably the viral budding compartment. A short exposure (<20 min)

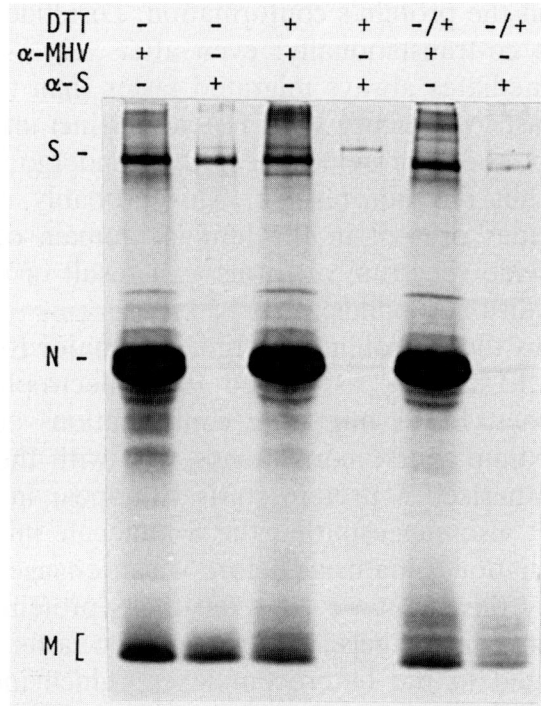

Fig. 3. Effects of in vivo reduction. MHV-infected cells were labeled for 10 min in the absence or presence of 5 mM DTT. In the latter case the cells were treated with 5 mM DTT for 5 min before labeling. As a control, MHV-infected cells were labeled for 10 min in the absence of DTT and chased for 10 min in the presence of 5 mM DTT. The cell lysates were split and the viral proteins were precipitated with the polyclonal anti-MHV serum and with the monoclonal anti-S serum. MHV structural proteins are indicated (*S, N, M*)

of the cells to a reducing milieu resulted in almost complete absence of ER staining by the conformation-specific anti-S monoclonal antibodies. However, the fraction of S protein that co-localized with the M protein was still recognized by the antibody. Thus, in vivo reduction with DTT affects the conformation of S present within the ER while the protein outside the ER appears to be much more resistant to DTT.

Do M/S complexes still form during in vivo reduction? As mentioned above, only the oxidized S protein occurs in complexes with M under normal conditions. Thus, under reducing conditions one would not expect the formation of M/S complexes to take place. To confirm this assumption we labeled MHV-infected cells under reducing conditions and analyzed the complex formation by co-immunoprecipitation. While under normal conditions much of M can be co-precipitated with S already after a 10-min pulse labeling, no co-precipitation was observed in the presence of DTT (Fig. 3). This result indicated either that the complexes were not formed or that existing complexes were no longer recognized by the anti-S monoclonal antibody. To rule out the latter possibility we prelabeled MHV-infected cells for 10 min in the absence of DTT and then chased for 10 min in its presence. As shown in Fig. 3, a significant fraction of the M/S complexes formed during the pulse were still recognized by the conformation-specific antibody after DTT treatment. We therefore conclude that M/S complexes are no longer formed under reducing conditions. Apparently, the folding of S is a prerequisite for its association with M.

Differential effects of in vivo reduction on transport of the coronaviral envelope proteins

The endoplasmic reticulum controls the exit of proteins to the Golgi complex. Only properly folded molecules are allowed to leave, misfolded proteins are generally retained in the ER. This "quality control" still functions during in vivo reduction, as was shown for the influenza HA protein, the reduced form of which is unable to leave the ER [3]. As a general consequence of this finding, those proteins that require disulfide bond formation for their proper folding will accumulate in the ER under reducing conditions. Accordingly, we found that the reduced MHV S protein stably stayed in the ER and was re-oxidized upon DTT removal.

The M protein of MHV does not form luminal disulfide bonds, as can be deduced from its structure; no cysteines are present in the luminally exposed part of the protein [1, 10]. This enabled us to study whether transport of such proteins to the Golgi complex still occurs during DTT treatment. Taking advantage of its well-characterized O-glycosylation pattern [7, 14], the M protein allows us to follow its intracellular transport biochemically. Pulse-chase analysis under reducing conditions (Fig. 4) showed that M was efficiently transported out of the ER and reached the trans side of the Golgi complex; the slower migrating forms appearing during the chase are indicative of modifications occurring in this part of the Golgi complex. Thus, transport of M is independent of disulfide bond formation. Moreover, the behavior of M under reducing conditions shows that cellular processes such as glycosylation and ER-to-Golgi transport are not disturbed.

Another interesting observation from these experiments was that the M protein was transported to the Golgi complex faster under reducing than under normal conditions (Fig. 4; compare M in lanes 3 and 7). In contrast to its glycosylation in the presence of DTT, after a 30 min chase in the absence of DTT a large fraction of M had still not been modified by Golgi enzymes, suggesting that it was retained somewhere before the Golgi complex. As concluded from the effects of DTT on the M protein

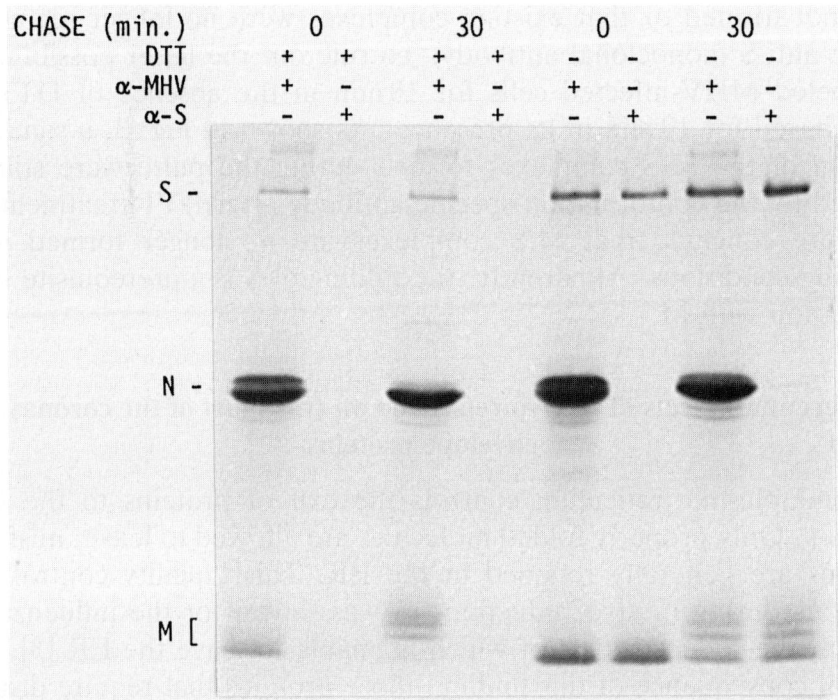

Fig. 4. Effects of in vivo reduction on transport of the MHV-M protein. Parallel cultures of MHV-infected cells were labeled for 10 min and chased for 30 min in the presence or absence of 5 mM DTT. In the former case the cells had been treated with 5 mM DTT for 5 min before labeling. The cell lysates were split and the viral proteins were precipitated with polyclonal anti-MHV serum and with monoclonal anti-S serum. MHV structural proteins are indicated (*S, N, M*)

expressed by a recombinant vaccinia virus, the reducing agent itself did not affect the transport kinetics. M was transported to the Golgi complex at the same rate in the presence and absence of DTT. Thus under conditions of in vivo reduction of the S protein, M is no longer retained in a pre-Golgi compartment. These data strongly suggest that complex formation between S and M plays an important role in the retention of both proteins at the site of virus budding.

Envelope protein interaction and viral budding

The data obtained so far raise several questions. One is where the two envelope proteins associate. The presence of unglycosylated M protein in complexes with S strongly suggests that the proteins interact in the ER. In this instance the proteins may either be transported to the budding compartment as small oligomers of a discrete composition or in the form of larger aggregates. It cannot yet be excluded, however, that the two proteins are transported individually to the budding compartment before they associate. Clearly, complex formation can occur in the ER as budding takes place in this compartment late in infection, probably as the result of the abundant co-accumulation of the envelope proteins.

Like other RNA viruses that assemble intracellularly, coronaviruses lack a matrix protein; the nucleocapsid must interact directly with one or both envelope proteins. There are no indications that the RNA genome is involved in the budding process, and protein-protein interactions probably are the driving force. Two models can be envisioned. The envelope proteins may form large rafts, which are composed solely of M and S, in the plane of the budding compartment's membrane. In this model the nucleocapsids would interact with the envelope proteins present in these preformed patches. Alternatively, the nucleocapsid may be the organizing factor, and the nucleocapsid protein would selectively recruit the envelope proteins for formation of the viral membrane. Our findings support the first model in which S and M form large complexes by lateral interactions, the specificity of which would exclude non-viral proteins. Because we did not observe co-immunoprecipitation of the nucleocapsid protein in our assays, the interaction between the viral membrane proteins is apparently capsid-independent. An obvious way to obtain more conclusive information on these and other issues related to coronaviral budding is by co-expressing the structural genes in the absence of other viral components. Such experiments are currently in progress.

Acknowledgements

We thank Dr. H. Vennema for helpful discussions. We are grateful to Dr. J. Fleming for providing the monoclonal antibodies to the M and S proteins.

References

1. Armstrong JH, Niemann H, Smeekens S, Rottier P, Warren G (1984) Sequence and topology of a model intracellular membrane protein, E1 glycoprotein, from a coronavirus. Nature 308: 751–752
2. Braakman I, Hoover-Litty H, Wagner KR, Helenius A (1991) Folding of influenza hemagglutinin in the endoplasmic reticulum. J Cell Biol 114: 401–411
3. Braakman I, Helenius J, Helenius A (1992) Manipulating disulfide bond formation and protein folding in the endoplasmic reticulum. EMBO J 11: 1717–1722
4. Griffiths G, Rottier P (1992) Cell biology of viruses that assemble along the biosynthetic pathway. Semin Cell Biol 3: 367–381
5. Holmes KV, Doller EW, Sturman LS (1981) Tunicamycin resistant glycosylation of a coronavirus glycoprotein: demonstration of a novel type of viral glycoprotein. Virology 115: 334–344
6. Hurtley SM, Helenius A (1989) Protein oligomerization in the endoplasmic reticulum. Annu Rev Cell Biol 5: 277–307

 7. Krijnse Locker J, Griffiths G, Horzinek MC, Rottier PJM (1992) O-glycosylation of the coronavirus M protein: differential localization of sialyltransferases in N- and O-linked glycosylation. J Biol Chem 267: 14094–14101
 8. Pettersson R (1991) Protein localization and virus assembly at intracellular membranes. In: Compans RW (ed) Protein traffic in eukaryotic cells. Curr Top Microbiol Immunol 170: 67–106
 9. Rottier PJM, Horzinek MC, van der Zeijst BAM (1981) Viral protein synthesis in mouse hepatitis virus strain A59-infected cells: effects of tunicamycin. J Virol 40: 350–357
10. Rottier PJM, Welling GW, Welling-Wester S, Niesters HGM, Lenstra JA, van der Zeijst BAM (1986) Predicted membrane topology of the coronavirus protein E1. Biochemistry 25: 1335–1339
11. Rottier PJM, Rose JK (1987) Coronavirus E1 glycoprotein expressed from cloned cDNA localizes in the Golgi region. J Virol 61: 2042–2045
12. Simons K, Garoff H (1980) The budding mechanism of enveloped animal viruses. J Gen Virol 50: 1–21
13. Tooze J, Tooze S, Warren G (1984) Replication of coronavirus MHV-A59 in sac- cells: determination of the first site of budding of progeny virions. Eur J Cell Biol 33: 281–293
14. Tooze J, Tooze S, Warren G (1988) Site of addition of N-acetylgalactosamine to the E1 glycoprotein of mouse hepatitis virus-A59. J Cell Biol 106: 1475–1487
15. Vennema H, Rottier PJM, Heijnen L, Godeke GJ, Horzinek MC, Spaan WJM (1990) Biosynthesis and function of the coronavirus spike protein. In: Cavanagh D, Brown TDK (eds) Coronaviruses and their diseases. Plenum Press, New York, pp 9–19

Authors' address: Dr. P. J. M. Rottier, Institute of Virology, Yalelaan 1, 3584 CL Utrecht, The Netherlands.

Arch Virol (1994) [Suppl] 9: 329–338

_Archives_____
Virology
of
© Springer-Verlag 1994
Printed in Austria

Assembly and entry mechamisms of Semliki Forest virus

H. Garoff[1], **J. Wilschut**[2], **P. Liljeström**[1], **J. M. Wahlberg**[1], **R. Bron**[2], **M. Suomalainen**[1], **J. Smyth**[1], **A. Salminen**[1], **B. U. Barth**[1], **H. Zhao**[1], **K. Forsell**[1], and **M. Ekström**[1]

[1] Department of Molecular Biology, Huddinge, Sweden
[2] Department of Physiological Chemistry, University of Groningen,
Groningen, The Netherlands

Summary. The alphavirus Semliki Forest (SFV) is an enveloped virus with a positive single-stranded RNA genome. The genome is complexed with 240 copies of a capsid protein into a nucleocapsid structure. In the membrane the virus carries an equal number of copies of a membrane protein heterodimer. The latter oligomers are grouped into clusters of three. These structures form the spikes of the virus and carry its entry functions, that is receptor binding and membrane fusion activity. The membrane protein heterodimer is synthesized as a p62E1 precursor protein which upon transport to the cell surface is cleaved into the mature E2E1 form. Recent studies have given much new information on the assembly and entry mechanism of this simple RNA virus. Much of this work has been possible through the construction of a complete cDNA clone of the SFV genome which can be used for in vitro transcription of infectious RNA. One important finding has been to show that a spike deletion variant and a capsid protein deletion variant are budding-negative when expressed separately but can easily complement each other when transfected into the same cell. This shows clearly that enveloped viruses use different budding strategies: one which depends on a nucleocapsid-spike interaction as exemplified by SFV and another one which is based on a direct core-lipid bilayer interaction as shown before to be the case with retroviruses. Another important finding concerns the activation process of the presumed fusion protein of SFV, the E1 subunit. In the original p62E1 heterodimer E1 is completely inactive. Activation proceeds in several steps. First p62 cleavage activates the potential for low pH inducible fusion. Next the low pH which surrounds incoming virus in endosomes induces dissociation of the heterodimeric structure. This is followed by a rearrangement of E1 subunits into homotrimers which are fusion active.

Introduction

The positive, single-stranded RNA genome of alphaviruses (e.g., Semliki Forest virus (SFV) and Sindbis virus) is packaged into a nucleocapsid (NC). This is surrounded by a lipid membrane which contains the viral spike proteins. Alphaviruses mature by budding at the plasma membrane (PM) and enter new cells by fusion of viral and cell membranes in endosomes. We are interested in the mechanisms of alphavirus assembly and entry, and use SFV infection of baby hamster kidney (BHK) cells as our model system. SFV directs the synthesis of a nucleocapsid (C) protein and two membrane proteins (p62 and E1). The membrane proteins associate into p62E1 heterodimers soon after synthesis in the ER, and are then transported to the PM where they form virus spikes. During this process the p62 subunit is proteolytically processed into the smaller E2 form [7].

Virus assembly

The viral 42 S RNA genome is dispensable for particle formation

The assembly of SFV starts with the formation of viral NCs in the cell cytoplasm. This is probably initiated by the binding of C protein to one or more regions of the viral (+) 42 S RNA sequence; these regions constitute the encapsidation signal. For Sindbis virus, this signal has been localized to the 5′ part of the viral genome, the coding region of the first nonstructural protein [28]. After this specific recognition event, NC formation can be envisaged as proceeding via more C protein making additional (and less specific) interactions with the genomic RNA, as well as via C-C interactions, until a T = 4 icosahedral NC structure is formed. Each NC contains 240 C protein units in complex with one 42 S RNA molecule [5]. An important question is whether an interaction between C and the encapsidation signal is absolutely required for assembly of alphavirus particles. In the cases of hepadnaviruses and retroviruses, it has clearly been shown that the viral genome is not required for particle assembly [3, 9]. In order to determine whether alphavirus particles can be made in the absence of genomic RNA, we have expressed the structural proteins of SFV from a T7-promoter plasmid using a vaccinia virus-T7 polymerase recombinant [22]. The results showed that particles were released from cells. These particles appeared similar to wild-type (wt) SFV by electron microscopy (EM) and were also of wt density when analyzed on sucrose density gradients. However, when assembly of virus subunits into particles in transfected cells was compared to that of wt virus (i.e. assembly in the presence of viral genome) by pulse/chase analysis, the efficiency with which assembly occurred was found to be only 10% of that of wt. This suggests that the viral genome with its packaging sequence is not absolutely required for virus particle formation. However, its presence apparently increases the efficiency of the assembly process considerably. Indeed, particle production increased with about 50% in transfected cells when the structural proteins were coexpressed with an SFV recombinant RNA that lacked the normal structural region but retained the nonstructural region with an intact encapsidation signal [22]. At present we do not know whether the virus-like particles produced from the vaccinia construct contain any RNA. However, it is likely that assembly of viral NCs can in this case proceed via nonspecific C-RNA interactions, as foreign nucleic acids have also been shown to be encapsidated in vitro [29].

Mapping of the functional domains of the C protein

The COOH-terminal half of the 33 kD C protein is folded into a serine protease [5]. This is responsible for cleavage of the C polypeptide from the structural polyprotein after translation. The NH_2-terminal half of the C polypeptide is very rich in Arg, Lys and Pro residues and is therefore thought to interact with RNA during NC assembly [7]. Using an infectious clone of SFV [14], we have tested this hypothesis by analyzing a mutant from which the Arg-, Lys- and Pro-rich domain was deleted. The deletion variant was found to direct efficient synthesis of all viral subunits, but assembly of NCs and virus particles was not observed. Thus, it was clearly demonstrated that the NH_2-terminal domain of the C protein is required for NC assembly. It is probable that this domain interacts with the viral RNA, thereby facilitating the occurrence of intermolecular C-C interactions. The phenotype of this deletion mutant also showed that the COOH-terminal serine protease unit can function independently of the NH_2-terminal half of the molecule.

The spike-NC interaction is the driving force for alphavirus budding

Since it became clear in the early 1970s that alphavirus spike proteins are transmembrane molecules, it has been hypothesized that the driving force for virus budding is provided by the interaction between the cytoplasmic domains of the spike protein and the NC surface [8]. As the lipid bilayer is bound to the spikes through their transmembrane peptides, successive spike-NC interactions would force the membrane to curve around the NC resulting in virus budding. Although a very attractive model, it has never been tested by direct biochemical experiments. The importance of elucidating this question is underlined by the fact that for retroviruses, budding has been found to be the result of an interaction between the internal core of the virus and the lipid bilayer, and not between the core and the spikes. Indeed, several recent studies have demonstrated efficient budding of spike-less retrovirus particles, using in vitro-engineered proviral DNA constructs from which the complete *env* gene has been deleted [9, 32]. This prompted us to test the original alphavirus budding model by examining viral particle formation for two SFV mutants that were designed to contain deletions of either the capsid or the spike gene region [22]. Biochemical analysis of BHK cells that had been transfected with the spike-gene deletion variant showed normal intracellular NC assembly but no virus particle formation. Furthermore, budding virus could not be detected by EM analysis. The capsid-gene deletion variant was also unable to release enveloped particles from cells. However, when both constructs were cotransfected into cells, virus particle formation was regenerated to 30% of that of wt. Thus, in the case of alphaviruses, the driving force for budding comes from an interaction between the spike proteins and the NC. Therefore, enveloped viruses utilize at least two different mechanisms for budding; one of which is used by retroviruses and is based on an interaction between the internal core of the virus and the lipid bilayer, and a second which is used by alphaviruses and which is based on an interaction betwen the viral NC and the viral spike proteins.

The spike cytoplasmic tail-NC interaction remains unproven

The cytoplasmic tail of E2 is considered to be the most likely site of interaction of the spike with the NC. However, any evidence that has been cited as proof for such an interaction is largely circumstantial, and comes from experiments where NC and spike proteins have been cross-linked [8], or from experiments where solubilized spikes have been shown to bind to NCs in vitro [11]. Cross-linking at best indicates that spike proteins and NC are in close proximity, and in the case of the in vitro spike-NC binding assays the specificity of the interaction remains to be determined.

The recently reported ability of an anti-idiotypic monoclonal antibody (Mab, F13) that was raised against a Mab specific for the cytoplasmic tail of E2 to bind to NCs, has been cited as evidence for an interaction between the E2-cytoplasmic tail and the NC [23]. We previously reported that a synthetic E2-tail peptide binds to isolated NCs in vitro [17], and confirmation of the specificity of the E2-tail/NC interaction was based on competition studies using the F13 Mab. However a re-evaluation of the specificity of F13 has led to the conclusion that this Mab binds to an unrelated epitope and therefore the results of these experiments can no longer be considered proof of an interaction between the E2-tail and the NC [21].

To analyze the role of the cytoplasmic tail of the SFV spike in budding, we have mutagenesized the conserved Cys_4, Tyr_9, Cys_{24} and Cys_{25} residues of the 31 amino acid-long E2 tail, and the two conserved Arg residues which constitute the tail of the other spike subunit, E1 [7]. Analysis of the virus mutants showed that the two conserved Arg residues in the E1 tail had no role in virus assembly (nor in virus entry nor multiplication in cell culture), whereas all mutations in the E2 tail inhibited virus particle formation [2] (Zhao et al., unpubl.). As these E2 mutants produced spike oligomers that were processed and transported to the cell surface in a fashion similar to wt, the effect of the mutations appeared to be at the level of virus budding, possibly through defective binding of spikes to the NC. This interpretation was supported by EM analysis of cells transfected with infectious mutant RNA; in these cells there was no evidence of budding virus particles, or of binding of NCs to the PM. Therefore the E2 tail is clearly involved in budding, but whether or not it actually binds to the NC during budding still remains to be proven.

Control of budding site

Typically, alphaviruses bud at the PM of vertebrate cells. A simple explanation as to how this is controlled, is that the viral spike proteins are transported to the cell surface where they accumulate and function as receptors for NCs. We have tested this model by analyzing the fate of the virus spike protein when expressed from a C protein-deletion variant of the SFV genome [33]. Our analyses have demonstrated that, in contrast to wt, spikes expressed from the C-deletion variant did not accumulate on the cell surface after intracellular transport as expected, but were rapidly shed from the cell surface as fragments derived from the external spike domains. The fact that this was not observed during budding of wt virus suggests that normally no sizeable pool of free spikes exists on the PM, but that spikes are captured very efficiently by NCs to form budding complexes. This result is in agreement with the earlier finding that Sindbis spikes are efficiently immobilized in the PM of infected cells, as measured by photobleaching recovery [12]. At present there are no data to suggest how such an efficient spike capturing could take place. The mechanism could be as simple as over production of NCs which then efficiently bind all spikes arriving at the PM. Alternatively, a more complicated control mechanis might exist, one involving co-localization of spikes and NCs to the sites of budding.

An important piece in the puzzle of what controls budding could lie in the small 6 kD protein which is encoded by the membrane protein-coding region of the alphavirus genome [7]. This protein carries the signal sequence for the E1 subunit in its COOH-terminal region, but this is probably not the sole function of this protein. We have shown that it associates with the spike-protein complex in the ER, and is transported with it to the cell surface [15]. However, either during or just prior to budding it is largely excluded from the viral membrane. Interestingly, if the 6kD-coding region is deleted from the viral genome, it is the budding efficiency that is affected; this drops to about 10% of that of wt virus [14]. One possibility is that the 6kD protein modulates the maturation of the spike so that it becomes competent for budding.

Virus entry

The sole purpose for virus assembly is to export the virus genome out from the cell, enclosed in a package that promotes entry of the genome into a new target cell. In the case of alphaviruses the entry process involves binding to receptors on the host cell, endocytosis through clathrin-coated pits and vesicles, and acid-induced membrane fusion in endosomes [16, 27]. We have recently studied three important aspects of SFV entry; these are the mechanisms by which SFV spike proteins become fusion-competent in acidic endosomes, deactivation of the fusion function which is necessary to prevent aberrant fusion within the cell after virus entry, and inhibition of the fusion activity of spike proteins during virus assembly.

Activation and deactivation of the SFV spike fusion function

Since the E2E1 heterodimeric spike protein of SFV is able to fuse with cell membranes only after entering acidic endosomes, it was of interest to study the conformational changes that appear in spike tertiary and quaternary structures during virus entry [13]. For this purpose we used ^{35}S-methionine-labelled SFV, which was allowed to bind to target cells (BHK) at 4°C, and then to enter cells by incubation at 37°C. Analysis of cell lysates after increasing times of incubation with virus, showed that within a few minutes, the original E2E1 heterodimer was reorganized into E1 homotrimers and E2 monomers [24–26]. The trimers were found to be trypsin resistant and also to expose an unique epitope recognized by a mAb (antiE1″) that did not detect the original E2E1 heterodimer. To study the possible involvement of the E1 homotrimer in membrane fusion, we utilized an in vitro fusion assay [4, 26]. Pyrene phospholipid-labelled SFV was allowed to fuse with unilamellar lipid vesicles of uniform size. Pyrene-labelled virus has a very characteristic pyrene excimer fluorescence of 480 nm, at an excitation wave length of 343 nm. Upon low pH-induced fusion of virus and liposomes, this fluorescence decreases as a consequence of lipid mixing and the resulting dilution of the pyrene lipids. Monitoring the excimer fluorescence allows the fusion reaction to be followed. With this system fusion occurred with an efficiency of 53%; the optimum pH and temperature conditions were found to be 5.55 and 37°C respectively, and cholesterol was required in the target membrane. All these parameters have previously been demonstrated to be optimal for alphavirus fusion both in vitro and in vivo [18, 30]. Moreover, we could also demonstrate that all conformational changes occurring in the spike proteins in our in vivo studies also occurred in the in vitro fusion assay. These conformational changes also appeared upon incubation of SFV at low pH in the absence of target membranes. However, this treatment resulted in complete inactivation of the fusion activity of the virus. This suggests a mechanism for SFV fusion in which the acidic milieu of the endosome triggers the formation of E1 trimers. These interact with target membranes, initiate membrane fusion and finally achieve a non-active conformation. By using suboptimal conditions for in vitro fusion (i.e., higher pH and lower incubation temperature, pH 5.85 and 20°C) it was possible to make a very clear correlation between the kinetics of appearance of conformational changes in the spike protein, kinetics of membrane binding, kinetics of membrane fusion and kinetics of inactivation of spike protein fusion ability. These results showed that the most rapid process, occurring immediately upon acidification, was the formation of E1 trimers. This was followed by binding of SFV to liposomes and then, after a considerable time lag, by fusion of the SFV envelope with the liposomal membranes. Finally, after an additional time lag, E1 lost its ability to cause fusion i.e. become deactivated. In these experiments the conformational changes were monitored by incubating virus and liposomes at suboptimal fusion conditions for various times, after which the samples were neutralized to stop further conformational changes. Spike proteins were then analyzed for trimer formation, trypsin resistance and anti-E1″ reactivity. The binding of SFV to liposomes was monitored by flotation experiments following similar incubations. The loss of the fusion ability of the E1 homotrimer was followed by first incubating the virus at suboptimal fusion conditions in the absence of liposomes, and then after various times of incubation adding liposomes and a pre-titrated volume of buffer in order to obtain optimal conditions for fusion. Residual fusion activity was then monitored. The results clearly supported a mechanism for fusion that involves the E1 trimer. The time lag observed before fusion occurs at suboptimal conditions can be explained if several E1 trimers must be recruited in order to initiate binding and lipid mixing between the virus and the target membrane. The fact that loss of fusion ability occurs with the slowest kinetics substantiates the theory that the

E1 trimer, after performing its membrane fusion function, undergoes some additional conformational changes in order to prevent any further and possibly deleterious intracellular membrane fusions. The direct involvement of the E1 homotrimer in the fusion reaction was further supported by additional in vitro and in vivo assays, in which addition of the antiE1″ mAb very efficiently inhibited fusion if present during incubation of SFV and liposomes under acidic conditions.

This mechanism of SFV spike-catalyzed membrane fusion is reminiscent of the situation with influenza virus: it has previously been shown that the influenza virus hemagglutinin (HA) trimer undergoes acid-induced conformational changes which are followed by spike protein binding to the target membrane, fusion and finally loss of the fusion activity of the hemagglutinin molecule [20, 31]. However, a problem with the SFV model is that a fusion peptide has never been identified. In the case of the influenza hemagglutinin photo-affinity lipid-labelling techniques have clearly demonstrated the involvement of the NH_2-terminal peptide of the HA_2 subunit in acid-induced membrane fusion reaction [10]. Such studies would also be valuable in the case of the alphavirus spike protein.

Inhibition of entry functions during virus assembly

It is evident that all viruses must have a control mechanism which ensures that entry functions such as receptor binding, membrane fusion and virus dissociation do not interfere with virus assembly. In the case of alphaviruses, even though activation of viral entry functions requires the acidic conditions within the endosomes, this is probably not sufficient as a control mechanism, because the pH of distal portions of the exocytic compartment also reach pH values as low as those of the endosomes [1]. Therefore alphaviruses must have additional control mechanisms built into their multiplication process in order to prevent premature activation of entry functions in infected cells. The p62 cleavage represents such a mechanism. We have generated SFV variants that are not cleaved by the host protease because their protease cleavage sites have been violated by mutagenesis of the p62 gene [19]. These viruses formed mature particles with wt efficiency, but could not infect new cells. This "noninfectivity" was due to deficient binding and fusion when assayed under normal conditions. The block was, however, possible to overcome both in vivo and in vitro, by lowering the fusion pH from pH 5.5 (which is optimal for wt virus) to 4.5. At this low pH the p62E1 heterodimer of the variant viruses dissociated and E1 homotrimers formed. Thus, under normal conditions the p62E1 spike of the variant viruses cannot undergo the low pH-induced conformational changes and is therefore unable to fuse.

Since the p62E1 heterodimer of wt virus is known to be transported to the very distal parts of the biosynthetic pathways before cleavage takes place, the acid stability of p62E1 offers a likely explanation as to how SFV controls (inhibits) activation of E1-mediated entry functions while virus assembly is occurring [6].

References

1. Anderson RGW, Orci L (1988) A view of acidic intracellular compartments. J Cell Biol 106: 539–543
2. Barth B-U, Suomalainen M, Liljeström P, Garoff H (1992) Alphavirus assembly and entry: The role of the cytoplasmic tail of the E1 spike subunit. J Virol 66: 7560–7564
3. Birnbaum F, Nassal M (1990) Hepatitis B virus nucleocapsid assembly: Primary structure requirements in the core protein. J Virol 64: 3319–3330
4. Bron R, Wahlberg JM, Garoff H, Wilschut J (1993) Membrane fusion of Semliki Forest virus in a model system: correlation between fusion kinetics and structural changes in the envelope glycoprotein. EMBO J 12: 693–701
5. Choi H-K, Tong L, Minor W, Dumas P, Boege U, Rossmann MG, Wengler G (1991) Structure of Sindbis virus core protein reveals a chymotrypsin-like serine proteinase and the organization of the virion. Nature 354: 37–43
6. de Curtis I, Simons K (1988) Dissection of Semliki Forest virus glycoprotein delivery from the trans-Golgi network to the cell surface in permeabilized BHK cells. Proc Natl Acad Sci USA 85: 8052–8056
7. Garoff H, Kondor-Koch C, Riedel H (1982) Structure and assembly of alphaviruses. Curr Top Microbiol Immunol 99: 1–50
8. Garoff H, Simons K (1974) Location of the spike glycoproteins in the Semliki Forest virus membrane. Proc Natl Acad Sci USA 71: 3988–3992
9. Gheysen D, Jacobs E, de Foresta F, Thiriart C, Francotte M, Thines D, De Wilde M (1989) Assembly and release of HIV-1 precursor Pr55[gag] virus-like particles from recombinant Baculovirus-infected cells. Cell 59: 103–112
10. Harter C, James P, Bächi T, Semenza G, Brunner J (1989) Hydrohobic binding of the ectodomain of influenza hemagglutinin to membranes occurs through the "fusion peptide. J Biol Chem 264: 6459–6464
11. Helenius A, Kartenbeck J (1980) The effects of octylglucoside on the Semliki Forest virus membrane. Evidence for spike protein-nucleocapsid interaction. Eur J Biochem 106: 613–618
12. Johnson DC, Schlesinger MJ (1981) Fluorescence photobleaching recovery measurements reveal differences in envelopment of Sindbis and vesicular stomatitis viruses. Cell 23: 423–431
13. Kielian M, Helenius A (1985) pH-induced alterations in the fusogenic spike protein of Semliki Forest virus. J Cell Biol 101: 2284–91
14. Liljeström P, Lusa S, Huylebroeck D, Garoff H (1991) In vitro mutagenesis of a full-length cDNA clone of Semliki Forest virus: the 6000-molecular-weight membrane protein modulates virus release. J Virol 65: 4107–4113
15. Lusa S, Garoff H, Liljeström P (1991) Fate of the 6K membrane protein of Semliki Forest virus during virus assembly. Virology 185: 843–846
16. Mash M, Wellsteed J, Kern H, Harms E, Helenius A (1982) Monensin inhibits Semliki Forest virus penetration into culture cells. Proc Natl Acad Sci USA 79: 5297–5301

17. Metsikkö K, Garoff H (1990) Oligomers of the cytoplasmic domain of the p62/E2 membrane protein of Semliki Forest virus bind to the nucleocapsid in vitro. J Virol 64: 4678–4683
18. Phalen T, Kielian M (1991) Cholesterol is required for infection by Semliki Forest virus. J Cell Biol 112: 615–623
19. Salminen A, Wahlberg JM, Lobigs M, Liljeström P, Garoff H (1992) Membrane fusion process of Semliki Forest virus II: Cleavage dependent reorganization of the spike protein complex controls virus entry. J Cell Biol 116: 349–357
20. Stegmann T, White JM, Helenius A (1990) Intermediates in influenza membrane fusion. EMBO J 9: 4231–4241
21. Suomalainen M, Garoff H (1992) Alphavirus spike-nuclecapsid interaction and network antibodies. J Virol 66: 5106–5109
22. Suomalainen M, Liljeström P, Garoff H (1992) Spike protein-nucleocapsid interactions drive the budding of alphaviruses. J Virol 66: 4737–4747
23. Vaux DJT, Helenius A, Mellman I (1988) Spike-nucleocapsid interaction in Semliki Forest virus reconstructed using network antibodies. Nature 336: 36–42
24. Wahlberg J, Garoff H (1992) Membrane fusion process of Semliki Forest virus I: Low pH-induced rearrangement in spike protein quaternary structure preceeds virus penetration into cells. J Cell Biol 116: 339–348
25. Wahlberg JM, Boere WA, Garoff H (1989) The heterodimeric association between the membrane proteins of Semliki Forest virus changes its sensitivity to mildly acidic pH during virus maturation. J Virol 63: 4991–4997
26. Wahlberg JM, Bron R, Wilschut J, Garoff H (1992) Membrane fusion of Semliki Forest virus involves homotrimers of the fusion protein. J Virol 66: 7309–7318
27. Wang K-S, Kuhn RJ, Strauss EG, Ou S, Strauss JH (1992) High-affinity laminin receptor is a receptor for Sindbis virus in mammalian cells. J Virol 66: 4992–5001
28. Weiss B, Nitschko H, Ghattas I, Wright R, Schlesinger S (1989) Evidence for specificity in the encapsidation of Sindbis virus RNAs. J Virol 63: 5310–5318
29. Wengler G, Boege U, Wengler G, Bischoff H, Wahn KI (1982) The core protein of the alphavirus Sindbis virus assembles into core-like nucleoproteins with the viral genome RNA and with other single-stranded nucleic acids in vitro. Virology 118: 401–410
30. White J, Helenius A (1980) pH-dependent fusion between the Semliki Forest virus membrane and liposomes. Proc Natl Acad Sci USA 77: 3273–3277
31. White JM, Wilson IA (1987) Anti-peptide antibodies detect steps in a protein conformational change: low pH-activation of the influenza virus hemagglutinin. J Virol 105: 2887–2896
32. Wills JW, Craven RC, Weldon RA Jr, Nelle TD, Erdie CR (1991) Suppression of retroviral MA deletions by the amino-terminal membrane-binding domain of p60^{src}. J Virol 65: 3804–3812
33. Zhao H, Garoff H (1992) The role of cell surface spikes for alphavirus budding. J Virol 66: 7089–7095

Authors' address: Dr. H. Garoff, Department of Molecular Biology, Novum, S-141 57 Huddinge, Sweden.

Arch Virol (1994) [Suppl] 9: 339–348

Archives
of
Virology
© Springer-Verlag 1994
Printed in Austria

The interactions of the flavivirus envelope proteins: implications for virus entry and release

F. X. Heinz, G. Auer, K. Stiasny, H. Holzmann, C. Mandl, F. Guirakhoo[*], and **C. Kunz**

Institute of Virology, University of Vienna, Vienna, Austria

Summary. Viral membrane proteins play an important role in the assembly and disassembly of enveloped viruses. Oligomerization and proteolytic cleavage events are involved in controlling the functions of these proteins during virus entry and release. Using tick-borne encephalitis virus as a model we have studied the role of the flavivirus envelope proteins E and prM/M in these processes. Experiments with acidotropic agents provide evidence that the virus is taken up by receptor-mediated endocytosis and that the acidic pH in endosomes plays an important role for virus entry. The envelope glycoprotein E undergoes irreversible conformational changes at acidic pH, as indicated by the loss of several monoclonal antibody-defined epitopes, which coincide with the viral fusion activity in vitro. Sedimentation analysis reveals that these conformational changes lead to aggregation of virus particles, apparently by the exposure of hydrophobic sequence elements. None of these features are exhibited by immature virions containing E and prM rather than E and M. Detergent solubilization, sedimentation, and crosslinking experiments provide evidence that prM forms a complex with protein E which prevents the conformational changes necessary for fusion activity. The functional role of prM before its endoproteolytic cleavage by a cellular protease thus seems to be the protection of protein E from acid-inactivation during its passage through acidic trans Golgi vesicles in the course of virus release.

Introduction

Upon entry of enveloped viruses into cells a membrane fusion event takes place, either at the plasma membrane (e.g. herpes viruses) or, after uptake by receptor-mediated endocytosis, at the endosomal mem-

---

*Present address: ORAVAX Inc., Cambridge, Massachusetts, U.S.A.

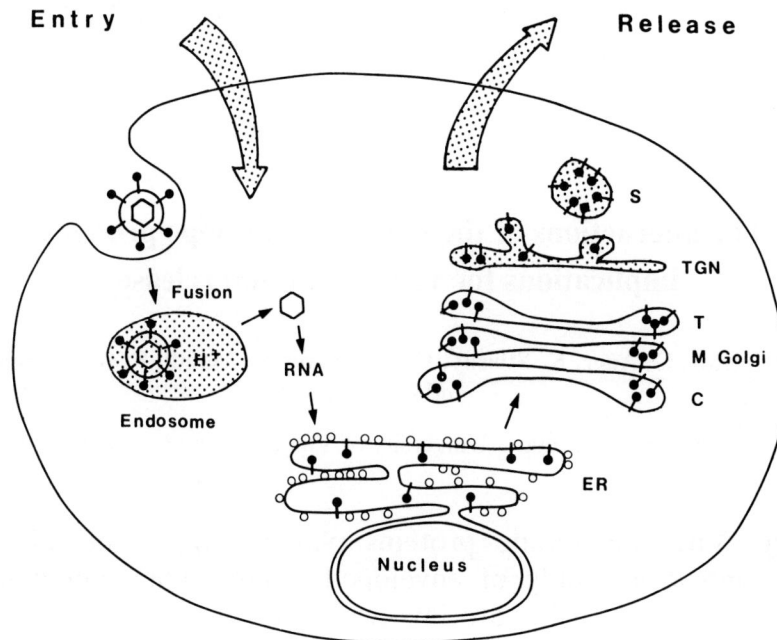

Fig. 1. Pathways of enveloped virus entry and release. Shaded areas indicate that viral envelope glycoproteins encounter acidic intracellular compartments during both endocytosis and exocytosis. *T/M/C-Golgi* Trans, medial, cis-Golgi, *TGN* trans Golgi network, *S* secretory vesicle

brane (e.g. myxoviruses) (for reviews see [10, 13]). Fusion activity is mediated by viral envelope proteins, which in all known cases exist as homo- or heterooligomeric complexes. Many of these are synthesized as inactive precursors that can be assembled into virus particles. Fusion competence is acquired through proteolytic cleavage by a cellular protease. In those case where virus uptake occurs via receptor-mediated endocytosis the acidic pH in endosomes plays a crucial role, because it induces a conformational change in viral envelope proteins that triggers their fusion activity (Fig. 1) [22].

It is a general problem of this latter class of viruses that their potential fusion proteins also encounter an acidic environment during late stages of the biosynthetic pathway in the trans-Golgi network and secretory vesicles (Fig. 1) [1]. This could lead to irreversible conformational changes and fusion activity at the wrong time and at the wrong place. For influenza viruses there is now convincing evidence that this unwanted effect is counteracted by the ion channel activity of the M2 protein [15], which elevates the pH in acidic vesicles and allows the transfer of native hemagglutinin molecules to the plasma membrane [3]. For Semliki Forest virus (an alphavirus) acid pH stabilization of the E1 fusion pro-

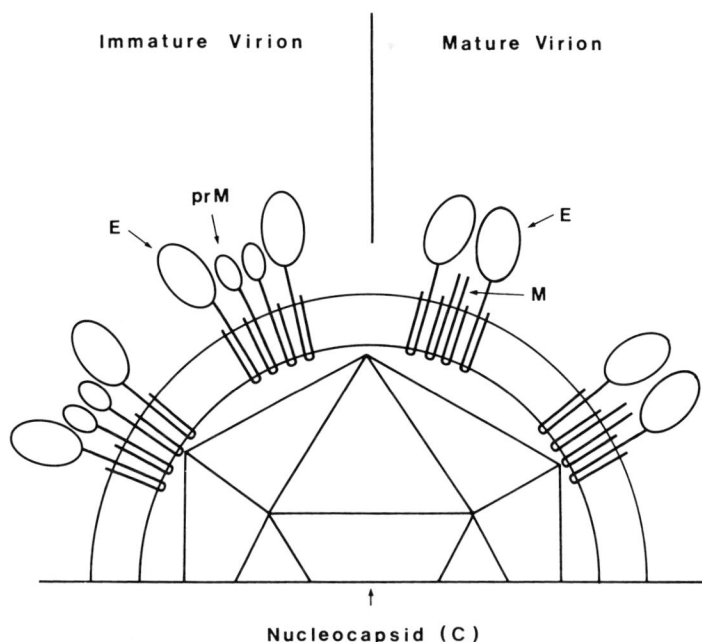

Fig. 2. Schematic picture of mature and immature flavivirions. *E* Envelope protein, *M* membrane protein, *prM* precursor of M

tein was shown to be mediated by oligomerization with the precursor (p62) of a second envelope glycoprotein (E2) [20].

We have been studying related mechanisms with another virus family, *Flaviviridae*, using tick-borne encephalitis (TBE) virus as a model. Flaviviruses are small enveloped viruses that contain a capsid protein (C) and two membrane-associated proteins (E and M) (Fig. 2) (for review of the molecular biology of flaviviruses see [2]).

The E protein represents the viral hemagglutinin and is believed to mediate both receptor-binding and fusion activity. Virion assembly takes place intracellularly and first yields "immature virions" containing a precursor of the M protein (prM) (Fig. 2). As with other viral proteins, this precursor is cleaved by a cellular protease after the consensus sequence R-X-R/K-R [19]. We have studied the function of these envelope proteins during virus entry and release.

Role of acidic vesicles in virus uptake

There is considerable evidence that mosquito-borne flaviviruses are taken up by receptor-mediated endocytosis [4, 11] but direct fusion at the plasma membrane has also been described in certain instances [7]. To study the involvement of acidic vesicles during entry of TBE virus into cells we have investigated the effect of the acidotropic agent ammonium chloride and of bafilomycin A1 on the early phase of the viral life cycle. Bafilomycin A1 is a specific inhibitor of the vacuolar type of H^+-ATPases and thus inhibits the acidification of intracellular vesicles [23]. Both agents exhibited a strong inhibitory effect (~90%) on the generation of new virus progeny when present at the time of infection. Addition at 1 h p.i. resulted in only 35% inhibition and at 4 h p.i. an inhibitory effect was not observed.

When ammonium chloride or bafilomycin A1 were present in the late phase of the viral life cycle only (i.e. added at 24 h p.i.) there was no reduction in the number of virus particles released from the cell. However, in both cases these particles were in the form of immature virions, containing prM rather than M. A similar inhibition of prM processing by acidotropic amines also has been observed with mosquito-borne flaviviruses [16]. These data suggest that acidic compartments play an important role both in the early and in the late phase of TBE virus replication.

Fusion activity of TBE virus

Fusion activity of purified virus preparations was studied by "Fusion from without" (FFWO) of C6/36 mosquito cells [5]. Fusion was observed by the formation of polykaryocytes (Fig. 3a) which, however, was strictly pH-dependent. As shown in Fig. 3b fusion started to occur at pH 6.4 and reached maximal levels at pH 6.0.

In the case of other enveloped viruses, fusion-active proteins have been shown to undergo conformational changes upon exposure to acidic pH (for review see [22]). We have addressed this question for the TBE virus E protein by analyzing the ELISA-reactivity of a panel of 19 distinct monoclonal antibodies (mabs) before and after acidification. The corresponding epitopes were mapped previously and assigned to specific sequence elements of the E protein by a number of immunochemical analyses and the sequence determination of mab-resistant mutants. Together with the assignment of the 12 absolutely conserved cysteines to six disulfide bridges in the E protein of West Nile virus [14] these data have led to a working structural model of protein E, which shows the arrangement of the polypeptide chain into distinct protein domains (A, B, and C) (Fig. 4) [12]. As indicated in Fig. 4 irreversible acid pH-induced conformational changes predominantly affect sites within the discontinuous domain A and the isolated epitope i2. These structural changes start to occur at pH 6.4 and are complete at pH 6.0 and thus parallel the viral fusion activity. As revealed by sedimentation analysis, exposure to acid pH leads to the aggregation of purified virus particles. Similar to related events in other viral fusion proteins the structural

a

b

Fig. 3. a Fusion from without (FFWO) of C6/36 mosquito cells by purified preparations of TBE virus. Giemsa staining. **b** pH-dependency of FFWO

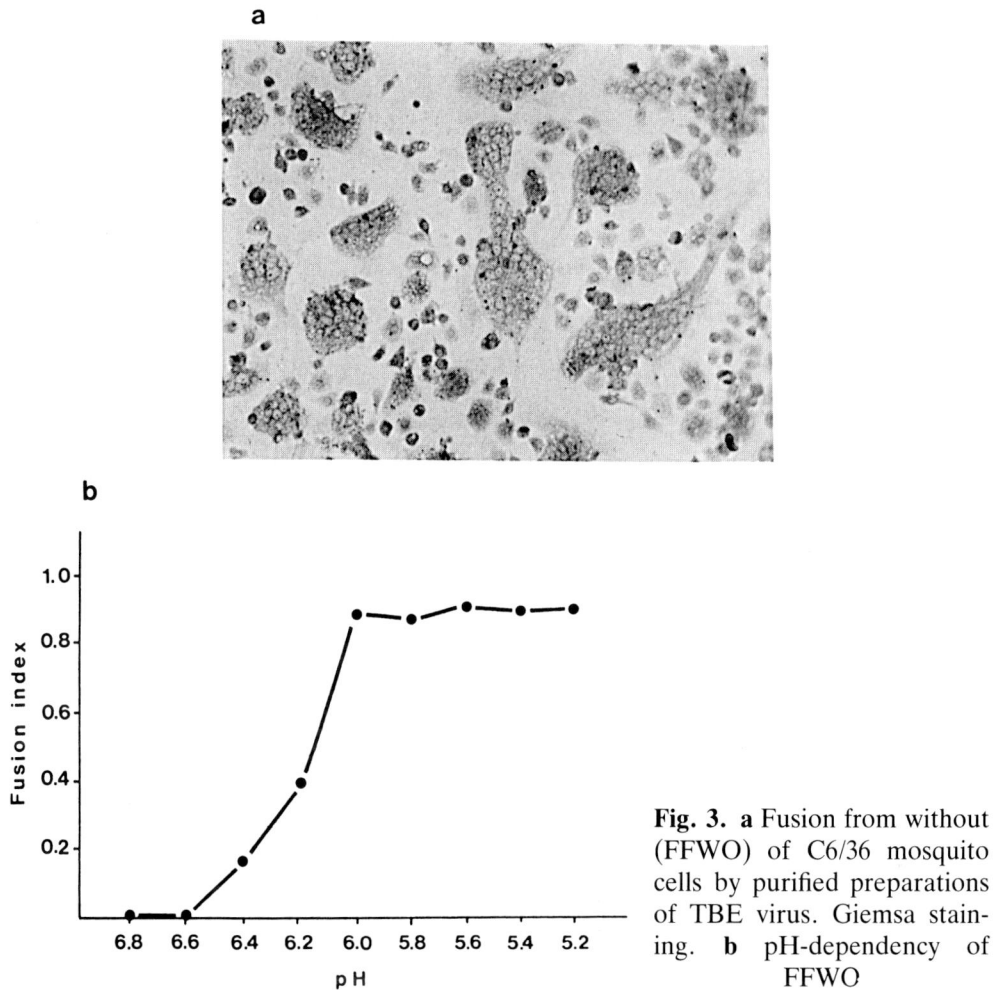

alterations in protein E therefore seem also to be associated with the exposure of hydrophobic sequence elements.

Domain A of protein E (cf. Fig. 4) also contains the most highly conserved sequence among all flavivirus E proteins. By the use of anti-peptide sera Roehrig et al. [18] have demonstrated that this part of the protein becomes more accessible after low pH treatment. Since it also contains the tetrapeptide GLFG which is also present at the fusion-active N-terminus of the influenza virus HA2, this sequence was proposed to form part of an internal fusion element [17].

One could also hypothesize that the cleavage of prM liberates an N-terminal fusion sequence in the M protein. This is unlikely however, since the N-terminus of protein M (about 40 amino acids extend from the lipid bilayer) is neither significantly conserved nor hydrophobic and does not contain sequences found in other fusion proteins.

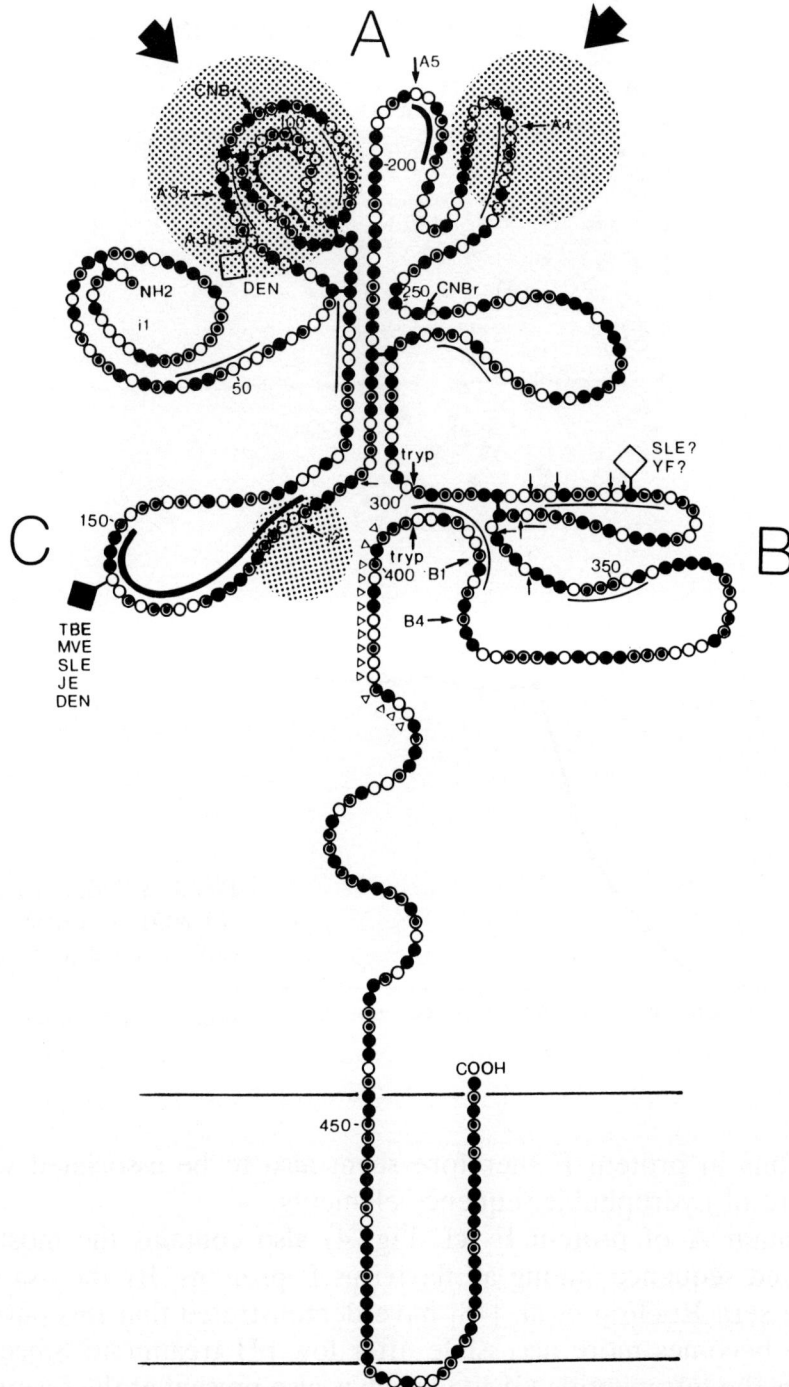

Fig. 4. Structural model of TBE virus protein E, reproduced from Mandl et al. [12]. Open circles represent hydrophilic amino acid residues, dotted circles show intermediate amino acid residues, and solid circles show hydrophobic amino acid residues. Cysteine residues forming disulfide bridges are connected by solid lines. A solid diamond represents the carbohydrate side chain of TBE virus. A line of solid triangles indicates the conserved putative fusion sequence within domain *A*. Shaded areas represent sites undergoing irreversible acid-pH-induced conformational changes. Large arrows indicate sites of impaired mab-binding in immature virions

Properties of immature virions

To assess the functional role of the prM protein we analyzed the properties of immature virions generated by growing the virus in the presence of ammonium chloride. Preparations of mature and immature virus were standardized to the same specific protein contents and used for the determination of specific activities. Similar to observations with mosquito-borne flaviviruses [6, 21], immature TBE virions exhibited a strongly reduced specific infectivity and HA activity as compared to mature virions and did not show any fusion activity in the FFWO assay [5]. This suggests that in the presence of prM the E protein exists in a different functional state in these particles. Also, antigenic analysis of immature virions with the 19 protein E-specific mabs revealed strongly reduced reactivities in the same two areas within domain A which were also shown to undergo low-pH-induced conformational changes (Fig. 4). A similar reduced reactivity of certain E protein-specific mabs was also observed with immature forms of MVE virus [6].

The presence of prM not only prevents acid pH induced structural changes at certain sites but apparently also the exposure of hydrophobic sequences, because acid-pH-induced aggregation was not demonstrable with immature virions (unpubl. obs.).

Oligomeric structure of envelope proteins

Viral envelope proteins that control fusion processes usually exist as oligomeric complexes, such as the infuenza virus HA trimer or the alpha-virus (p62)E2-E1 heterodimer. We have therefore analyzed the oligomeric structure of the TBE virus envelope glycoproteins by crosslinking, detergent solubilization, and sedimentation analyses. After solubilization of mature virions with nonionic detergents such as Triton X-100 or octylglucoside, which usually preserve the structure of native protein complexes, protein E sediments as a homogeneous peak which crosslinks into a dimer [8]. Also, a soluble crystallizable membrane anchor-free form of protein E generated by limited trypsin digestion of purified virions exists as a dimer [9]. Sedimentation and crosslinking experiments performed with immature virions, on the other hand, provide convincing evidence that E and prM form a heterooligomeric complex (unpubl. res.) similar to those described for West Nile virus [21].

Conclusions

The inhibitory effect of ammonium chloride and bafilomycin A1 on TBE-virus entry and the prevention of prM cleavage during virus release

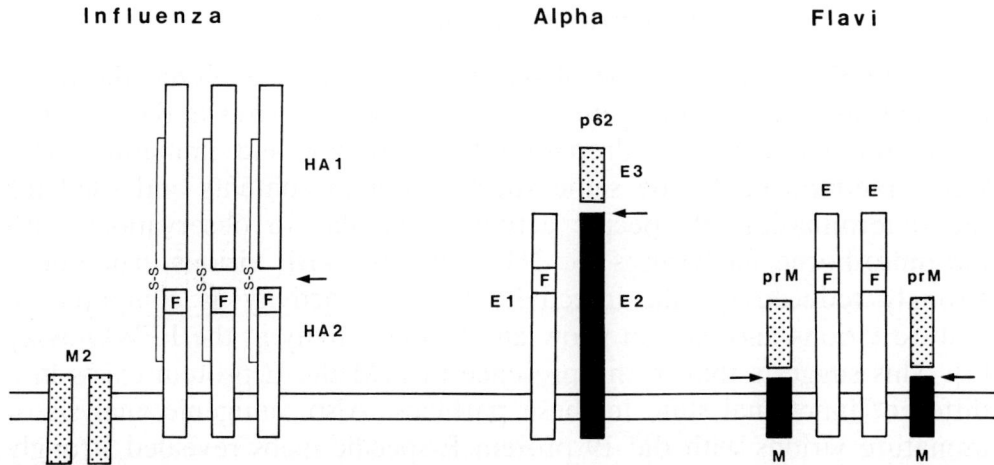

Fig. 5. Schematic representation of proteolytic cleavage activation, protection and oligomerization-control of fusion-active proteins in influenza virus, alphaviruses and flaviviruses. Arrows indicate proteolytic cleavage sites. *F* indicates (putative) fusion-active sequence

is consistent with a role of acidic vesicles in both the early and late phases of the viral life cycle. The function of the prM protein seems to be the protection of protein E from irreversible acid pH-induced conformational changes during exocytosis, changes that are necessary to trigger fusion activity in the endosome after receptor-mediated endocytosis.

As depicted schematically in Fig. 5, different virus families have apparently evolved different mechanisms for activating the fusogenic potential of their envelope glycoproteins by proteolytic cleavage and for protecting them against low pH-induced conformational changes in the exocytic pathway.

In the case of influenza virus the protein that is proteolytically cleaved (HA) is itself the fusion active protein. The fusion activity has been mapped to the N-terminus of HA2. Another protein, M2, functions as an ion channel, elevates the pH in the exocytic pathway and thus protects acid-sensitive hemagglutinins from inactivation.

Alphaviruses and flaviviruses appear to be significantly different. In both cases the fusion-active proteins, E1 and E, respectively, are synthesized as a heterodimer with a precursor protein (p62 and prM, respectively) which prevents those acid pH-induced conformational changes necessary for fusion from occurring. The proteolytic cleavage of these protective proteins by a cellular protease renders the other protein potentially fusogenic. Again, in contrast to the N-terminal fusion sequence in the influenza virus HA2, fusion active sites both in E1 of alphaviruses and in E of flaviviruses seem to reside in highly conserved glycine-rich internal sequence elements.

References

1. Anderson RGW, Orci L (1988) A review of acidic intracellular compartments. J Cell Biol 106: 539–543
2. Chambers TJ, Hahn CS, Galler R, Rice CM (1990) Flavivirus genome organization, expression, and replication. Annu Rev Microbiol 44: 649–688
3. Ciampor F, Bayley PM, Nermut MV, Hirst EMA, Sugrue RJ, Hay AJ (1992) Evidence that the amantadine-induced, M2-mediated conversion of influenza A virus hemagglutinin to the low pH conformation occurs in an acidic trans Golgi compartment. Virology 188: 14–24
4. Gollins SW, Porterfield JS (1986) pH-dependent fusion between the flavivirus West Nile and liposomal model membranes. J Gen Virol 67: 157–166
5. Guirakhoo F, Heinz FX, Mandl CW, Holzmann H, Kunz C (1991) Fusion activity of flaviviruses: comparison of mature and immature (prM-containing) tick-borne encephalitis virions. J Gen Virol 72: 1323–1329
6. Guirakhoo F, Bolin RA, Roehrig JT (1992) The Murray Valley encephalitis virus prM protein confers acid resistance to virus particles and alters the expression of epitopes within the R2 domain of E glycoprotein. Virology 191: 921–931
7. Hase T, Summers PL, Eckels KH (1989) Flavivirus entry into cultured mosquito cells and human peripheral blood monocytes. Arch Virol 104: 129–143
8. Heinz FX, Kunz C (1980) Isolation of dimeric glycoprotein subunits from tick-borne encephalitis virus. Intervirology 13: 169–177
9. Heinz FX, Mandl CW, Holzmann H, Kunz C, Harris BA, Rey F, Harrison SC (1991) The flavivirus envelope protein E: isolation of a soluble form from tick-borne encephalitis virus and its crystallization. J Virol 65: 5570–5583
10. Kielian M, Jungerwirth S (1990) Mechanisms of virus entry into cells. Mol Biol Med 7: 17–31
11. Kimura T, Gollins SW, Porterfield JS (1986) The effect of pH on the early interaction of West Nile virus with P 388 D1 cells. J Gen Virol 67: 2423–2433
12. Mandl CW, Guirakhoo F, Holzmann H, Heinz FX, Kunz C (1989) Antigenic structure of the flavivirus envelope protein E at the molecular level, using tick-borne encephalitis as a model. J Virol 63: 564–571
13. Marsh M, Helenius A (1989) Virus entry into animal cells. In: Maramorosch K, Murphy FA, Shatkin A (eds) Advances in virus research, vol 36. Academic Press, San Diego, pp 107–151
14. Nowak T, Wengler G (1987) Analysis of disulfides present in the membrane proteins of the West Nile flavivirus. Virology 156: 127–137
15. Pinto LH, Holsinger LJ, Lamb RA (1992) Influenza virus M_2 protein has ion channel activity. Cell 69: 517–528
16. Randolph VB, Winkler G, Stollar V (1990) Acidotropic amines inhibit proteolytic processing of flavivirus prM protein. Virology 174: 450–458
17. Roehrig JT, Hunt A, Johnson AJ, Hawkes RA (1989) Synthetic peptides derived from the deduced amino acid sequence of the E-glycoprotein of Murray Valley encephalitis virus elicit antiviral antibody. Virology 171: 49–60
18. Roehrig JT, Johnson AJ, Hunt AR, Bolin RA, Chu MC (1990) Antibodies to dengue 2 virus E-glycoprotein synthetic peptides identify antigenic conformation. Virology 177: 668–675

19. Strauss JH, Strauss EG, Hahn CS, Hahn YS, Galler R, Hardy WR, Rice CM (1987) Replication of alphaviruses and flaviviruses: proteolytic processing of poly-proteins. In: Brinton MA, Rueckert RR (eds) Positive strand RNA viruses. Alan R. Liss, New York, pp 209–225
20. Wahlberg JM, Boere WAM, Garoff H (1989) The heterodimeric association be-tween the membrane protein of Semliki Forest virus changes its sensitivity to low pH during virus maturation. J Virol 63: 4991–4997
21. Wengler G, Wengler G (1989) Cell-associated West Nile flavivirus is covered with E + Pre-M protein heterodimers which are destroyed and reorganized by proteolytic cleavage during virus release. J Virol 63: 2521–2526
22. White J (1990) Viral and cellular membrane fusion proteins. Annu Rev Physiol 52: 675–697
23. Yoshimori T, Yomamoto A, Moriyami Y, Futai M, Tashiro Y (1991) Bafilomycin A1, a specific inhibitor of vacuolar-type H^+-ATPase, inhibits acidification and protein degradation in lysosomes of cultured cells. J Biol Chem 266: 17707–17712

Authors' address: Dr. F. X. Heinz, Institute of Virology, Kinderspitalgasse 15, A-1095 Wien, Austria.

Arch Virol (1994) [Suppl] 9: 349–358

Archives
of
Virology
© Springer-Verlag 1994
Printed in Austria

Coronavirus polyprotein processing

S. R. Weiss[1], **S. A. Hughes**[1], **P. J. Bonilla**[1], **J. D. Turner**[1], **J. L. Leibowitz**[2], and **M. R. Denison**[3]

[1] Department of Microbiology, University of Pennsylvania School of Medicine, Philadelphia, Philadelphia, [2] Department of Pathology and Laboratory Medicine, University of Texas Health Sciences Center, Houston, Texas, [3] Department of Pediatrics, Vanderbilt University School of Medicine, Nashville, Tennessee, U.S.A.

Summary. MHV gene 1 contains two ORFs in different reading frames. Translation proceeds through ORF 1a into ORF 1b via a translational frame-shift. ORF 1a potentially encodes three protease activities, two papain-like activities and one poliovirus 3C-like activity. Of the three predicted activities, only the more amino terminal papain-like domain has been demonstrated to have protease activity. ORF 1a polypeptides have been detected in infected cells by the use of antibodies. The order of polypeptides encoded from the 5′ end of the ORF is p28, p65, p290. p290 is processed into p240 and p50. Processing of ORF1a polypeptides differs during cell free translation of genome RNA and in infected cells, suggesting that different proteases may be active under different conditions. Two RNA negative mutants of MHV-A59 express greatly reduced amounts of p28 and p65 at the non-permissive temperature. These mutants may have defects in one or more viral protease activities. ORF 1b, highly conserved between MHV and IBV, potentially contains polymerase, helicase and zinc finger domains. None of these activities have yet been demonstrated. ORF 1b polypeptides have yet been detected in infected cells.

Introduction

The coronavirus mouse hepatitis virus (MHV) contains the largest known viral RNA genome, a 31 kilobase (kb) single stranded positive sense RNA [1, 2]. The replication strategy includes the replication of full-length genome RNA via a full-length negative strand RNA, as well as synthesis of a nested set of six positive-stranded subgenomic mRNAs [3]. These mRNAs are thought to be synthesized via a leader priming mechanism. Recent data have identified subgenomic negative strand

RNAs [4], but the mechanism of synthesis of these RNAs and their possible role in expression of mRNAs is not yet understood.

Gene 1 of the coronaviruses is presumed to encode the viral polymerase. The complete nucleotide sequences for gene 1 are now available for the coronavirus, avian infectious bronchitis virus (IBV) [5] and the JHM [2] and A59 ([6]; unpubl. res.) strains of MHV. The general structures of the genes for the avian and murine viruses are similar; most of the specific information described in this manuscript will be for MHV-A59. Gene 1 contains two large ORFs of approximately 14 (ORF 1a) and 8 (ORF 1b) kb. These ORFs overlap by about 75 nucleotides. ORF 1b is translated in the −1 frame with respect to ORF 1a. Translation is thought to begin at nucleotide 210 of ORF 1a and proceed through the end of ORF 1a where a translational frame shift allows translation of ORF 1b [2, 5–7]. Translation of ORF 1a of MHV-A59 predicts a polypeptide of 4 469 amino acids while a fusion polypeptide translated from ORFs 1a and 1b would predict a polypeptide of 7 202 amino acids [6]. While these very large polypeptides have not yet been detected in coronavirus-infected cells, we describe below the detection of smaller polypeptides presumably processed from these very large precursors by proteases encoded in gene 1. We will discuss the structure of gene 1 and the functional domains predicted from analysis of the sequence and then describe our recent work on the detection of the polypeptides encoded in gene 1 of MHV-A59.

Structure of gene 1, the putative polymerase gene

Figure 1 is a schematic diagram of the functional domains predicted to be encoded in gene 1 of MHV. There are two predicted papain-like protease domains and one polio 3C-like protease domain within ORF 1a ([2]; Bonilla et al., unpubl. data). The two papain-like domains contain the expected cysteine and histidine catalytic residues at similar spacing to the cellular papain-like enzymes; however, sequence comparison suggests that the viral and cellular enzymes are only distantly related [8]. An "X" domain located downstream of the first papain domain, is also found in IBV gene 1 and conserved among other groups of viruses, such as the alpha and rubi viruses [8]. The function of this domain has not yet been demonstrated. The coronavirus 3C-like protease domain sequence predicts a protease similar to the poliovirus 3C protease, but different from the chymotrypsin-like serine proteases in that the catalytic serine residue is replaced by a cysteine. The coronavirus 3C-like protease domain also differs from the poliovirus enzyme and from chymotrypsin like serine proteases in the following ways: 1) there is a tyrosine substituted for glycine in the putative substrate binding region of the coro-

ORF 1a   **ORF 1b**

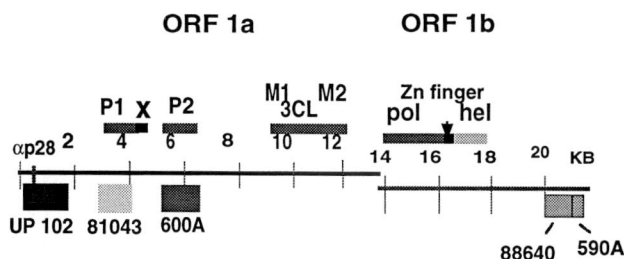

Fig. 1. Predicted functional domains encoded within gene 1 of MHV. ORF 1a and ORF 1b are shown with the predicted functional domains ([2] Bonilla et al., unpubl.). *P1* and *P2* designate the putative papain-like protease domains. *X* Conserved domain discussed in the text; *M1* and *M2* predicted membrane spanning domains; *3CL* polio 3C-like protease domain; *Pol* predicted polymerase domain; *hel* predicted dNTP binding (helicase) domain. Below the lines are shown the regions of genome against which gene 1 specific antisera (as discussed in the text) are directed

navirus sequences [2]; and 2) the predicted coronavirus enzymes lack the third catalytic residue of aspartic/glutamic acid [2]. The MHV 3C-like protease domain is bounded on either side by a potential membrane spanning domain [2]. Of these three predicted protease domains in MHV, only the more amino terminal papain-like enzyme has been demonstrated to have activity ([9]; see below).

ORF 1b is predicted to contain polymerase, dNTP binding (or helicase) and zinc finger domains [2, 5]. An unusual feature of the predicted polymerase domain in both MHV and IBV [2, 6] is that the usual GDD (glycine, aspartic acid, aspartic acid) core polymerase motif is replaced by SDD (serine, aspartic acid, aspartic acid). The NTP binding domain has several clusters of conserved amino acids. The most conserved "A" site motif (GKS) and "B" motifs (aspartic acid residue preceeded by three out five hydrophobic residues) thought to be the Mg^{+2} binding domain are present in the coronavirus sequences [10]. This predicted NTP binding domain is also thought to contain helicase activity because NTP binding proteins have homology with some bacterial helicases [10]. Between the polymerase and NTP binding domains is a cysteine rich putative zinc finger domain. This is thought to interact with nucleic acids during viral replication [2, 11].

Predictions of actual polypeptides to be generated from gene 1 were made by examining putative cleavage sites for the poliovirus 3C protease [2, 11]. Analysis of glutamine/glycine, glutamine/serine and glutamine/ alanine dipeptides (cleavage sites for the poliovirus 3C protease) and surrounding sequences both in IBV and MHV suggests that cleavages may occur to generate separate polypeptides containing the 3C-like protease, each of the potential membrane spanning domains, the polymerase and helicase/zinc finger domains. However, none of this has yet been confirmed experimentally. This analysis also predicts several other similar cleavage sites in other portions of ORF 1a and ORF 1b, for which functional domains have not yet been predicted.

ORF 1b is thought to be translated via a frame shift. Indeed, several labs have described a conserved "slippery sequence" (UUUUUAAAC in MHV) at the end of ORF 1a followed by a pseudoknot structure, both thought to be elements in frame shifting. Both the MHV [2, 6] and the IBV [7] sequences have been demonstrated to have frame shifting activity both during in vitro translation and in eukaryotic cells. We have also demonstrated that prokaryotic ribosomes will also frame shift when translating this sequence. The coronavirus frame shift is about 40% efficient [2, 6, 7]. It is interesting that ORF 1a, containing the proteases necessary to process the polymerase polypeptides, is probably expressed at a higher level than ORF 1b, which is believed to encode the actual polymerase activity.

Comparison of gene 1 of MHV-JHM, MHV-A59 and IBV

The above predicted functional domains are highly conserved between MHV-JHM and MHV-A59 and also between MHV and IBV. In fact, ORF 1b is 52.8% conserved at the amino acid level between JHM and IBV and 94.4% between JHM and A59 [2]. ORF 1a is less conserved than ORF 1b when comparing either A59 and JHM or MHV and IBV. ORF 1a of IBV and MHV have considerable differences, particularly near the 5′ end of the ORF. In fact IBV contains only one of the papain domains and is almost 2 kb shorter than MHV ORF 1a [2, 5]. There are some interesting differences between the ORF 1a sequences of JHM and A59. The A59 sequence has several regions containing small deletions and insertions resulting in frame shifts of up to 40 amino acids. There is also an 18 amino acid in frame deletion in the 5′ portion of the gene (Bonilla et al., in prep.). These frame shifts and deletions are all outside of the predicted functional domains, with the exception of the presence of one frame shift in the predicted membrane spanning domain upstream of the 3C-like protease domain. These differences between A59 and JHM in ORF 1a suggest that these regions are perhaps dispensable and that not all of the 21 kb of gene 1 is essential for function.

Antisera directed against proteins encoded in gene 1

Very little is known about the proteins encoded in gene 1 of the coronaviruses. This is probably due to the lack of antisera specific for gene 1 encoded polypeptides. We have made a considerable effort toward raising antisera against polypeptides encoded in this gene and have begun to analyze these proteins. Figure 1 illustrates the regions of the genome encoding proteins against which these antisera were raised. Anti p28 serum is an anti-peptide serum directed against 14 amino acids encoded by the JHM genome (nucleotides 287–329). UP102 is directed against a viral/bacterial fusion protein expressed in the pET 3b vector [12], containing an approximately 2 kb fragment of the A59 genome as shown in Fig. 1. The antisera 81043, 600A, 88640 and 590A are directed against bacterial/viral tripartite proteins containing viral sequences encoded by the A59 genome (see Fig. 1) inserted into the plasmid pGE374 between truncated recA and lacZ sequences [13]. The fusion protein antisera were raised in rabbits, which had been inoculated with both native and denatured fusion proteins.

Synthesis of ORF 1 proteins by in vitro translation

Denison and Perlman [14] have shown that translation of MHV genome RNA results in the synthesis of p220 and p28. In the presence of leupeptin, an inhibitor of serine and cysteine proteases there is less p28 detected and a new polypeptide, p250 is observed. This suggests that p250 is a precursor to p220 and p28 [14]. As expected, p28 is immunoprecipitated with anti-p28 and UP102 antisera. The p220 is immunoprecipitated by UP102 and 81043, but not by anti-p28 antiserum because the p28 sequences are not present in this polypeptide. The precursor, p250, is immunoprecipitated by anti-p28, as well as UP102 and 81043 antisera, consistent with this polypeptide containing the p28 sequences. Baker et al. [9] have demonstrated that the cleavage of p250 into p220 and p28 is carried out by the first papain-like protease, encoded in ORF 1a. They showed that in vitro translation of a synthetic RNA representing the first 5.3 kb of the JHM genome resulted in the cleavage of p28. They also showed that translation products of a shorter, 3.9 kb RNA did not include p28, suggesting that the protease activity is encoded between 3.9 and 5.3 kb, and that the first papain-like protease was responsible for cleavage. They also concluded that this was a *cis*-acting protease only. We have carried out similar experiments with A59 in which we in vitro translated RNAs representing various portions of the 5′ end of ORF 1a. As shown in Table 1, we observed that RNAs representing from 4.2 kb to 5.2 kb of ORF 1a were capable of synthesis of p28. However 3.6 kb and 1.9 kb transcripts were not. This confirms the results of Baker [9] that the first papain-like protease is likely to be responsible for p28 synthesis. Furthermore, the conserved "X" domain, which is encoded downstream of the first papain-like protease, would not be present in the translation products of the 4.2 kb RNA; this demonstrates that the "X" domain is not necessary for cleavage of p28.

Table 1. Cleavage of P28 during in vitro translation of
RNAs translated from MHV-A59 ORF 1a

ORF 1a (nucleotide)	Location (amino acid)	P28 Cleavage
1985	591	−
3690	1160	−
4242	1344	+
4664	1484	+
4934	1574	+
5220	1670	+

We have used antisera to detect polypeptides encoded in ORF 1b of
the genome in the products of in vitro translation [15]. We detected
products of ORF 1b only when translations were allowed to proceed for
long time periods, that is, longer than 60 min: this suggests that the
ORF 1b products are synthesized by initiation of translation in ORF 1a
followed by frame shifting into ORF 1b. In summary the two antisera
directed against proteins encoded in the 3′ end of ORF 1b (88640 and
590A, see Fig. 1) detected several polypeptides of 90 kd, 74 kd, 53 kd,
44 kd, 32 kd. We did not detect any larger proteins, even when leupeptin
was included during the translation. We do not known if these polypeptides are similar to those in infected cells because we have not yet
been able to detect the ORF 1b polypeptides during infections. In
preliminary experiments, translation of a full-length RNA transcribed
from an ORF 1b cDNA resulted in the synthesis of many polypeptides;
this was unexpected since a protease domain in ORF 1b has not yet been
predicted from its sequence. The observation of many polypeptides in
the cell free translation products of ORF 1b could due to sub optimal
conditions for the in vitro translation of very long mRNAs; alternatively
there may actually be a protease activity encoded in ORF 1b.

Synthesis of ORF1a proteins in MHV-A59 infected cells

We have used the antisera described above to detect the polypeptide
products of ORF 1a in infected cells (Fig. 2). Anti p28 antiserum immunoprecipitates p28; we do not observe larger polypeptides immunoprecipitated with this antiserum, suggesting that p28 is rapidly cleaved
from its precursor. UP102 detects p28 and also p65, a polypeptide not
observed during in vitro translation of genome RNA; the observation
that p65 is detected with UP102 serum, but not with anti p28 serum,
suggests that p65 is encoded downstream of p28. Pulse chase experiments support this conclusion. During pulse-chase experiments, the

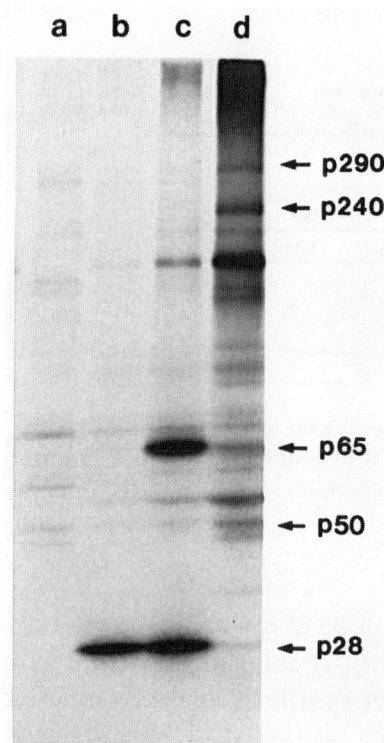

Fig. 2. Immunoprecipitation of MHV-A59 ORF1a products. Immunoprecipitates of [^{35}S] methionine labeled, (*a*) mock infected and (*b,c,d*) MHV-A59 infected murine fibroblast DBT cells were analyzed by 5–18% gradient polyacrylamide gel electrophoresis. *a,c* Antiserum UP102; *b* anti-p28 antiserum; *d* 81043 antiserum. The molecular weights of the specific precipitation products are indicated to the right of the gel with arrows

UP102 antiserum detects a high molecular weight polypeptide before p65 is apparent, suggesting that cleavage of p65 is not as rapid as p28 cleavage. The 81043 and 600 antisera detect p290 and p240, while only 81043 detects p50. These data, along with previously published kinetic data [16] suggest that p290 is a precursor of p240 and p50 and that p50 represents the amino terminus of p290 [16].

The products of ORF 1a observed in infected cells differ from those found in the products of cell free translation of genome RNA (see Fig. 3). Most strikingly, p65 is not observed during in vitro translation. Furthermore the proteins encoded downstream of differ. While in vitro translation products include p250 and p220, the intracellular polypeptides include p290, p240 and p50. As shown in Fig. 3, p290 would represent nearly all of ORF 1a. We do not know why these differences are observed between in vitro translation products and intracellular polypeptides. However, it is possible that there is temporal regulation of

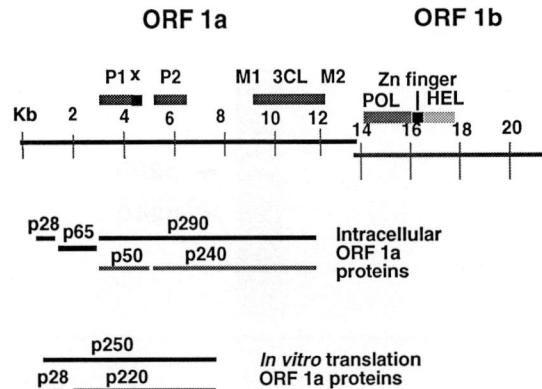

Fig. 3. Model of ORF1a translation and processing in MHV-A59 infected cells and in *in vitro* translation of MHV-A59 genome RNA. The predicted sizes of the polypeptides and the possible cleavage sites are all shown to scale. General alignment of polypeptides is based on apparent molecular weights and antisera specificity

processing during infection. We can only observe ORF 1a polypeptides late in infection and in vitro translation products might be more representative of polypeptides synthesized early in infection. Also, while we have shown that ORF 1b products are synthesized in vitro [15], this may not be very efficient. It is possible that there are not high enough levels of the downstream proteases to process the ORF 1a proteins properly. Thus, p65 may be cleaved either by the second papain protease or by the 3C-like protease and these are simply not present in high enough concentrations during in vitro translation. Another explanation for the differences in vivo and in vitro is the possibility of cellular enzymes participating in the processing of viral polypeptides. We feel this is unlikely as expression of the 5.2 kb ORF 1a construct via a vaccinia virus expression system in mouse cells results in the synthesis of p28 but not p65 (data not shown). Even in eukaryotic cells, p65 is not synthesized in the absence of infection.

Temperature sensitive mutants with possible processing defects

We have identified two RNA negative mutants of MHV-A59 than may have processing defects in the synthesis of ORF 1a polypeptides. Both NC11 (group B mutant) and LA16 (group A/B mutant) [17] synthesize greatly reduced levels of p28 and barely detectable amounts of p65 at the non-permissive temperature. Furthermore, p28 and p65 are stable at the non-permissive temperature. Since, in preliminary experiments these mutants do appear to synthesize the p290 and p240 ORF 1a polypeptides, we suggest that these viruses may be defective in processing of p28 and p65. It is interesting that both the synthesis of p28 and p65 appear to be defective in these mutants because, as we have discussed above, these proteins are apparently processed by different protease activities. These results, in combination with the kinetics of accumulation of p28 and p65, suggest the possibility that p28 cleavage is necessary for p65 cleavage to occur.

Acknowledgements

We thank Dr. R. Baric for the temperature sensitive mutants, Dr. S. Perlman for the anti-p28 peptide antibody and Ms X. Wang for excellent technical help. We thank Dr. P. Zoltick for construction of fusion proteins and for raising of the gene 1 antisera and Dr. S. Kraft and Mr. A. Pekosz for sequencing portions of MHV-A59 ORF 1a. This work was supported by Public Health Service grants AI-17418 (SRW) and AI-26603 (MLD) and National Multiple Sclerosis Society grant RG 2203-A-5 (JLL). Dr. P. Bonilla was supported by Public Health Service training grant NS-07180. Mr. S. Hughes was supported in part by Merck, Sharp and Dohme and by Public Health Service training grant NS-07180.

References

1. Pachuk CJ, Bredenbeek PJ, Zoltick PW, Spaan WJM, Weiss SR (1989) Molecular cloning of the gene encoding the putative polymerase of mouse hepatitis virus strain A59. Virology 171: 141–148
2. Lee HJ, Shieh CK, Gorbalenya AE, Koonin EV, LaMonica N, Tuler J, Bagdzhadzhyan A, Lai MMC (1991) The complete sequence of the murine coronavirus gene 1 encoding the putative protease and RNA polymerase. Virology 180: 567–582
3. Spaan WJM, Cavanagh D, Horzinek MC (1988) Coronaviruses. Structure and genome expression. J Gen Virol 69: 2939–2952
4. Sethna PB, Hung S-L, Brian DA (1989) Coronavirus subgenomic minus-strand RNAs and the potential for RNA replicons. Proc Natl Acad Sci USA 86: 5626–5630
5. Boursnell MEG, Brown TDK, Foulds IJ, Green PH, Tomley FM, Binns MM (1987) Completion of the sequence of the genome of the coronavirus avian infectious bronchitis virus. J Gen Virol 68: 57–77
6. Bredenbeek PJ, Pachuk CJ, Noten AFH, Charite J, Luytjes W, Weiss SR, Spaan WJM (1990) The primary structure and expression of the second open reading frame of the polymerase gene of the coronavirus MHV-A59; a highly conserved polymerase is expressed by an efficient ribosomal frameshifting mechanism. Nucleic Acids Res 18: 1825–1832
7. Brierly I, Boursnell MEG, Binns MM, Bilimoria B, Blok VC, Brown TDK, Inglis SC (1987) An efficient ribosomal frame-shifting signal in the polymerase encoding region of the coronavirus IBV. EMBO J 6: 3779–3785
8. Gorbalenya AE, Koonin EV, Lai MMC (1991) Putative papain-related thiol proteases of positive strand RNA viruses. FEBS Lett 288: 201–205
9. Baker SC, Shieh CK, Chang MF, Vannier DM, Lai MMC (1989) Identification of a domain required for autoproteolytic cleavage of murine coronavirus gene A polyprotein. J Virol 63: 3693–3699
10. Gorbalenya AE, Blinov VM, Donchenko AP, Koonin EV (1989) An NTP-binding domain is the most conserved seq uence in a highly diverged monophyletic group of proteins involved in positive strand RNA viral replication. J Mol Evol 28: 256–268

11. Gorbalenya AE, Koonin EV, Donchenko AP, Blinov VM (1992) Coronavirus genome: prediction of putative functional domains in the non-structural polyprotein by comparative amino acid sequence analysis. Nucleic Acids Res 17: 4847–4861
12. Studier WF, Rosenberg AH, Dunn JJ, Dubendorf JW (1990) Use of T7 polymerase to direct the expression of cloned genes. Methods Enzymol 185: 60–89
13. Zoltick PW, Leibowitz JL, DeVries JR, Weinstock GM, Weiss SR (1989) A general method for the induction and screening of antisera for cDNA-encoded polypeptides: antibodies specific for a coronavirus putative polymerase encoding gene. Gene 85: 413–420
14. Denison MR, Perlman S (1986) Translation and processing of mouse hepatitis virus virion RNA in a cell-free system. J Virol 60: 12–18
15. Denison MR, Zoltick PW, Leibowitz JL, Pachuk CJ, Weiss SR (1991) Identification of polypeptides encoded in open reading frame 1b of the putative polymerase gene of the murine coronavirus mouse hepatitis virus A59. J Virol 65: 3076–3082
16. Denison MR, Zoltick PW, Hughes SA, Giangreco B, Olson AL, Perlman S, Leibowitz JL, Weiss SR (1992) Intracellular processing of the N-terminal ORF 1a proteins of the coronavirus MHV-A59 requires multiple proteolytic events. Virology 189: 274–284
17. Shaad MC, Stohlman SA, Egbert J, Lum K, Fu K, Wei T, Baric RS (1990) Genetics of mouse hepatitis virus transcription: Identification of cistrons which may function in positive and negative strand RNA synthesis. Virology 177: 634–645

Authors' address: Dr. S. R. Weiss, Department of Microbiology, 319 Johnson Pavilion, University of Pennsylvania, Philadelphia, PA 19104-6076, U.S.A.

Archives
Virology

© Springer-Verlag 1994
Printed in Austria

Processing of dengue type 4 and other flavivirus nonstructural proteins

C.-J. Lai, M. Pethel, L. R. Jan, H. Kawano, A. Cahour, and **B. Falgout**

Molecular Viral Biology Section, Laboratory of Infectious Diseases,
National Institute of Allergy and Infectious Diseases, National
Institutes of Health, Bethesda, Maryland, U.S.A.

Summary. Dengue type 4 (DEN4) and other flaviviruses employ host and viral proteases for polyprotein processing. Most proteolytic cleavages in the DEN4 nonstructural protein (NS) region are mediated by the viral NS2B-NS3 protease. The N-terminal third of NS3, containing sequences homologous to serine protease active sites, is the protease domain. To determine required sequences in NS2B, deletions were introduced into DEN4 NS2B-30%NS3 cDNA and the expressed polyproteins assayed for self-cleavage. A 40 amino acid segment within NS2B was essential. Sequence analysis of NS2B predicts that this segment constitutes a hydrophilic domain surrounded by hydrophobic regions. Hydophobicity profiles of other flavivirus NS2Bs show similar patterns. Cleavage of DEN4 NS1-NS2A requires an octapeptide sequence at the NS1 C terminus and downstream NS2A. Comparison of the analogous octapeptide sequences among flaviviruses indicates a consensus cleavage sequence of (P8)/Met/Leu-Val-Xaa-Ser-Xaa-Val-Ala(P1), where Xaa are non-conserved amino acids. The effects on cleavage of amino acid substitutions in this consensus sequence were analyzed. Most substitutions of the conserved residues interfered with cleavage, whereas substitutions of non-conserved residues had little or no effect. These findings indicate that the responsible enzyme recognizes well-defined sequences at the cleavage site.

Introduction

The four dengue viruses and several other arthropod-borne flaviviruses, such as yellow fever virus and Japanese encephalitis virus, are important etiologic agents of human diseases. The flavivirus genome is a positive-sense RNA that codes for a long polyprotein in one open reading frame.

The order of the individual proteins has been determined as NH_2-anch/C-pre/M-E-NS1-NS2A-NS2B-NS3-NS4A-NS4B-NS5-COOH.

Analysis of amino acid sequences near the intergenic junctions suggests that flaviviruses employ a combination of host and viral proteases to process their polyproteins. The first type of cleavage occurs after a long hydrophobic region generating the N terminus of preM, E, NS1 and NS4B. This is mediated by a signal peptidase which is a resident enzyme associated with the membranes of endoplasmic reticulum. The second type of cleavage occurs after a dibasic amino acid sequence, such as Lys-Arg, Arg-Lys or Arg-Arg, and is followed by a small amino acid Gly, Ser or Ala. These cleavages are believed to take place in the cytosol generating the N-terminus of NS2B, NS3, NS4A and NS5. In addition, a third type of cleavage involves the processing of the NS1/NS2A site because this region contains neither a long hydrophobic sequence nor a dibasic amino acid motif. To elucidate the mechanisms of the latter two types of cleavage, we have studied expression of dengue type 4 virus (DEN4) polyprotein in the nonstructural region using vaccinia virus as a vector or the recently developed vT7/pTM-1 transient expression system.

Cleavage of NS1-NS2A requires a hydrophobic N terminus, an octapeptide sequence at the NS1 C terminus and downstream NS2A

We have previously shown that proper synthesis of NS1 requires the N terminal hydrophobic signal of NS1 and downstrean NS2A [1]. The requirement for a signal sequence suggests that cleavage of NS1-NS2A polyprotein occurs at some intracellular site along the exocytic pathway. Deletion analysis has shown that the N terminal 70% of NS2A appears to be sufficient to mediate cleavage, whereas deletion of additional residues from the C terminus yields mostly uncleaved NS1-NS2A [2]. The cleavage defect cannot be corrected by co-infection with a recombinant virus producing NS1-NS2A capable of cleavage. Thus, it appears that NS2A functions *in cis*. Available evidence suggests two plausible mechanisms for cleavage of NS1-NS2A: (1) NS2A may be a protease but only autocatalyzes NS1-NS2A, and (2) NS2A is essential for recognition of the NS1/NS2A cleavage site by a responsible protease by altering the conformation of the polyprotein, or by directing the polyprotein to the correct intracellular site.

There is also evidence indicating that a minimum length of 8 amino acids at the NS1 C terminus preceding the cleavage junction is required for cleavage [3]. Based on a comparison of this 8 amino acid sequence with the analogous sequences of more than 14 other flaviviruses, the consensus cleavage sequence is (P8) Met/Leu-Val-Xaa-Ser-Xaa-Val-Xaa-Ala (P1) in which Met or Leu at position P8, Val at P7, Ser at P5,

Val at P3 and Ala at P1 appear to be strictly conserved. Amino acids at positions P2, P4 and P6 vary. To investigate whether any other portion of NS1 is required, a recombinant virus that encodes a fusion protein containing a signal sequence, a portion of preM, the C terminal 8 amino acids of NS1 and all of NS2A was constructed. The fusion protein produced in recombinant infected cells was cleaved at the predicted site. This indicates that the NS1 sequence is not required for cleavage except for the C terminal 8 amino acids (Falgout and Lai, unpubl. obs.).

Mutational analysis of the octapeptide sequence motif at the NS1 C terminus

In order to provide experimental evidence for the NS1-NS2A cleavage model, we studied in vivo cleavage of DEN4 NS1-NS2A polyprotein encoded by a series of mutant NS1-NS2A cDNA constructs [4]. Site-directed mutagenesis was performed to construct mutant NS1-NS2A that contained a single amino acid substitution at one of the eight positions of the consensus cleavage sequence or at the immediate downstream position. Three to eight different amino acid substitutions were made at each position. The cleavage phenotype of these mutant NS1-NS2A polyproteins was analyzed by detection of the cleaved NS1 product. The results, summarized in Fig. 1, showed that most substitutions at positions P8, P7, P5, P3, and P1, occupied by conserved amino acids yield low levels of cleavage ranging from no cleavage (nearly all substitutions for Ala at P1, or Val at P3 and P7) to approximately 20% of wild type cleavage (Thr substitution for Ser at P5 or Val substitution for Met at position P8). Exceptions are the Pro and Ala substitutions for Ser at position P5 and the Leu substitution for Met at P8. Ala and Pro differ in polarity from Ser, but like Ser they have small side chains and both appear to be recognized equally efficiently at position P5 by the responsible protease. The observation that the Leu substitution for Met at position P8 has only a minimal effect on cleavage is predicted as position P8 is occupied by Met in the DEN4 sequence or by Leu in other flavivirus sequences. For the most part, substitutions for conserved amino acids even with amino acids having similar side chains, yield cleavage-deficient NS1-NS2A polyproteins. This provides evidence that amino acids at these positions are essential for cleavage. This finding indicates that the responsible enzyme recognizes a well-defined sequence at the cleavage site with a highly stringent specificity.

Although amino acids at positions P2, P4, and P6 of the cleavage sequence are not conserved among flaviviruses, evidence indicates that not all substitutions at these positions of DEN4 produce a cleavage-competent NS1-NS2A polyprotein. Examples are the substitutions of

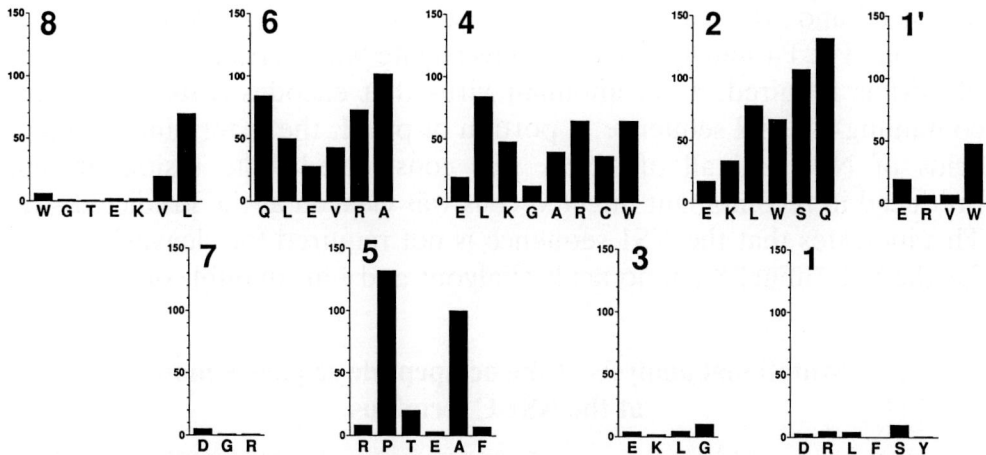

Fig. 1. Efficiency of cleavage at the DEN4 NS1/NS2A junction containing amino acid substitutions. Cleavage of DEN4 NS1-NS2A was analyzed by immunoprecipitation of ^{35}S-methionine labeled lysates prepared from recombinant virus infected or pTM-1 transfected cells. The percent cleavage of the wild type or mutant NS1-NS2A polyprotein was detected by measuring the radioactivity in the NS1 and NS1-NS2A bands separated on polyacrylamide gels. The cleavage efficiency of the wild type NS1-NS2A polyprotein measured in each experiment was used to normalize the cleavage efficiency for each mutant. Amino acids are designated using the single-letter code. Cleavage at DEN4 NS1/NS2A junction requires an 8-amino acid sequence at the NS1 C terminus which in the order of P8 to P1 is Met/Leu-Val-Lys-Ser-Gln-Val-Thr-Ala. The amino acid position immediately following the cleavage site is occupied by Gly

Gly or Glu for Gln (P4) and Glu for Lys (P6) that permit only a low level of cleavage. These findings indicate that these amino acids are not completely unimportant for cleavage. Nonetheless, it is clear that amino acids at positions P2, P4 and P6 are less important determinants for NS1-NS2A cleavage than the other five positions in the consensus cleavage sequence. Gly at position P1′ immediately following the DEN4 NS1/NS2A cleavage site is not a conserved amino acid. Three other amino acids, Tyr, Phe or Asp, occupy this position in the sequence of other flaviviruses. The results in Fig. 1 revealed that amino acid substitutions at P1′ yielded low to moderate cleavage to approximately 50% of the wild type NS1-NS2A cleavage level. These findings suggest that the amino acid at position P1′ also plays a role in cleavage.

Processing of most nonstructural proteins is mediated by NS2B-NS3 viral proteinase

Molecular modeling studies have revealed that the first 180 amino acids of flavivirus NS3 contained sequence motifs homologous to the catalytic triad and the substrate binding pocket of a serine protease. On the basis of this analysis, it was proposed that flavivirus NS3 is a viral proteinase

[5, 6]. This proposal has been experimentally tested by in vitro translation-processing or in vivo expression of polyproteins of DEN4 [7, 8], DEN2 [9], and yellow fever [10]. In the case of DEN4, recombinant vaccinia viruses have been constructed for expression of various portions of the nonstructural polyprotein. A polyprotein consisting of NS2A, NS2B, and N terminal 184 amino acids of NS3 is cleaved at the NS2A/NS2B and NS2B/NS3 junctions, whereas a similar protein containing only the N terminal 77 amino acids of NS3 is not cleaved. This finding is consistent with the proposal that the N terminal 180 amino acids constitute a protease domain. Cells infected with recombinants expressing poly-protein NS3-NS4A-NS4B or NS3-NS4A-NS4B-NS5 produce only the uncleaved polyprotein precursor. Polyprotein NS2A-ΔNS2B-NS3 containing a large deletion of NS2B is also defective for cleavage. The cleavage defect of the NS2B deletion polyprotein can be complemented by co-infection with a recombinant expressing NS2B [7]. Thus, in addition to NS3, NS2B is also required for cleavage and can act *in trans*. These studies do not rule out the possibility that NS2B can also act *in cis*. In fact, others have shown that in vitro processing of the NS2A/NS2B and NS2B/NS3 sites of DEN2 or yellow fever virus polyprotein is not affected by dilution [9, 10]. The insensitivity of cleavage to dilution is viewed as evidence for a *cis* acting mode.

The dengue virus NS2B/NS3 cleavage junction is unique in that the cleavage site follows Gln-Arg instead of a dibasic sequence. In order to determine whether additional amino acids surrounding the NS2B/NS3 junction are required for cleavage, we performed deletion analysis similar to that used to establish the octapeptide sequence at the NS1-NS2A junction. Our findings indicate that the Gln-Arg sequence at the DEN4 NS2B/NS3 junction is sufficient for mediating cleavage. Mutagenesis studies showed that substitution of Asn for Gln at position P2 slightly reduces the cleavage efficiency. However, cleavage of a poly-protein that contains negatively charged Glu substituting for Thr at position P3 is severely reduced. Thus, it appears that the amino acid at position P3 of the NS2B/NS3 site can influence cleavage. We have also obtained evidence indicating that cleavages at the NS3/NS4A and NS4B/NS5 junctions are also mediated by NS2B-NS3 viral proteinase [8]. However, it is observed consistently that processing of the NS3/NS4A junction, regardless whether NS2B is present *in cis* or *in trans*, is inefficient because processing intermediates NS3-NS4A and NS3-NS4A-NS4B accumulates. Recombinants specifying polyprotein NS4A-NS4B-NS5 or NS4B-NS5 produce authentic NS5 when complemented with recombinants expressing NS2B and NS3. These results indicate that cleavage at NS4B/NS5 can be mediated by NS2B-NS3 proteinase *in trans*.

Identification of domains required for NS2B-NS3 protease activity

In contrast to NS3, little is known about the sequence in NS2B required for protease activity. As a first step in obtaining such information, we performed deletion analysis of NS2B to delineate sequence requirements for NS2B-NS3 proteinase activity. A DEN4 DNA segment that codes for NS2B and 184 amino acids of NS3, approximately 30% of its full length, was used for expression using the vT7/pTM-1 transient expression system. This polyprotein is self-cleaved producing NS3 and NS2B that can be detected by precipitation with hyperimmune mouse ascitic fluid (HMAF). PCR was performed to introduce nucleotide substitutions creating unique restriction enzyme sites in NS2B without altering the encoded amino acid sequence. Seven new sites were made, including NheI at nt 4 125, StuI at nt 4 163, AflII at nt 4 194, ApaI at nt 4 218, NruI at nt 4 388, SpeI at nt 4 420 and Bst EII at nt 4 475 of the DEN4 sequence. In addition, there is a naturally occurring BglII site at nt 4 277 (Fig. 2). Oligonucleotides were used to bridge two adjacent restriction enzyme sites creating a total of seven individual deletions in NS2B.

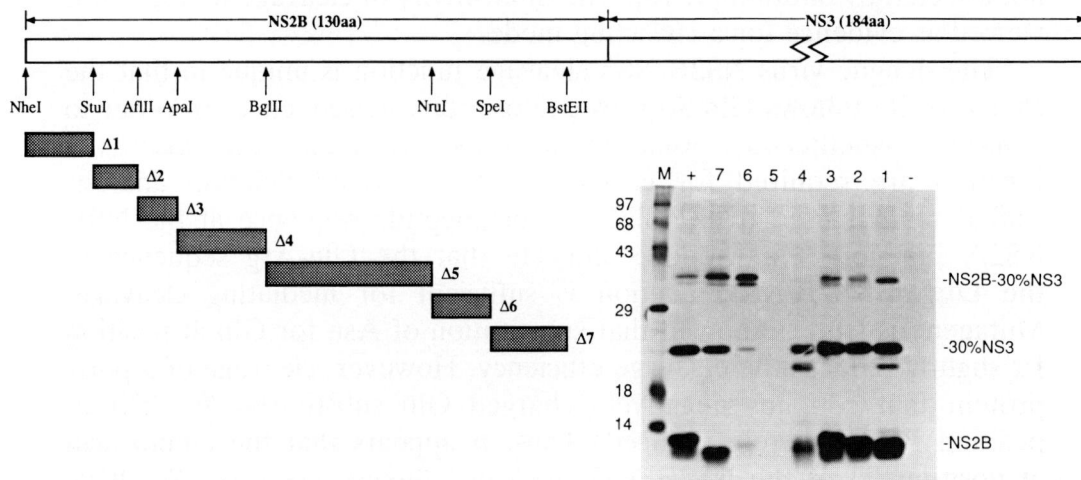

Fig. 2. NS2B deletions and analysis of their cleavage phenotypes. The map of wild type DEN4 NS2B-30%NS3 DNA that codes for NS2B (130 amino acids) and 30%NS3 (184 amino acids) is shown at the top. The relative positions of the restriction enzyme sites used in the construction of NS2B deletions are indicated. CV-1 cells were infected with vT7 and transfected with recombinant pTM-1 DNA. Infected cells were labeled with ^{35}S-methionine for 2h and the lysate was prepared. Radio-labeled DEN4 proteins were immunoprecipitated and analyzed by polyacrylamide gel electrophoresis. Recombinant plasmids used for transfection are: lane −, pTM-1 as a negative control; lanes *1* to *7*, deletion mutants Δ1 to Δ7; lane +, pTM-1-NS2B-30%NS3; lane *M* at the left contains molecular size markers in kDa. The positions of uncleaved NS2B-30% NS3, cleaved 30% NS3 and NS2B are shown

These deletions, designated Δ1 to Δ7, remove from 4 to 33 amino acids. The cleavage phenotype of Δ1 to Δ4 was similar to that of wild type as seen from the intense signals of cleavage products NS2B and 30%NS3. The polyprotein containing Δ7 was also cleaved but with a slightly reduced efficiency. Deletion Δ6 severely reduced cleavage and only a low level of the cleavage products was detected. Neither the uncleaved Δ5 NS2B-30%NS3 precursor nor the cleavage products was detected using (HMAF). However, the uncleaved precursor was readily detected by an anti-NS3 serum, suggesting that Δ5 interferes with the ability of HMAF to immunoprecipitate the uncleaved polyprotein precursor presumably by altering the conformation of the molecule. It was also shown that the defective cleavage of the polyprotein was complemented during co-transfection with pTM-NS2A-NS2B. Thus, the cleavage defect of Δ5 was not due to the uncleavability of the misfolded Δ5 polyprotein. These data suggest that the NS2B sequences important for the NS3 protease function have been removed by Δ5.

The finding that two neighboring mutants Δ5 and Δ6 are defective for cleavage suggests that the sites of these deletions include either end of a domain of NS2B essential for protease activity. Additional experiments were designed to map the boundaries of this putative domain. To map the right hand boundary, mutants 6a, 6b and 6c were constructed to progressively remove amino acids from Δ6 (Fig. 3). Analysis of the cleavage phenotype showed that 6a like Δ6 was defective for cleavage, but the cleavage phenotype of 6b which adds one more amino acid to 6a was similar to wild type. This result mapped the right hand boundary to Thr at amino acid 1 435 of the DEN4 sequence. Similarly, to map the left hand boundary, mutants 4a to 4c were constructed removing more amino acids than Δ4 (Fig. 4). The phenotype analysis of these mutants showed that 4a to 4e was cleaved efficiently at the NS2B/NS3 junction. On the other hand, 4d and 4e like Δ5 were defective. The uncleaved polyprotein from each of these mutants, although not detected by HMAF, was precipitated using the NS3-specific sera. Defective cleavage of 4d and 4e was corrected by complementation with added NS2A-NS2B. These data indicate that the left hand boundary of the essential domain is Leu at amino acid 1 395.

Deletion analyses described above have identified a sequence of 40 amino acids within NS2B that is essential for the protease activity of NS3 [11]. Analysis of the NS2B hydrophobicity profile reveals that the 40 amino acid domain constitutes a hydrophilic region surrounded by hydrophobic domains. Additional analysis indicates that deletion of almost the entire hydrophobic domain upstream (combination of deletions Δ1 to Δ4) or downstream (Δ7) or both hydrophobic domains except amino acids at the cleavage site has little or no effect on cleavage.

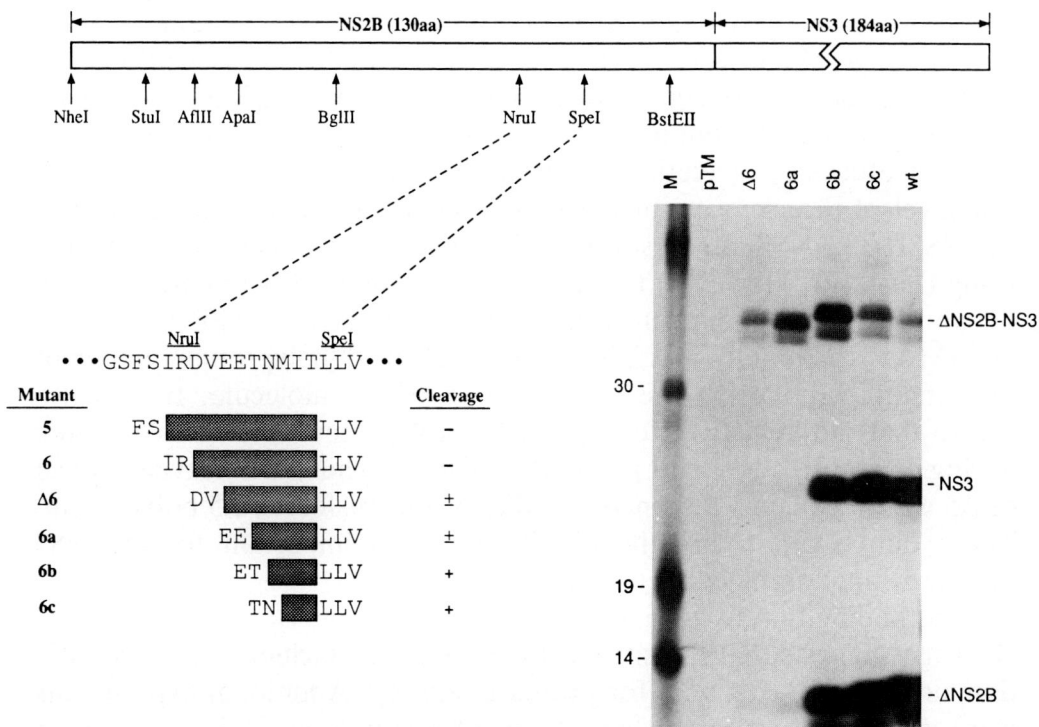

Fig. 3. Deletion constructs and mapping the right boundary of NS2B required for NS3 protease activity. The map position of the series of deletions is depicted. The expanded region shows the wild type amino acid sequence (*wt*) and residues removed by the indicated deletion mutants. Cleavage data of mutants 5 and 6 are not shown. Radiolabeled lysates were prepared and analyzed as in Fig. 2. Recombinant plasmids used for transfection are indicated at the top of gel lanes

Thus, the minimum length of NS2B essential for the NS3 protease activity appears to be the identified 40 amino acid domain. Alignment of this 40 amino acid sequence of DEN4 with the corresponding sequences of nine other mosquito-borne flaviviruses reveals five strictly conserved amino acids that include Trp_{1403}, Ala_{1407}, Gly_{1411}, Ser_{1413}, and Gly_{1424} of the DEN4 sequence. This finding suggests that this region of NS2B is also required for protease activity of other flaviviruses. These studies and others cited above have provided evidence that flavivirus NS3 requires the cooperation of NS2B for cleavage after a dibasic amino acid or Gln-Arg sequence in the nonstructural protein region. How NS2B affects the NS3 protease function remains an intriguing question. We have previously suggested one possibility that NS2B may interact with NS3 to form a complex yielding the NS3 conformation necessary for the protease activity. Direct demonstration that NS2B and NS3 form a complex has been achieved in our recent study in which we showed that

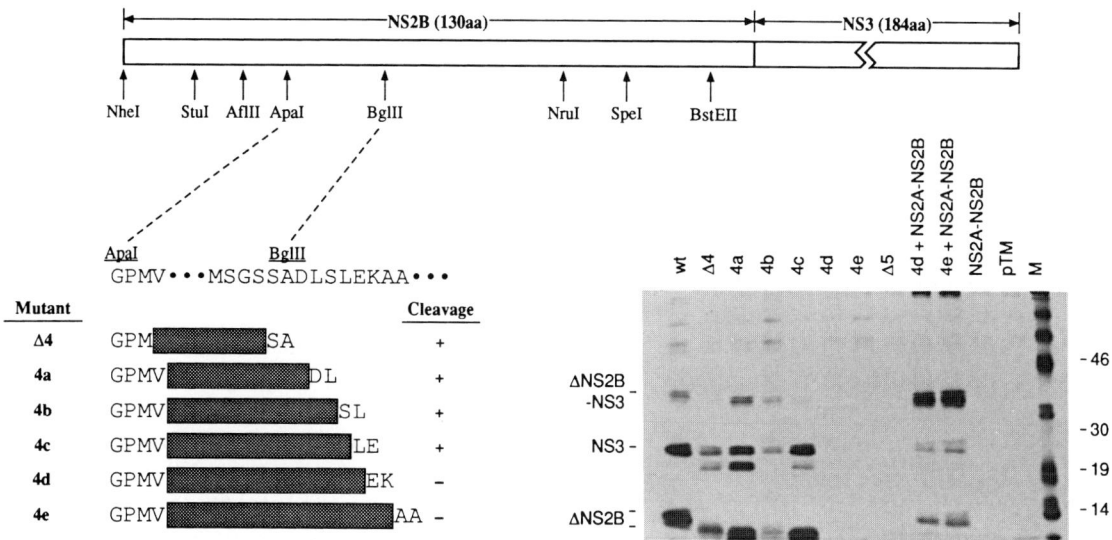

Fig. 4. Deletion constructs and mapping the left boundary of NS2B required for NS3 protease activity. The map position of the series of deletions is depicted. The expanded region shows the wild type amino acid sequence and residues removed by the indicated deletion mutatants. Radio-labeled lysates were prepared and analyzed as in Fig. 2. Recombinant plasmids used for transfection are indicated at the top of gel lanes

Japanese encephalitis virus NS2B and NS3 were co-precipitated using NS2B- or NS3-specific serum. Co-immunoprecipitation of NS2B and NS3 was also observed for DEN2 [12] and yellow fever virus (Rice, pers. comm.). At present, the nature of molecular interactions involved in the cleavage site recognition or in the formation of NS2B-NS3 complex has not been elucidated. It is interesting to point out that there are 14 charged amino acids and 6 polar amino acids within the 40 amino acid domain of DEN4 NS2B suggesting that these residues may play an important role in charge-charge interactions. Additional experiments using point mutants of NS2B constructed by site-directed mutagenesis will be needed to determine whether mutants which abolish complex formation also eliminate protease activity.

References

1. Falgout B, Chanock R, Lai C-J (1989) Proper processing of dengue virus non-structural glycoprotein NS1 requires the N-terminal hydrophobic signal sequence and the downstream nonstructural protein NS2A. J Virol 63: 1852–1860
2. Falgout B, Lai C-J (1990) Synthesis of dengue virus nonstructural protein NS1 requires the N-terminal signal and the downstream nonstructural protein NS2A. In: Brinton MA, Heinz FX (eds) New aspects of positive-strand RNA viruses. Americal Society for Microbiology, Washington, pp 192–195

3. Hori H, Lai C-J (1990) Cleavage of dengue virus NS1-NS2A requires an octapeptide sequence at the C-terminus of NS1. J Virol 64: 4573–4577

4. Pethel M, Falgout B, Lai C-J (1992) Mutational analysis of the octapeptide sequence motif at the NS1-NS2A cleavage junction of dengue type 4 virus. J Virol 66: 7225–7231

5. Bazan JF, Flettrick RJ (1989) Detection of a trypsin-like serine protease domain in flaviviruses and pestiviruses. Virology 171: 637–639

6. Gorbalenya AE, Donchenko AP, Koonin EV, Blinov VM (1989) N-terminal domains of putative helicases of flavi- and pestiviruses may be serine proteases. Nucleic Acids Res 17: 3889–3897

7. Falgout B, Pethel M, Zhang Y-M, Lai C-J (1991) Both nonstructural proteins NS2B and NS3 are required for the proteolytic processing of dengue virus nonstructural proteins. J Virol 65: 2467–2475

8. Cahour A, Falgout B, Lai C-J (1992) Cleavage of the dengue virus polyprotein at the NS3/NS4A and NS4B/NS5 junctions is mediated by viral protease NS2B-NS3, whereas NS4A/NS4B may be processed by a cellular protease. J Virol 66: 1535–1542

9. Preuschat F, Yao C-W, Strauss JH (1990) In vitro processing of dengue 2 nonstructural proteins NS2A, NS2B and NS3. J Virol 64: 4364–4374

10. Chambers TJ, Weir RC, Grakoui A, McCourt DW, Bazan JF, Fletterick RJ, Rice CM (1990) Evidence that the N-terminal domain of yellow fever virus NS3 is a serine protease responsible for site-specific cleavage in the viral polyprotein. Proc Natl Acad Sci USA 87: 8898–8902

11. Falgout B, Miller R, Lai C-J (1993) Deletion analysis of dengue virus type 4 nonstructural protein NS2B: identification of domains required for NS2B-NS3 protease activity. J Virol 67: 2034–2042

12. Arias CF, Preugschat F, Strauss JH (1993) Dengue 2 virus NS2B and NS3 form a stable complex that can cleave NS3 within the helicase domain. Virology 193: 888–899

Authors' address: Dr. C.-J. Lai, Laboratory of Infectious Diseases, National Institute of Allergy and Infectious Diseases, National Institutes of Health, Bethesda, MD 20892, U.S.A.

Arch Virol (1994) [Suppl] 9: 369–377

Archives
of
Virology
© Springer-Verlag 1994
Printed in Austria

Nuclear targeting of Semliki Forest virus nsP2

M. Rikkonen, J. Peränen, and **L. Kääriäinen**

Institute of Biotechnology, University of Helsinki, Helsinki, Finland

Summary. The Semliki Forest virus-specific nonstructural protein nsP2 is transported into the nuclei of both infected and transfected BHK cells. The pentapeptide sequence P ^{648}R RRV is an essential part of the nuclear localization signal (NLS) of nsP2, the middle arginine being the most critical residue for nuclear targeting. Host DNA and RNA syntheses are rapidly inhibited in virus-infected cells, and nsP2 could be involved in these processes. It has been postulated that the inhibition of cellular replication could be due to viral NTPase activity. We have expressed and purified nsP2 in *E. coli* using the highly efficient T7 based expression system. Purified nsP2 was shown to have ATPase and GTPase activities, and these specific activities were increased in the presence of single-stranded RNA, a typical feature of RNA helicases. The role of nsP2 in the nucleus was studied by creating a mutant virus SFV-RDR, which contained an altered NLS (PRDRV). The mutation affected neither the processing nor the stability of nsP2, but it did render nsP2 completely cytoplasmic. SFV-RDR was shown to be fully infectious, and no difference could be seen in the expression of viral proteins. In addition, the inhibition of host DNA synthesis was almost equally efficient in both wild-type and mutant-infected cells. The pathogenic properties of the mutant will be further studied.

Introduction

Semliki Forest virus (SFV) belongs to the genus *Alphavirus* of the family *Togaviridae*. SFV is an enveloped virus, and its genome is composed of one single-stranded 42S RNA which is about 11 500 nucleotides in length. Upon release into the cytoplasm, the genomic 42S RNA directs the synthesis of one large polyprotein P1234 [15]. P1234 is further processed by viral proteinases [5, 29] into four mature nonstructural (ns) proteins (nsP1–nsP4), all required in the viral replication. The first step of viral replication is the synthesis of negative-strand 42S RNA which is complementary to the genomic 42S RNA. The minus-strand is then used as a template for the synthesis of new positive-strand 42S RNA and subgenomic 26S RNA.

Despite years of work on alphaviruses, the molecular mechanisms of viral RNA synthesis are not well known. However, several functions in viral RNA replication have been assigned to individual nonstructural proteins. One of these, nsP2, appears to have a vital function in RNA synthesis, because temperature-sensitive mutants with mutations in nsP2 coding sequence display an RNA-negative phenotype [11, 15]. NsP2 is involved in the regulation of the synthesis of the subgenomic 26S RNA which codes for the structural proteins [11, 15]. It is also needed for the cessation of transcription of negative-strand RNA [25]. NsP2 has also proteinase activity responsible for the proteolytic processing of the ns-polyprotein [5]. This activity has been mapped in the C-terminal part of nsP2 which shares homology with papain proteinases [12]. The N-terminal half of nsP2, which is highly conserved between alphaviruses, is homologous to many putative helicases of diverse origin [9]. Among these, amino acid comparisons have revealed six conserved motifs, which also can be found in nsP2. It has been suggested that all these proteins carry out NTP-dependent reactions in the replication, recombination, and processing of RNA or DNA. Motif I is the classical nucleotide binding site (consensus sequence G-X-X-X-X-G-K-S/T) found in many ATP- or GTP-binding proteins [2, 30]. We have recently shown that about 50% of nsP2 is transported into the nucleus during viral infection, whereas other ns-proteins, all derived from the same polyprotein, are found in the cytoplasm [20]. Indirect immunofluorescence revealed that nsP2 is often concentrated in the nucleoli. In the nucleus most of nsP2 is associated with the nuclear matrix, which has an essential role in the organization and replication of DNA, transcription, and processing of RNA [18]. When nsP2 was expressed alone, in the absence of other viral proteins, it was transported almost quantitatively into the nucleus and nucleolus. This implied that the protein itself must have information for its nuclear targeting. The putative nuclear function of nsP2 is probably unrelated to the viral RNA synthesis and protease activities.

Nuclear localization signal of nsP2

Nuclear localization signals (NLS) have been identified from a number of proteins in yeast, *Xenopus*, mammals, plants and viruses [8]. Although no consensus sequence has emerged, in general, NLSs contain short stretches of basic residues often flanked by a proline. In order to define the NLS of nsP2, two approaches were taken: construction of in-frame deletion mutants and cloning of chimeric β-galactosidase-nsP2 fusions [24]. The deletions analysis showed that two distinct regions (amino acids 25–110 and 634–661) affected the nuclear localization of nsP2. The latter includes a pentapeptide sequence P ^{648}RRRV which resembles some of the known NLSs. In vitro mutagenesis of this sequence showed that it is an essential part of the functional NLS. Substitution of all three arginines with aspartic acids rendered nsP2 cytoplasmic. The replacement of the first arginine resulted in a diffuse nuclear and cytoplasmic distribution of the altered protein. The second arginine seems to be the most critical one, because the mutant nsP2 was restricted entirely to the cytoplasm. Substitution of the third arginine had no effect on nuclear transport. Regions from nsP2 having this pentapeptide were able to target β-galactosidase into the nucleus, although the activity of the PRRRV signal was affected by the context within which it was presented. The N-terminal region, identified by deletion analysis, seemed not to be important for nuclear transport, because the C-terminal half of nsP2 alone was effectively targeted into the nucleus. Neither were N-terminal sequences able to promote nuclear transport of a cytoplasmic marker protein [24].

What is the function of nsP2 in the nucleus? As alphaviruses can grow in enucleated cells [7], the nuclear location of nsP2 is not vital for viral replication per se. However, the virus might benefit from nonvital functions such as inhibition of host macromolecular syntheses. It is known that in alphavirus-infected cells host DNA, RNA, and protein syntheses are rapidly inhibited [27]. It has been suggested that inhibition of DNA synthesis is due to a virus-specific nucleoside triphosphate phosphohydrolase (NTPase), which releases γ-phosphate from all rNTPs and dNTPs. It has been reported that NDPs, particularly rADP and dADP, are strong inhibitors of TMP kinase. Thus, nucleoside diphosphates accumulating in the cell due to the NTPase activity could block the TMP kinase leading to the decrease in TTP pool and eventually to inhibition of DNA synthesis [27]. In order to see whether nsP2 has NTPase activity, as also suggested by sequence homology, we have purified nsP2 after expression in *Escherichia coli*.

Expression, purification and enzymatic activity of nsP2

NsP2 was produced in *E. coli* using the highly efficient T7 based expression system [28]. NsP2 was cloned into tightly regulated pBAT-HF vector (Peränen and Rikkonen, unpublished results) and expressed in JM 109 (DE3) cells (Promega), in which the T7 RNA polymerase is under *lac*UV5 promoter, and can be induced with isopropyl-β-D-tiogalactose (IPTG). For rapid protein purification, we introduced a short sequence encoding six histidine residues at the 5′ end of the coding region for nsP2. Histidines have high affinity for certain metal ions such as nickel [21], and the histidine-tagged proteins can be purified by immobilized metal affinity chrornatography. Proteins fused to a stretch of histidines bind to Ni^{2+}-nitrilotriacetic acid, and can be eluted under nondenaturing conditions by application of imidazole which competes in binding to the Ni^{2+} [14]. As can be seen in Fig. 1A, nsP2 was efficiently produced, and most of it was soluble. NsP2 was purified essentially free of host proteins in a single step (lane 10). The identity of the expected size nsP2 was verified by immunoblotting, which also showed some smaller degradation products (Fig. 1B). Because our antibody has been made against β-galactosidase-nsP2 fusion protein [19], it also recognizes authentic β-galactosidase.

NsP2 preparation (Fig. 1A, lane 10) was tested for ATPase activity, which was measured as the release of $^{32}P_i$ from γ-^{32}P-ATP, as previously described [4]. From these preliminary experiments it was calculated that the specific activity of nsP2 is about 1.3 pmol/μg/s (in the absence of RNA). In the presence of single-stranded RNA (polyU), the activity was increased (2.1 pmol/μg/s), which is typical of RNA helicases. The high basal activity, as compared to other known helicases, also has been described for plum pox potyvirus RNA helicase Cl [16]. When the elution fraction from control cells (without nsP2 gene) or cells expressing another SFV-specific protein, nsP3, was tested, no activity could be seen (data not shown), suggesting strongly that SFV nsP2 has ATPase activity. The activity of nsP2 was about 1–2 orders of magnitude higher than that described for eIF-4A (an RNA helicase; [1]), and one fifth of that of p68, another RNA helicase [13]. NsP2 also had GTPase activity with about the same specific activity as for ATP (data not shown). Preliminary experiments have shown that α-^{32}P-ATP and α-^{32}P-GTP can be cross-linked to nsP2 (data not shown).

Mutant virus with nonnuclear nsP2

Because nsP2 has NTPase activity, it could be the virus-specific factor causing the inhibition of host DNA synthesis. To see whether the nuclear localization of nsP2 is essential for this function, we created a mutant virus in which the second arginine of the NLS of nsP2 was substituted with aspartic acid. This was done by changing the wild type nsP2 gene of the cDNA clone of SFV ([17]; a kind gift from P. Liljeström and H. Garoff, Karolinska Institute, Sweden) with the mutant one. From the cDNA clone, infectious RNA can be transcribed in vitro followed by transfection into cells to obtain viral replication. After transfection, the distribution of viral proteins was monitored by indirect immunofluorescence (Fig. 2). As expected, mutant nsP2 was totally cytoplasmic, as opposed to the nuclear and cytoplasmic distribution

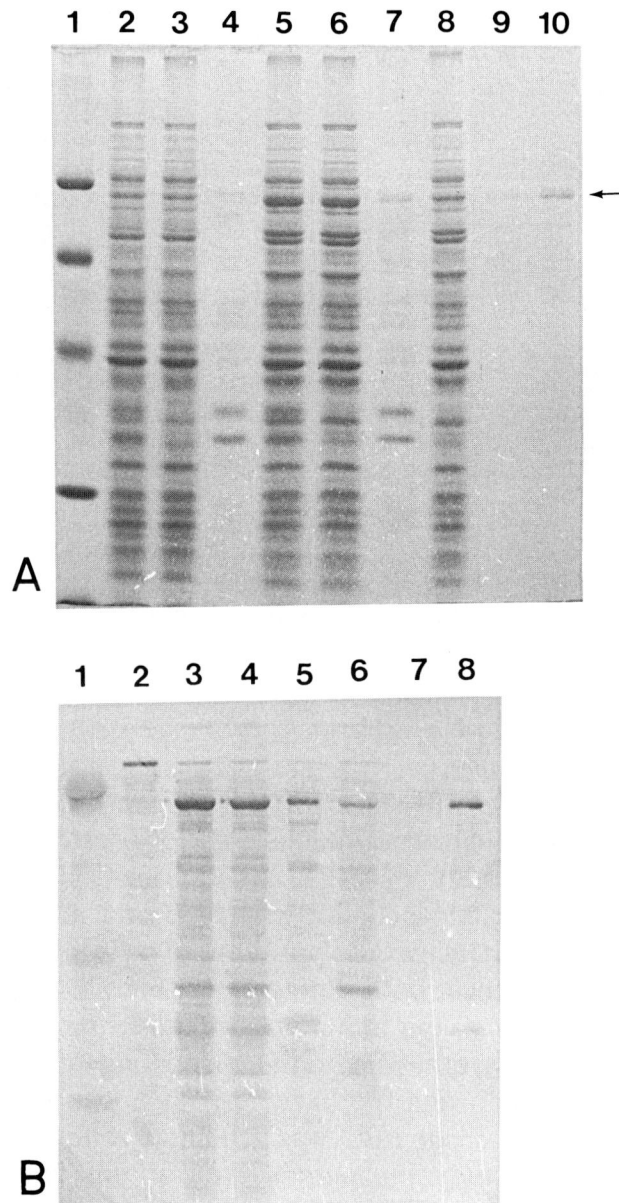

Fig. 1. A Expression of nsP2 in *E. coli*. NsP2 was purified on nondenaturing Ni^{2+}-NTA-agarose column. Lanes: *1* molecular mass marker (from the top 94, 67, 43, 30, and 20.1 kilodaltons); *2* and *5* total cell lysate; *3* and *6* soluble proteins (column load); *4* and *7* insoluble pellet fraction; *8* flow through; *9* wash with 20 mM imidazole; *10* elution with 100 mM imidazole. *2–4* are from control cells (without nsP2 gene), *5–10* from cells expressing nsP2 after induction with 50 µM IPTG (arrow points to the position of nsP2). **B** Immunoblotting of the same fractions as in **A** with anti-nsP2 antisera. Lanes: *1* molecular mass marker; *2* total cell lysate from control cells; *3* total cell lysate; *4* soluble proteins; *5* insoluble pellet fraction; *6* flow through; *7* wash with 20 mM imidazole; *8* elution with 100 mM imidazole; *3–8* from cells expressing nsP2

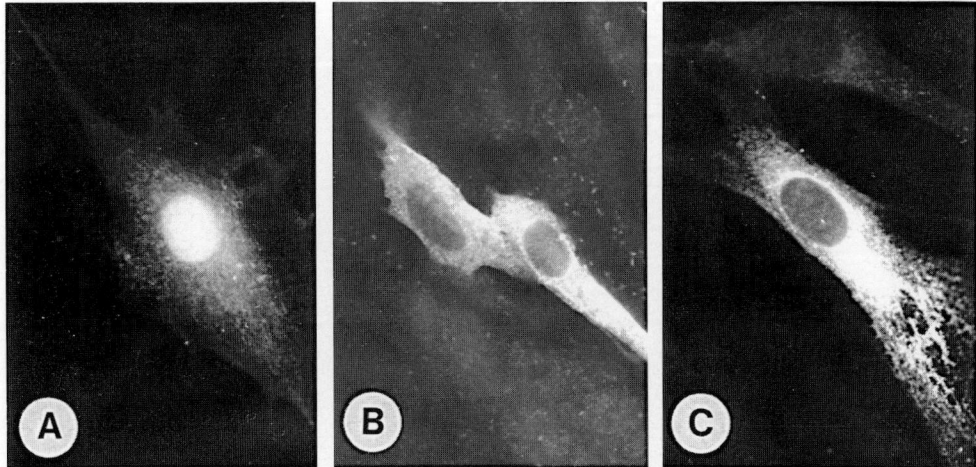

Fig. 2. Localization of SFV proteins in BHK cells transfected with RNA transcribed in vitro from wild-type cDNA clone of SFV (**A**), or from a cDNA clone with a mutation in the NLS of nsP2 (**B** and **C**). Cells were stained for indirect immunofluorescence with affinity purified anti-nsP2 antisera (**A** and **B**), or with anti-E2 antisera (**C**)

of the wild type nsP2. No difference could be seen in the localization of other viral proteins (Fig. 2, and data not shown). From such a transfection, a mutant virus stock (SFV-RDR) was propagated, and this virus was used in further experiments. First, we tested the effect of the mutation on the stability of nsP2. As can be seen in Fig. 3A, no difference was found between the wild type and mutant viruses. The processing of the nonstructural polyprotein, in cells infected with the two viruses, was also identical both during early (Fig. 3B) and late infection (data not shown). Thus, the mutation introduced into the NLS of nsP2 affected neither the synthesis nor the processing of the nonstructural polyprotein P1234, allowing the biogenesis of infectious SFV. Only the intracellular distribution of nsP2 was rendered completely cytoplasmic. This enabled us to study the putative role of nuclear localization on cellular functions.

The inhibition of DNA synthesis in alphavirus-infected cells is rapid [27]. In Sindbis virus infected cells, host DNA replication is about 10% of that of mock-infected cells at 5–7h p.i. [3]. In western equine encephalitis virus infected cells (m.o.i. 10), DNA synthesis is reduced to about 35% at 3h p.i., and to 10% at 5h p.i. [26]. Consistent with these results, host DNA replication was decreased rapidly in wild type SFV-infected cells (Table 1). This occurred also in cells infected with SFV-RDR, although the decrease was somewhat slower. However, this is not so surprising if the hypothesis of the mechanism of inhibition suggested by Simizu and coworkers is correct. Although the localization of nsP2 is

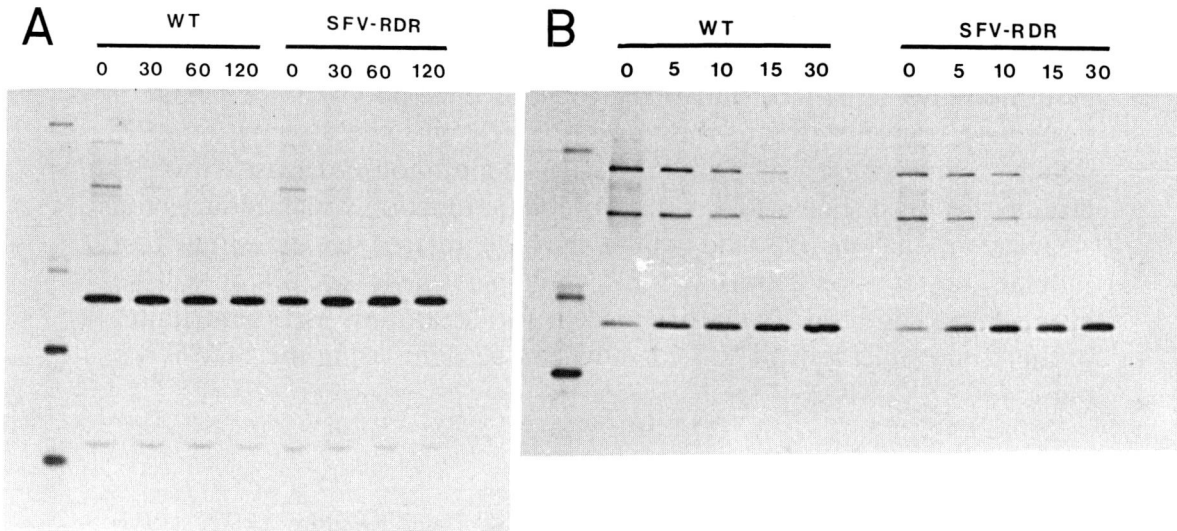

Fig. 3. A Stability analysis of nsP2 in BHK cells infected with either wild-type SFV or SFV-RDR. Cells were labeled with [^{35}S] methionine for 30 min at 2.5 h p.i., and chased for 0, 30, 60, or 120 min. NsP2 was detected by immunoprecipitation with anti-nsP2 antisera. Molecular mass marker is in the left (from the top 200, 92.5, 69, and 46 kilodaltons) **B** Processing of nsP2 early in infection. BHK cells were infected with either wild-type SFV or SFV-RDR and labeled with [^{35}S] methionine for 10 min at 2 h p.i. followed by a chase of 0, 5, 10, 15, or 60 min. Labeled proteins were immunoprecipitated with anti-nsP2 antisera

Table 1. Inhibition of DNA synthesis in SFV-infected cells

| | h after infection | | | |
	2	3	4	5
Wild type SFV	78.2	33.8	5.5	3.6
SFV-RDR	78.6	45.6	7.7	5.1

BHK cells were infected with wild-type SFV or SFV-RDR at m.o.i. 10, or mock-infected. At time indicated, cells were exposed to [^{3}H] thymidine for 30 min, and DNA synthesis was measured as incorporation of radioactivity into TCA-insoluble fraction

changed, the NTPase activity of mutant nsP2 seems to be unaltered, and NDPs still accumulate in the cell. The slight difference found between the two viruses may well reflect some other inhibitory aspect of nsP2.

Also cellular transcription is efficiently inhibited in alphavirus-infected cells. The next question we want to ask is whether nsP2 plays a role in this process. NsP2 is often localized in the nucleoli of infected cells [20]. It has also been shown to be associated with ribosomes where it can be cross-linked to RNA by UV [22]. Interestingly, foot-and-mouth-disease virus [10] and plant tobacco etch potyvirus [23] have a proteinase transported into the nucleus which, like the alphavirus nsP2, cleaves the replicase polyprotein. The foot-and-mouse-disease virus 3C protein has been shown to have a proteolytic function which induces the cleavage of histone H3 and which may be related to the shutoff of cellular transcription reported for several picornaviruses [6]. It will be interesting to determine whether nuclear localization of nsP2 contributes to the pathogenicity of SFV. This now can be done using the SFV-RDR mutant.

References

1. Abramson RD, Dever TE, Lawson TG, Ray BK, Thach RE, Merrick WC (1987) The ATP-dependent interaction of eukaryotic initiation factors with mRNA. J Biol Chem 262: 3826–3832
2. Argos P, Leberman R (1985) Homologies and anomalies in primary structural patterns of nucleotide binding proteins. Eur J Biochem 152: 651–656
3. Atkins GJ (1976) The effect of infection with Sindbis virus and its temperature-sensitive mutants on cellular protein and DNA synthesis. Virology 71: 593–597
4. Clark R, Lane DP, Tjian R (1981) Use of monoclonal antibodies as probes of simian virus 40 T antigen ATPase activity. J Biol Chem 256: 11854–11858
5. Ding M, Schlesinger MJ (1989) Evidence that Sindbis virus nsP2 is an autoprotease which processes the virus nonstructural polyprotein. Virology 171: 280–284
6. Falk MM, Grigera PR, Bergmann IE, Zibert A, Muthaup G, Beck E (1990) Foot-and-mouth disease virus protease 3C induces specific proteolytic cleavage of host cell histone H3. J Virol 64: 748–756
7. Follet EAC, Pringle RC, Pennington TH (1975) Virus development in enucleate cells: echovirus, poliovirus, pseudorabies virus, reovirus, respiratory syncytial virus and Semliki Forest virus. J Gen Virol 26: 183–196
8. Garcia-Bustos J, Heitman J, Hall MN (1991) Nuclear protein localization. Biochim Biophys Acta 1071: 83–101
9. Gorbalenya AE, Koonin EV (1989) Viral proteins containing the purine NTP-binding sequence pattern. Nucleic Acids Res 17: 8413–8440
10. Grigera PR, Sagedahl A (1986) Cytoskeletal association of an aphtovirus-induced polypeptide derived from P3ABC region of the viral polyprotein. Virology 154: 369–380
11. Hahn YS, Strauss EG, Strauss JH (1989) Mapping of RNA⁻ temperature-sensitive mutants of Sindbis virus: assignment of complementation groups A, B and G to nonstructural proteins. J Virol 63: 3142–3150
12. Hardy WR, Strauss JH (1989) Processing of the nonstructural polyproteins of Sindbis virus: nonstructural proteinase is in the C-terminal half of nsP2 and functions both in *cis* and in *trans*. J Virol 63: 4653–4664
13. Hirling H, Scheffner M, Restle T, Stahl H (1989) RNA helicase activity associated with the human p68 protein. Nature 339: 562–564
14. Hoffmann A, Roeder RG (1991) Purification of his-tagged proteins in non-denaturing conditions suggests a convenient method for protein interaction studies. Nucleic Acids Res 19: 6337–6338

15. Kääriäinen L, Takkinen K, Keränen S, Söderlund H (1987) Replication of the genome of alphaviruses. J Cell Sci [Suppl] 7: 231–250
16. Laín S, Martín MT, Riechmann JL, García JA (1991) Novel catalytic activity associated with positive-strand RNA virus infection: nucleic acid stimulated ATPase activity of the plum pox potyvirus helicaselike protein. J Virol 65: 1–6
17. Liljeström P, Lusa S, Huylebroeck D, Garoff H (1991) In vitro mutagenesis of a full-length cDNA clone of Semliki Forest virus: the small 6000-molecular-weight membrane protein modulates virus release. J Virol 65: 4107–4113
18. Nelson WG, Pienta KJ, Barrack ER, Coffey DS (1986) The role of the nuclear matrix in the organization and function of DNA. Annu Rev Biophys Chem 15: 457–475
19. Peränen J, Takkinen K, Kalkkinen N, Kääriäinen L (1988) Semliki Forest virus-specific nonstructural protein, nsP3, is a phosphoprotein. J Gen Virol 69: 2165–2178
20. Peränen J, Rikkonen M, Liljeström P, Kääriäinen L (1990) Nuclear localization of Semliki Forest virus-specific nonstructural protein nsP2. J Virol 64: 1888–1896
21. Porath J, Carlsson J, Olsson I, Belfrage G (1975) Metal chelate affinity chromatography, a new approach to protein fractionation. Nature 258: 598–599
22. Ranki M, Ulmanen I, Kääriäinen L (1979) Semliki Forest virus-specific nonstructural protein is associated with ribosomes. FEBS Lett 108: 299–302
23. Restrepo MA, Freed DD, Carrington JC (1990) Nuclear transport of plant potyviral proteins. Plant Cell 2: 987–998
24. Rikkonen M, Peränen J, Kääriäinen L (1992) Nuclear and nucleolar targeting signals of Semliki Forest virus nonstructural protein nsP2. Virology 189: 462–473
25. Sawicki DL, Sawicki SG (1993) Alphavirus nsP2 and nsP4 proteins play a role in the cessation of transcription of minus strand RNA. Arch Virol [Suppl] 9: 393–405
26. Simizu B, Wagatsuma M, Oya A, Hanaoka F, Yamada M (1976) Inhibition of cellular DNA synthesis in hamster kidney cells infected with western equine encephalitis virus. Arch Virol 51: 251–261
27. Simizu B (1984) Inhibition of host cell macromolecular synthesis following Togavirus infection. In: Fraenkel-Conrat H, Wagner RR (eds) Comprehensive virology, vol 19. Plenum Press, New York, pp 465–499
28. Studier FW, Moffatt BA (1986) Use of bacteriophage T7 RNA polymerase to direct expression of cloned genes. Methods Enzymol 185: 60–89
29. Takkinen K, Peränen J, Keränen S, Söderlund H, Kääriäinen L (1990) The Semliki-Forest-virus-specific nonstructural protein nsP4 is an autoproteinase. Eur J Biochem 189: 33–38
30. Walker JE, Saraste M, Runswick MJ, Gay NJ (1982) Distantly related sequences in the α- and β-subunits of ATP synthase, myosin, kinases and other ATP-requiring enzymes and a common nucleotide binding fold. EMBO J 8: 945–951

Authors' address: Dr. Marja Rikkonen, Institute of Biotechnology, P.O. Box 45 (Valimotie 7), University of Helsinki, SF-00014 Helsinki, Finland.

RNA replication

Arch Virol (1994) [Suppl] 9: 381–392

Archives
Virology
© Springer-Verlag 1994
Printed in Austria

Replication and translation of cowpea mosaic virus RNAs are tightly linked

J. Wellink, H. van Bokhoven*, **O. Le Gall****, **J. Verver**, and **A. van Kammen**

Department of Molecular Biology, Agricultural University, Wageningen,
The Netherlands

Summary. The genome of cowpea mosaic virus (CPMV) is divided among two positive strand RNA molecules. B-RNA is able to replicate independently from M-RNA in cowpea protoplasts. Replication of mutant B-transcripts could not be supported by co-inoculated wild-type B-RNA, indicating that B-RNA cannot be efficiently replicated in trans. Hence replication of a B-RNA molecule is tightly linked to its translation and/or at least one of the replicative proteins functions in cis only. Remarkably also for efficient replication of M-RNA one of its translation products was found to be required in cis. This 58K protein possibly helps in directing the B-RNA-encoded replication complex to the M-RNA. In order to identify the viral polymerase the CPMV B-RNA-specific proteins have been produced individually in cowpea protoplasts using CaMV 35S promoter based expression vectors. Only protoplasts transfected with a vector containing the 200K coding sequence were able to support replication of co-transfected M-RNA. Despite this, CPMV-specific RNA polymerase activity could not be detected in extracts of these protoplasts using a poly(A)/oligo(U) assay. These results indicate that, in contrast to the poliovirus polymerase, the CPMV polymerase is not able to accept oligo(U) as a primer and in addition support the concept that translation and replication are linked.

Introduction

The cowpea mosaic virus (CPMV) genome is composed of two plus strand RNA molecules. Both RNAs are translated into large polypro-

---

* Present address: Department of Human Genetics, University Hospital, University of Nijmegen, Nijmegen, The Netherlands.
** Present address: Station de Pathologie Végétale, INRA, Villenave D'Ornon Cédex, France.

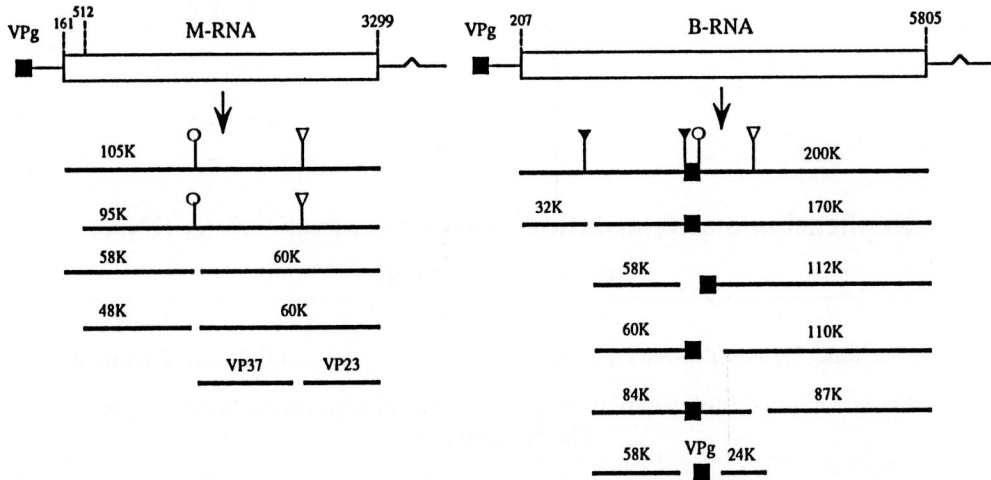

Fig. 1. Diagram of the expression of the CPMV RNAs. The open reading frames are represented by bars, proteins by single lines and VPg by ■. Cleavage sites are indicated by ○ gln/met, ▽ gln/gly, and ▼ gln/ser. All proteins have been identified in infected cells

teins that are subsequently cleaved by the viral 24K proteinase into several stable intermediate and final cleavage products (Fig. 1; [1, 2]). The larger B RNA replicates independently from the smaller M RNA in cowpea protoplasts and encodes all the proteins involved in replication. The M RNA is translated into two carboxycolinear proteins of 105K and 95K which are cleaved into the 58K and 48K proteins and the capsid proteins VP37 and VP23. The 48K protein and the capsid proteins have been shown to be indispensable for virus cell-to-cell movement (see chapter by Goldbach et al.).

The genetic organization and expression strategy of CPMV and picornaviruses are strikingly similar. Furthermore the 58K, 24K and 87K proteins of CPMV show considerable amino acid sequence homology to the poliovirus 2C, 3C and 3D proteins respectively ([3, 4] see also Fig. 4). A striking difference is that picornaviruses have encoded their genetic information on a single RNA molecule. Experimental evidence obtained by several groups strongly suggests that translation and replication of picornaviral RNA are linked and/or that at least one of the non-structural proteins is cis-acting in replication [5, 6]. Considering the similarities in genome organization and expression strategy between CPMV and poliovirus it is possible that CPMV B RNA translation and replication also are linked. If so, the question arises how replication of M RNA is achieved by the B RNA encoded replicative proteins. To study this, mutations were introduced in cDNA sequences of B and M RNA and T7 transcripts from the mutant cDNAs were used for

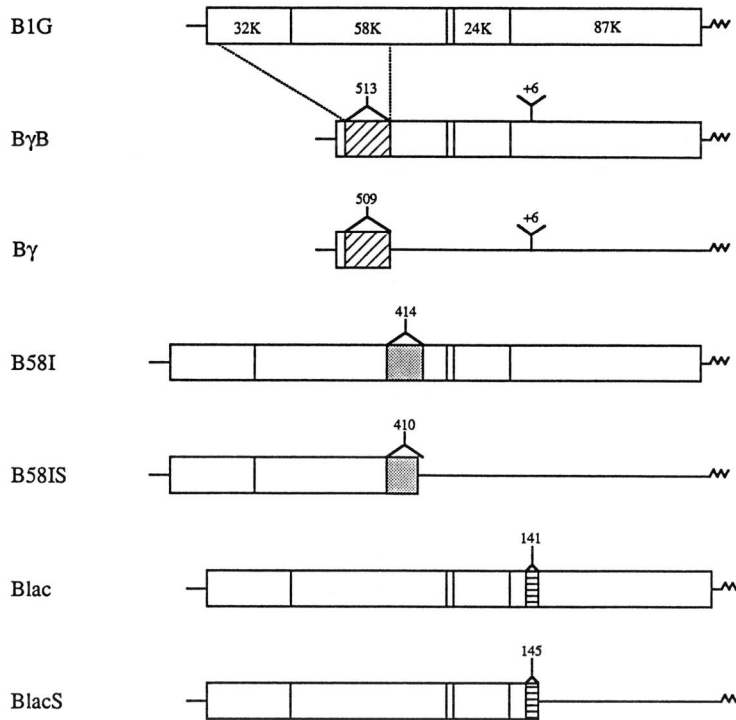

Fig. 2. Schematic representation of the B RNA insertion mutants. Open bars denote open reading frames in the transcripts. Insertions originate from parts of the coding sequences of CaMV gene 1 (hatched) lacZ (horizontal lines) and CPMV M RNA (dotted) with sizes indicated above the mutant transcripts. Cowpea protoplasts were inoculated with these mutant B RNAs together with wt B RNA. Infectivity of the mutant B RNAs was determined by Northern blot analysis using probes specific for the inserts

transfection of cowpea protoplasts. In an effort to determine which of the CPMV B RNA encoded proteins possesses polymerase activity we have expressed individual proteins in insect cells and cowpea protoplasts. Extracts of these cells were assayed for polymerase activity using poly(A) oligo(U) assays.

B RNA is not replicated in trans

To study whether B RNAs defective in replication can be replicated by proteins provided by a co-inoculated wildtype (wt) B RNA, B RNA mutants were created that carry insertions at different positions in the open reading frame ([7]; Fig. 2). Insertions of heterologous sequences were used to facilitate detection of the mutant RNAs among the excess of cotransfected wt B RNA on Northern blots. To exclude that the recognition of these B RNA mutants as templates for replication would be obstructed by shielding of recognition sites on the RNA by defective replicative proteins, frame shift mutants were constructed as well (Fig. 2).

Upon inoculation of cowpea protoplasts with these mutant B RNAs together with wt B RNA no replication of the mutant RNAs could be detected using Northern blot analysis. The explanation that the insertions create cis acting defects by disrupting RNA structures involved in replication is not very likely. Only the 5' and 3' non-coding regions of M and B RNA show nucleotide sequence homology and large regions of the M RNA coding region are not essential for replication ([7], also see below). It thus appears that CPMV B RNA, like poliovirus RNA, is not or very inefficiently replicated in trans. In this respect CPMV B components are very similar to poliovirus DI particles.

Evidence that the N-terminus of the 58K protein of M RNA is involved in replication

Previous studies had shown that large regions of the M RNA coding region were not essential for replication [7]. However, mutants in which translation of the 105K protein was disrupted replicated very poorly [8, 9]. To study the possible involvement of the 105K protein in replication of M RNA, mutant M RNAs were constructed as shown in Fig. 3. These mutant M RNAs were transfected into cowpea protoplasts together with B RNA. At 42 h after transfection protoplasts were assayed by immunofluorescent staining with anti CPMV serum or anti 48K serum to determine the number of cells in which M RNA encoded proteins have accumulated (Fig. 3) and by Northern blotting to measure the accumulation of the mutant M RNAs in these cells.

The deletion in the 58K coding region in M58ΔH RNA resulted in complete loss of RNA replication for this mutant M RNA (Fig. 3), emphasizing the importance of the 58K-specific region for M RNA replication. Evidence that the protein, rather than the nucleotide sequence of this part of the M RNA, is involved in the replication of M RNA is provided by M RNA mutants MΔBg1 and M58GS, which respectively contain insertions of 4 and 12 nucleotides at the same position (Fig. 3). It appeared that for M58GS RNA, with the in frame insertion, replication was maintained at wt levels whereas for MΔBg1 RNA, where the out frame insertion prevented synthesis of the 58K protein, the replication level was reduced about 50 fold as compared to M RNA. MΔ48 RNA, in which almost the complete 48K coding sequence has been deleted, replicated to similar levels as wt M RNA (Fig. 3). These results strongly suggest that the N-terminal part of the 58K protein is involved in replication of M RNA [9]. This 58K protein

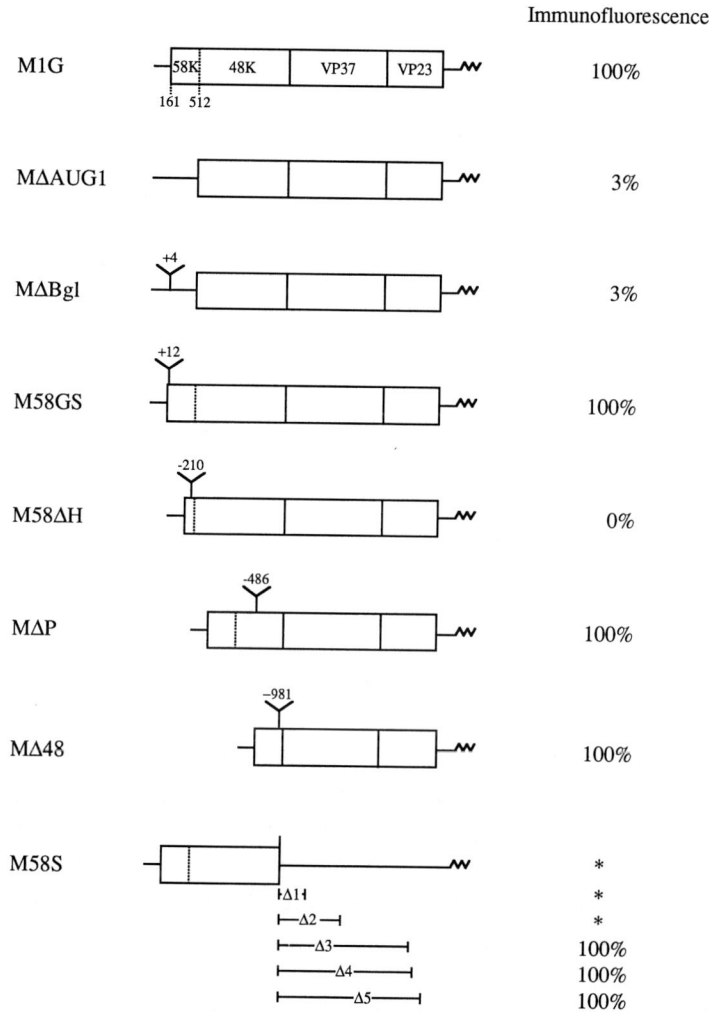

Fig. 3. Schematic representation of mutant M RNAs. Immunofluorescence of protoplasts inoculated with these transcripts together with B RNA was determined at 42 h p.i. using anti-CPMV and anti-48K serum. Immunofluorescence for M1G RNA was arbitrarily set at 100% and values for the other transcripts were correlated to this value. The immunofluorescence signal for M58S, M58SΔ1 and M58SΔ2 was too weak to be quantitated (*)

probably functions in cis only since replication of the defective M RNAs could not be rescued by wt M RNA.

Remarkably, the N-terminus of the 58K protein is not conserved between comoviruses, except for the presence of many hydrophobic and aromatic amino acid residues [10]. Hydrophobic and aromatic amino acid motifs are often found in members of the highly heterologous "family" of RNA-binding proteins [11]. Therefore, it is tempting to speculate that the N-terminal domain of the 58K protein of M RNA is involved in RNA-binding. The results indicate that replication of M RNA depends on translation of the 58K polypeptide from the very same RNA molecule, suggesting that the translator ribosomes transport the N-terminal domain of the 58K protein, contained in the 105K polyprotein, to the 3' end of the RNA molecule. Then a ribonucleoprotein complex occasionally may be formed, consisting of the 105K polyprotein, viral RNA and possibly ribosomal factor(s). This complex is then recognized by the B RNA-encoded replicative machinery to start negative strand RNA synthesis. For B RNA, the observed linkage between translation and replication may be effected similarly by transportation of the replicative proteins to the 3' end of the RNA ([5, 6]; see Discussion). Each M RNA molecule must be translated many times in order to produce enough coat proteins to encapsidate the RNAs (120 times for one B and one M RNA). Interestingly, the virus prevents (over)production of the 105K (58K) protein by synthesizing the 95K (48K) protein (Fig. 1). The 48K protein has a function in cell-to-cell movement, and probably is a structural component of the tubule wall (see chapter by Goldbach et al.). In vivo labelling experiments have shown that the 48K protein is produced in larger amounts than the 58K protein in infected cells [12].

The mutant M58SΔ5 RNA, which has a large deletion in the capsid coding region, excludes a possible involvement of the coat proteins in M RNA replication. Complicating the interpretation of results with coat protein mutants is that accumulation of these mutant RNAs in protoplasts is greatly reduced due to the absence of encapsidation. However, immunofluorescence data reveal that M58SΔ5 RNA is replicated as efficiently as wt M RNA (Fig. 3). Remarkably the M58S, M58SΔ1 and M58SΔ2 RNAs replicated very poorly. It is likely that the stability of these RNAs is reduced by the presence of a large 3' non-translated region. A decreased stability of mRNAs caused by a premature translational stop has also been described for a frameshift mutant of the soybean Kunitz trypsin inhibitor gene [13] and for certain pseudogenes in transgenic tobacco [14, 15]. Alternatively, for the mutant M58S and M58SΔ1 and Δ2 RNAs a linkage between translation and replication is hampered due to their large 3' non-translated regions, which interfere in the formation of a ribonucleoprotein complex as described above.

The 5' and 3' non-coding regions of B- and M RNA are exchangeable

Except for the N-terminal part of the 58K protein, none of the other M RNA coding sequences and proteins derived there from seem to be involved in replication. However, it is possible that the 5' and 3' non-coding regions of M RNA contain sequences that, in cooperation with the N-terminus of the 58K protein, allow in trans replication of M RNA.

Comparison of the non-coding regions of M RNA with those of B RNA reveals a high degree of homology at the 5′ terminal 44 nucleotides (89%) and 3′ terminal 65 nucleotides (82%), which can be folded into very similar secondary structures [16]. Proximal to the coding region the nucleotide sequences of B- and M RNA have widely diverged. Particularly striking is a section of 100 nucleotides in the 3′ non-coding region of M RNA that is lacking at the 3′ end of B RNA. To study whether these deviant nucleotides in the non-coding regions of M RNA are required for in trans replication, two mutant M RNAs were constructed that have exchanged the 5′ and 3′ ends of M RNA for the corresponding sequences of B RNA, which flank an RNA molecule that is only replicated in cis. Both mutant M RNAs still replicated efficiently in cowpea protoplasts. Therefore we consider it very unlikely that the 5′ and 3′ non-coding regions of M RNA contain signals specific for in trans replication of this RNA.

Synthesis of putative CPMV polymerases in *E. coli* and insect cells

Thus far it has not been possible to isolate from infected plants a CPMV polymerase fraction that is dependent on exogenous template. As an alternative approach to obtain such template dependent CPMV specific polymerase activity, and in order to be able to study individual CPMV replicative proteins, two heterologous expression systems have been employed, *E. coli* [17] and insect cells [18–20].

Extracts of cells producing the putative CPMV polymerase and its precursors were assayed for polymerase activity using assays with poly(A) or CPMV RNA as template and oligo(U) as primer. However, none of the extracts contained such activity, despite a wide variety of conditions used for the assay. In sharp contrast, polioviral polymerase produced in both expression systems as well exhibited easily detectable activity in the same assays [17, 19, 20]. There are two main possibilities to explain the lack of activity of the CPMV polymerase: 1) CPMV polymerase requires a plant (host) factor for activity; 2) CPMV polymerase cannot use oligo(U) as a primer in RNA synthesis (despite the high degree of homology to the poliovirus polymerase) or is not able to function on exogenous RNA (template + primer).

Synthesis of putative CPMV poymerases in cowpea protoplasts

In order to study whether the CPMV polymerase needs a plant factor for activity, a transient expression system was developed for cowpea protoplasts [21]. Transient expression vectors were constructed that contained the 87K, 110K, 170K and complete 200K coding regions of CPMV B RNA under control of the 35S promoter of cauliflower mosaic

A   CPMV B-RNA

B   POLIOVIRUS

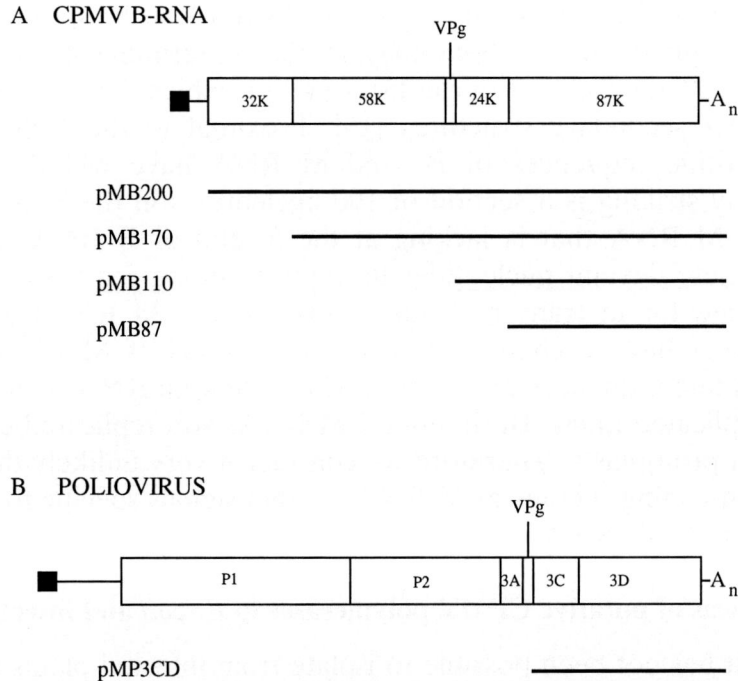

Fig. 4. Schematic diagram of the expression vectors for cowpea protoplasts. **A** Genetic organization of CPMV B RNA. VPg at the 5′ end of the RNA is drawn as ■. The single open reading frame is represented by an open bar in which the polypeptide domains are indicated. Underneath are shown the expression vectors containing CPMV B cDNA sequences. The sequences represented by solid lines were cloned in the polylinker of pMON999. **B** Genetic organization of poliovirus RNA and the position of the cDNA sequence inserted in pMON999 to create expression vector pMP3CD

virus (Fig. 4). In cowpea protoplasts transfected with these vectors large amounts of the expected CPMV-specific proteins were synthesized, as determined by Western blotting and as were visualized by immunofluorescent staining (Table 1). These proteins exhibited the same characteristic activities (proteolytic processing, and induction of cytopathic structures) that were manifested by the viral proteins produced in protoplasts transfected with B RNA. To determine directly whether these proteins possessed the activities that are required for viral RNA replication, cowpea protoplasts were co-transfected with the transient expression vectors and with M RNA transcripts. Replication of M RNA was supported in protoplasts that transiently expressed the entire 200K coding sequence of B RNA (Table 1). It was therefore concluded that all the viral replicative proteins, including the viral polymerase, are functionally intact upon their synthesis in protoplasts. M RNA was not replicated in cells in which the 170K, 110K or 87K protein were synthesized individually, indicating that expression of the complete 200K

Table 1. Immunofluorescent staining of cowpea protoplasts

Protoplasts[a]	Immunofluorescence[b]	
	anti-110K	anti-CPMV
B-RNA	40	20
pMB200	20	10
pMB170	20	0
pMB110	50	0
pMB87	50	0
pMGUS	0	0

[a] Cowpea protoplasts were transfected with B-RNA transcripts or expression vector DNA (see Fig. 4) in the presence of M-RNA transcripts

[b] Immunofluorescent staining was with the indicated antisera 18 h after transfection. The numbers represent the percentage of fluorescing protoplasts. Staining with anti-CPMV serum can only be observed in protoplasts in which M-RNA is replicated

coding region of B RNA is required for CPMV RNA replication (Table 1).

In spite of the in vivo RNA replicase activity of transiently expressed 200K protein in protoplasts, it was not possible to demonstrate in a poly(A)/oligo(U) assay in vitro RNA polymerase activity in extracts of these protoplasts (Table 2). In contrast, extracts of protoplasts in which poliovirus polymerase has been expressed (pMP3CD) did contain such activity (Table 2). Again, these results demonstrate that the polymerases of CPMV and poliovirus have different requirements for in vitro activity. Because it was unequivocally demonstrated that expression in a homologous system yielded functional replicative proteins, it is not likely that the lack of in vitro activity for the CPMV polymerase is caused by the absence of an essential host factor in the assay. Use of feasible natural templates for the CPMV polymerase, i.e. CPMV RNA and negative strand transcripts of B RNA with or without an oligo-ribonucleotide primer, did not result in detectable RNA-synthesizing activity.

Discussion

Thus far all efforts to obtain a template dependent CPMV polymerase have failed [17, 19–21]. In our opinion the most likely explanation for this is the inability of the CPMV polymerase to function in poly(A)/oligo(U) assays that are used to detect such activity. Possibly the CPMV

Table 2. Poly(A)/oligo(U) polymerase activity in cell extracts

Extract[a]	Poly(A)/oligo(U)	No template/primer
AcHB3CD	126 000[b]	160
HeLa	7 200	150
pMP3CD	1 500	140
pMB200	160	170
pMB170	160	150
pMB110	130	160
pMB87	100	130
pMGUS	150	150

[a] Extracts of protoplasts were prepared at 18 h after transfection. AcHB3CD is a crude extract of insect cells containing poliovirus 3D^{pol} and HeLa is a partially purified preparation of 3D^{pol} from poliovirus-infected HeLa cells

[b] Polymerase activity is expressed as the amount of [^{3}H]-UTP label incorporated after 30 min incubation

polymerase cannot use oligo(U) as a primer or cannot function on any added template/primer combination because translation and replication are linked. The ability of poliovirus polymerase to use poly(A)/oligo(U) may be a fortuitous property that is not shared by the CPMV polymerase. The oligo(U) primed activity of the poliovirus polymerase has no specificity towards poliovirus RNA as template and certainly does not mimic the initiation of replication events that occur in vivo. VPg precursors might have a role in this process [22]. Thus far we have not been able to test whether the CPMV polymerase can use VPg-pU as a primer.

The results described in the first part of this chapter suggest that replication and translation of the CPMV RNAs are linked. Also complete replication of poliovirus (e.g. negative strand and positive strand synthesis) has only been accomplished with extracts of uninfected HeLa cells in which the input poliovirus RNA was first translated [23]. Thus it may be that in vitro CPMV replicase activity can only be observed under conditions that first allow translation of the RNA template.

At present the actual link between translation and replication is not clear. It seems reasonable to assume that it involves the formation of the complex that initiates negative strand synthesis [5]. One of the replicative proteins involved in this process may only function in cis. This may be difficult to reconcile with replication of M RNA (by B RNA-encoded proteins) unless one assumes that the M RNA- encoded 58K (105K) protein can convert a cis acting protein to a trans acting one. Alternatively as proposed in a previous section of this chapter, ribosomes are involved because they "transport" either the replicative proteins or an essential host factor (of ribosomal origin?) to the 3′ end of the RNA. It may also be that RNA has to interact with ribosomes in order to make it a suitable template for replication [6]. Irrespective of the actual mechanism, the advantage of this translation-linked mode of replication is that it results in a positive selection of replication-competent mutants.

Acknowledgements

The authors wish to thank Dr. C. Hemenway of Monsanto Company for use of the expression vector pMON999 and G. Heitkönig for preparation of the manuscript. This work was supported by the Netherlands Foundation for Chemical Research (SON) with financial aid from the Netherlands Organization for Scientific Research (NWO).

References

1. Peters SA, Voorhorst WGB, Wellink J, Van Kammen A (1992) Processing of VPg-containing polyproteins encoded by the B RNA from cowpea mosaic virus. Virology 191: 90–97
2. Peters SA, Voorhorst WGB, Wery J, Wellink J, Van Kammen A (1992) A regulatory role for the 32K protein in proteolytic processing of cowpea mosaic virus polyprotcins. Virology 191: 81–89
3. Franssen H, Leunissen J, Goldbach R, Lomonossoff G, Zimmern D (1984) Homologous sequences in non-structural proteins from cowpea mosaic virus and picornaviruses. EMBO J 3: 855–861
4. Argos P, Kamer G, Nicklin MJH, Wimmer W (1984) Similarity in gene organisation and homology between proteins of animal picornaviruses and a plant comovirus suggest common ancestry of these virus families. Nucleic Acids Res 12: 7251–7267
5. Bernstein HD, Sarnow P, Baltimore D (1986) Genetic complementation among poliovirus mutants derived from an infectious cDNA clone. J Virol 60: 1040–1049
6. Hagino-Yamayishi U, Nomoto A (1989) In vitro construction of poliovirus defective interfering particles. J Virol 63: 5386–5392
7. Van Bokhoven H, Le Gall O, Kasteel D, Verver J, Wellink J, Van Kammen A (1993) Cis and trans acting elements in cowpea mosaic virus RNA replication. Virology 195: 377–386
8. Wellink J, Van Kammen A (1989) Cell-to-cell transport of cowpea mosaic virus requires both the 58K/48K proteins and the capsid proteins. J Gen Virol 70: 2279–2286
9. Holness CL, Lomonossoff GP, Evans D, Maule AJ (1989) Identification of the initiation codons for translation of cowpea mosaic virus middle component RNA using site-directed mutagenesis of an infectious cDNA clone. Virology 172: 311–320
10. Chen X, Bruening G (1992) Nucleotide sequence and genetic map of cowpea severe mosaic virus RNA2 and comparisons with RNA2 of other comoviruses. Virology 187: 682–692
11. Kenan DJ, Query CC, Keene JD (1991) RNA recognition: towards identifying determinants of specificity. Trends Biochem Sci 16: 214–220
12. Rezelman G, Van Kammen A, Wellink J (1989) Expression of cowpea mosaic virus M RNA in cowpea protoplasts. J Gen Virol 70: 3043–3050

13. Jofuka KD, Schipper RD, Goldberg RG (1989) A frameshift mutation prevents Kunitz trypsine inhibitor mRNA accumulation in soybean embryos. Plant Cell 1: 427–435

14. Voelker TA, Moreno J, Chrispeels MJ (1990) Expression analysis of a pseudogene in transgenic tobacco: a frameshift mutation prevents mRNA accumulation. Plant Cell 2: 255–261

15. Vancanneyt G, Rosahl S, Wilmitzer L (1990) Translatability of a plant mRNA strongly influences its accumulation in transgenic plants. Nucleic Acids Res 18: 2917–2921

16. Eggen R, Verver J, Wellink J, Pley K, Van Kammen A, Goldbach R (1989) Analysis of sequences involved in cowpea mosaic virus RNA replication using site specific mutants. Virology 173: 456–464

17. Richards OC, Eggen R, Goldbach R, Van Kammen A (1989) High level synthesis of cowpea mosaic virus RNA polymerase and protease in Escherichia coli. Gene 78: 135–146

18. Van Bokhoven H, Wellink J, Usmany M, Vlak JM, Goldbach R, Van Kammen A (1990) Expression of plant virus genes in animal cells: high level synthesis of cowpea mosaic virus B-RNA-encoded proteins with baculovirus expression vectors. J Gen Virol 71: 2509–2517

19. Van Bokhoven H, Mulders M, Wellink J, Vlak JM, Goldbach R, Van Kammen A (1991) Evidence for dissimilar properties of comoviral and picornaviral RNA polymerases. J Gen Virol 72: 567–572

20. Van Bokhoven H, Van Lent JWM, Custers R, Vlak JM, Wellink J, Van Kammen A (1992) Synthesis of the complete 200K polyprotein of cowpea mosaic virus B RNA in insect cells. J Gen Virol 73: 2775–2784

21. Van Bokhoven H, Verver J, Wellink J, Van Kammen A (1993) Protoplasts transiently expressing the 200K coding sequence of cowpea mosaic virus B-RNA support replication of M-RNA. J Gen Virol 79 (in press)

22. Takeda N, Kuhn RJ, Yang C, Takegami T, Wimmer E (1986) Initiation of poliovirus plus-strand RNA synthesis in a membrane complex of infected HeLa cells. J Virol 60: 43–53

23. Molla A, Paul AV, Wimmer E (1991) Cell-free de novo synthesis of poliovirus. Science 254: 1647–1651

Authors' address: Dr. J. Wellink, Department of Molecular Biology, Agricultural University, Dreijenlaan 3, 6703 HA Wageningen, Netherlands.

Arch Virol (1994) [Suppl] 9: 393–405

_Archives_____

Virology

© Springer-Verlag 1994
Printed in Austria

Alphavirus positive and negative strand RNA synthesis and the role of polyproteins in formation of viral replication complexes

D. L. Sawicki and **S. G. Sawicki**

Department of Microbiology, Medical College of Ohio, Toledo, Ohio, U.S.A.

Summary. The genome of alphaviruses is translated into polyproteins that are processed into a viral replicase that produces both negative and positive strands. In infected cells, negative strand synthesis is short-lived and occurs only early, whereas positive strand synthesis is stable and occurs both early and late. Analysis of temperature sensitive mutants indicated: nsP1 functioned in the initiation of transcription; nsP3 acted to form initial transcription complexes; and nsP2 and nsP4 first recognized positive strands as templates and then made negative strands the preferred templates. While nsP4 and nsP1 individually rescued early defects in transcription, nsP2 and nsP3 acted initially in *cis*. We interpret our results to suggest nsP1234 was cleaved to nsP4, nsP1 and nsP23, bound a positive strand and synthesized a negative strand. Cleavage of P23 or other modifications to nsP2 and nsP4 convert the initial transcription complex to a stable complex that synthesizes positive strands. Negative strand synthesis is unstable because of the failure to form initial transcription complexes after host factors that are part of the replicase are depleted or the half-life of polyprotein precursors like P23 is shortened.

Introduction

A superfamily of animal and plant viruses has been defined that share amino acid homologies with the nonstructural proteins (nsP) nsP1, nsP2 and nsP4 of the alphavirus Sindbis virus [1, 2]. The Sindbis nsPs are translated initially as two polyproteins of 200 kDa and 250 kDa, the former when translation terminates at an opal stop codon that is located at the end of the nsP3 open reading frame and the latter from ribosomal readthrough of the opal stop codon [3]. Cleavage by a papain-like protease [3] activity of nsP2 gives rise to nsP1, nsP2 and nsP3 from the 200 kDa polyprotein and to nsP1, nsP2, nsP3 and nsP4 or fusion protein nsP34 from the 250 kDa polyprotein. Thus, nsP4 and polyprotein nsP34 are underproduced relative to nsP1–nsP3.

From our current understanding of the functions of the Sindbis nsPs, it is clear that most of them act in both positive and negative strand synthesis. The nsP1 contains sequence motifs for a methyltransferase and guanylyltransferase activity that would cap positive strand RNA; cloned and expressed Sindbis [4] and Semliki Forest virus [5] nsP1 exhibit methyltransferase activity in the absence of other nonstructural proteins. As described below, we also found that nsP1 provides an essential function for alphavirus negative strand synthesis [6]. The nsP2 sequence encodes two enzymatic activities. The N-half of the Sindbis protein has sequence motifs for helicase and NTP-binding activity that are conserved within the superfamily; hepatitis E, rubella and furoviruses have a helicase domain partly resembling that of coronaviruses [7]. Using linker insertion mutagenesis, it has been demonstrated [8] that brome mosaic virus (BMV) helicase functions to elongate both positive and negative strand RNA. The C-half of the Sindbis nsP2 sequence is encoded in genomes of animal virus members of the superfamily and contains a papain-like protease activity that has been shown to be responsible for processing the viral nonstructural polyproteins and allowing RNA replication [3, 9]. To date, the absence of protease sequences in genomes of plant virus members of the superfamily such as tobacco mosaic virus (TMV) indicates their nsPs function as polyproteins. For alphaviruses, nsP2 also plays an essential role in the internal initiation of transcription of the subgenomic mRNA [10, 11] and in the regulation of negative strand synthesis [12]. Recently, it was found that nsP2 contains a nuclear localization signal and may contribute to the shut down of host transcription in infected cells [13]. Flanking the papain protease domain, the N-terminus of alphavirus nsP3 contains a sequence motif of unknown function found also in the rubella and hepatitis E genome and related to a domain in coronaviruses [9]. A role of this "X" domain in the regulation of polyprotein processing has been suggested [9]. For alphaviruses, only the N-half of the nsP3 sequence is essential and nsP3 is phosphorylated by a casein kinase II-like activity [14]. As detailed below, it has been found that nsP3 is a component of transcriptionally active replication complexes [15] and plays a role in the formation of the replication complex for negative strand synthesis [16]. Alphavirus nsP4 contains conserved sequences that encode the RNA-dependent RNA polymerase domain [17], assigned to subgroup III by Koonin et al. [2]. It functions in elongation of both positive and negative strand RNAs [18], in promoter recognition [19], and in host range [20].

Alphavirus RNA synthesis

Alphaviruses are unique among positive strand RNA viruses because they regulate negative strand synthesis and form a stable positive strand

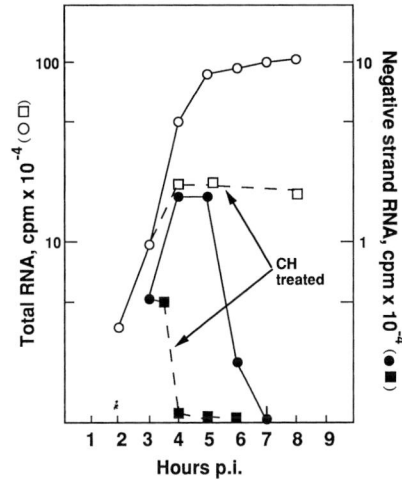

Fig. 1. Kinetics of alphavirus RNA synthesis. Cultures of chicken embryo fibroblast cells infected with Sindbis HR at an multiplicity of infection of 100 at 37°C were given 60 min pulses of ^{3}H-uridine in medium also containing 20 µg of actinomycin D/ml. Beginning at 3 h p.i., duplicate cultures were treated with 100 µg of cycloheximide (CH)/ml and pulsed with ^{3}H-uridine in medium containing CH and actinomycin D. Total labeled RNA, greater than 95% of which is positive strand genomes or 26S mRNAs, represents the acid-insoluble incorporation in the absence (○) or presence (□) of CH. Labeled negative strand RNA in the absence (●) or presence (■) of CH

[26]

polymerase [21]. Genome-length negative strands are templates for replication of the genome and transcription of subgenomic mRNA, and the same template strand can switch between the two activities [22]. While the alphavirus positive strand polymerase is stable, the negative strand activity in unstable. Synthesis of negative strands occurs early and, once synthesized, the negative strands form stable templates (Fig. 1) whose accumulation results in proportionally increased numbers of replication complexes and increased rates of positive strand synthesis. Addition of protein synthesis inhibitors such as cycloheximide selectively and rapidly stop negative strand synthesis, leaving positive strand synthesis to continue at the rate ongoing at the time of addition of the drug (Fig. 1). After the early phase there is a selective cessation of negative strand synthesis, when about 5000 negative strand templates/cell have accumulated [6]. This results in the number of replication complexes becoming constant when the rate of positive strand synthesis is maximal. Cessation is not mediated by the encapsidation of positive strand templates or by a transactive repressor [19], and occurs even if the non-structural proteins are overproduced late in infection [23]. The regulation of negative strand synthesis may be a common replication feature of the

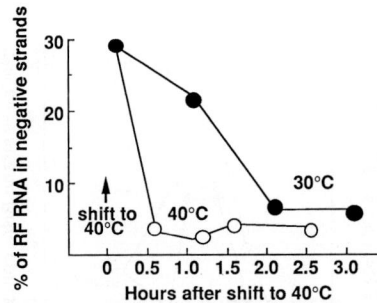

Fig. 2. The phenotype of *ts*11, a Sindbis RNA-negative mutant exhibiting a rapid and selective inhibition of negative strand RNA synthesis after shift to 40°C early in infection. Cultures infected with *ts*11 at 30°C were maintained at 30°C or were shifted to 40°C at 3 h p.i., when the overall rate of RNA synthesis was 10% of maximum. The synthesis of negative strand RNA was determined using pulses of ^{3}H-uridine of 60 min at 30°C or 30 min at 40°C and quantitating the % of total incorporation in double-stranded cores (RFs) of viral replicative intermediates that was in negative strand RNA [26]

superfamily since TMV [25] and BMV [8] produced negative strands only early in infection.

These findings led us to propose the following model [21]. An initial replication complex associates with a positive strand template and synthesizes a negative strand. The negative strand then becomes a preferred template of the same replication complex and converts it to positive strand synthesis. A corollary of this model is that only a positive strand is capable of initiating the formation of a replication complex. Moreover, continued negative strand synthesis requires continued formation of new complexes. This suggests different viral functions are required for the formation of replication complexes which result in the synthesis of negative strands and for their conversion later to positive strand synthesis. To identify such proteins, we screened for and characterized two classes of Sindbis mutants conditionally-lethal for RNA synthesis.

Viral functions necessary for negative strand synthesis early in infection

In the first class were two temperature-sensitive (*ts*) mutants, *ts*11 and *ts*4, whose phenotype upon shift to 40°C reproduced the effect of translation inhibition. That is, they rapidly and selectively shut off the synthesis of negative strand RNA at 40°C, in contrast to the normal cessation that occured later at 30°C (Fig. 2) (not shown here, positive strand synthesis continued at the rate ongoing at the time of shift). The responsible mutations were located by swapping mutant nonstructural genes into the infectious Sindbis clone [24], from which RNA was transcribed and transfected into cells. The recombinant viruses that resulted

Reactivation of negative strand synthesis

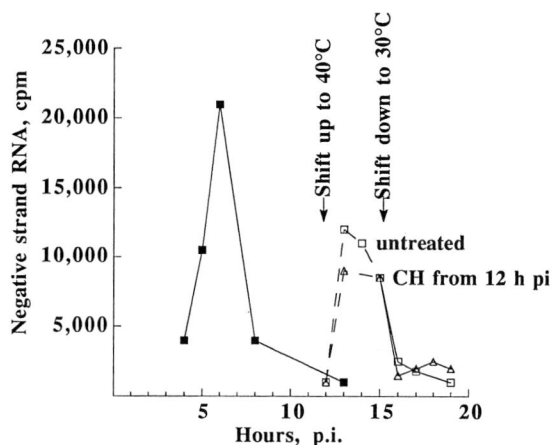

Fig. 3. Reactivation of negative strand synthesis at 40°C late in infection after its cessation at 30°C is a property of Sindbis mutant *ts*24. Cultures infected with *ts*24 maintained at 30°C (■) were pulse labeled with ³H-uridine in the presence of actinomycin D for 60 min periods during the first 13 h p.i. At 12 h p.i., duplicate cultures were shifted to 40°C, incubated in the presence (△) or absence (□) of 100 μg of cycloheximide (CH)/ml and pulse-labeled with ³H-uridine at 40°C for 60 min periods. Negative strand RNA synthesis was determined [26] (modified from Sawicki and Sawicki [10], reproduced by courtesy of Academic Press)

were assayed for retention of mutant phenotype and were sequenced. The mutation in *ts*11 mapped to amino acid 348 of nsP1 [6, 11], which is downstream of the methyltransferase domain located within the first 200 amino acids of nsP1 [4]. The mutation in *ts*4 mapped to amino acid 268 of nsP3 [16] and downstream of the "X" domain in nsP3. Thus, both nsP1 and nsP3 apparently provide essential functions early in infection to form a replication complex and/or to initiate negative strand synthesis. It was interesting that nsP3 played an essential role in negative strand synthesis because an nsP3 sequence is not present in the genomes of plant viruses of the superfamily. This suggests that for transcription nsP3 may be required for proper folding of nsP2 or for a function that also requires the thiol protease sequence, or that it may be required only in animal cells because a host factor provides the function in plant cells.

Viral functions necessary for the regulation of negative strand synthesis

The second class of three mutants had *ts* defects in the shut off of negative strand synthesis and allowed negative strand synthesis to continue at 40°C or turned negative strand synthesis back on once it had stopped. As shown in Fig. 3, shifting *ts*24 infected cells to 40°C late in infection, after negative strand synthesis had ceased at 30°C, led to its reactivation, even

under conditions when no new proteins were synthesized [26]. These
negative strands accumulated but did not increase the rate of positive
strand synthesis, even though they were present in replicative intermediates and served as templates for positive strand synthesis. Thus,
there was no net increase in the number of replication complexes. The
*ts*24 mutation was not conditionally lethal and mapped to amino acid 194
of nsP4 [19]. It introduced a positive amino acid into a linear stretch of
19 uncharged polar or nonpolar amino acids, in a sequence upstream of
the conserved replicase Gly-Asp Asp (GDD) domain and near the
region affecting host range. The other two mutants were *ts*17 and *ts*133,
whose individual single mutations mapped within the protease domain of
nsP2 at amino acids 517 and 700, respectively [11, 12]. Both mutations
inactivated protease activity, inhibited polyprotein cleavage, and blocked
RNA synthesis. They also inhibited subgenomic mRNA synthesis and
reactivated negative strand synthesis [12]. Because the nsP4 mutation
altered activities of preformed replication complexes, we asked if the
loss of subgenomic mRNA synthesis and reactivation of negative strand
synthesis by the nsP2 mutations affected the same preformed complexes.

Cells infected at 30°C were shifted to 40°C late in infection and
continuously incubated with cycloheximide to monitor the activities of
replication complexes formed earlier in infection. Negative strand synthesis reactivated at 40°C but shut off again when cultures were returned
to 30°C within 3 h [12]. Similar results were found for subgenomic mRNA
synthesis, which was inhibited at 40°C and recovered upon return of
cultures to 30°C in the absence of new protein synthesis. Thus, reactivation of negative strand synthesis was the property of stable nsP2 or
nsP4 proteins that were produced early in infection and present in
preformed complexes [12]. These results did not support the hypothesis
[27] which proposed that cleavage of polyprotein nsP34 and release
of nsP4 reactivated negative strand synthesis because, if this model
was correct, reactivation would have required translation of polyprotein
molecules and would not have occurred in the presence of cycloheximide.

Reactivation did not cause an increase in the rate of positive strand
synthesis because there was no formation of additional replication complexes [12]. Thus, we argue that reactivation was due to *template switching*: replacement of the preferred negative strand template with a positive
strand template. Reversibility by a temperature shift of ten degrees
suggested that mutations in nsP2 and nsP4 induced reversible changes
in protein interactions. These interactions affected changes either in
protein conformation that modified the structure or affinity of the positive strand polymerase, such that it resembled the negative strand polymerase and was able to recognize the 3' promoter on a positive strand
template, or in the association of a host factor(s) with the replication

complex. Reactivation and its reversal thus resulted from the modification of one polymerase activity into another polymerase activity and in this way mimicked conversion events postulated to occur in replication complexes early in infection [21]. The rate of reactivation of negative strand synthesis was higher for viruses carrying the nsP4 mutation compared to the nsP2 mutation, but the individual rates were additive when both mutations were in the same genome and when their proteins were components of the same replication complex [15]. This suggests that nsP2 and nsP4 function differently in negative strand synthesis but both act, nevertheless, to fix the negative strand as the stable template of the replication complex. Modifications that cause the replication complex to favor positive strand synthesis would not affect essential helicase and elongation activities required for both positive and negative strand synthesis. They would result in the conversion of the replication complex from negative to positive strand synthesis and lead to the overproduction of positive strands and to cessation of negative strand synthesis if no new replication complexes were formed.

Are functions of nsP1 and nsP3 required early for negative strand synthesis also required late?

Mutation in nsP4 resulted in the failure to turn off negative strand synthesis late in infection and in the ability to reactivate negative strand synthesis by preformed replication complexes [16]. Recombinant viruses containing the nsP4 mutation and either the nsP1 or the nsP3 mutation were constructed (Fig. 4). To summarize our results, in the absence of any other mutation the nsP4 mutant reactivated negative strand synthesis when shifted to 40°C in the presence of cycloheximide, while parental virus and nsP1 or nsP3 mutants had background levels [16]. The double mutant construct containing the nsP1 mutation did not reactivate negative strand synthesis. Thus, the function in negative strand synthesis provided by nsP1 was required for negative strand synthesis by preformed replication complexes as well as newly formed ones and dominated that provided by nsP4. On the other hand, the double mutant with the nsP3 mutation reactivated negative strand synthesis to a level comparable to that observed with the nsP4 mutation alone. We take this to mean that the nsP3 function was required only early in infection. We also determined the composition of the late replication complexes to verify that nsP3 was a component of replication complexes reactivating negative strand synthesis. Replication complexes solubilized from the 15 000 × **g** pellet membrane fraction of infected cells and transcriptionally active [18] were incubated with antibodies specific for each nonstructural protein. When the complexes were denatured before immunoprecipita-

The defect in negative-strand RNA synthesis in nsP1 functions early and late, but the defect in nsP3 functions only early.

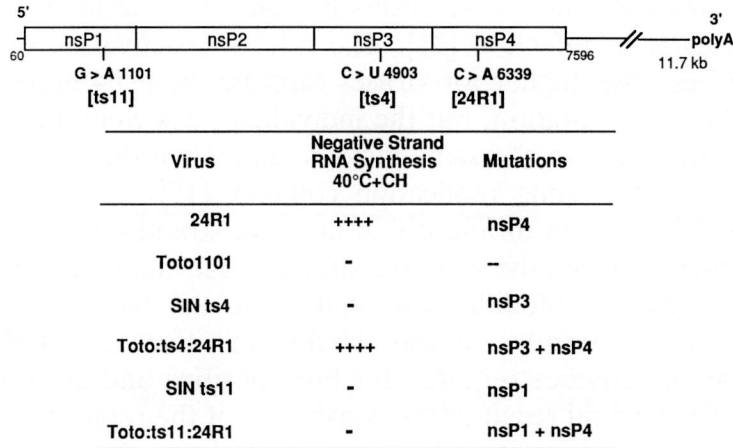

Virus	Negative Strand RNA Synthesis 40°C+CH	Mutations
24R1	++++	nsP4
Toto1101	-	–
SIN ts4	-	nsP3
Toto:ts4:24R1	++++	nsP3 + nsP4
SIN ts11	-	nsP1
Toto:ts11:24R1	-	nsP1 + nsP4

Fig. 4. Reactivation potential of mutant genomes containing the *ts*11 mutation in nsP1 [6] plus the 24R1 mutation [19] in nsP4 (Toto:*ts*11A1:24R1) or the *ts*4 mutation in nsP3 [16] plus the 24R1 mutation in nsP4 (Toto:*ts*4:24R1). Double mutants were constructed by swapping cDNAs of the mutant nonstructural genes into the infectious Toto1101 cDNA [24], from which infectious RNA was transcribed and transfected into CEF cells. The resulting virus stocks were screened for mutant phenotypes as described [19]. Data are from Wang et al. [16]

tion, only the individual nsP was precipitated by specific antibody to it (Fig. 5). When not pre-denatured, each of the antibodies, and especially those specific for nsP3, coprecipitated the four proteins as well as a 120 kDa protein of possible host origin (Fig. 5), consistent with their being present in the same complex. We interpret such findings to mean that significant amounts of nsP3 are associated with replication complexes reactivating negative strand synthesis late in infection. We argue that the reason the nsP3 function is not required late in infection for negative strand synthesis by preformed replication complexes is because this function acts during the formation of replication complexes and is not required once the complex has formed.

As shown in Table 1, we compared the ability of the mutants to complement each others' defects. Complementation measures the efficiency with which the proteins separate after polyprotein processing and then reassociate. This would not occur if the replication complex were formed by proteins processed from the same polyprotein, as the mutations would be cis-acting. Because, with the exception of the Sindbis HR mutants, alphavirus mutants do not show complementation, these replication complexes likely are composed of proteins processed from the same polyprotein. For Sindbis HR mutants, nsP1 and nsP4 were

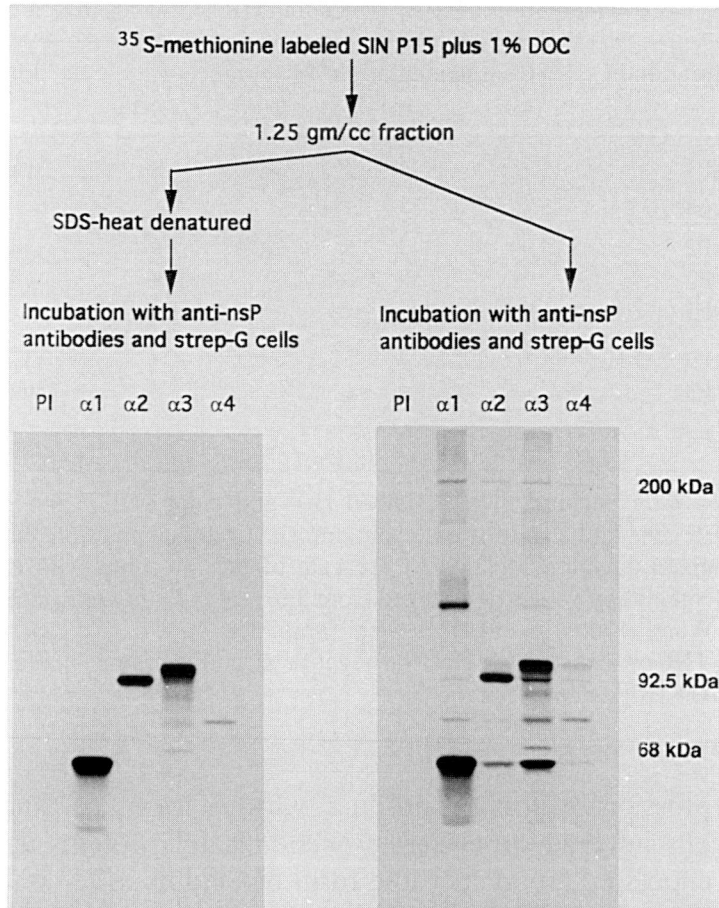

Fig. 5. Immunoprecipitation of Sindbis replication complexes solubilized from P15 membranes. P15 membranes from infected cells pulse-labeled with ^{35}S-methionine 2–4 h p.i. and chased from 4–6 h p.i. were collected and solubilized with 1% sodium deoxycholate to obtain the 1.25 gm/cc replication complexes [15]. Equal amounts of complexes were treated with 1% lithium dodecylsulfate in 1 mM EDTA, heated to 100°C for 3 min and fast cooled (left panel) or were left untreated (right panel) before both were aliquoted and incubated either with preimmune serum (PI) or with mono-specific nsP antisera (the gift of J. H. Strauss [3]), followed by addition of Strep. G cells [15]. Immunoprecipitates were analyzed by electrophoresis on 5 to 10% polyacrylamide gels (modified from Barton et al. [15]. Reproduced by courtesy of ASM

trans-active but nsP2 and nsP3 were not as readily dissociated, even when produced by mutants with no defects in polyprotein cleavage. One interpretation is that nsP2 and nsP3 function initially as a cis-active pair during the formation of a replication complex that then engages in negative strand synthesis. If so, increased amounts or stability of polyprotein nsP23 is predicted to produce increased levels of RNA synthesis. Such a finding was observed in a recent study [28] using

Table 1. Complementation analysis of Sindbis HR RNA-negative mutants

Virus	Mutation in	Complementation[a] by mutants of				Interpretation
		nsP1	nsP2	nsP3	nsP4	
*ts*11	nsP1	−	+	+	+	trans-active
*ts*17	nsP2	+	−	−	+	
*ts*18	nsP2	+	−	−	+	cis-active
*ts*21	nsP2	+	−	−	+	with nsP3
*ts*24	nsP2	+	−	−	+	
*ts*7	nsP3	+	−	−	+	cis-active
*ts*4	nsP3	−[b]	−	−	+	with nsP2
*ts*6	nsP4	+	+	+	−	trans-active

[a] Analyses were performed as described [10] and employed 35 mm petri dishes containing CEF cells and 20 pfu of each mutant/cell. Complementation values are the yield of the mixed infections divided by the sum of the infections with each mutant alone. Complementation values (+) ranged from 3 to ~800. Data are from Sawicki and Sawicki [10], Wang et al. [16] and D. Sawicki (unpubl.)

[b] Sindbis HR *ts*4 exhibits a defect in negative strand synthesis similar to that of nsP1 mutant *ts*11 [16]

vaccinia expression vectors to produce various forms of Sindbis virus proteins. Cells infected with vectors encoding polyproteins nsP123 and nsP34 that expressed an uncleavable form of Sindbis nsP23 polyproteins and individual nsP1, nsP3 and nsP4 proteins yielded increased levels of negative and positive strand RNA, as compared to levels obtained from expression of fully cleavable polyprotein nsP123 plus nsP34.

Model for alphavirus transcription and the regulation of positive and negative strand synthesis

In summary, we suggest the following model (Fig. 6): early in infection, the nsP1234 polyprotein is cleaved at the 3/4 site to form an active nsP4 replicase and at the 1/2 site to release nsP1 and to form the polyprotein nsP23 or a cis-active complex of nsP2 + nsP3. These viral proteins, perhaps in association with host factors, bind a positive strand to form the initial replication complex and synthesize a negative strand. The nsP1 functions in the initiation of transcription, while nsP3 is needed to form the complex or the active replicase. In such a model, only genomes, not negative strands, are capable of initiating the formation of replication complexes. Once the negative strand is synthesized, it becomes the preferred template of the replication complex as a result of modifications to nsP2 and nsP4 that affect promoter recognition and that

Early

nsP1234

(nsP1)(nsP2) (nsP3)(nsP4)   (nsP1)(nsP2) (nsP3)

nsP1   nsP4   nsP1   nsP23

nsP23

nsP123

Negative strand synthesis = formation of RC$_{initial}$ = translation of nsP1234

RC$_{initial}$ = RC(+RNA)

49S (+)

RC$_{initial}$ (+) ⟶ RC$_{stable}$ (-) ⟶ 49S (+) RNA
                                         26S (+) RNA
            conversion   *nsP2

nsP1, nsP4   nsP1, nsP4
nsP23   nsP2 + nsP3
HF(s)   HF(s), p120

49S (-) RF

RC$_{reactivate}$ (+) ⟵ RC$_{stable}$ (-)
       reactivation   *nsP4
                          *nsP2

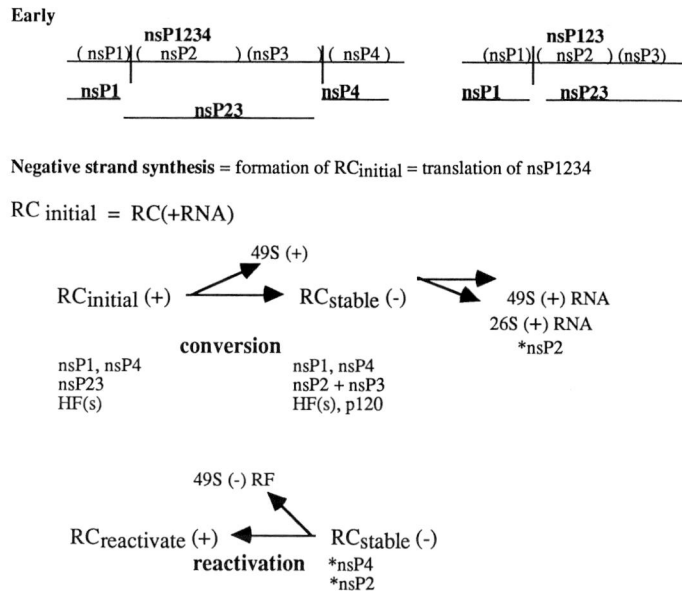

Fig. 6. Model for the regulation of alphavirus positive and negative strand transcription and the role of polyproteins in the formation of viral replication complexes.* To date, Sindbis mutants *ts* for continued synthesis of 26S mRNA have mapped to nsP2 [10–12] and for reactivation of negative strand RNA synthesis have mapped to nsP4 or nsP2 [12, 19]. Data are from Wang et al. [16], Sawicki and Sawicki [12], Sawicki et al. [19], and Strauss and Strauss [3]

convert the replication complex to positive strand synthesis. Reactivation of negative strand synthesis by preformed replication complexes indicates such modifications are reversible and suggests they may involve changes in protein interactions or conformation, possibly from a cleavage of polyprotein nsP23 or the binding or loss of factors. Finally, activities that decrease the half-life of polyprotein nsP23 or prevent its formation or that limit essential host factors lead to a failure to form new replication complexes and a failure to synthesize negative strands late in infection, as is observed in alphavirus-infected cells.

References

1. Goldbach R (1990) Genome similarities between positive-strand RNA viruses from plants and animals. In: Brinton MA, Heinz FX (eds) New aspects of positive-strand RNA viruses. ASM Press, Washington, pp 3–11
2. Koonin EV, Gorbalenya AE, Purdy MA, Rozanov MN, Reyes GR, Bradley DW (1992) Computer-assisted assignment of functional domains in the nonstructural polyprotein of hepatitis E virus: delineation of an additional group of positive-strand RNA plant and animal viruses. Proc Natl Acad Sci USA 89: 8259–8263
3. Strauss JH, Strauss EG (1991) Alphavirus proteinases. Semin Virol 1: 347–356

4. Mi S, Stollar V (1991) Expression of Sindbis virus nsP1 and methyltransferase activity in *E. coli.* Virology 184: 423–427
5. Kääriäinen L, Peränen J (1992) Pers. comm.
6. Wang YF, Sawicki SG, Sawicki DL (1991) Sindbis nsP1 functions in negative strand synthesis. J Virol 65: 985–988
7. Gorbalenya AE, Blinov VM, Dochenko AP, Koonin EV (1989) An NTP-binding motif is the most conserved sequence in a highly diverged monophyletic group of proteins involved in positive strand RNA viral replication. J Mol Evol 28: 256–268
8. Kroner PA, Young BM, Ahlquist P (1990) Analysis of the role of brome mosaic virus 1a protein domains in RNA replication, using linker insertion mutagenesis. J Virol 64: 6110–6120
9. Gorbalenya AE, Koonin EV, Lai MMC (1991) Putative papain-related thiol proteases of positive strand RNA viruses. FEBS Lett 288: 210–205
10. Sawicki DL, Sawicki SG (1985) Functional analysis of the A complementation group mutants of Sindbis HR virus. Virology 144: 20–34
11. Hahn YS, Strauss EG, Strauss JH (1989) Mapping of RNA-negative temperature-sensitive mutants of Sindbis virus: assignment of complementation group A, B and G to nonstructural proteins. J Virol 63: 3142–3150
12. Sawicki DL, Sawicki SG (1992) A second nonstructural protein functions in the regulation of alphavirus negative strand RNA synthesis. J Virol 67: 3605–3610
13. Peränen J, Rikkonen M, Liljeström P, Kääriäinen L (1990) Nuclear localization of Semliki Forest virus-specific nonstructural protein nsP2. J Virol 64: 1888–1896
14. Li G, LaStarza M, Hardy WR, Strauss JH, Rice CM (1990) Phosphorylation of Sindbis virus nsP3 in vivo and in vitro. Virology 179: 416–427
15. Barton DJ, Sawicki SG, Sawicki DL (1991) Solubilization and immunoprecipitation of alphavirus replication complexes. J Virol 65: 1496–1506
16. Wang YF, Sawicki SG, Sawicki DL (1993) Initiation of alphavirus negative strand synthesis requires two different nonstructural proteins, nsP1 and nsP3. J Virol (submitted)
17. Hodgman TC (1988) A new superfamily of replicative proteins. Nature 332: 22–23
18. Barton DJ, Sawicki SG, Sawicki DL (1988) Demonstration in vitro of temperature-sensitive elongation of RNA in Sindbis virus mutant ts6. J Virol 62: 3597–3602
19. Sawicki DL, Barkhimer DB, Sawicki SG, Rice CM, Schlesinger S (1990) Temperature sensitive shutoff of alphavirus minus strand RNA synthesis maps to a nonstructural protein, nsP4. Virology 174: 43–52
20. Lemm JA, Durbin RK, Stollar V, Rice CM (1990) Mutations which alter the level or structure of nsP4 can affect the efficiency of Sindbis virus replication in a host-dependent manner. J Virol 64: 3001–3011
21. Sawicki DL, Sawicki SG (1987) Alphavirus plus and minus strand RNA synthesis. In: Brinton M, Ruckert R (eds) Positive-strand RNA viruses. Alan R. Liss, New York, pp 251–259
22. Sawicki DL, Kääriäinen L, Lambek C, Gomatos PG (1978) Mechanism for control of synthesis of Semliki Forest virus 26S and 42S RNA. J Virol 25: 19–27
23. Sawicki SG, Sawicki DL (1986) The effect of overproduction of nonstructural proteins on alphavirus plus-strand and minus-strand RNA synthesis. Virology 152: 507–512
24. Rice CM, Levis R, Strauss JH, Huang HV (1987) Production of infectious RNA transcripts from Sindbis virus cDNA clones: mapping of lethal mutations, rescue of a temperature-sensitive marker, and in vitro mutagenesis to generate defined mutants. J Virol 61: 3809–3819

25. Ishikawa M, Mesi T, Ohno T, Okada Y (1991) Specific cessation of minus strand RNA accumulation at an early stage of Tobacco mosaic virus infection. J Virol 65: 861–868
26. Sawicki SG, Sawicki DL (1986) The effect of loss of regulation of minus strand RNA synthesis on Sindbis virus replication. Virology 151: 339–349
27. deGroot RJ, Hardy WR, Shirako Y, Strauss JH (1990) Cleavage-site preferences of Sindbis virus polyproteins containing the nonstructural proteinase. Evidence for temporal regulation of polyprotein processing in vivo. EMBO J 9: 2631–2638
28. Lemm J, Rice CM (1993) Roles of nonstructural polyproteins and cleavage products in regulating Sindbis virus RNA replication and transcription. J Virol 67: 1916–1926

Authors' address: Dr. D. L. Sawicki, Department of Microbiology, Medical College of Ohio, P.O. Box 10008, Toledo, OH 43699-0008, U.S.A.

Arch Virol (1994) [Suppl] 9: 407–416

Archives
of
Virology
© Springer-Verlag 1994
Printed in Austria

Nodavirus RNA replication: mechanism and harnessing to vaccinia virus recombinants

L. A. Ball, B. Wohlrab, and **Y. Li**

Department of Microbiology, University of Alabama at Birmingham,
Birmingham, Alabama, U.S.A.

Summary. In order to harness RNA replication for the amplification of mRNAs expressed from recombinant vectors and vaccines, we constructed a VV recombinant that expressed the RNA replicase encoded in the larger genomic segment of the nodavirus FHV. When both termini of the VV-derived transcript were correct, the encoded enzyme replicated its own mRNA, and replication dominated the RNA synthetic capacity of the cell. The smaller genomic segment of FHV could also be replicated by the enzyme when supplied *in trans*, either by coinfection with another VV recombinant or by transfection of an appropriate plasmid. However, two requirements had to be fulfilled for replication of the smaller FHV RNA segment. The first was the prior replication of the larger genomic segment, which was interpreted as a mechanism to achieve sufficient replicase synthesis before the onset of coat protein synthesis. The second was the presence in the smaller genomic RNA of an internal region between about nucleotides 525–620. Work is in progress to elucidate the reasons for these requirements for RNA 2 replication.

Introduction

RNA replication provides a powerful tool to amplify mRNA and may be usefully harnessed to drive mRNA expression from DNA-based recombinant vectors and vaccines. Furthermore, several important questions concerning the molecular mechanisms of RNA replication can be addressed with DNA-based expression systems by genetic manipulation of cDNAs that encode RNA replicases and their corresponding templates. With these considerations in mind, we undertook the expression of a functional RNA replicase and competent templates for RNA replication from vaccinia virus (VV) recombinants [1]. The RNA replicase we chose for this work was the enzyme encoded by the larger genomic RNA segment of the nodavirus flock house virus (FHV). Nodaviruses are small, non-enveloped, isometric, riboviruses with bipartite, positive-sense, RNA genomes. Their larger genomic RNA segment (RNA 1) encodes an RNA replicase that replicates both its own mRNA and the smaller genomic RNA segment (RNA 2), which encodes a precursor to the viral coat proteins. Although the subunit composition of the RNA-dependent RNA polymerase is unknown, it is clear that RNA 1 encodes the entire viral contribution to this enzyme, since RNA 1 transfected alone into cells replicates autonomously [10]. Moreover, the large protein that it encodes (protein A) [6] contains a clear polymerase motif.

The 5′ ends of the FHV genomic RNAs have cap O structures, whereas the 3′ ends, which are not polyadenylylated, are blocked by an entity which has yet to be defined [4, 6]. During RNA replication, a third viral RNA (RNA 3) is synthesized in infected cells, but it is not subsequently encapsidated in virus particles. RNA 3 represents the 3′ terminal 387 nucleotides of RNA 1, and is synthesized by the RNA replicase, probably by a mechanism that involves internal initiation on the negative strand of RNA 1. RNA 3 directs the synthesis of protein B, a 10 kd non-structural viral protein of unknown function. Sequence determination of RNA 1 has established that the sequences encoding proteins A and B overlap in different reading frames [6] (for reviews see [11] and the article by Dasgupta et al. in this volume). The RNAs whose replication we have studied to date are the two natural templates of the FHV replicase, the FHV genomic RNAs 1 and 2, and internal deletions of these molecules.

The basic concept, illustrated in Fig. 1, is to construct a VV recombinant which uses a VV promoter to express an RNA that can serve both as a message for an RNA-dependent RNA polymerase and as a template for replication by the same enzyme. Because of the template specificity of the RNA replicase, mRNA amplification should be selective and restricted to those RNAs that can serve as templates for the enzyme. Use of this approach for the amplification and expression of heterologous mRNAs would therefore depend on providing them with the cis-acting signals necessary for replication; experiments designed to identify these signals are presented below. Although this idea was developed using VV as the DNA vector, it may also be applicable to other DNA-based expression vectors.

An attractive aspect of this expression system in the context of recombinant vaccines is that the VV recombinant need only carry out early transcription to yield message/template RNAs for the replicase and for the heterologous target sequence (not shown in Fig. 1). Thereafter, amplification and expression of the RNAs should be independent of VV replication. Since the requirements for attenuation to achieve safety in a recombinant vaccine virus can conflict with the requirements for high-

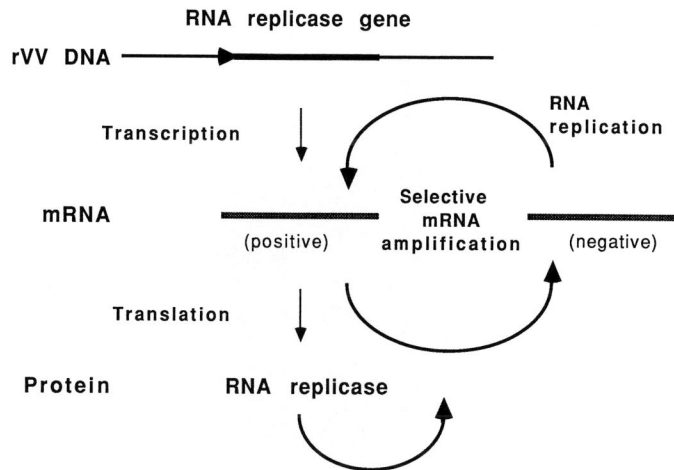

Fig. 1. Schematic representation of the harnessing of RNA replication to a DNA-based vector such as vaccinia virus

level antigen expression to achieve vaccine efficacy, uncoupling these features by the approach shown in Fig. 1 may prove advantageous.

Construction of VV recombinants

We constructed VV recombinants that contained full-length cDNAs of FHV RNAs 1 and 2 [7], and which expressed, from a VV promoter, RNA transcripts with no additional nucleotides at either terminus. At the 5' end, this was achieved by positioning the VV 7.5k promoter so that its major initiation site during early transcription [8] corresponded to the first nucleotide in the FHV sequences [4, 6]. At the 3' end we inserted a cDNA copy of the antigenomic ribozyme of hepatitis delta virus (HDV), positioned so that the site of autolytic cleavage of the corresponding RNA transcript immediately followed the 3' nucleotide of the FHV sequences [15]. These DNA sequence arrangements were recombined into VV at the thymidine kinase locus and recombinants were isolated and purified by conventional methods [3].

The FHV 1 VV recombinant initiated autonomous RNA replication

Baby hamster kidney (BHK 21) cells were infected with the VV recombinant that contained the cDNA to FHV RNA 1, and total cytoplasmic RNAs were examined by metabolic labeling with [³H] uridine, followed by RNA extraction and gel electrophoresis. By far the major products of RNA synthesis in the infected cells were FHV RNAs 1 and 3 (Fig. 2, lane 3). These replication products dominated the synthesis of other RNAs in the infected cells, including tRNAs, ribosomal RNAs, and VV

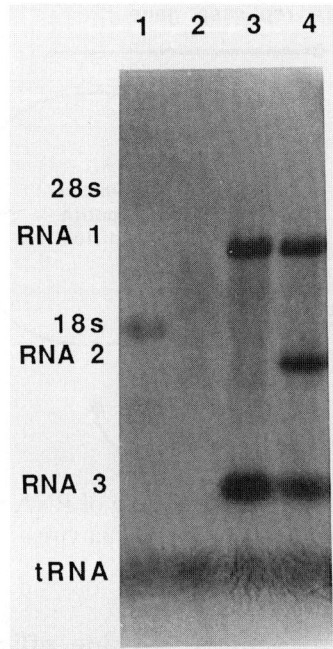

Fig. 2. Analysis of total cytoplasmic RNAs from cells infected with VV recombinants. BHK 21 cells were labeled by incorporation of [^{3}H] uridine for 2 h at 24 h post-infection at 28°. Labeled RNAs were purified from cytoplasmic extracts by extraction with phenol-chloroform and precipitation with ethanol. Samples derived from 5×10^4 cells were resolved by electrophoresis on an agarose-formaldehyde gel [13] and visualized by fluorography. *1* RNAs from uninfected cells; *2* RNAs from cells infected with a VV recombinant that expressed FHV replicase from an mRNA with long 5′ and 3′ terminal extensions; *3* RNAs from cells infected with a VV recombinant that expressed FHV replicase from an mRNA with no terminal extensions; *4* RNAs from cells infected as for *3*, and transfected with a circular DNA plasmid that contained FHV 2 cDNA between inverted tandem copies of the HDV antigenomic ribozyme

late mRNAs. Cells infected with a VV recombinant in which the FHV cDNA transcript had long 5′ and 3′ extensions supported no detectable RNA replication (Fig. 3, lane 2), although synthesis of protein A could be clearly seen (data not shown). RNA replication was resistant to actinomycin D and 10–50% as active as replication initiated by authentic FHV RNA 1.

The FHV RNA replicase could replicate transcripts *in trans*

FHV RNA 2 transcripts were delivered in two ways: by coinfection of cells with a second VV recombinant that expressed a perfect FHV RNA 2 transcript (constructed as described for the FHV 1 recombinant), or by transfection of cells with a DNA plasmid that contained the FHV 2 cDNA sequence sandwiched between inverted tandem cDNA copies of

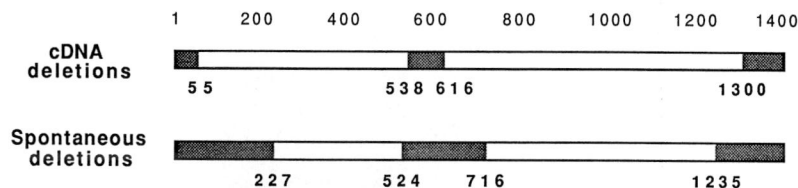

Fig. 3. Schematic linear representations of the regions of FHV RNA 2 required for replication. The upper line summarizes the results from analysis of constructed cDNA deletions; the lower line summarizes the results from analysis of spontaneous deletions of RNA 2. Open areas can be deleted

the HDV ribozyme. Since the ribozyme functions only in positive-sense RNA, the copy at the 3′ end cleaved only the positive strand, whereas the copy at the 5′ end cleaved only the negative strand. It was unnecessary to deliberately include a promoter site in plasmids of this design, since the VV DNA-dependent RNA polymerase was sufficiently promiscuous to initiate transcription within the vector sequences, and transcription from anywhere, in either direction, ultimately yielded a perfectly-terminated, replicable, RNA molecule. Replication of an FHV RNA 2 transcript derived in this way is shown in Fig. 2, lane 4.

Cis-acting sequences required for replication of FHV RNA 2

To identify replaceable regions of FHV RNA 2, it was necessary to determine what parts of the molecule were essential for replication. This question was approached in two ways: by constructing deletions of FHV 2 cDNA and determining whether the corresponding deleted RNAs could be replicated [2], and by analyzing a series of spontaneous RNA 2 deletions that arose during repeated replication at high RNA concentration [14]. Both approaches showed that in addition to the RNA termini, an internal region of 100–200 nucleotides starting at about position 525 was important for replication of FHV RNA 2 (Fig. 3). This region was conserved in each of more than 30 different spontaneous deletions of RNA 2, despite large deletions on either side. Moreover, its removal from the FHV 2 cDNA yielded RNA transcripts that failed to replicate. The role of this internal region is unknown at present.

FHV RNA 1 must replicate before FHV RNA 2

Since the termini of FHV RNAs 1 and 2 were found to be critical for the ability of these molecules to replicate [2], it was possible to construct a plasmid that expressed an FHV RNA 1 transcript that could function as an mRNA for the replicase subunit, protein A, but was unable to replicate. For these experiments, plasmids that contained a promoter site

L. A. Ball et al.

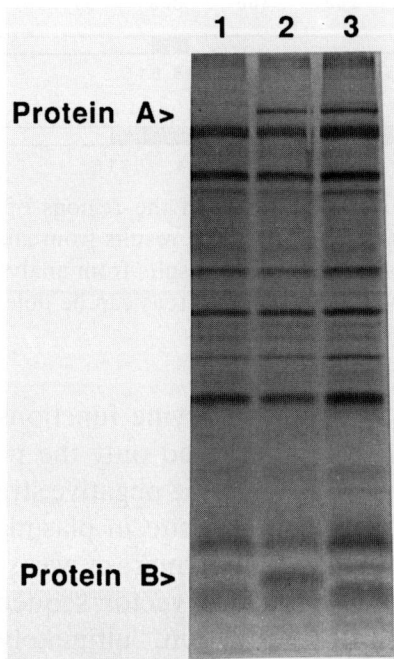

Fig. 4. Analysis by SDS-polyacrylamide gel electrophoresis [12] of the proteins labeled with [^{35}S] methionine in VV-T7 infected BHK 21 cells that were transfected with plasmids expressing mRNAs for FHV replicase. *1* Proteins from infected, untransfected cells; *2* proteins from infected cells that were transfected with plasmid FHV 1 (1,0); *3* proteins from infected cells that were transfected with plasmid FHV 1 (26, 5). Labeled proteins were visualized by autoradiography of the dried gel

recognized by the bacteriophage T7 DNA-dependent RNA polymerase were used, and their behavior was examined after transfection into cells infected with the VV-T7 polymerase recombinant constructed by Fuerst et al. [9]. These T7 transcription plasmids were designated FHV 1 (or 2) (N, M), where N and M represent the lengths of the 5′ and 3′ extensions, respectively, at the ends of the authentic FHV 1 (or 2) sequence as transcribed by T7 RNA polymerase. One example of such a plasmid, FHV 1 (26, 5), directed synthesis of an FHV RNA 1 molecule that had 26 extra nucleotides at the 5′ end, but lacked five nucleotides from the 3′ end. When transfected into cells that were infected with the VV-T7 polymerase recombinant, this plasmid made functional mRNA for FHV protein A (Fig. 4, lane 3). However, no RNA replication occurred, due to the absence of a competent template (Fig. 5, lane 1), and no protein B was made (Fig. 4, lane 3). This contrasted with the behavior of plasmid FHV 1 (1,0), which directed the synthesis of an FHV RNA 1 molecule with one extra nucleotide at its 5′ end and a

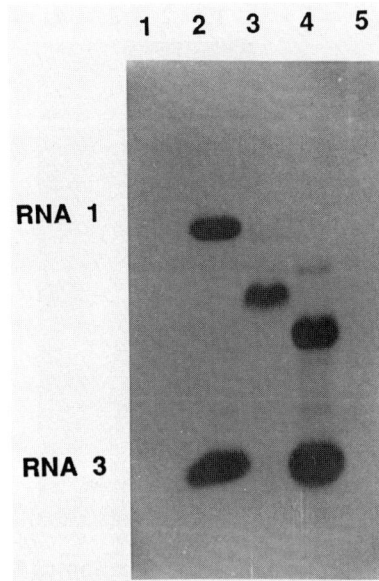

Fig. 5. Analysis of the products of RNA replication by electrophoresis on an agarose-formaldehyde gel. BHK 21 cells were infected with VV-T7 and transfected with plasmids as indicated. The products of RNA replication were labeled by 2 h incorporation of [^{3}H] uridine in the presence of actinomycin D (10 μg per ml) at 24 h post-infection at 28°, and then processed as described for Fig. 2. Cells were transfected with: *1* FHV 1 (26, 5) DNA only; *2* as *1*, plus FHV 1 (1,0)687R688 DNA; *3* As *1*, plus FHV 1 (1,0)746Δ1830 DNA; *4* As *1*, plus FHV 1 (1,0)313Δ941, 1261Δ2283 DNA; *5* As *1*, plus FHV 2 (0,0) DNA. Labeled RNAs were visualized by fluorography

perfect 3' end. Cells transfected with this plasmid made both proteins A and B (Fig. 4, lane 2) and carried out abundant replicative synthesis of RNAs 1 and 3 (data not shown).

The replicase enzyme expressed from plasmid FHV 1 (26, 5), although unable to replicate its own mRNA, could replicate FHV RNA 1 molecules with near-perfect termini that were supplied by transcription from another plasmid. To prevent the latter RNA molecules from contributing to the pool of functional protein A, a codon for arginine was inserted between the nucleotides that encoded amino acid residues 687 and 688, yielding plasmid FHV 1 (1,0)687R688. This amino acid position is in the heart of the polymerase motif, and the insertion rendered protein A catalytically inactive (data not shown). Thus FHV 1 (1,0)687R688 provided RNA that was inactive as mRNA for functional protein A, but active as a template for replication, whereas FHV 1 (26, 5) provided RNA that was active as mRNA for functional protein A but incompetent for replication. Cotransfection of these plasmids showed that the protein

414   L. A. Ball et al.

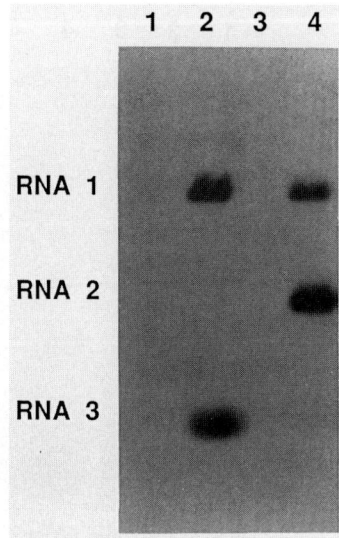

Fig. 6. Facilitation of RNA 2 replication by replication of RNA1. Cells were infected and labeled as described for Fig. 5. Shown are the products of RNA replication in cells transfected with: *1* FHV 1 (26, 5) DNA only; *2* As *1*, plus FHV 1 (1,0)687R688 DNA; *3* As *1*, plus FHV 2 (0,0) DNA; *4* As *1*, plus FHV 1 (1,0)687R688 DNA, plus FHV 2 (0,0) DNA. RNAs labeled in the presence of actinomycin D were resolved by electrophoresis on an agarose-formaldehyde gel and visualized by fluorography

A made by RNA from one plasmid could replicate the RNA expressed from the other; i.e. that the enzyme was active *in trans* on RNA 1 molecules other than those from which it was translated (Fig. 5, lane 2). Deleted versions of FHV RNA 1 could also be replicated by the enzyme expressed from FHV 1 (26, 5). A cDNA deletion that lacked nucleotides 747–1829 of the FHV 1 sequence yielded a replicable RNA molecule, although little, if any, RNA 3 was made during replication (Fig. 5, lane 3). Similarly, a transcript of a spontaneous double deletion of FHV RNA 1 that lacked nucleotides 314–940 and 1261–2283 [5] was replicated efficiently, although in this case RNA 3 was synthesized in abundance (Fig. 5, lane 4).

Unexpectedly, FHV RNA 2 could not be replicated by the enzyme expressed from FHV 1 (26, 5) (Fig. 5, lane 5), although it could be replicated efficiently during self-replication of FHV 1 (1,0) transcripts (data not shown). Indeed, it appeared that RNA 1 replication was necessary before RNA 2 replication could occur, since providing the replicase with a competent RNA 1 template sufficed to facilitate the replication of RNA 2 (Fig. 6, lane 4). These results indicated that some event or product of RNA 1 replication (other than protein A) is essential for the replication of RNA 2. This effect may be understood in the context of the genome replication strategy of the virus, as it may help to ensure sufficient replication of RNA 1 and the concomitant accumulation of RNA replicase activity before the onset of replication of RNA 2, which is the mRNA for the structural proteins of the virus. However, the mechanistic basis of the requirement for RNA 1 replication before the onset of RNA 2 replication is unknown.

Acknowledgements

We thank Drs. R. Dasgupta and P. Kaesberg for providing the cDNA clones of FHV RNAs 1 and 2; J. Ball for help with one of the constructions; and T. Harper, S.-X. Wu, and R. Rueckert for helpful discussions. This work was supported by Public Health Service grant R37 AI18270 from the National Institute of Allergy and Infectious Diseases and by the WHO/UNDP Programme for Vaccine Development.

References

1. Ball LA (1992) Cellular expression of a functional nodavirus RNA replicon from vaccinia virus vectors. J Virol 66: 2335–2345
2. Ball LA, Li Y (1993) Cis-acting requirements for the replication of flock house virus RNA 2. J Virol 67: 3544–3551
3. Ball LA, Young KKY, Anderson K, Collins PL, Wertz GW (1986) Expression of the major glycoprotein G of human respiratory syncytial virus from recombinant vaccinia virus vectors. Proc Natl Acad Sci USA 83: 246–250
4. Dasgupta R, Ghosh A, Dasmahapatra B, Guarino LA, Kaesberg P (1984) Primary and secondary structure of black beetle virus RNA 2, the genomic messenger for BBV coat protein precursor. Nucleic Acids Res 12: 7215–7223
5. Dasgupta R (1993) Personal communication
6. Dasmahapatra B, Dasgupta R, Ghosh A, Kaesberg P (1985) Structure of the black beetle virus genome and its functional implications. J Mol Biol 182: 183–189
7. Dasmahapatra B, Dasgupta R, Saunders K, Selling B, Gallagher T, Kaesberg P (1986) Infectious RNA derived by transcription from cloned cDNA copies of the genomic RNA of an insect virus. Proc Natl Acad Sci USA 83: 63–66
8. Davison AJ, Moss B (1989) Structure of vaccinia virus early promoters. J Mol Biol 210: 749–769
9. Fuerst TR, Niles EG, Studier FW, Moss B (1986) Eukaryotic transient-expression system based on recombinant vaccinia virus that synthesizes bacteriophage T7 RNA polymerase. Proc Natl Acad Sci USA 83: 8122–8126
10. Gallagher TM, Friesen PD, Rueckert RR (1983) Autonomous replication and expression of RNA 1 from black beetle virus. J Virol 46: 481–489
11. Hendry DA (1991) Nodaviridae of invertebrates. In: Kurstak E (ed) Viruses of invertebrates. Marcel Dekker, New York, pp 227–276
12. Laemmli UK (1970) Cleavage of structural proteins during the assembly of the head of bacteriophage T4. Nature 227: 680–685
13. Lehrach H, Diamond D, Wozney JM, Boedtker H (1977) RNA molecular weight determination by gel electrophoresis under denaturing conditions, a critical reexamination. Biochemistry 16: 4743–4751
14. Li Y, Ball LA (1993) Non-homologous RNA recombination during negative strand synthesis of flock house virus RNA. J Virol 67: 3854–3860

15. Perrotta AT, Been MD (1991) A pseudoknot-like structure required for efficient self-cleavage of hepatitis delta virus RNA. Nature 350: 434–436

Authors' address: Dr. L. A. Ball, BHSB 70, UAB, Birmingham, AL 35294-0005, U.S.A.

Arch Virol (1994) [Suppl] 9: 417–427

Archives
Virology
© Springer-Verlag 1994
Printed in Austria

Molecular characterization of Borna virus RNAs

J. M. Pyper, L. Brown, and **J. E. Clements**

Division of Comparative Medicine, The Johns Hopkins University School of Medicine, Baltimore, Maryland, U.S.A.

Summary. Borna disease virus is cell-associated in infected animals. Antibodies in animals are directed against BDV proteins of 38/39, 24, and 14.5 kD. cDNA clones that encode these proteins hybridize to five mRNAs of 10.5, 3.6, 2.1, 1.4, and 0.85 kb. The 10.5, 3.6, 2.1, and 0.85 kb RNAs are 3′ co-terminal; the 1.4 kb RNA is contained within the 10.5, 3.6, and 2.1 kb species but is not 3′ co-terminal. A negative strand 10 kb RNA is also present in infected cells. To determine which of the large 10 kb species represents the genomic RNA, strand-specific probes were used for Northern analyses of RNA from infectious particles isolated by Freon extraction of BDV-infected rat brain. RNA purified from these particles contained both positive and negative sense 10 kb species. Treatment of particles with RNaseA before isolation of RNA resulted in detection of only negative strand species, suggesting that BDV is a negative strand RNA virus. However, the genomic organization of BDV is unlike any known negative strand RNA virus.

Introduction

Borna disease is a rare but severe neurological disease that infects horses and sheep in parts of Germany and Switzerland. It was first described nearly a century ago in the town of Borna, Saxony [1]. In 1927 Zwick et al. [2] determined that a virus caused the disease, but despite years of work, the viral agent has not yet been classified. Although Borna disease is rare, the immunopathological nature of the disease and the wide range of vertebrates that can be infected (avians to primates) make it an interesting disease model. Further, the ability of the virus to replicate in neuronal cells in both the central and peripheral nervous systems make it an attractive model for central nervous system (CNS) disease. In natural infections Borna disease has only been identified in horses and sheep, but similar behavioral diseases in humans have been linked to the agent by results of serologic studies [3, 4].

Borna disease

Infected animals exhibit severe neurological symptoms, such as ataxia, excitation, abnormal posture, lethargy, and blindness. Examination of the brains of infected animals shows encephalomyelitis, with the most severe lesions located in the cerebral cortex and the hippocampus. Infectivity can be transmitted to uninfected animals by intracranial inoculation of brain homogenate from an infected animal. Thus far it has been impossible to isolate cell-free infectious virus from infected animals (or from cells infected in culture); all infectivity is cell-associated. This has hindered efforts to classify the virus.

The dependence of the virus on infection of neurons has been shown by experiments in which virus is inoculated into the footpad of a rat. CNS disease results after transport through nerve processes. However, viral transport to the CNS can be blocked if the sciatic nerve is transected prior to or soon after inoculation [5]. Transmission of infection has also been demonstrated by inoculating the nasal epithelium [6, 7]. In contrast, intravenous inoculation with the same inoculum usually fails to cause disease [5]. These results have demonstrated that the most efficient route of infection in the host is via the nerves.

Immunopathology of Borna disease

The outcome of infection with the etiologic agent of Borna disease, herein called Borna disease virus (BDV) for convenience, depends on the age of the animal at the time of inoculation. Adult rats inoculated with BDV develop disease characterized by an initial phase of frenzied aggressive behavior and followed by a chronic passive phase. There is a high titer of virus in the brains of these acutely infected animals. In contrast, neonatal rats do not develop clinical signs of disease even though there are high titers of BDV in their brains. This persistent tolerant infection (PTI) is accompanied by subtle changes in behavior [8]. The age-dependent difference in response to BDV has been shown to be due to the state of the host's immune system [9, 10]. Neither adult animals immunosuppressed with cyclosporine A [11] nor athymic adult rats [12] develop the clinical disease seen in immunocompetent adults. Characteristic lesions in the brains of animals with Borna disease are immunopathological in nature and are caused by CD4+ T cells [13]. Immunocompetent animals respond to infection by producing antibodies specific to BDV, but these antibodies are not protective and do not neutralize infectivity either in vivo or in vitro. However, these antibodies have been useful both diagnostically and for characterization of the prominent BDV proteins produced during infection.

BDV is similar to lymphocytic choriomeningitis (LCM) virus in that both viruses can establish a persistent tolerant infection (PTI) in neonatal animals. However, BDV persists only in neurons and astrocytes [14] whereas the LCM virus persists in many cell types (neurons, meningeal and choroid plexus cells, lymphocytes, macrophages, hepatocytes, etc.) [15]. Additionally, LCM-PTI mice can be cured of their infection by adoptive immunization with activated, immunologically specific CTL from adult mice [16] whereas BDV-PTI rats have not been cured using a similar strategy.

Borna disease virus

Culture of BDV

Although cell-free virus cannot be isolated from infected animals, a homogenate from infected brain is infectious for other animals and for some cells in culture. Cultured fetal rabbit glial cells (FRGs) can be directly infected by these homogenates and FRGs have been used to titrate virus in brain and other organs. However, these cells do not produce significant quantities of virus because of their limited growth capacity in culture. The MDCK cell line cannot be directly infected using brain homogenates, but a persistently infected MDCK line can be established after cocultivation with infected FRG cells. Clonal cell lines can then be isolated which are uniformly infected. No cell-free virus is produced in either the BDV-infected FRGs or in persistently infected MDCKs.

Physical characteristics of BDV

Homogenates from either PTI rat brains or persistently infected cell lines have been used to assess physical characteristics of BDV. From early filtration experiments the size of the particle has been estimated to be approximately 85–125 nm [17, 18]. Density gradient centrifugation concentrated infectivity at ~1.2 g/ml (ranging from 1.18 to 1.22 g/ml) [19, 20].

BDV-specific proteins have been characterized using immune sera. From both Western analyses and immunoprecipitations of labeled proteins from infected MDCK cells, prominent BDV-specific proteins of 38/39, 24, and 14.5 kD can be detected [21–23]. The large amounts of the 38/39 and 24 kD proteins made it possible to purify enough material to obtain limited amino acid sequence from proteolytic fragments.

Molecular characterization of BDV

Several laboratories used molecular techniques to extend the characterization of BDV. Because it was not known whether BDV was an RNA or DNA virus, initial approaches utilized cDNA cloning of mRNA from infected cells using subtractive libraries. In our laboratory, an expression

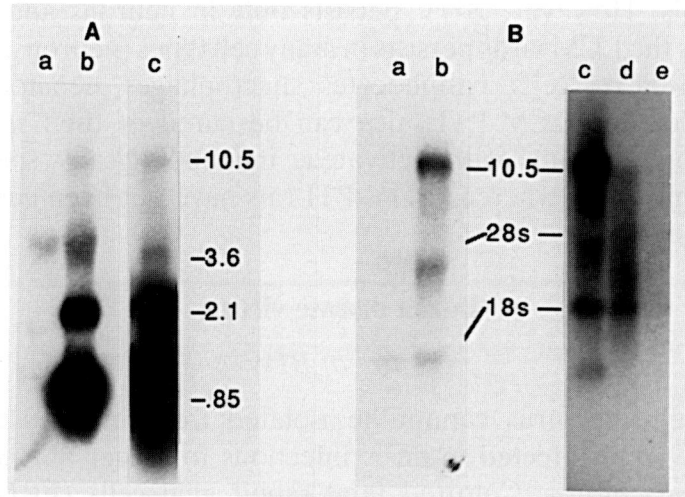

Fig. 1. Northern analysis determination of polarity of BDV transcripts. **A** Detection of mRNAs using a BDV anti-sense riboprobe generated from clone B8. *a* 7.5 μg RNA from uninfected rat brain; *b* 7.5 μg RNA from BDV-infected rat brain; *c* 2.0 μg poly A+ RNA from BDV-infected rat brain. **B** Detection of negative-sense transcripts using a BDV sense riboprobe generated from clone B8. *a* 7.5 μg RNA from uninfected rat brain; *b* 7.5 μg RNA from BDV-infected rat brain; *c* 5 μg of total RNA from BDV-infected rat brain; *d* 3 μg of RNA from oligo(dT) column wash; *e* 3 μg of polyA+ RNA

library was screened using monoclonal antibodies to the 38/39 kD proteins [24, 25]. Other laboratories screened their libraries using subtractive probes [26, 27] or oligonucleotide probes based on the amino acid sequence of the proteolytic fragments of purified proteins [23]. The three techniques led to the isolation of cDNA clones encoding the 24 kD protein.

Analysis of RNA species using probes specific for the 24 kD ORF

The initial molecular clone characterized was a 700 bp clone (B8) isolated from the MDCK subtractive library [24, 25]. Radiolabeled probes made from this clone hybridized only to RNA from infected cells and not to DNA. This demonstrated that BDV was an RNA virus; similar results were reported by Lipkin et al. [27]. Using strand-specific probes and sequence information (see below), it was shown that there were four polyadenylated transcripts of positive orientation (10.5, 3.6, 2.1, and 0.85 kb) and three of negative orientation (10.0, 3.5, and 1.7 kb) (Fig. 1). The data suggested that the mRNAs could be described as a 3' coterminal nested set similar to the organization seen in the coronavirus superfamily [25].

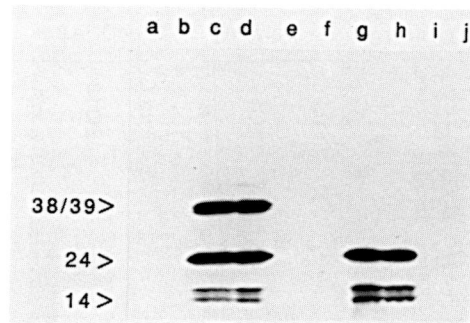

Fig. 2. Identification of BDV-specific proteins encoded by the B8 clone. Proteins were in vitro translated either from polyA+ RNA isolated from rat brain or from in vitro transcribed RNA and then immunoprecipitated with anti-BDV sera from rat or rabbit. In vitro translations contained: polyA+ RNA from uninfected rat brain (*a* and *b*); polyA+ RNA from BDV-infected rat brain (*c* and *d*); in vitro transcribed anti-sense RNA from clone B8 (*e* and *f*); in vitro transcribed sense RNA from clone B8 (*g* and *h*); no RNA (*i* and *j*). Proteins were immunoprecipitated using polyclonal anti-BDV sera from rat (*a, c, e, g,* and *i*) or rabbit (*b, d, f, h,* and *j*)

Analysis of the nucleotide sequence of the B8 clone predicted that it would encode two proteins [24, 25, 28]. The predicted amino acid sequence contained peptide sequences previously determined for the p24 proteolytic fragments. In vitro transcription and translation of B8 followed by immunoprecipitation with rat sera from an infected animal showed that 24, 14.5, and 13 kD proteins were encoded by B8 (Fig. 2).

cDNAs encoding the 38 kD protein

In this laboratory a second subtractive library was also generated using BDV-infected rat brain as the source of mRNA for cDNA cloning [28]. This library was screened using oligonucleotides derived from the 5′ end of B8 as probes. One clone, E20/1, contained most of the B8 sequence but also extended 5′ of B8 by 140 nt. Sequence analysis of this clone showed that the additional 140 nt contained a sequence corresponding to the amino acid sequence from a proteolytic fragment of the 38 kD protein.

Clones encoding sequences 5′ of E20/1 were obtained by applying the RACE technique [29] to gel-purified primer extension products [28]. A cDNA clone, pcDNA2.1, encodes the entire 38 kD protein. The predicted amino acid sequence of this clone contained eight peptide sequences previously identified from proteolytic fragments of the 38 kD protein, including those previously identified in E20/1. This was confirmed using in vitro transcription and translation, followed by immu-

J. M. Pyper et al.

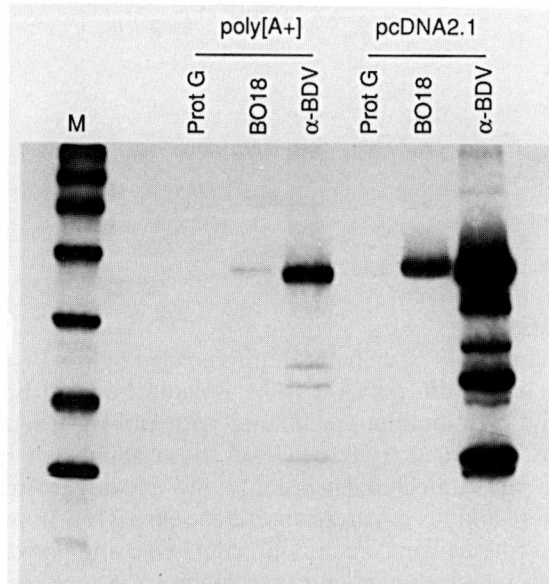

Fig. 3. Immunoprecipitation of the in vitro translated 38 kD BDV-specific protein. Proteins were in vitro translated using poly A+ RNA from rat brain (*poly[A⁺]*) or RNA in vitro transcribed from pcDNA2.1 in the sense orientation (*pcDNA2.1*). Proteins were immunoprecipitated with Protein G Sepharose only (*Prot G*); BO18 (monoclonal antibody to the 38 kD protein) and Protein G Sepharose (*BO18*); or polyclonal anti-BDV rabbit sera and Protein G Sepharose (*α-BDV*)

noprecipitation using a monoclonal antibody specific for the 38 kD protein (Fig. 3).

Comparison of mRNAs detected using probes specific for the 24 kD or 38 kD ORFs

Earlier analyses of RNA species used oligonucleotide probes derived from clone B8, which encodes the 14 and 24 kD proteins. We extended this analysis using oligonucleotide probes specific for the 38 kD open reading frame (ORF). Using a Northern transfer of a 30 cm formaldehyde-agarose gel to improve resolution of RNA species, we were able to detect an additional mRNA species of 1.4 kb, which was specifically hybridized by probes derived from the 38 kD ORF. Interestingly, these probes failed to hybridize to the 0.85 kb mRNA (Fig. 4A). These results suggest that the organization of the BDV mRNAs is more complex than originally postulated and that they cannot be accurately and simply described as a nested set of 3' coterminal transcripts. One possible arrangement of mRNAs is shown schematically in Fig. 4B. It remains to be seen whether additional mRNAs will be detected when probes specific to other protein coding regions are obtained.

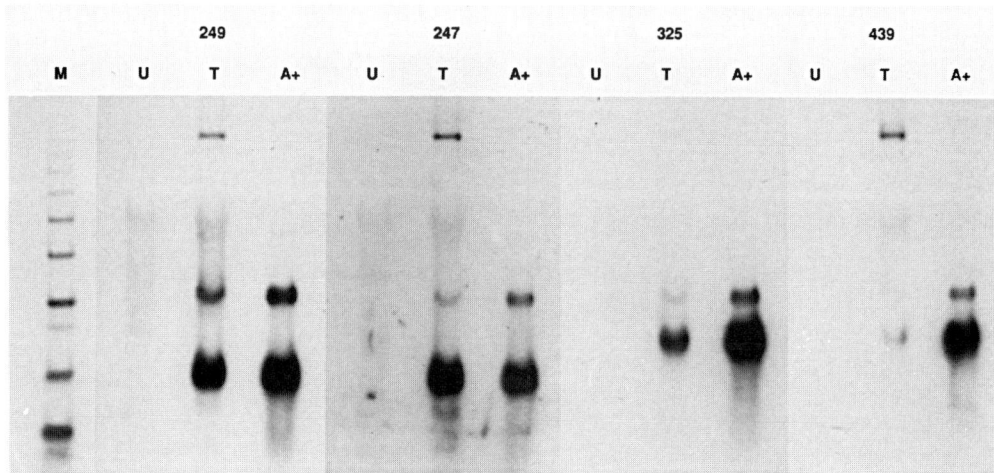

mRNA Transcripts for the BDV Structural Genes

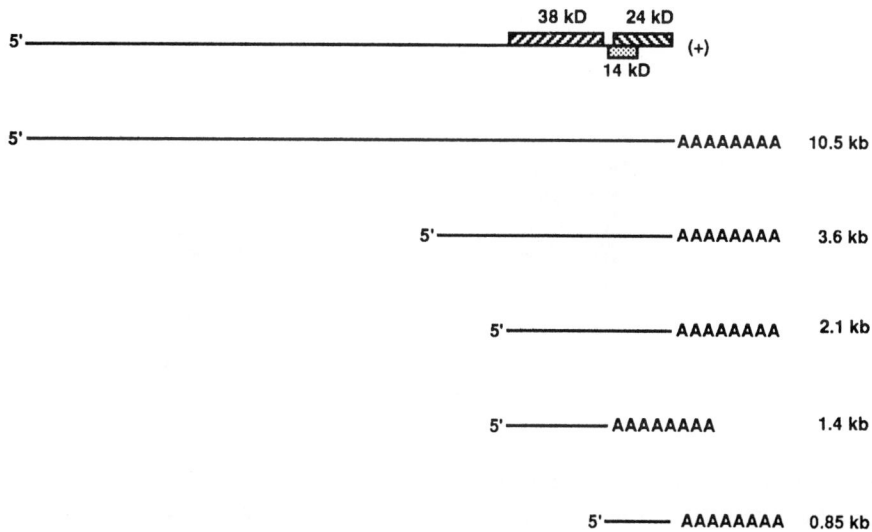

Fig. 4. BDV mRNAs. A Northern analysis of BDV RNAs. RNA samples were electropheresed in a 30 cm agarose-formaldehyde gel and transferred to nylon. The blot was sequentially probed with oligonucleotides located within the 24 kD ORF (249 and 247) or within the 38 kD ORF (325 and 439). U 10 µg total RNA from uninfected rat brain; T 10 µg total RNA from BDV-infected rat brain; A+ 0.5 µg polyA+ RNA from BDV-infected rat brain. B mRNA transcripts for the structural genes of BDV. The top line shows the locations of the three ORFs for the structural proteins relative to the full-length BDV sense strand RNA. Below are shown the polyadenylated mRNA species which have been identified

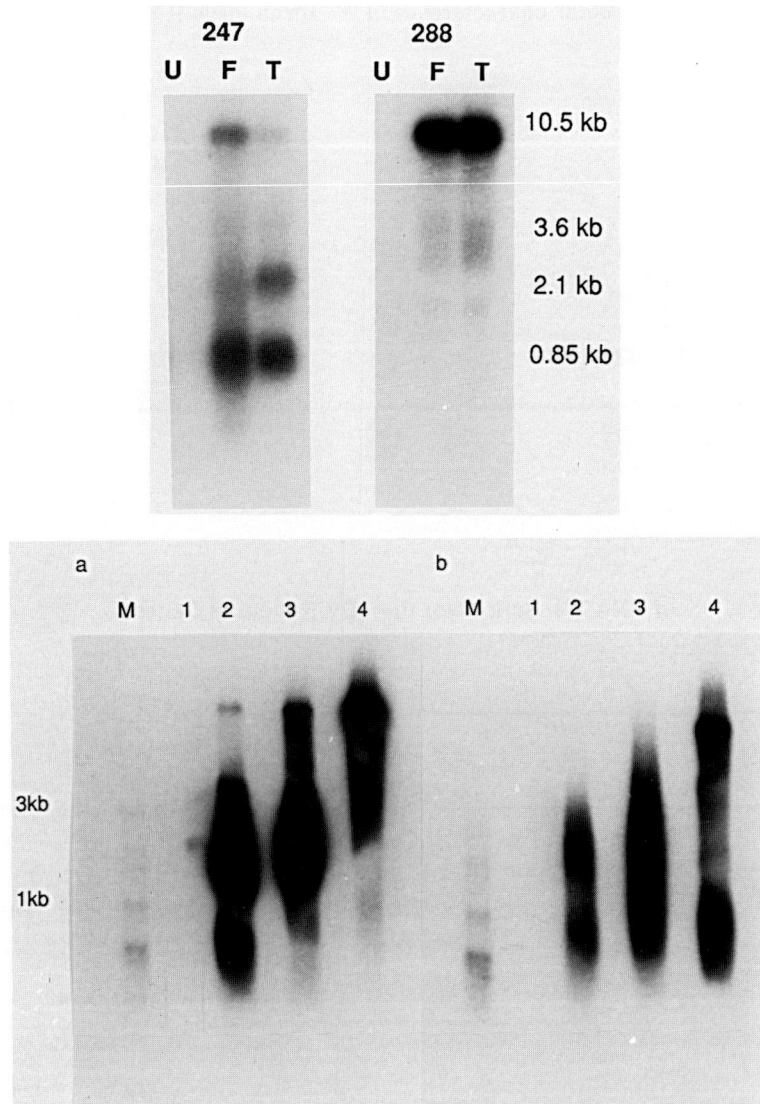

Fig. 5. Analysis of RNA contained in particles isolated from the aqueous phase after Freon extraction of BDV-infected rat brain. Northern blots were analyzed using strand-specific probes. **A** A Northern blot was sequentially probed with complementary oligonucleotides located in the 24 kD ORF. Oligonucleotide 247 detects sense transcripts; 288 detects anti-sense transcripts. *U* 10 μg RNA from uninfected rat brain; *F* 10 μg RNA prepared after Freon-extraction of BDV-infected rat brain; *T* 10 μg total RNA from BDV-infected rat brain. **B** Northern analysis of Freon extracted rat brain homogenate treated with RNase A prior to preparation of RNA. Rat brain homogenate was extracted with Freon; aliquots were used to prepare RNA directly or they were treated with RNase A at a concentration of 20 μg/ml or 40 μg/ml for 2 h at room temperature prior to preparation of RNA. Panels **a** and **b** are transfers from duplicate gels. **a** was hybridized with a sense RNA probe, **b** with an antisense RNA probe. *1* 10 μg RNA from uninfected rat brain; *2* RNA from Freon-extracted BDV rat brain treated with 40 μg/ml RNase A prior to RNA preparation (25 μg RNA); *3* RNA from Freon-extracted BDV rat brain treated with 20 μg/ml RNase A prior to RNA preparation (25 μg RNA); *4* RNA prepared directly from Freon-extracted BDV rat brain (10 μg RNA)

Analysis of genomic RNA

Because the original observations of a nested set of 3′ coterminal mRNAs (based on probes derived from the 24 kD ORF) appeared similar to the organization of coronavirus mRNAs, a working hypothesis has been that the genomic RNA is the positive stranded 10.5 kb species. This has been complicated by the fact that in infected cells the negative strand 10 kb RNA species is present at much higher levels. In order to address this issue directly, we have conducted experiments to isolate infectious particles from either lysates of infected tissue culture cells or from homogenates of BDV-infected rat brain using a Freon extraction protocol [20]. RNA prepared from these extracts contained BDV-specific RNA of varying sizes and polarities, although it did appear to be enriched for the 10.5 kb positive strand species (Fig. 5A). However, it was not clear that all these species were within the particles or whether they were nonspecifically copurified with the particles. We then treated a Freon extract from BDV-infected rat brain with RNase A prior to preparing RNA, using the guanidinium isothiocyanate method of RNA isolation [30] to immediately inactivate the RNase. Strand-specific riboprobes were then hybridized to duplicate Northern blots of the RNA. Although this treatment did not remove all small BDV species, it did clearly show that the only 10 kb species detected after RNase treatment was a negative strand species (Fig. 5B). Because infectious particles are present in the Freon extracted material and the only RNase-resistant genomic sized species detectable is of negative orientation, it seems likely that BDV is a negative strand RNA virus [20].

Although these data strongly suggest that BDV is a negative-strand virus, its classification remains unclear. No other negative-strand RNA viruses are known to have a nested mRNA organization. Also, our clones suggest that the ORFs encoding the probable structural proteins are located at the 3′ end of the positive-strand RNA. Such an organization has not been described for negative-stranded RNA viruses. Further characterization of BDV depends on the isolation of cDNAs encoding additional BDV proteins. Isolation of a clone encoding the polymerase gene(s) may allow its classification because the polymerases are highly conserved between related viruses.

References

1. Heinig H (1969) Die Borna'sche Krankheit der Pferde und Schafe. In: Roehrer H (ed) Handbuch der Virusinfektionen bei Tieren, vol 4. VEB Fischer, Jena, pp 83–148
2. Zwick W, Seifried O, Witte J (1927) Experimentelle Untersuchungen ueber die seuchenhafte Gehirn- und Rueckenmarks-entzuendung der Pferde (Bornasche Krankheit). Z Infektionskrkh 30: 42–136

3. Bode L, Riegel S, Ludwig H, Amsterdam J, Lange W, Koprowski H (1988) Borna disease virus specific antibodies in partients with HIV infection and with mental disorders. Lancet 2: 689

4. Rott R, Herzog S, Fleischer B, Winokur A, Amsterdam J, Dyson W (1985) Detection of serum antibodies to Borna disease virus in patients with psychiatric disorders. Science 228: 755–756

5. Carbone KM, Duchala CS, Griffin JW, Kincaid AL, Narayan O (1987) Pathogenesis of Borna disease in rats: evidence that intra-axonal spread is the major route for virus dissemination and the determinant for disease incubation. J Virol 61: 3431–3440

6. Morales JA, Herzog S, Kompter C, Frese K, Rott R (1988) Axonal transpot of Borna disease virus along olfactory pathways in spontaneously and experimentally infected rats. Med Microbiol Immunol 177: 51–68

7. Shankar V, Kao M, Hamir AN, Sheng H, Koprowski H, Dietschold B (1992) Kinetics of virus spread and changes in levels of several cytokine mRNAs in the brain after intranasal infection of rats with Borna disease virus. J Virol 66: 992–998

8. Dittrich W, Bode L, Ludwig H, Kao M, Schneider K (1989) Learning deficiencies in Borna disease virus-infected but clinically healthy rats. Biol Psychol 26: 818–828

9. Narayan O, Herzog S, Frese K, Scheefers H, Rott R (1983) Behavioral disease in rats caused by immunopathological responses to persistent Borna virus in the brain. Science 220: 1401–1403

10. Narayan O, Herzog S, Frese K, Scheefers H, Rott R (1983) Pathogenesis of Borna disease in rats: immune-mediated ophthalmoencephalopathy causing blindness and behavioral abnormalities. J Infect Dis 148: 305–315

11. Stitz L, Soeder D, Deschl U, Frese K, Rott R (1989) Inhibition of immuneme-diated meningoencephalitis in persistently Borna virus-infected rats by cyclosporine A. J Immunol 143: 4250–4256

12. Herzog S, Wonigeit K, Frese K, Hedrich HJ, Rott R (1985) Effect of Borna disease virus infection on athymic rats. J Gen Virol 66: 503–508

13. Richt JA, Stitz L, Wekerle H, Rott R (1989) Borna disease, a progressive menin-goencephalomyelitis as a model for CD4+ T cell-mediated immunopathology in the brain. J Exp Med 170: 1045–1050

14. Carbone KM, Trapp BD, Griffin JW, Duchala CS, Narayan O (1989) Astrocytes and Schwann cells are virus-host cells in the nervous system of rats with Borna disease. J Neuropathol Exp Neurol 48: 631–644

15. Buchmeier MJ, Welsh RM, Dutko FJ, Oldstone MBA (1980) The virology and immunobiology of LCM virus infection. Adv Immunol 30: 275–331

16. Ahmed R (1988) Curing of a congenitally acquired chronic virus infection: Acqui-sition of T-cell competence by a previously "tolerant" host. In: Lopez C (ed) Immunobiology and pathogenesis of persistent virus infections. ASV Publ., Washington, pp 159–167

17. Elford WG, Galloway IA (1935) Filtration of the virus of Borna disease through graded collodion membranes. Br J Exp Pathol 14: 196–205

18. Danner K, Mayr A (1979) In vitro studies on Borna virus. II. Properties of the virus. Arch Virol 61: 261–271

19. Pauli G, Ludwig H (1985) Increase of virus yields and releases of Borna disease virus from persistently infected cells. Virus Res 2: 29–33

20. Richt JA, Clements JE, Herzog S, Pyper J, Wahn K, Becht H (1993) Analysis of virus-specific RNA species in Freon-113 preparations of the Borna disease virus. Med Microbiol Immunol (in press)

21. Haas B, Becht H, Rott R (1986) Purification and properties of an intranuclear virus-specific antigen from tissues infected with Borna disease virus. J Gen Virol 67: 235–241

22. Schaedler R, Diringer H, Ludwig H (1985) Isolation and characterization of a 14500 molecular weight protein from brains and tissue cultures persistently infected with Borna disease virus. J Gen Virol 66: 2479–2484

23. Thierer J, Riehle H, Grebenstein O, Binz T, Herzog S, Thiedemann N, Stitz L, Rott R, Lottspeich F, Niemann H (1992) The 24 K protein of Borna disease virus. J Gen Virol 73: 413–416

24. VandeWoude S, Richt JA, Zink MC, Rott R, Narayan O, Clements JE (1990) A Borna virus cDNA encoding a protein recognized by antibodies in humans with behavioral diseases. Science 250: 1278–1281

25. Richt JA, VandeWoude S, Zink MC, Narayan O, Clements JE (1991) Analysis of Borna disease virus-specific RNAs in infected cells and tissues. J Gen Virol 72: 2251–2255

26. delaTorre JC, Carbone KM, Lipkin WI (1990) Molecular characterization of the Borna disease agent. Virology 179: 853–856

27. Lipkin WI, Travis GH, Carbone KM, Wilson MC (1990) Isolation and characterization of Borna disease agent cDNA clones. Proc Natl Acad Sci USA 87: 4184–4188

28. Pyper JM, Richt JA, Brown L, Rott R, Narayan O, Clements JE (1993) Genomic organization of the structural proteins of Borna disease virus revealed by a cDNA clone encoding the 38 kD protein. Virology 195: 229–238

29. Frohman MA, Dush MK, Martin GR (1988) Rapid production of full-length cDNAs from rare transcripts: Amplification using a single gene-specific oligonucleotide primer. Proc Natl Acad Sci USA 85: 8998–9002

30. Chomczynski P, Sacchi N (1987) Single-step method of RNA isolation by acid guanidinium thiocyanate-phenol-chloroform extraction. Anal Biochem 162: 156–159

Authors' address: Dr. J. M. Pyper, Retrovirus Laboratory, Traylor G-60, Johns Hopkins University School of Medicine, 720 Rutland Avenue, Baltimore, MD 21205, U.S.A.

Arch Virol (1994) [Suppl] 9: 429–439

_Archives_____

V̇ïrology

© Springer-Verlag 1994
Printed in Austria

Genomic organization and expression of astroviruses and caliciviruses

M. J. Carter

School of Biological Sciences, University of Surrey, Guildford, U.K.

Summary. Astroviruses and caliciviruses are two families defined initially by their characteristic morphology. Many of these viruses have been difficult to grow in culture. Molecular biology has now provided a valuable insight into the nature of these viruses, and in many respects knowledge of genome structure now outstrips that of more classical virological features. However these advances have allowed a more detailed approach to virus classification and have led to the establishment of the *Astroviridae* as a distinct virus family.

Introduction

Both astroviruses (family *Astroviridae*) and caliciviruses (family *Caliciviridae*) were initially defined by their morphology under the electron microscope. Caliciviruses are 34–38 nm and covered by cup-like depressions in which stain accumulates (*calix*, a cup). The particle margin is jagged where cups are seen edge-on [3, 4]. Astroviruses are smaller (28 nm), with a smooth margin [22]. A proportion of the virions bear a prominent five or six-pointed star-like motif (*astro*, a star). The morphologies of astroviruses and caliciviruses are each unique (Fig. 1) and until recently there was no satisfactory explanation of how either is formed.

Both virus families are mostly associated with diarrhoea, but some members may cause skin lesions, or respiratory or hepatic infections. Some caliciviruses grow readily in vitro, but despite this they remained relatively uncharacterized. Astroviruses were more difficult to grow and have only recently been recognized as a distinct family. A few strains of a human virus were adapted to growth in continuous cell lines by passage in primary human embryo cells [10]. Caliciviruses remain the better characterized, and it now appears that they may share some features with the astroviruses.

A

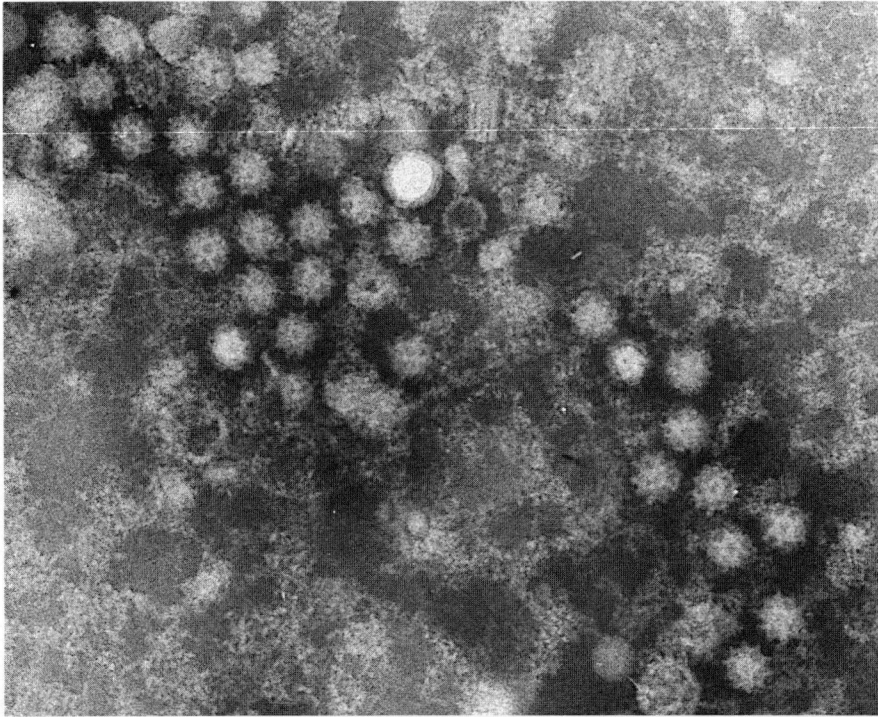

B

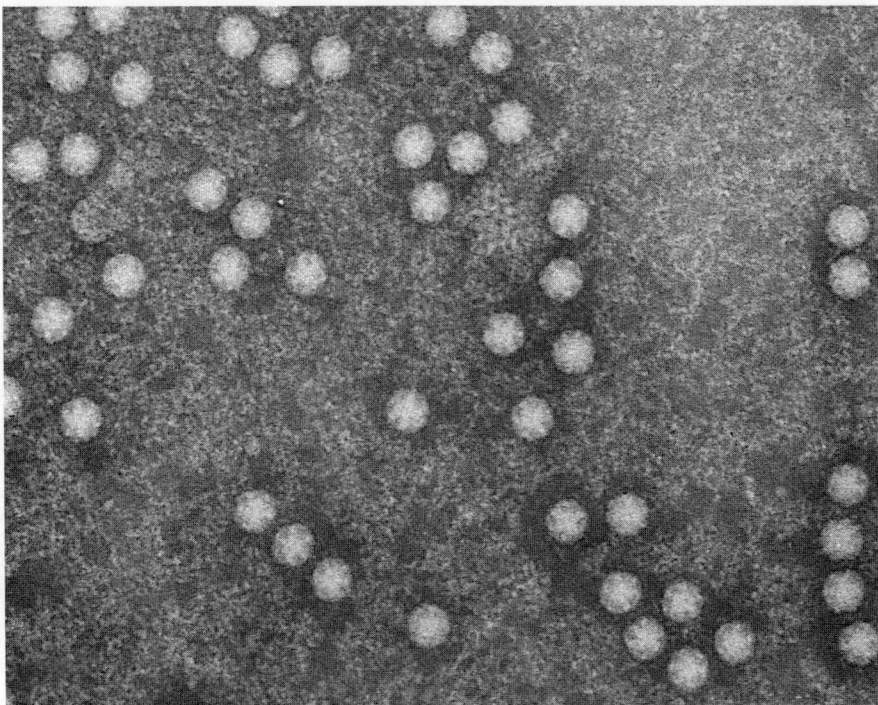

Fig. 1. Caliciviruses and astroviruses. Virions stained with phosphotungstic acid, pH 7.0, and are reproduced at a magnification of ×2000000. **A** Caliciviruses. **B** Astroviruses. Electron micrographs were kindly provided by Prof. C. R. Madeley

Genome structure of caliciviruses

The best characterized caliciviruses are listed in Table 1. Complete sequences have been determined for feline calicivirus (FCV) [6], for Norwalk virus (NV) [EMBL Database Accession No. M87661], for rabbit haemorrhagic disease virus (RHDV) [11] and for the candidate virus of enterically transmitted human non-A non B hepatitis, hepatitis E virus (HEV) [18]. Partial sequence has been obtained for San Miguel sea-lion virus (SMSV) [15] and for European brown hare syndrome virus (EBHSV) (G. Meyers, pers. comm.). The genomes of these viruses are organized in a similar manner (Fig. 2). The virions contain a single type of capsid protein which is encoded towards the 3' end [5, 17], non-structural proteins are encrypted by the 5' end [14]. FCV, NV and HEV carry genes for non-structural proteins in a separate open reading frame (ORF1), and the capsid protein is specified by a second region (ORF2), separated from ORF1 by a termination codon and a frameshift. However in RHDV and also EBHSV, these two genes are fused in frame (ORF1/2) and translation forms a single large polyprotein. Finally, most caliciviruses contain a small, third ORF at the extreme 3' end (ORF3). This is lacking in HEV, in which a third gene is located internally, spanning the juncton of ORFs 1 and 2.

Calicivirus RNA synthesis

Early reports suggested that up to eight RNA species were synthesized in infected cells and that sequential open reading frames are present in the genome [2, 16]. This has been partially confirmed by sequencing studies, but there are too few ORFs to account for all these mRNAs, and a re-examination of virus-specific mRNA synthesis found only two were present (T. D. K. Brown, pers. comm.). Multiple RNAs were observed in FCV infection, a rapidly cytocidal process, and most

Table 1. Best characterized caliciviruses

Name		Serotypes
Vesicular exanthema of swine	(VESV)	13
San Miguel sealion	(SMSV)	16
Feline calicivirus	(FCV)	1
Small round structured viruses	(SRSV)	5
e.g. Norwalk virus	(NV)	
Rabbit haemorrhagic disease	(RHDV)	1
European brown hare syndrome	(EBHSV)	1
Hepatitis E virus	(HEV)	1

Calicivirus Coding Strategy

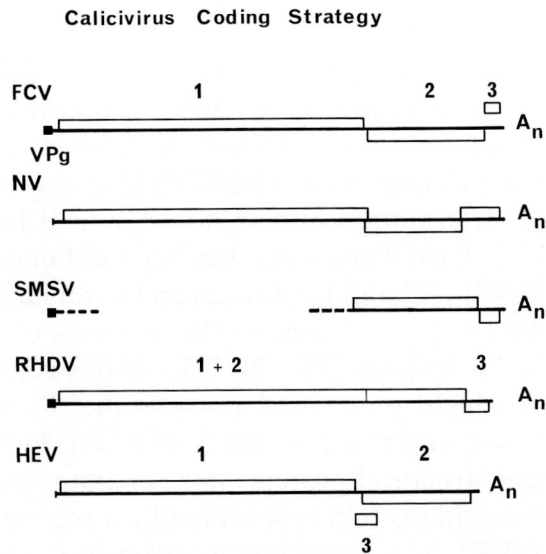

Fig. 2. Coding strategy of the caliciviruses. Open reading frames are illustrated as open boxes. The poly A tract at the 3' end is present in all cases. The 5' end is covalently linked to a protein VpG (*VP*$_g$), illustrated where its presence has been demonstrated. VpG is probably also present in NV RNA, although HEV could be capped. Genome lengths (bp) are: *FCV* (feline calicivirus) 7 690; *NV* (Norwalk virus) 7 644; *RHDV* (rabbit haemorrhagic disease virus) 7 437; *HEV* (hepatitis E virus) 7 194. *SMSV* San Miguel sea-lion virus

could thus have derived from degradation, or artefactual origin. The prominent bands marked in Fig. 3 result from ribosomal RNA acting as a "snowplough" and concentrating the more diffuse background material into pseudo-bands. These features are removed if mRNA is selected before analysis. This effect was probably compounded [2] by the addition of different RNA as carrier, leading to localized overloading in other positions and an increase in the number of apparent bands.

Both of the RNAs detected in the cell are encapsidated. These comprise the genome (approx. 7.5 kb), and a subgenomic RNA, 2.4–2.7 kb in size. The smaller molecule is 3' co-terminal with the genome, extending to the 5' end of ORF2 and serves as an mRNA for the capsid protein. This mRNA is produced by RHDV and EBHSV, in which ORF2 is not separated from ORF1. The subgenomic RNA is presumably the major source of capsid protein made by these viruses, although processing from the ORF1/2 fusion polyprotein may also be possible. Subgenomic mRNA produced to increase expression of a section of a continuous ORF rather than to express a cryptic or modified sequence, is a feature unique to the caliciviruses. It is unclear how the subgenomic RNA is produced. It is likely that the subgenomic RNA is transcribed

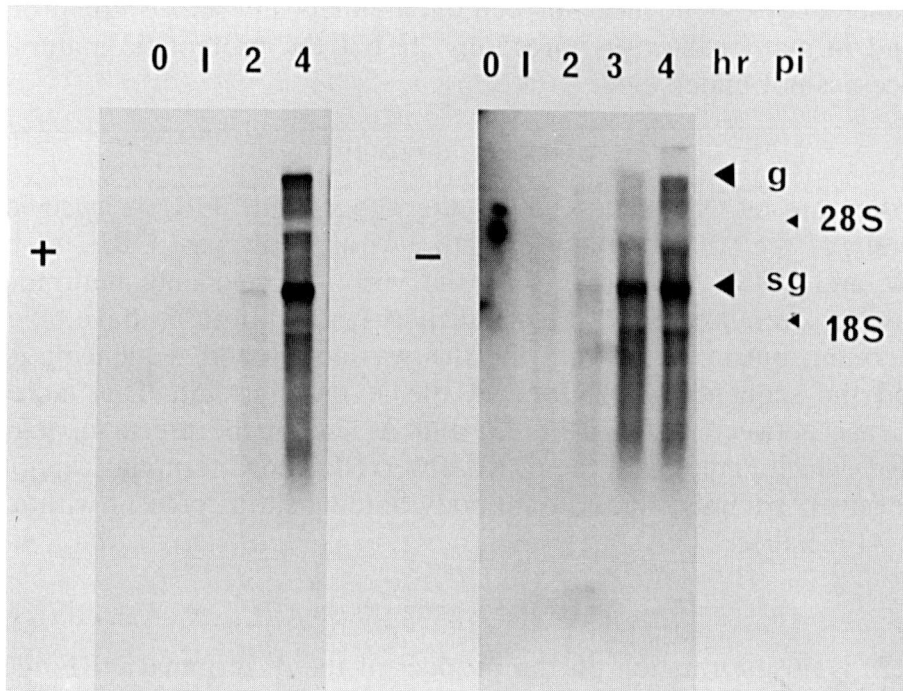

Fig. 3. Calicivirus RNA synthesis. Total cell RNA was extracted from cells at the times shown following infection, Northern blotted and probed with strand-specific cDNA probes from the 3′ terminal 300 bases. Designation of (+) and (−) given for each panel refers to the polarity of the probe used. Bands present between the genomic (g) and subgenomic (sg) RNAs are caused by ribosomal RNA (*18S, 28S*)

from a genome-sized RNA template, but since both RNAs are encapsidated, and both have been found inside the cell in their negative forms (Fig. 3), and each may be replicated independently [2]. However the existence of the negative form of subgenomic RNA has been disputed and an alternative derivation of this species by partial degradation has been suggested [17].

Protein translation and maturation

Non-structural proteins

Proteins specified by ORF1 resemble picornavirus non-structural proteins, regions have thus been designated 3C-like (cysteine protease), 2C-like (helicase) and 3D-like (polymerase) [14]. However, this leaves about 80 Kd of protein for which no function has yet been assigned. Nomenclature used is assigned by sequence comparison and does not imply maturational relationships. ORF1 is translated as a polyprotein which is subsequently cleaved to produce the final virus products. There is general correspondence between the numbers and sizes of the proteins found in many calicivirus infections [1] but the proteolytic maturation process is not understood.

The capsid protein

In the case of NV and RHDV, the subgenomic mRNA specifies a protein of the size observed in mature virus. However, ORF2 of both FCV and SMSV are larger than the virion protein and maturational cleavage is required [1, 7]. Most calicivirus capsid proteins have blocked N termini; but in FCV strain F9, this was not so and sequencing identified the sequence ADDGSIT at the N terminus [5]. This indicates cleavage between residues E(123) and A(124) of the precursor leaving 63 Kd, which agrees with the 62 Kd observed. The C terminus cannot be extensively trimmed since an antibody binding site is present within the last 34 residues.

The ORF3 gene product

There is no information on the product of the 3′ terminal ORF3 gene. The protein specified by this gene is the best conserved among different strains of FCV (92–98%) and shows 27% identity between FCV and RHDV [6], rising to 47% similarity when conservation of amino acid character is considered [12]. Although there is more variation between the ORF3 protein specified by NV and SMSV, all such calicivirus proteins are relatively small and have a predominance of basic residues. This high degree of conservation argues strongly for a role at the protein level, but no product from this area has been identified. Sera from FCV-vaccinated cats do not recognize the product of the ORF3 gene when it is expressed as a beta-galactosidase fusion protein in *E. coli* (I. D. Milton, unpubl. obs.). ORF3 expression by frameshifting has been suggested, fusing ORF3 to the capsid protein [17]. However no such chimaeric protein has been found and it may be that a small RNA detected in the cell [2], but not confirmed [17], could serve as mRNA for this ORF.

The ORF3 product specified by the internal gene of HEV is distinct from that encoded by the other caliciviruses. It is recognized by antibodies from infected animals. A third mRNA has been detected in infected liver, slightly larger than that encoding the capsid protein, and possibly functions as mRNA for ORF3 [23].

Classification

The caliciviruses were first classified on the basis of their distinctive morphology. This limited the number of firmly established members, because classical morphological features are not always clear. NV for instance has remained only a candidate calicivirus because of the "fuzzy" appearance of the virion. However, despite differences in ORF1/2 structure RHDV, EBHSV and NV can now be considered caliciviruses because of their gene order, and the similarities of their non-structural polypeptides. However, the status of HEV is questionable. HEV virions do display (albeit poorly) many of the morphological features of the caliciviruses [19], and the gene order is largely calicivirus-like; but there is no convincing similarity between the sequence of HEV ORF1 proteins and those of other caliciviruses (10–20%), and the internal ORF3 has no counterpart in other caliciviruses. The product of an internal ORF in FCV [5] was not recognized by immunize cat serum, and there is no evidence that such a gene is real (I. D. Milton, unpubl. obs.). Furthermore, recent computer analysis has grouped functional motifs in the ORF1 sequence of HEV away from other caliciviruses, and motifs indicative of capping functions have been identified [8]. If HEV RNA is capped rather than protein-linked, such differences may imply that HEV belongs to a separate taxonomic grouping, irrespective of the morphology of the virion.

Astroviruses: genome structure

The best characterized astroviruses are human astrovirus (HuAV, 5–6 serotypes), bovine astrovirus (BAV, at least 2 serotypes), porcine astrovirus (PAV), and ovine astrovirus (OAV). The best characterized of all is HuAV, a complete genome sequence should be available soon, but information is already available from the 3' end and from some internal regions [20]. These show few similarities to members of any other virus family, and no small 3' ORF equivalent to ORF3 of the caliciviruses is present. However, the genome does not comprise a single unbroken ORF and a separate gene at the 3' end, could encode a protein of 90 Kd (S. Monroe, pers. comm. 1992). A second ORF extends towards the 5' end and contains both a polymerase motif (T. Lewis, pers. comm.) and a nuclear addreassing sequence (EMBL Database Accession No. Z 16420).

RNA synthesis

Astrovirus genome expression is similar to that of the caliciviruses and two RNAs are synthesized [13], the genome and a subgenomic species of a size suitable for the expression of the 3' ORF. At first this was not confirmed by Northern blotting [20] but this discreapancy has now been resolved: Unlike the caliciviruses, the astrovirus subgenomic RNA may not be packaged, and does not persist at the late times studied. Blotting studies easily demonstrated both RNAs when samples were analysed

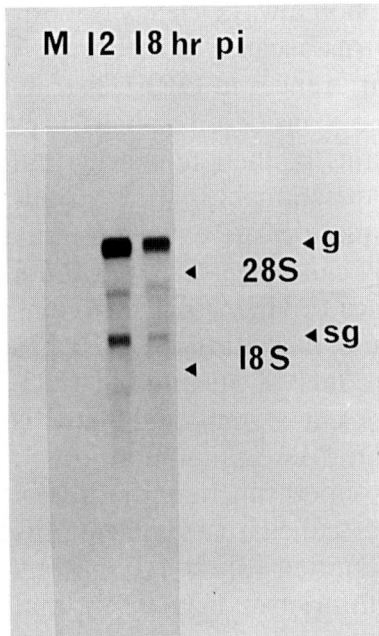

Fig. 4. Astrovirus RNA synthesis. RNA was extracted from CaCo-2 cells infected with human astrovirus serotype 1 at 12 and 18 h pi for Northern blotting and detection with a cDNA probe prepared from the 3′ end of the genome. *g* Genomic and *sg* subgenomic RNAs. The position of ribosomal RNA causing "snowploughing" between the two virus RNAs is indicated

earlier in infection (Fig. 4). Both RNAs may exist in both positive and negative senses (S. Monroe, pers. comm.).

Protein translation and maturation

Non structural proteins

Little is known about astrovirus non-structural proteins but these are probably encoded as a polyprotein towards the 5′ which contains a polymerase motif and a nuclear targeting signal. It is not known if this is functional, but a discrete immune-fluorescence has been observed in the nucleus [22].

Structural proteins

The large ORF at the 3′ terminus is thought to specify the structural proteins. In support of this hypothesis we have identified a short run of basic amino acids in this gene which is similar to basic stretches found in the capsid proteins of two coronaviruses (Table 2). All astroviruses contain at least two proteins of approximately 30 Kd, as well as a smaller molecule whose size varies between viruses 13–27 Kd [22]. A fourth small protein (5.2 Kd) has been observed but not confirmed [9]. These account for 80–90 Kd and an unstable protein of this size has been observed in infection [13]. This could be a precursor to the virion

Table 2. Comparison of basic residues in the presumptive structural gene of human astrovirus type 1 with similar sequences in coronavirus capsid proteins

GRSRSKSRARSQSRGR	Human astrovirus type 1
*** .** ****** *	
SRSRSRNRSQSRGR	Transmissible gastroenteritis virus
*** *** *	
SRSTSRASS	Bovine coronavirus nucleocapsid

* Residues identical to those in astrovirus type 1 sequence; (.) residues conserved in character with astrovirus type 1

proteins, and its size is consistent with the coding capacity of the 3′ ORF.

Classification

Astroviruses have been variously suggested as possible members of the family *Picornaviridae*, based on their structural proteins, or the family *Caliciviridae*, based on intracellular RNA synthesis. However, their unique morphology, and the observation of protein migration to the nucleus, suggest that they are distinct from members of both these virus families. Accordingly the astroviruses are now considered as representing a distinct family, the *Astroviridae* (Monroe et al., in press). Perhaps both astroviruses and caliciviruses belong to a larger grouping of non-enveloped positive-stranded RNA viruses with distinctive gene order and mode of expression.

Note added in proof

The complete sequences of two serotypes of human astrovirus (types 1 and 2) have now been determined. These both show that the non-structural gene (ORFf1) is present as two sections (ORF 1a and 1b), linked by motifs indicative of ribosomal frameshifting. Furthermore the viral protease has a serine at the active site, rather than a cysteine as in the picorna- and caliciviruses. These features are similar to those observed in the luteoviruses of plants and further support the classification of astroviruses as a novel family of animal viruses.

Acknowledgements

I am grateful to those workers who allowed me to present their data at Clearwater and in this review. In particular, Dr. J. Neill at Pioneer Hibred International, Iowa, U.S.A.; Dr. G. Meyers and Prof. J. Thiel at the Bundesforschungsanstalt für Virus-krankheiten der Tiere, Tübingen, Germany; Dr. F. Parra at the University of Oviedo,

Spain, and Dr. C. Rossi at the Instituto Zooprofilattico Sperimentale, Brescia, Italy, for their help with the calicivirus section of this presentation. Also Dr. S. Monroe at U.S. Centres for Disease Control, Atlanta, U.S.A. for his help with the astroviruses. Finally, I thank my own colleagues at Newcastle, Dr. M. Willcocks and Dr. I. Milton for their assistance.

References

1. Carter MJ (1989) Feline calicivirus protein synthesis investigated by Western blotting. Arch Virol 108: 69–79
2. Carter MJ (1990) Transcription of feline calicivirus RNA. Arch Virol 114: 143–152
3. Carter MJ, Madeley CR (1987) Caliciviridae. In: Nermut MV, Steven AC (eds) Animal virus structure. Elsevier, Amsterdam, pp 121–128 (Perspectives in Medicine and Virology, vol 3)
4. Carter MJ, Milton ID, Madeley CR (1991) Caliciviruses. Rev Med Virol 1: 177–186
5. Carter MJ, Milton ID, Turner PC, Meanger J, Bennett M, Gaskell RM (1992) Identification and sequence determination of the capsid protein gene of feline calicivirus. Arch Virol 122: 223–235
6. Carter MJ, Milton ID, Meanger J, Bennett M, Gaskell RM, Turner PC (1992) The complete sequence of a feline calicivirus. Virology 190: 443–448
7. Fretz M, Schaffer FL (1978) Calicivirus proteins in infected cells: evidence for a capsid polypeptide precursor. Virology 89: 318–321
8. Koonin EV, Gorbalenya AE, Purdy MA, Rozanov MN, Reyes GR, Bradley DW (1992) Computer-assisted alignment of functional domains in the non-structural polyprotein of hepatitis E virus: delineation of an additional group of positive-strand RNA plant and animal viruses. Proc Natl Acad Sci USA 89: 8259–8263
9. Kurtz JB (1989) Astroviruses. In: Farthing MJG (ed) Viruses and the gut. Proceedings of the Ninth BSG Smith Kline and French International Workshop. Smith Kline and French Laboratories Ltd, Welwyn Garden City, pp 84–87
10. Lee TW, Kurtz JB (1981) Serial propagation of astrovirus in tissue culture with the aid of trypsin. J Gen Virol 57: 421–424
11. Meyers G, Wirblich C, Thiel H-J (1991) Rabbit haemorrhagic disease virus – molecular cloning and sequencing of a calicivirus genome. Virology 184: 664–676
12. Milton ID, Vlasak R, Nowotny N, Rodak L, Carter MJ (1992) Genomic 3' terminal sequence comparison of three isolates of rabbit haemorrhagic disease virus. FEMS Microbiol Lett 93: 37–42
13. Monroe SS, Stine SE, Gorelkin L, Herrmann JE, Blacklow NR, Glass RI (1991) Temporal synthesis of proteins and RNAs during human astrovirus infection in cultured cells. J Virol 65: 641–648
14. Neill JD (1990) Nucleotide sequence of a region of the feline calicivirus genome which encodes picorna virus-like RNA dependent RNA polymerase, cysteine protease and 2C polypeptides. Virus Res 17: 145–160
15. Neill JD (1992) Nucleotide sequence of the capsid protein gene of two serotypes of San Miguel sea-lion virus: identification of conserved and non-conserved amino acid sequences among calicivirus capsid proteins. Virus Res 24: 211–222
16. Neill JD, Mengeling WL (1988) Further characterization of the virus-specific RNAs in feline calicivirus infected cells. Virus Res 11: 143–152
17. Neill JD, Reardon IM, Heinrikson RL (1991) Nucleotide sequence and expression of the capsid protein gene of feline calicivirus. J Virol 65: 5400–5447

18. Tam AW, Smith MM, Guerra ME, Huang C-C, Bradley DW, Fry KE, Reyes GR (1991) Hepatitis E virus (HEV): molecular cloning and sequencing of the full-length viral genome. Virology 125: 120–131

19. Ticehurst J, Rhodes LL, Krawczynski K, Asher LVS, Engler WF, Mensing TL, Caudill JD, Sjogren MH, Hoke CH, LeDuc JW, Bradley DW, Binn LN (1992) Infection of owl monkeys (*Aotus trivirgatus*) and cynomolgus monkeys (*Macaca fascicularis*) with hepatitis E virus from Mexico. J Infect Dis 165: 835–845

20. Willcocks MM, Carter MJ (1992) The 3′ terminal sequence of a human astrovirus. Arch Virol 124: 279–289

21. Willcocks MM, Carter MJ, Laidler FR, Madeley CR (1990) Growth and characterization of human faecal astroviruses in a continuous cell line. Arch Virol 113: 73–82

22. Willcocks MM, Carter MJ, Madeley CR (1992) Astroviruses. Rev Med Virol 2: 97–106

23. Yarbough PO, Tam AW, Fry KE, Krawczynski K, McCaustland KA, Bradley Dw, Reyes GR (1991) Hepatitis E virus: identification of type-common epitopes. J Virol 65: 5790–5797

Author's address: Dr. M. J. Carter, School of Biological Sciences, University of Surrey, Guildford GU2 5XH, U.K.

Arch Virol (1994) [Suppl] 9: 441–448

Archives
of
Virology
© Springer-Verlag 1994
Printed in Austria

Lelystad virus belongs to a new virus family, comprising lactate dehydrogenase-elevating virus, equine arteritis virus, and simian hemorrhagic fever virus

J. J. M. Meulenberg, M. M. Hulst, E. J. de Meijer, P. L. J. M. Moonen, A. den Besten, E. P. de Kluyver, G. Wensvoort, and **R. J. M. Moormann**

Central Veterinary Institute, Department of Virology, Lelystad, The Netherlands

Summary. Lelystad virus (LV) is an enveloped positive-stranded RNA virus, which causes abortions and respiratory disease in pigs. The complete nucleotide sequence of the genome of LV has been determined. This sequence is 15.1 kb in length and contains a poly(A) tail at the 3′ end. Open reading frames that might encode the viral replicases (ORFs 1a and 1b), membrane-associated proteins (ORFs 2 to 6) and the nucleocapsid protein (ORF7) have been identified. Sequence comparisons have indicated that LV is distantly related to the coronaviruses and toroviruses and closely related to lactate dehydrogenase-elevating virus (LDV) and equine arteritis virus (EAV). A 3′ nested set of six subgenomic RNAs is produced in LV-infected alveolar lung macrophages. These subgenomic RNAs contain a leader sequence, which is derived from the 5′ end of the viral genome. Altogether, these data show that LV is closely related evolutionarily to LDV and EAV, both members of a recently proposed family of positive-stranded RNA viruses, the Arteriviridae.

Introduction

A new pig disease, causing reproductive failures in sows and respiratory problems in piglets, was first observed in 1987 in the United States [1]. In subsequent years the disease spread through the United States to Canada. In late 1990 it appeared in Germany, after which it spread rapidly through Western Europe [2]. As the disease spread, it acquired more and more names, three of which are most prominent: swine infertility and respiratory syndrome (SIRS), porcine reproductive respiratory syndrome (PRRS), porcine epidemic abortion and respiratory syndrome (PEARS).

The causal agent of the disease, Lelystad virus (LV), was first isolated by Wensvoort et al. [3]. In experimentally induced infections, LV
causes reproductive failure in sows, resulting in abortions and mummified,
stillborn and weak piglets [3]. In addition, respiratory disease was observed in fattening pigs and in piglets. In the United States, the SIRS
virus, which is antigenically and structurally closely related to LV, was
identified as the causal agent of the disease [4].

LV is a small, enveloped, single-stranded RNA virus that replicates
in vitro only in primary cultures of porcine alveolar macrophages. Titres
in these macrophage cultures reach a maximum of $10^{6.5}$ $TCID_{50}$/ml. The
buoyant density of LV ranges from 1.14 g/ml on a sucrose gradient to
1.19 g/ml in a caesium-chloride gradient. In ultra-thin sections of LV-
infected macrophages, LV virions appear as 45 to 55 nm large spherical
particles that consist of a 30 to 35 nm large nucleocapsid which is surrounded by a lipid bilayer membrane [5].

In this report we will present recent data concerning the organization
and expression of the genome of Lelystad virus.

Genome organization

Recently, the nucleotide sequence of the genome of LV has been determined [6]. RNA was isolated from LV-infected alveolar macrophages
and this was used to construct a cDNA library. This library was screened
with a radioactive-labeled probe synthesized from LV genomic RNA.
A set of overlapping LV-specific cDNA clones was isolated and a consecutive sequence of 15 088 was obtained. A poly (A) sequence was
present at the 3' end of the genome. Eight open reading frames (ORFs)
that might encode virus-specific proteins were identified (Fig. 1). ORF1a
and ORF1b comprise about 80% of the viral genome and are predicted
to encode the viral RNA polymerase. Their amino acid sequences
contain elements conserved in RNA polymerases of the torovirus Berne
virus (BEV) [7], equine arteritis virus (EAV) [8], lactate dehydrogenase-
elevating virus (LDV) [9], and coronaviruses [10–12]. A putative serine
protease domain and two papain-like protease domains were identified
in the coding region of ORF1a (Fig. 2). The four characteristic domains,
identified in ORF1b of the viruses mentioned above were also present in
ORF1b, at the same relative position. They were: (I) the polymerase
motif containing the core sequence S/GDD, identified in the RNA
polymerase of all positive-strand RNA viruses; (II) a cysteine- and
histidine-rich Zinc finger domain; (III) a nucleoside triphosphate binding
or helicase motif; and (IV) a conserved domain of unknown function.

ORF1a and ORF1b overlap over a small region of 16 nucleotides and
occupy different reading frames. We assume that the expression of

A

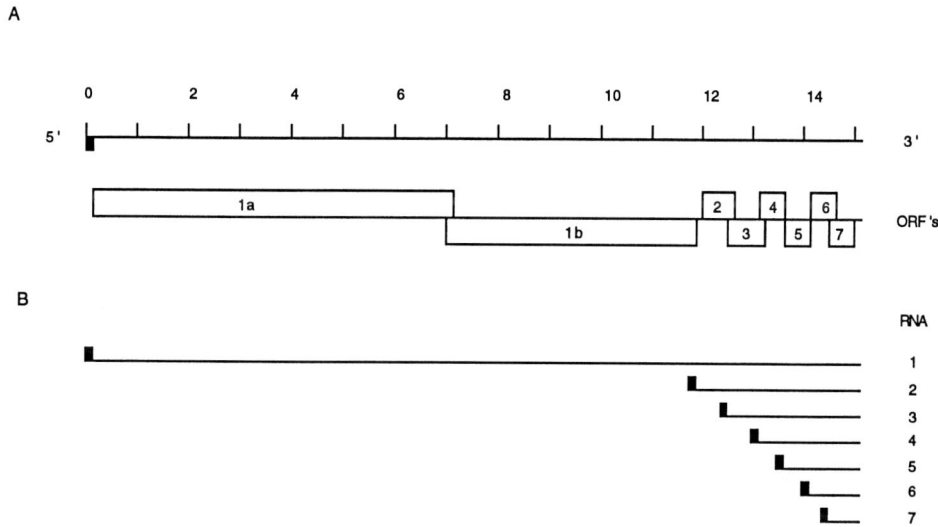

Fig. 1. A Organization of the LV genome. The ORFs, identified in the nucleotide sequence are shown. **B** shows the subgenomic set of RNAs, encoding the ORFs. The leader sequence is indicated by a solid box

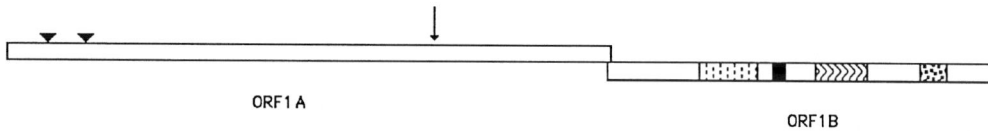

Fig. 2. Conserved domains in the polymerase encoding regions, ORF1a and ORF1b, of Lelystad virus. Two putative papain protease domains and a putative serine protease domain in ORF1a are indicated with black triangles and an arrow, respectively. The four conserved domains in ORF1b are indicated by shaded boxes. From left to right they are: polymerase motif, zinc finger domain, nucleoside triphosphate binding domain and a conserved domain of unknown function

ORF1b involves ribosomal frameshifting. A heptanucleotide slippery sequence (UUUAAAC) and a putative pseudoknot structure, which are both required for efficient ribosomal frameshifting during translation of the RNA polymerase ORF1b of BEV, EAV and the coronaviruses [10, 13], were identified in the overlapping region of ORF1a and ORF1b of LV [6].

The calculated sizes of the gene products of ORFs 2 to 6 range from 13.8–30.6 kDa and they show features reminiscent of membrane proteins [6]. They all contain putative N-linked glycosylation sites and N- and C-terminal hydrophobic sequences that may function as a signal sequence and a membrane anchor, respectively. The polypeptide encoded by ORF7 was extremely basic and the degree of identity with nucleocapsid proteins of other viruses (see below) suggests a similar function for this polypeptide.

Sequence comparisons of LV, LDV, and EAV

Comparison of the amino acid sequences encoded by the ORFs identified in the genome of LV with those of other viruses indicate that LV is distantly related to coronaviruses and toroviruses and closely related to LDV and EAV. Apart from the conserved regions in the polymerase encoding regions, mentioned earlier, no extensive identity between the amino acid sequences encoded by the ORFs of LV and coronaviruses or toroviruses was identified, using the FASTA sequence comparison program [14]. In contrast, the RNA polymerase-encoding regions (ORF1a and ORF1b) and the ORFs encoding the putative envelope proteins (ORFs 2 to 6) and the nucleocapsid protein (ORF7) of LV and LDV share a high percentage (29 to 67%) identical amino acids (Fig. 3). The amino acid sequences encoded by ORFs 1b to 7 of LV and LDV can be aligned without the introduction of large gaps. This is not the case for ORF1a of these viruses. Although the N-terminal amino acids (1–500) and the C-terminal amino acids (1 100–2 380) of ORF1a of LV and LDV are highly identical (46 and 40%, respectively), the sequence of the amino acids between were not significantly identical (as determined with FASTA). Comparison of the amino acid sequences encoded by ORF1a and ORF1b of LV and EAV also indicated that ORF1b is more conserved than is ORF1a. The amino acid sequences encoded by ORF1b of LV and EAV were 36% identical over the complete coding region. However, the amino acid sequence encoded by ORF1a of EAV is about 600 residues shorter than the amino acid sequence encoded by ORF1a of LV and is only 25% identical to the C-terminal portion of the amino acid sequence encoded by ORF1a of LV. Further experiments must be done to establish whether these differences in ORF1a also result in different functions of the RNA replicases of these viruses. Comparison of the amino acid sequences of ORFs 2 to 7 of LV and EAV showed that the amino acid sequences encoded only by ORFs 6 and 7 of these viruses share some amino acid identity (23 and 20%, respectively; Fig. 3).

In summary, the analysis of the amino acid sequences indicates that LV is more similar to LDV than to EAV. It is conjectured that such relationships indicate evolutionary development.

Subgenomic RNAs

Analogous to RNA synthesis during replication of EAV and LDV, multiple subgenomic RNAs are synthesized in LV-infected alveolar macrophages. A set of oligonucleotides was used in Northern blot

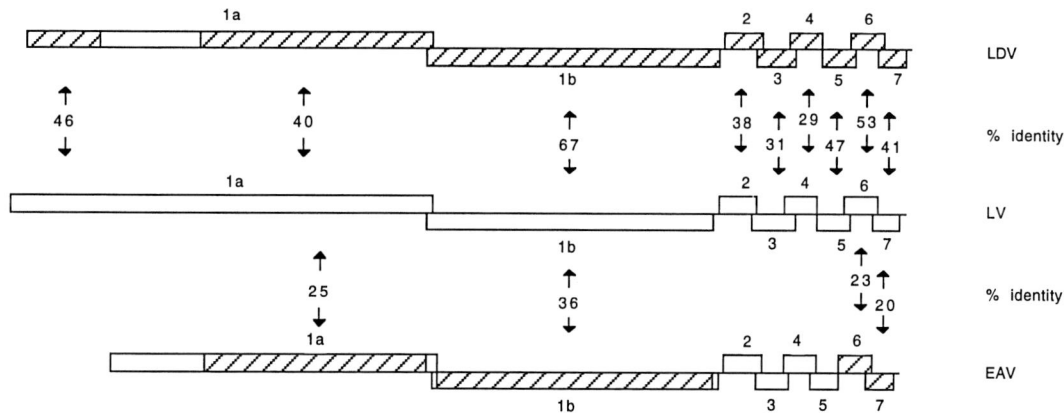

Fig. 3. Comparison of the viral proteins of LV, EAV and LDV. The percentage identity between the amino acid sequences encoded by ORFs of LV, LDV ([9], M. Brinton, pers. comm.) and EAV [8] are indicated. The boxes represent the open reading frames, identified in the nucleotide sequence of the genome of these viruses. The regions of the ORFs of LDV and EAV, which have identity with the ORFs of LV are shaded

hybridization analyses to identify and characterize LV subgenomic RNAs. These LV-specific oligonucleotides were located in the unique part of the various ORFs and at the extreme 5′ and 3′ end of the LV genome (Fig. 4A). We were able to identify six subgenomic RNAs by northern analyses of RNA isolated from LV-infected alveolar macrophages using oligonucleotide 25U101R, which is complementary to the 5′ noncoding region of the viral genome. This oligonucleotide hybridized to the viral genomic RNA of about 15 kb (RNA1) and to six smaller RNAs of 3.3 (RNA2), 2.7 (RNA3), 2.2 (RNA4), 1.7 (RNA5), 1.1 (RNA6), and 0.7 (RNA7) kb. Hence, these six subgenomic RNAs possess a common leader sequence, which is derived from the 5′ end of the viral genome. The hybridization patterns of the other oligonucleotides demonstrated a correlation between ORFs 1a to 7 and RNAs 1 to 7 (Fig. 4B). Oligonucleotides specific for ORF1a and ORF1b (5U94R and 41U101R, respectively) hybridized only to the largest RNA, RNA1, oligoncleotide 60R219, specific for ORF2, hybridized to the two largest RNAs, RNA1 and RNA2, etc. This indicated that these subgenomic RNAs, together with RNA1 form a 3′ co-terminal nested set (Figs. 1 and 4B). Besides the seven RNAs, we observed two additional bands at about 1.5 and 4.3 kb. They appear to reflect aspecific binding of the probes to the ribosomal 18S and 28S RNA.

A

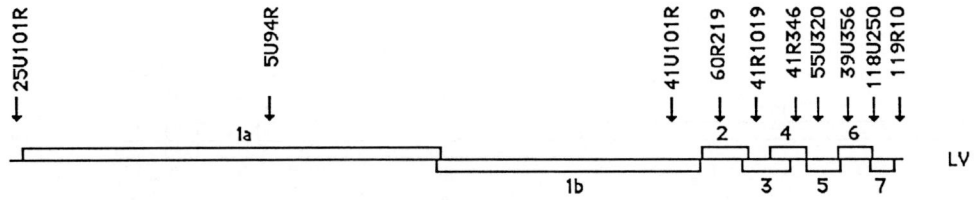

B

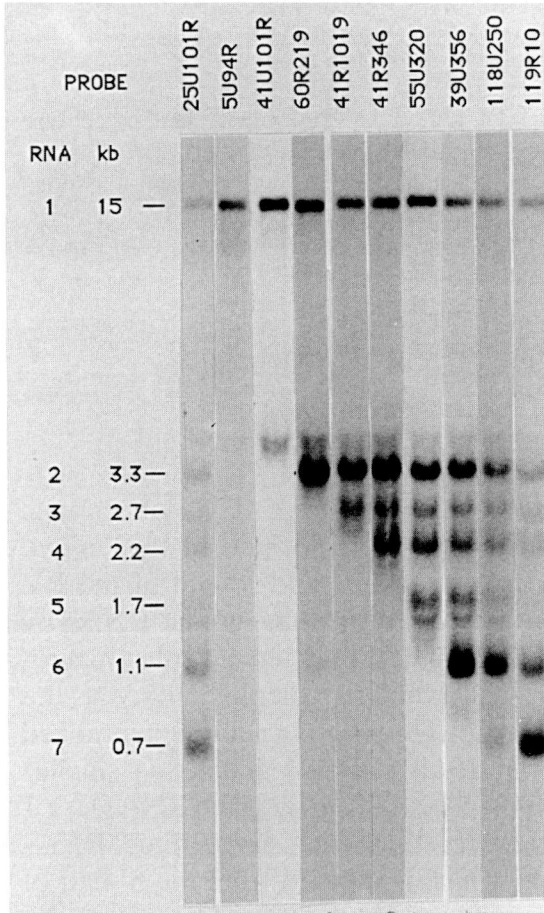

Fig. 4. Northern hybridization of RNA, isolated from LV-infected alveolar lung macrophages. The RNA was separated on a 0.8% denaturing formaldehyde agarose gel. The blots were sequentially hybridized with the indicated LV-specific oligonucleotides B. The location of these oligonucleotides in the genome is shown in A

A new family of positive-strand RNA viruses

The genomic data presented here indicate that LV is closely related to EAV and LDV and distantly related to coronaviruses and toroviruses. The size and organization of the genome of LV, LDV and EAV is similar, they all produce a coterminal nested set of 6–7 subgenomic RNAs and the ORFs of these viruses share extensive sequence homology. Preliminary sequence data indicate that simian hemorrhagic virus

(SHFV) is also closely related to LV, LDV, and EAV (E. Godeny, pers. comm.). Besides the similarities, mentioned above, these four viruses have other structural and biological properties in common [6, 15].

LDV and EAV have been classified in the family *Togaviridae* and SHFV in the family *Flaviviridae*. Recently, however, it has been suggested that LDV, EAV, SHFV, together with LV should be grouped in a new virus family, called the *Arteriviridae* [15]. We agree that these four viruses share enough characteristics and are sufficiently different from the togaviruses or flaviviruses to warrant inclusion in a new virus family. However, of these viruses only EAV causes arteritis, whereas the other three viruses cause diseases with different clinical symptoms. We therefore feel that an appropriate name for this new group of viruses still has to be found. We are in favor of a name, related to the macrophage tropism of these viruses in their natural hosts.

Acknowledgements

We thank Drs. P. Plagemann and M. Brinton for providing additional sequence data. Part of this work was supported by Boehringer Ingelheim, Germany, and the Produktschap voor Vee en Vlees (PVV), the Netherlands.

References

1. Hill H (1990) Overview and history of mystery swine disease (swine infertility respiratory syndrom). In: Proceedings Mystery Swine Disease Committee Meeting, October 6, 1990, Denver, Colorado. Livestock Conservation Institute, Madison, pp 29–31
2. Paton DJ, Brown IH, Edwards S, Wensvoort G (1991) Blue ear disease of pigs. Vet Rec 128: 617
3. Wensvoort G, Terpstra C, Pol JMA, Ter Laak EA, Bloemraad RA, de Kluyver EP, Kragten C, van Buiten L, den Besten A, Wagenaar F, Broekhuijsen JM, Moonen PLJM, Zetstra T, de Boer EA, Tibben HJ, de Jong MF, van 't Veld P, Groenland GJR, van Gennep JA, Voets MTh, Verheijden JHM, Braamskamp J (1991) Mystery swine disease in the Netherlands: the isolation of Lelystad virus. Vet Q 13: 121–130
4. Collins JE, Benfield DA, Christianson WT, Harris L, Hennings J, Shaw DP, Goyal SM, McCullough S, Morrison RB, Joo HS, Gorcyca D, Chladek D (1992) Isolation of swine infertility and respiratory syndrome virus (Isolate ATCC VR 2332) in North America and experimental reproduction of the disease in gnotobiotic pigs. J Vet Diagn Invest 4: 117–126
5. Pol JAM, Wagenaar F (1992) Morphogenesis of Lelysted virus in porcine lung alveolar macrophages. Am Ass Swine Practit 4: 29
6. Meulenberg JJM, Hulst MM, de Meijer EJ, Moonen PLJM, den Besten A, de Kluyver EP, Wensvoort G, Moornmann RJM (1992) Lelystad virus, the causative agent of porcine epidemic abortion and respiratory syndrome (PEARS), is related to LDV and EAV. Virology 192: 62–72
7. Snijder EJ, den Boon JA, Bredenbeek PJ, Horzinek MC, Rijnbrand R, Spaan WJM (1990) The carboxyl-terminal part of the putative Berne virus polymerase is

expressed by ribosomal frameshifting and contains sequence motifs which indicate that toro- and coronaviruses are evolutionary related. Nucleic Acids Res 18: 4535–4542

8. den Boon JA, Snijder EJ, Chirnside ED, de Vries AAF, Horzinek MC, Spaan WJM (1991) Equine arteritis virus is not a togavirus but belongs to the coronavirus superfamily. J Virol 65: 2910–2920

9. Kuo L, Harty JT, Erickson L, Palmer GA, Plagemann PGWW (1991) A nested set of eight RNAs is formed in macrophages infected with lactate dehydrogenase-elevating virus. J Virol 65: 5118–5123

10. Boursnell MEG, Brown TDK, Foulds IJ, Green PF, Tomley FM, Binns MM (1987) Completion of the sequence of the genome of the coronavirus avian infectious bronchitis virus. J Gen Virol 68: 57–77

11. Brierly I, Boursnell MEG, Binns MM, Bilimoria B, Blok VC, Brown TDK, Inglis SC (1987) An efficient ribosomal frameshifting signal in the polymerase-encoding region of the coronavirus IBV. EMBO J 6: 3779–3785

12. Lee H, Shieh C, Gorbalenya AE, Koonin EV, Monica NL, Tuler N, Bagdzhadzhyan A, Lai MMC (1991) The complete sequence (22 kilobases) of murine coronavirus gene 1 encoding the putative proteases and RNA polymerase. Virology 180: 567–582

13. Brierley I, Diggard P, Inglis SC (1989) Characterization of an efficient coronavirus ribosomal frameshifting signal: requirement for an RNA pseudoknot. Cell 57: 537–547

14. Pearson WR, Lipman DJ (1988) Improved tools for biological sequence comparison. Proc Natl Acad Sci USA 85: 2444–2448

15. Plagemann PGW, Moennig V (1991) Lactate dehydrogenase-elevating virus, equine arteritis virus, and simian hemorrhagic fever virus: a new group of positive-strand RNA viruses. Adv Virus Res 41: 99–192

Authors' address: Dr. J. J. M. Meulenberg, Central Veterinary Institute, Department of Virology, Houtribweg 39, 8221 RA Lelystad, The Netherlands.

Virus receptors

Arch Virol (1994) [Suppl] 9: 451–459

Archives
Vˢ̣ᵗᵒ̣ᶠirology

© Springer-Verlag 1994
Printed in Austria

Recognition of cellular receptors by bovine coronavirus

B. Schultze and **G. Herrler**

Institut für Virologie, Philipps-Universität Marburg, Marburg,
Federal Republic of Germany

Summary. Bovine coronavirus (BCV) initiates infection by attachment to cell surface receptors the crucial component of which is N-acetyl-9-O-acetylneuraminic acid. Inactivation of receptors by neuraminidase treatment and restoration of receptors by enzymatic resialylation of asialo-cells is described as a method to determine (i) the type of sialic acid that is recognized; (ii) the linkage specificity of the viral binding activity; (iii) the minimal amount of sialic acid required for virus attachment. Evidence is presented that both glycoproteins and glycolipids can serve as receptors for BCV provided they contain 9-O-acetylated sialic acid. A model is introduced proposing that after initial binding to sialic acid-containing receptors, the S-protein of BCV interacts with a specific protein receptor. This interaction may result in a conformational change that exposes a fusogenic domain and thus induces the fusion between the viral and the cellular membrane.

Introduction

The interactions between viruses and cell surface receptors are of major interest because they often determine the cell tropism of the respective virus. In addition they are a promising target for antiviral chemotherapy. Recently progress was made in elucidating the receptors for several coronaviruses. The picture emerging from the data reported appears rather complex. A member of the carcinoembryonic antigen family of proteins has been shown to serve as a receptor for mouse hepatitis virus (MHV) strain A59 [1]. Transmissible gastroenteritis virus (TGEV) and human coronavirus, strain 229E (HCV-229E), have been reported to use aminopeptidase N as a receptor for the infection of cells [2, 3]. Another variation in the recognition of receptors has been reported for bovine coronavirus (BCV). The strategy this virus uses to attach to the cell surface is similar to that of influenza C virus, i.e. the major receptor-determinant recognized is N-acetyl-9-O-acetylneuraminic acid (Fig. 1)

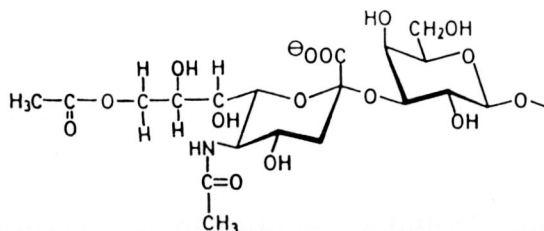

Fig. 1. Structure of N-acetyl-9-O-acetylneuraminic acid, the receptor determinant for BCV. The sugar is attached in an α2,3-linkage to galactose. Another common linkage type is the α2,6-linkage

[4–7]. This sugar is required on the cell surface not only to bind to erythrocytes in the hemagglutination reaction, but also to attach to and subsequently infect cultured cells [8, 9].

The importance of CEA and aminopeptidase as virus receptors has been established by two experimental techniques: (i) Monoclonal antibodies directed against these surface proteins are able to prevent virus infection. (ii) Receptor-negative cells become susceptible to infection after expression of the corresponding gene from permissive cells. These approaches are not applicable to viruses that use a sugar as a receptor determinant, as is the case with BCV.

Analysis of sialic acid as receptor determinant for viruses

Sialic acid is recognized as a receptor determinant by more viruses than any other determinant known. Both enveloped viruses (influenza viruses, some paramyxoviruses, some coronaviruses) and non-enveloped viruses (reoviruses, encephalomyocarditis virus, polyomavirus) have been reported to attach to glycoconjugates containing this type of sugar [10]. In order to determine the importance of sialic acid for binding of virus to cells, the approach shown in Fig. 2 has been used with success. Cells may contain different types of sialic acid, i.e. derivatives of neuraminic acid with different substitutents at the amino group (N-acetyl or N-glycolylneuraminic acid) or the various hydroxyl groups, e.g. 9-O-acetyl-N-acetylneuraminic acid. The sialic acid molecules may be attached to the glycoconjugates in different linkages, e.g. α2,3 or α2,6-linked to galactose (Fig. 1). Incubation of cells with neuraminidase results in the release of surface-bound sialic acid (Fig. 2). These asialo-cells are expected to be resistant to infection, if a virus requires sialic acid for attachment to cells. However, resistance to infection can also be due to a nonspecific effect, e.g. due to a change in the nagative charge of the cell surface. Such effects can be ruled out by showing that the virus specifically recognizes a certain type of sialic acid or a certain linkage type. For

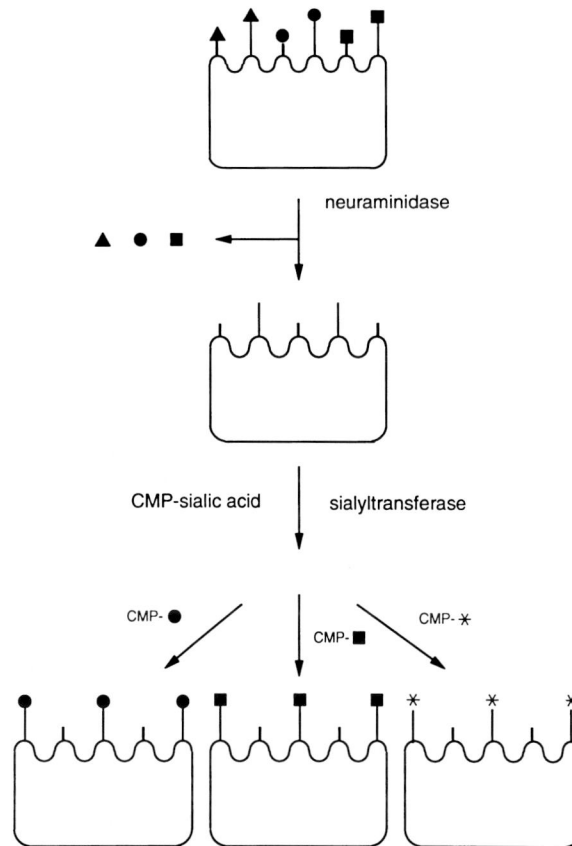

Fig. 2. Schematic presentation of the resialylation method. Sialic acid is released from the cell surface by treatment with neuraminidase. Asialo cells are resialylated by incubation with sialyltransferase and CMP-activated sialic acids. Because of the specificity of the enzyme and the choice of sialic acid, cells are obtained which contain only one type of sialic acid in a defined linkage type on the surface. Squares, circles, triangles, and asterisks represent different types of sialic acid. Long and short bars represent different linkage types, e.g. α2,3 and α2,6-linkage to galactose

this purpose, asialo-cells are incubated with sialyltransferase and CMP-activated sialic acid. Because of the specificity of the enzyme, sialic acids are attached to cell surface glycoconjugates in a defined linkage, e.g. α2,6-linked to galactose. The choice of CMP-activated sialic acid determines what type of sialic acid is transferred to the cell surface. In this way it is possible to obtain cells that contain only a single type of sialic acid attached in a single type of linkage (Fig. 2). The resialylated cells are tested either for agglutination by virus, in the case of erythrocytes, or for their susceptibility to infection, in the case of cultured cells.

Importance of 9-O-acetylated sialic acid as a receptor determinant for bovine coronavirus

Using the approach described above, it has been shown that BCV uses N-acetyl-9-O-acetylneuraminic acid as a receptor determinant not only for agglutination of erythrocytes, but also for infection of cultured cells [5, 8]. Sialic acid lacking a 9-O-acetyl group is not effective in this respect. Studies with erythrocytes indicate that hemagglutinating encephalomyelitis virus (HEV) and human coronavirus OC43 also require Neu5,9Ac$_2$ for attachment to cells [4, 5]. The resialylation method can be applied to compare the efficiency of two viruses in recognizing the same receptor determinant. For this purpose, aliquots of cells are resialylated using limiting concentrations of the activated sialic acid. This produces cells that have different amounts of sialic acid and, therefore, different numbers of receptors on their surface. As shown in Table 1, influenza C virus is more capable than is BCV in recognizing Neu5,9Ac$_2$ present in an α2,6-linkage; alternatively, if the same type of sugar is present in an α2,3-linkage, BCV requires a lower amount of sialic acid than does influenza C virus for agglutination of erythrocytes. Thus, the two viruses recognize Neu5,9Ac$_2$ in both linkage types but with different efficiencies.

Glycoproteins and glycolipids as receptors for BCV

As sialic acid is present on both glycoproteins and glycolipids, the question arises whether both types of glycoconjugates can serve as receptors for BCV. The sialyltransferase used for successful restoration of receptors on MDCK I cells was specific for the α2,6-linkage [8]. As this linkage is typically found on oligosaccharides of glycoproteins but not on gangliosides, the receptors generated by α2,6 sialyltransferase are glycoproteins. Thus, glycoproteins containing Neu5,9Ac$_2$ can serve as receptors for BCV. Hemagglutination-inhibition studies indicated that not only glycoproteins but also gangliosides can prevent BCV from agglutinating red blood cells. Bovine brain gangliosides are known to contain Neu5,9Ac$_2$ and, therefore, were analyzed for their ability to serve as virus receptors. If cells are incubated with gangliosides, the glycolipids attach to the cell surface via ionic interactions. Part of the bound gangliosides are incorporated into the lipid bilayer of the plasma membrane. As shown in Table 2, bovine brain gangliosides restored the susceptibility of neuraminidase-treated cells to infection to the same extent as resialylation. This result indicates that BCV can use both glycoproteins and glycolipids as receptors, provided they contain Neu5,9Ac$_2$.

Table 1. Comparison of the ability of bovine coronavirus (BVC) and influenza C virus (JHB/1/66) to agglutinate erythrocytes containing N-acetyl-9-O-acetylneuraminic acid in either of two different linkage types

Erythrocytes	Hemagglutination titer (units/ml)	
	BCV	C/JHB/1/66
Asialo	<2	<2
Resialylated		
Neu5Acα2,3Galβ1,4GlcNAc		
CMP-Neu5Ac, 0.50 nmol	64	<2
1.00 nmol	256	<2
2.00 nmol	2 048	128
Neu5Acα2,6Galβ1,4GlcNAc		
CMP-Neu5Ac, 0.25 nmol	<2	512
0.50 nmol	<2	512
1.00 nmol	<2	512
2.00 nmol	<2	512
4.00 nmol	128	512

Table 2. Restoration of receptors for bovine coronavirus on neuraminidase-treated MDCK I cells by resialylation or by incubation with bovine brain gangliosides

Treatment	Hemagglutination titer (units/ml) of the supernatant
None	256
Neuraminidase	<2
Neuraminidase, resialylated	32
Neuraminidase, gangliosides	32

The S protein is the receptor-binding protein of coronaviruses

Differences among coronaviruses have been reported not only with respect to their cellular receptors but also with respect to their viral receptor-binding proteins [11]. In the case of infectious bronchitis virus (IBV), transmissible gastroenteritis virus (TGEV), and related viruses, the S protein is responsible for attachment to cells. BCV and the serologically related viruses HEV and HCV-OC43 contain an HE protein missing from TGEV, IBV, and related viruses. The HE protein of BCV has acetylesterase activity [12, 13]. In addition, it has been suggested to function as a hemagglutinin, i.e. to be responsible for attachment to erythrocytes [14]. Surprisingly, the S protein of BCV was found recently to have hemagglutinating activity, too. Both proteins use Neu5,9Ac$_2$ as a receptor-determinant for attachment to cells [8]. However, the S protein is a more efficient hemagglutinin than is HE, because it requires less 9-O-acetylated sialic acid, i.e. fewer receptor determinants, on the cell surface for the agglutination of erythrocytes [8]. Therefore, the S protein is suggested to be the major receptor-binding protein of all coronaviruses.

Is one virus receptor sufficient for infection?

As described in the introductory section, different receptors have been described for different coronaviruses, a sugar as receptor determinant for BCV, and specific proteins as receptors for MHV, TGEV, and HCV-229E. It may appear surprising that coronaviruses apply such different strategies for binding to the cell surface. However, initiation of an infection is a complex process that comprises both attachment to the cell surface and, in the case of enveloped viruses, fusion between the membranes of the virus and the target cell. For some viruses a single type of receptor may be sufficient to induce both events. The hemagglutinin of influenza viruses, for example, binds to sialic acid-containing receptors and causes fusion after a conformational change which exposes a fusogenic epitope. The change in the conformation is triggered by the acid pH of the endosome after endocytotic uptake of the virus. However, the fusion activity of many viruses, including coronaviruses, is acid-independent. For these viruses fusion must be activated by factors other than pH.

The interaction of the viral fusion protein with a cell surface protein might be such a factor, as shown by our model of the inital events of an infection by BCV (Fig. 3). The first contact between the virus and the cell is established by binding of the S protein to 9-O-acetylated sialic acid (Fig. 3A). The initial attachment sites are most likely glycoproteins. On cells that lack appropriate glycoproteins, gangliosides containing Neu5,9Ac$_2$ can serve as receptors as well. In fact, it may be favorable for the virus to proceed from the glycoprotein receptor to glycoplipids in order to obtain closer contact between the viral and cellular lipid bilayer. The receptor-destroying enzyme, i.e. the acetylesterase of BCV, might be helpful in accomplishing this transition. Binding to sialic acid-containing receptors appears not to be sufficient for BCV to trigger the fusion event, because BCV does not cause hemolysis of erythrocytes, despite efficient agglutination. Therefore, we assume that the binding of the S protein to a second type of receptor (Fig. 3B) is required to induce a conformational change, one which results in exposure of a fusiogenic

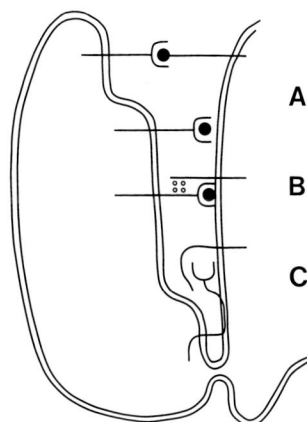

Fig. 3. Model of the initial events of an infection by BCV. Virus attaches to the cell surface by binding of the S protein to 9-O-acetylated sialic acid (solid circle) of glycoproteins or glycolipids (A). Interaction with a specific protein, which serves as a second receptor (B), results in a conformational change and exposes a fusogenic epitope of the S protein. This epitope interacts with the plasma membrane (C) and induces the fusion of the viral envelope with the cellular membrane

epitope (Fig. 3C). Interaction of the fusion-active S-protein with the target membrane destabilizes lipid bilayers and allows fusion between the virus and the cellular membrane.

The existence of a second receptor for BCV remains to be proven. It is interesting to note that coronaviruses, which recognize specific proteins as receptors, also have a sialic acid-binding activity. TGEV has been shown recently to agglutinate erythrocytes after pretreatment of the virus with neuraminidase to release inhibitory compounds from the viral surface [16]. As has been reported for infectious bronchitis virus, sialic acid α2,3-linked to cell surface compounds serves as a receptor determinant for TGEV on erythrocytes [16, 17]. Some murine corona-viruses are able to agglutinate mouse erythrocytes, suggesting that they recognize 9-O-acetylated sialic acid [18]. Thus, the ability to bind to two types of receptors is not restricted to BCV. However, the importance of the sugar-binding activity may vary among coronaviruses. Whereas BCV requires 9-O-acetylated sialic acid for efficient attachment to cultured cells, other viruses, such as MHV, may have optimized the recognition of the protein receptor in such a way that they are no longer dependent on the recognition of a carbohydrate receptor determinant.

Acknowledgements

We are grateful to Dr. Dr. R. Brossmer for providing CMP-Neu5,9Ac$_2$. This work was supported by grants from Deutsche Forschungsgemeinschaft (He 1168/2-1,2).

458   B. Schultze and G. Herrler

References

1. Dveksler GS, Pensiero MN, Cardelichio CB, Williams RK, Jiang G-S, Holmes KV, Dieffenbach CW (1991) Cloning of the mouse hepatitis virus (MHV) receptor: expression in human and hamster cell lines confers susceptibility. J Virol 65: 6881–6891

2. Delmas B, Gelfi J, L'Haridon R, Vogel LK, Sjöström H, Noren O, Laude H (1992) Aminopeptidase N is a major receptor for the enteropathogenic coronavirus TGEV. Nature 357: 417–420

3. Yeager CL, Ashmun RA, Williams RK, Cardellichio CB, Shapiro LH, Look AT, Holmes KV (1992) Human aminopeptidase N is a receptor for human coronavirus 229E. Nature 357: 420–422

4. Vlasak R, Luytjes W, Spaan W, Palese P (1988) Human and bovine coronaviruses recognize sialic acid containing receptors similar to those of influenza C virus. Proc Natl Acad Sci USA 85: 4526–4529

5. Schultze B, Gross H-J, Brossmer R, Klenk H-D, Herrler G (1990) Hemagglutinating encephalomyelitis virus attaches to N-acetyl-9-O-acetylneuraminic acid-containing receptors on erythrocytes: comparison with bovine coronavirus and influenza C virus. Virus Res 16: 185–194

6. Herrler G, Rott R, Klenk H-D, Müller H-P, Shukla AK, Schauer R (1985) The receptor-destroying enzyme of influenza C virus is neuraminate-O-acetylesterase. EMBO J 4: 1503–1506

7. Rogers GN, Herrler G, Paulson JC, Klenk H-D (1986) Influenza C virus uses 9-O-acetyl-N-acetylneuraminic acid as a high affinity receptor determinant for attachment to cells. J Biol Chem 261: 5947–5951

8. Schultze B, Herrler G (1991) Bovine coronavirus uses N-acetyl-9-O-acetylneuraminic acid as a receptor determinant to initiate the infection of cultured cells. J Gen Virol 73: 901–906

9. Herrler G, Klenk H-D (1987) The surface receptor is a major determinant of the cell tropism of influenza C virus. Virology 159: 102–108

10. Lentz TL (1990) The recognition event between virus and host cell receptor: a target for antiviral agents. J Gen Virol 71: 751–766

11. Spaan W, Cavanagh D, Horzinek MC (1988) Coronaviruses: structure and genome expression. J Gen Virol 69: 2939–2952

12. Vlasak R, Luytjes W, Leider J, Spaan W, Palese P (1988) The E3 protein of bovine coronavirus is a receptor-destroying enzyme with acetylesterase activity. J Virol 62: 4686–4690

13. Schultze B, Wahn K, Klenk H-D, Herrler G (1991) Isolated HE protein from hemagglutinating encephalomyelitis virus and bovine coronavirus has receptor-destroying and receptor-binding activity. Virology 180: 221–228

14. King B, Potts BJ, Brian DA (1985) Bovine coronavirus hemagglutinin protein. Virus Res 2: 1010–1013

15. Schultze B, Gross H-J, Brossmer R, Herrler G (1991) The S protein of bovine coronavirus is a hemagglutinin recognizing 9-O-acetylated sialic acid as a receptor determinant. J Virol 65: 6232–6237

16. Schultze B, Enjuanes L, Cavanagh D, Herrler G (1993) N-acetylneuraminic acid plays a critical role for the haemagglutinating activity of avian infectious bronchitis virus and porcine transmissible gastroenteritis virus. In: Laude H, Vautherot JF (eds) Coronaviruses: molecular biology and pathogenesis (in press)

17. Schultze B, Cavanagh D, Herrler G (1992) Neuraminidase treatment of avian infectious bronchitis coronavirus reveals a hemagglutinating activity that is dependent on sialic acid-containing receptors on erythrocytes. Virology 189: 792–794
18. Sugiyama K, Amano Y (1980) Hemagglutination and structural polypeptides of a new coronavirus associated with diarrhea in infant mice. Arch Virol 66: 95–105

Authors' address: Dr. G. Herrler, Institut für Virologie, Philipps-Universität, Biegenstrasse 10, D-35037 Marburg, Federal Republic of Germany.

Arch Virol (1994) [Suppl] 9: 461–471

Archives
V̌irology

© Springer-Verlag 1994
Printed in Austria

Mouse hepatitis virus receptors: more than a single carcinoembryonic antigen

K. Yokomori* and **M. M. C. Lai**

Howard Hughes Medical Institute and Department of Microbiology, University of Southern California School of Medicine, Los Angeles, California, U.S.A.

Summary. Mouse hepatitis virus (MHV), a murine coronavirus, has been shown to utilize carcinoembryonic antigen (CEA) as the receptor. We have demonstrated that MHV can utilize a different isoform of CEA, which is an alternatively spliced gene product that is expressed in different tissues, as a receptor. Furthermore, the CEA molecules from a resistant mouse strain (SJL) have different sequences and yet serve as functional viral receptors. Thus, MHV can use more than a single type of CEA molecule as the receptor. We have also shown that some mouse cell lines express functional CEA molecules and yet are resistant to infections by certain MHV strains. Biochemical studies of the infected cells indicate that MHV infections in these cell lines are blocked at the steps of virus entry. We conclude that MHV entry requires additional cellular factors other than CEA, the viral receptor. The significance of viral receptors and the additional cellular factors in regulating viral tropism is discussed.

Introduction

Viral infection of any cell type requires an initial interaction of virus with a specific cell surface molecule (viral receptor). This interaction usually leads to the entry of virus into cells by a mechanism which is not yet well understood. Different viruses utilize different cellular surface molecules as receptors. Depending on the distribution of these molecules in different cell types, different viruses infect different tissues. Thus, viral receptors play a crucial role in determining target cell specificity of virus. Viral receptors have been identified for several viruses, such as

*Present address: Department of Molecular and Cell Biology, University of California, Berkeley, California, U.S.A.

rhinoviruses [10], polioviruses [15], and human immunodeficiency virus (HIV) [14]. Many of these viral receptors have an immunoglobulin-like structure, with extracellular domains serving as virus-binding sites. However, virus binding, in many cases, is not sufficient for virus entry. For instance, HIV cannot infect several cell types that express CD4 molecules, the viral receptor [3, 14]. On the other hand, viruses occasionally infect cell types that do not express the prototype receptors. For example, HIV infects not only T cells, but also B cells, which do not express CD4 molecules [4, 7]. Thus, virus entry may require more than the presence of a specific receptor and also may have the flexibility of utilizing more than a single type of receptor. In this report, we present evidence that mouse hepatitis virus (MHV), a murine coronavirus, can utilize at least two different receptor molecules, which are differentially expressed in different tissues, and that MHV infection requires a second factor, which is more sensitively regulated and more discriminating than are the receptor molecules. MHV thus provides an excellent system for studying the mechanism of virus entry.

MHV infects the liver, intestine, brain or lymphoid tissue of mice, depending on the virus strain; for example, the JHM strain is neurotropic, causing either encephalitis or demyelinating diseases, while the A59 strain is relatively nonpathogenic [16]. The genetic background and age of mice also affect the susceptibility or resistance of mice to infection by different strains of MHV: for instance, BALB/C or C57BL mice are susceptible, while adult SJL mice are resistant to MHV-A59 or MHV-JHM and yet susceptible to MHV-3 infections [1]. MHV virions contain three or four structural proteins, among which the spike (S) protein mediates the interaction of MHV with the viral receptors on the surface of target cells [5]. Mutations of the S protein, such as in the case of neutralization-resistant variant viruses, often result in alterations of viral target cell specificity and pathogenicity [6]. Conceivably, viral receptors or associated cell surface molecules on different cell types are different, and variant forms of the S protein have different abilities to interact with different cell surface molecules, resulting in changes of target cell specificity. In this scenario, viral receptors in different cell types may show cell-specific variations. This possibility has so far not been demonstrated.

MHV can utilize two different receptors

The receptor for MHV has been identified as a member of the murine carcinoembryonic antigen (CEA) family, mmCGM1 [8, 19]. It is present in mouse liver and brush border membrane of gastrointestinal tract, which are targets of some MHV strains. However, it has not been

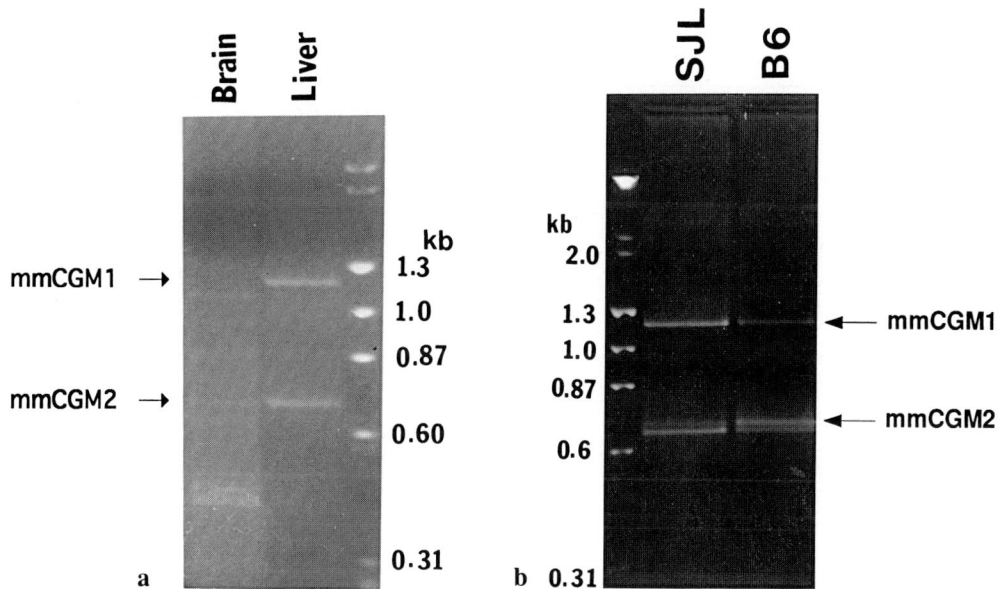

Fig. 1. Detection of MHV receptor-related molecules in different tissues and different mouse strains. RT-PCR was performed using two mmCGM1-specific primers: **a** Brain and liver RNAs from C57BL/6 mice. **b** Liver RNAs from SJL and C57BL/6 mice

detected in the brain, the target organ of neurotropic MHV strains, including MHV-JHM. Thus, it was not clear how MHV infects the central nervous system (CNS). We considered the possibility that brain tissue might contain a CEA-related molecule which is different from mmCGM1 but which can be used by MHV as an alternative receptor. To this end, we performed a reverse transcription-polymerase chain reaction (RT-PCR) amplification of CEA RNA from the mouse brain and liver tissue of C57BL/6 mice using two primers specific for mmCGM1 RNA. Figure 1a shows that the liver RNA yielded two PCR products. Sequence analysis of these two products showed that the larger one represents mmCGM1, the identified MHV receptor. The small one was an unexpected product, apparently an alternatively spliced product of the mmCGM1 gene with a deletion in the middle of the RNA (data not shown). The sequence and structure of this molecule are related to those of the reported mmCGM2 [18]. In contrast, the brain RNA yielded only the PCR product corresponding to mmCGM2. There were several other PCR products, which were misprimed PCR artifacts (Fig. 1a).

We then tested the possible viral receptor functions of these CEA-related molecules by cloning them into a mammalian expression vector, pECE, which has an SV40 T antigen promoter [9], and transfecting them into COS cells, which were originally resistant to MHV. Both the large

Table 1. Receptor functions of mmCGM molecules from SJL and C57BL/6 (B6) mice in COS cells[a]

	Virus titer (PFU/ml)	
Transfectant	JHM	A59
mmCGM1		
B6	1.1×10^3	4.4×10^3
SJL	2.1×10^2	2.0×10^3
mmCGM2		
B6	3.1×10^2	3.1×10^2
SJL	3.8×10^2	2.1×10^2
Vector (pECE)	0	2.0×10^1
None	0	3.3×10^1

[a] COS cells were transfected with various plasmids and infected with either JHM or A59 at 40 h posttransfection. Viruses were harvested from the supernatant 24 h after infection. The virus was plaque assayed on DBT cells

and small CEA molecules rendered COS cells susceptible to both MHV-A59 and MHV-JHM infection to roughly the same extent (Table 1). This result showed that both molecules can serve as MHV receptors. Because the small product (mmCGM2) is the major CEA-like molecule in the brain, it is likely the primary viral receptor used by MHV. In contrast, MHV probably can utilize either of the CEA molecules as the receptor in the liver. This result indicates that MHV can utilize at least two different receptor molecules, each of which is differentially expressed in different tissues.

The viral receptors in an MHV-resistant mouse strain are functional

We next examined whether the viral receptor plays a role in the genetic susceptibility or resistance of different mouse strains to MHV infections. Adult SJL mice have been known to be resistant to A59 and JHM infection (1) and the MHV receptor protein from this mouse strain has been reported to have a deletion and fail to bind MHV in an in vitro virus overlay binding assay [2, 20]. However, newborn SJL mice are susceptible to all MHVs, and MHV-3 can infect even adult SJL mice [1]; these findings are inconsistent with the possible structural defects of the receptors. We therefore re-examined this issue by studying the structure and receptor function of the CEA-like molecules from SJL mice. RT-PCR amplification of CEA RNAs was accomplished using two primers specific for the 5'- and 3'-ends, respectively, of mmCGM1. As shown in Fig. 1b, livers of SJL mice yielded two PCR products, each of which was identical in size to those from C57BL/6 mice. Sequence analysis of these two PCR products showed that both of the CEA-like molecules from SJL mice have the same length as that of the CEA molecules of C57BL/6 mice. However, there is approximately 10% sequence divergence, which is clustered mainly in the N-terminal one-third of the molecule, between the two mouse strains. Two of the potential N-glycosylation sites in C57BL/6 CEA molecules were mutated, but the SJL molecules gained another potential glycosylation site. As a result, the SJL CEA-like molecules lost a glycosylation site. This fact explains the apparently smaller size of the MHV-receptors in SJL mice, as reported previously [20]. To determine whether the SJL CEA molecules retain the MHV receptor functions, full-length mmCGM1 and mmCGM2 molecules in pECE plasmid were transfected into COS cells, and the cells were assayed for their susceptibility to infection with A59 and JHM viruses. Table 1 shows that COS cells transfected with SJL CEA molecules were infectable with both viruses and yielded roughly the same titers of viruses as those transfected with the corresponding C57BL/6 CEA molecules. These results indicate that MHV receptors in SJL mice are functional. Thus, we conclude that genetic resistance of SJL mice is not due to defectiveness of viral receptors, but rather due to another cellular factor required for viral infection.

Some murine cell lines are selectively resistant to certain MHV strains despite the presence of functional viral receptors

The results obtained with SJL receptors suggest that the establishment of MHV infection in mice requires more than expression of a functional viral receptor on target cells. To determine the possible additional cellular factors required for viral infection, we studied several murine cell lines derived from both susceptible and resistant mouse strains. We first examined the susceptibility of these cell lines to MHV infection. These cells were infected with A59 or JHM viruses, and the virus yields at various time points after infection were examined. Figure 2 shows that DBT cells, a murine astrocytoma cell line from BALB/C mice [11], yielded nearly the same titers of A59 and JHM. In contrast, MC7 and BXS cells from C57BL/6 mice [13] surprisingly were resistant to JHM infection, but allowed A59 virus infection, although the virus yield was slightly lower than that from DBT cells. Several other cell lines, including SSS, CSV and BC10, derived from either resistant SJL mice or susceptible BALB/C or C57BL/6 mice also showed a similar pattern of differential susceptibility to A59 and resistance to JHM strains, in

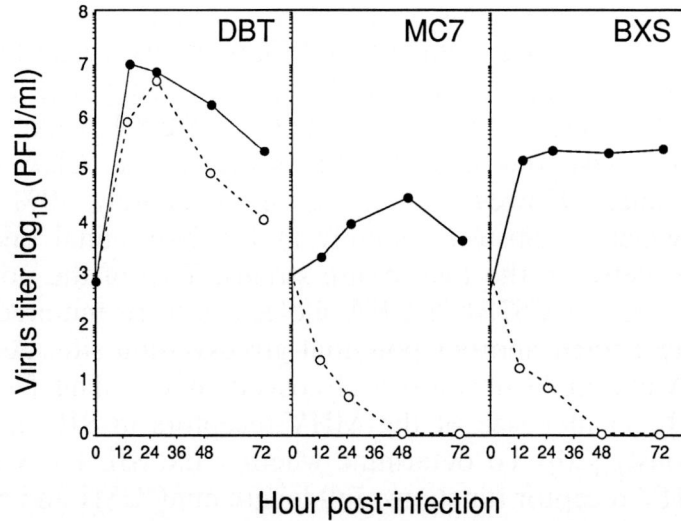

Fig. 2. Viral growth kinetics in different mouse cell lines. Virus infection was performed at a multiplicity of infection of 5. Virus yields were assayed at various time points post-infection; ● A59, ○ JHM

contrast to the patterns of viral resistance or susceptibility in parental mice (data not shown). Thus, there is no universal correlation between virus susceptibility in mice and in tissue culture. Most surprising were the significant differences observed between the susceptibility of cell lines to A59 and to JHM infection. Therefore, certain cellular factors essential for MHV infections may be regulated differently in animals and in tissue culture and can discriminate between different MHV strains. Because the MHV receptors have been shown to be functional for both A59 and JHM, these results suggest that a second cellular factor, more delicately regulated and more discriminating than the viral receptor, likely controls MHV infection.

The resistant cell lines are not defective in viral receptors

Several additional studies were done to determine the molecular basis of differential resistance of these cell lines to JHM infection. We first examined the expression of MHV receptors in these cells. Using mmCGM1-specific primers, RT-PCR amplification of the CEA-like RNAs revealed that all of the cell lines examined have a detectable level of mmCGM2-like molecules (Fig. 3). Although their levels of expression in these cell lines were not as high as that in mouse liver, they were comparable to that in the highly susceptible DBT cell line (data not shown). The mmCGM1 molecule was barely detectable in most of the cell lines examined. Thus, similar to the mouse brain, mmCGM2

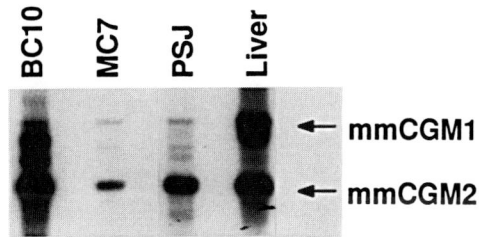

Fig. 3. Detection of MHV receptors in different mouse cell lines. RT-PCR was done as in Fig. 1, using RNA extracted from different mouse cell lines. Liver RNA was from C57BL/6 mice; the remainder represent RNAs from different murine cell lines tested

molecule is the major CEA molecule and probably the primary MHV receptor in these cell lines. We also have isolated mmCGM molecules from one of these cell lines by RT-PCR amplification and transfected them into COS cells. These molecules served as functional MHV receptors as efficiently as those from parental mice (data not shown). We conclude that the MHV receptor is expressed in these resistant cell lines and is functional. Therefore, these cells are likely defective in a second factor required for viral infection. This possibility was investigated by transfecting C57BL/6-derived mmCGM1 and mmCGM2 cDNA into these cell lines and determining whether resistance to JHM infections could be overcome by increased expression of viral receptors. The results showed that the resistance phenotype of these cells remained the same despite the expression of additional receptor molecules (data not shown). We conclude that the expression of receptor molecules in these cell lines is not sufficient to confer susceptibility to JHM virus infection.

Viral replication is blocked at an early step of JHM infection in resistant cell lines

To determine the steps of viral replication that are blocked in these resistant cell lines, viral macromolecular synthesis in various cell lines infected with JHM virus was examined by Northern blot analysis of viral RNA and polyacrylamide gel electrophoresis of ^{35}S-methionine-labeled viral proteins. Figure 4 shows that JHM viral RNA and proteins were detectable only in susceptible DBT cells. Thus, blockade of viral replication in resistant cell lines most likely occurs before initiation of viral RNA synthesis. Next, we transfected purified viral genomic RNA into these cell lines and examined virus yields from them. The results showed that both A59 and JHM viral genomic RNA led to the production of infectious virus particles from all cell lines transfected, although virus titers were, in general, lower for JHM than for A59 (data not shown).

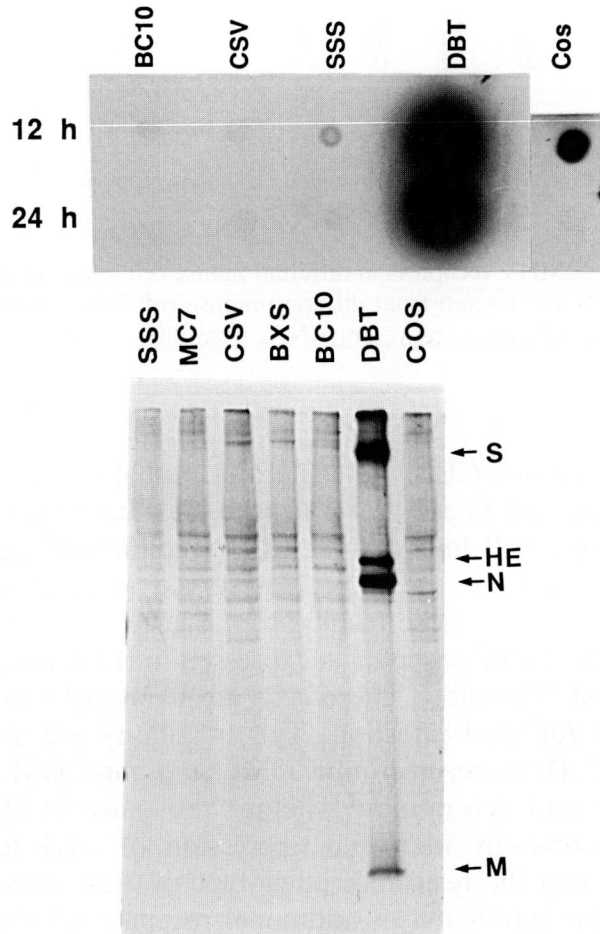

Fig. 4. Viral macromolecular synthesis in the different mouse cell lines infected with JHM virus. **a** Dot blot analysis of viral RNA from JHM-infected cells (isolated at 12 and 24 h post-infection). The probe used was ^{32}P-labeled RNA complementary to JHM mRNA 7. **b** Metabolic labeling with ^{35}S-methionine of JHM virus-infected cells. Viral proteins were precipitated with an antibody specific for JHM virus and separated by polyacrylamide gel electrophoresis. Viral proteins (*S*, *HE*, *N*, and *M*) are indicated

These results indicate that these cell lines are capable of supporting viral RNA synthesis and subsequent steps of viral replication and maturation, although we cannot rule out an additional possibility that JHM RNA replicates less efficiently. Therefore, blockade of viral replication in these cell lines likely occurs at an early step of viral replication, most probably at the level of virus entry.

Implications for virus entry mechanisms

The studies presented here revealed several interesting findings about virus entry. First, viruses can utilize multiple cell surface molecules as receptors; different tissues may express different viral receptors, so that receptor expression is a potential mechanism for regulation of viral tissue tropism. Conceivably, different receptor molecules can be utilized by different viruses at different efficiencies, thus accounting for the tissue tropism of different viral strains. In addition to mmCGM1 and mmCGM2 molecules, our recent preliminary studies have identified yet another species of CEA-like molecules, one which is expressed exclusively in the mouse brain. If this molecule is utilized only by neurotropic MHVs as the receptor, it may explain the target cell specificity of neurotropic MHV strains.

Second, the molecular basis of the viral resistance of certain mouse strains, such as the SJL mouse, does not reside in the structure or expression of the viral receptor. Although the receptor protein of SJL mice appears to be smaller than those from susceptible mice, and its in vitro virus-binding activity is deficient [20], these molecules serve as perfectly functional MHV receptors. Therefore, the genetic resistance of these mice is likely due to defects in some other cellular factors required for virus entry or to production of interfering factors. Because it has previously been shown that the primary macrophages obtained from SJL mice are also resistant to JHM infection in vitro [12, 17], the defect most likely resides in an intrinsic cellular factor required for viral infection.

Third, the resistance or susceptibility of some murine cell lines is different from those of the parental mouse strains, although the structure and expression of the viral receptors are comparable. Thus, a cellular factor other than the viral receptor is differentially expressed in animals and in tissue culture and is required for virus entry. Most interestingly, this additional factor can discriminate JHM from A59 infection, in contrast to CEA molecules, which are functional for both viruses. Thus, the expression of this second factor is more delicately regulated and its function is more discriminating than the primary receptor itself. This second factor likely works at the virus entry step. Although the nature of this factor is not yet known, preliminary data suggest that it functions through interaction with the viral spike protein, similar to the mechanism of action of viral receptors. Therefore, this second factor likely also interacts directly or indirectly with viral receptors. This factor, thus, provides a new tool for understanding the process of viral infection and the regulation of viral tissue tropism.

References

1. Barthold SW, Beck DS, Smith AL (1986) Mouse hepatitis virus nasoencephalopathy is dependent upon virus strain and host genotype. Arch Virol 91: 247–256
2. Boyle JF, Weismiller DG, Holmes KV (1987) Genetic resistance to mouse hepatitis virus correlates with the absence of virus-binding activity of target tissues. J Virol 61:185–189
3. Chesebro B, Buller R, Portis J, Wehrly K (1990) Failure of human immunodeficiency virus entry and infection in CD4-positive human brain and skin cells. J Virol 64: 215–221
4. Clapham PR, Weber JN, Whitby D, McIntosh K, Dalgleish AG, Maddon PJ, Deen KC, Sweet RW, Weiss RA (1989) Soluble CD4 blocks the infectivity of diverse strains of HIV and SIV for T cells and monocytes but not for brain and muscle cells. Nature 337: 368–370
5. Collins AR, Knobler RL, Powell H, Buchmeier MJ (1982) Monoclonal antibodies to murine hepatitis virus-4 (strain JHM) define the viral glycoprotein responsible for attachment and cell-cell fusion. Virology 119: 358–371
6. Dalziel RG, Lampert PW, Talbot PJ, Buchmeier MJ (1986) Site-specific alteration of murine hepatitis virus type 4 peplomer glycoprotein E2 results in reduced neurovirulence. J Virol 59: 463–471
7. De-Rossi A, Roncella S, Calabro ML, Ardrea E, Pasti M, Panozzo M, Mammano F, Ferrarini M, Chieco-Bianchi L (1990) Infection of Epstein-Barr virus-transformed lymphoblastoid B cells by the human immunodeficiency virus: evidence for a persistent and productive infection leading to B cell phenotypic changes. Eur J Immunol 20: 2041–2049
8. Dveksler GS, Pensiero MN, Cardellichio CB, Williams RK, Jiang GS, Holmes KV, Dieffenbach CW (1991) Cloning of the mouse hepatitis virus (MHV) receptor: expression in human and hamster cell lines confers susceptibility to MHV. J Virol 65: 6881–6891
9. Ellis L, Clauser E, Morgan DO, Edery M, Roth RA, Rutter WJ (1986) Replacement of insulin receptor tyrosine residues 1162 and 1163 compromises insulin-stimulated kinase activity and uptake of 2-deoxyglucose. Cell 45: 721–732
10. Greve JM, Davis G, Meyer AM, Forte CP, Yost SC, Marlor CW, Kamarck ME, McClelland A (1989) The major human rhinovirus receptor is ICAM-1. Cell 56: 839–847
11. Hirano N, Fujiwara K, Hino S, Matsumoto M (1974) Replication and plaque formation of mouse hepatitis virus (MHV-2) in mouse cell line DBT culture. Arch Ges Virusforsch 44: 298–302
12. Knobler RL, Tunison LA, Oldstone MBA (1984) Host genetic control of mouse hepatitis virus type 4 (JHM strain) replication. I. Restriction of virus amplification and spread in macrophages from resistant mice. J Gen Virol 65: 1543–1548
13. Lindsley MD, Thiemann R, Rodrigues M (1991) Cytotoxic T cells isolated from the central nervous system of mice infected with Theiler's virus. J Virol 65: 6612–6620
14. Maddon PJ, Dalgleish AG, McDougal JS, Clapham PR, Weiss RA, Axel R (1986) The T4 gene encodes the AIDS virus receptor and is expressed in the immune system and the brain. Cell 47: 333–348
15. Mendelsohn CL, Wimmer E, Racaniello VR (1989) Cellular receptor for poliovirus: molecular cloning, nucleotide sequence, and expression of a new member of the immunoglobulin superfamily. Cell 56: 855–865

16. Robb JA, Bond CW (1979) Pathogenic murine coronaviruses. I. Characterization of biological behavior in vitro and virus-specific intracellular RNA of strongly neurotropic JHMV and weakly neurotropic A59 viruses. Virology 94: 352–370
17. Stohlman SA, Frelinger JA (1978) Resistance to fatal central nervous system disease by mouse hepatitis virus, strain JHM. I. Genetic analysis. Immunogenetics 6: 277–281
18. Turbide C, Rojas M, Stanners CP, Beauchemin N (1991) A mouse carcinoembryonic antigen gene family member is a calcium-dependent cell adhesion molecule. J Biol Chem 266: 309–315
19. Williams RK, Jiang GS, Holmes, KV (1991) Receptor for mouse hepatitis virus is a member of the carcinoembryonic antigen family of glycoproteins. Proc Natl Acad Sci USA 88: 5533–5536
20. Williams RK, Jiang GS, Snyder SW, Frana MF, Holmes KV (1990) Purification of the 110-kilodalton glycoprotein receptor for mouse hepatitis virus (MHV)-A59 from mouse liver and identification of a nonfunctional, homologous protein in MHV-resistant SJL/J mice. J Virol 64: 3817–3823

Authors' address: Dr. M. C. Lai, Howard Hughes Medical Institute and Department of Microbiology, University of Southern California School of Medicine, 2011 Zonal Avenue, Los Angeles, CA 90033, U.S.A..

Arch Virol (1994) [Suppl] 9: 473–484

Archives
Virology
© Springer-Verlag 1994
Printed in Austria

Host-cell receptors for Sindbis virus

J. H. Strauss[1], **K.-S. Wang**[1,*], **A. L. Schmaljohn**[2,**], **R. J. Kuhn**[1,***],
and **E. G. Strauss**[1]

[1] Division of Biology, California Institute of Technology, Pasadena, California
[2] Department of Microbiology and Immunology, University of Maryland
School of Medicine, Baltimore, Maryland, U.S.A.

Summary. Sindbis virus has a very wide host range, infecting many species of mosquitoes and other hematophagous insects and infecting many species of higher vertebrates. We have used two approaches to study host cell receptors used by Sindbis virus to enter cells. Antiidiotype antibodies to neutralizing antibodies directed against glycoprotein E2 of the virus identified a 63-kDa protein as a putative receptor in chicken cells. In a second approach, monoclonal antibodies identified a 67 kDa protein, believed to be a high affinity laminin receptor, as a putative receptor in mammalian cells and in mosquito cells. We conclude that the virus attains its very wide host range by two mechanisms. In one mechanism, the virus is able to use more than one protein as a receptor. In a second mechanism, the virus utilizes proteins as receptors that are highly conserved across the animal kingdom.

Introduction

The receptors used by viruses to enter cells are an exciting area of study. Receptors determine in large part the host range of a virus and the pathology of the disease caused by it. Alphaviruses have an enormous host range [1–3]. They are vectored by mosquitoes and can grow in several tissues of their invertebrate host, and alphaviruses, including Sindbis virus, have also been isolated from other hematophagous insects such as mites. Alphaviruses also replicate in vertebrate hosts. Although birds are the major vertebrate host for Sindbis, it also replicates in mammals and reptiles. Sindbis and viruses closely related to it can cause

Present addresses: * Chiron Corporation, Emeryville, California; ** U.S. Army Medical Research Institute of Infectious Diseases, Frederick, Maryland; *** Department of Biological Sciences, Purdue University, West Lafayette, Indiana, U.S.A.

significant human disease, including a polyarthritis that can last for months [2, 4–6].

Early studies reported that proteins were used as receptors by Sindbis [7]. That a virus with such a wide host range would use a protein receptor is interesting; most viruses with protein receptors have narrow host ranges and often a limited tissue tropism [8–12]. A wide host range could result from using a highly conserved receptor that has domains invariant among the different organisms infected, or from using more than one protein as a receptor, or both. We are studying the nature of Sindbis receptors using two approaches. In one approach, anti-idiotypic antibodies have been characterized as potential antireceptor antibodies [13], and in a second approach, antibodies that bind to the surface of a susceptible cell and block virus binding and infection have been isolated and characterized [14].

A chicken receptor identified by anti-idiotypic antibodies

Sindbis virus has an icosahedral nucleocapsid surrounded by a lipoprotein envelope containing two virus encoded glycoproteins, called E1 and E2 [15]. The surface spikes that attach to a host cell consist of three E1–E2 heterodimers. We prepared anti-idiotypic (α-Id) antibodies against several neutralizing monoclonal antibodies (mAbs) that reacted with either Sindbis E2 or E1. Neutralizing antibodies may neutralize virus infectivity by binding to the part of the virus glycoprotein (the virus antireceptor) that interacts with a cellular receptor, and α-Id antibodies have the potential to act as antireceptor antibodies [16]. α-Ids made to three different neutralizing mAbs reactive with Sindbis E2 were able to partially block virus binding to chicken embryo fibroblasts (CEF). Of these, α-Id 49 made against mAb 49 was the most efficient. This α-Id blocked ^{35}S-labeled virus binding to cells by 50% (Fig. 1A) and blocked virus infection of cells to the same extent as measured by plaque assay (Fig. 1B). α-Id 49 bound to the surface of CEF as determined by fluorescence-activated cell sorting (FACS) assays (Fig. 2A). Thus this antibody appears to be an antireceptor antibody. This α-Id immunoprecipitated a protein of 63 kDa from CEF (Fig. 3B), and this 63-kDa protein is a putative chicken receptor for Sindbis. However, because α-Id 49 was only 50% effective at blocking virus binding, we concluded that the virus must also use a second receptor in chicken cells that is not recognized by α-Id 49. This is further supported by the finding that a variant of Sindbis selected to be resistant to mAb 49, called v49, is not affected in its binding to CEF by α-Id 49 (Fig. 1C); therefore, v49 must use a receptor different from the 63-kDa protein, most probably the second receptor used by the wild type virus.

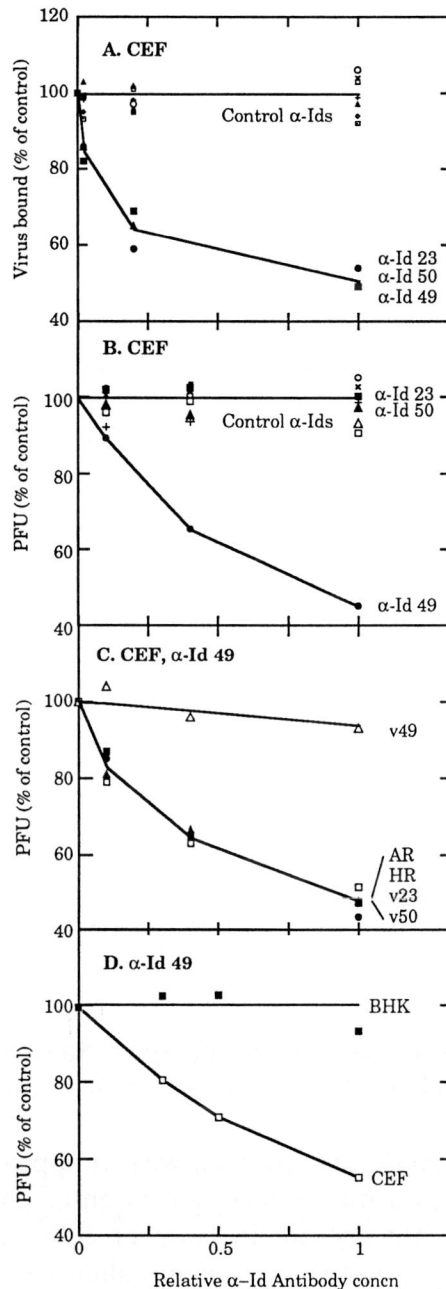

Fig. 1. α-Ids interfere with binding of Sindbis to CEF. **A** CEF monolayers were incubated with affinity purified α-Ids on ice for 60 min and then incubated with purified ^{35}S-labeled Sindbis for two hrs. After washing, cells were dissolved in SDS for assay of bound radioactivity. Control antibodies included α-Id 18, α-Id 33, α-Id K42, α-Id 53, and rabbit anti-mouse IgG. The maximum amount of antibody (= 1 on the abscissa) was 100 μg IgG. **B** CEF monolayers were incubated with different amounts of purified α-Ids at room temperature for 90 min, 200 PFU of Sindbis were added and incubation continued for 60 min. The inoculum was removed and the plates overlayed with Eagle's medium containing 1% agarose. Control α-Ids are the same as in **A**. The maximum amount of antibody was 250 μg IgG. **C** Plaque reduction by α-Id 49 for five different variants of Sindbis. Assay conditions are the same as in **B**. **D** Reduction of Sindbis plaques in CEF but not in BHK cells caused by α-Id 49. Assay conditions are the same as in **B**. Relative antibody concentration of 1 = 10% unpurified α-Id 49 antiserum

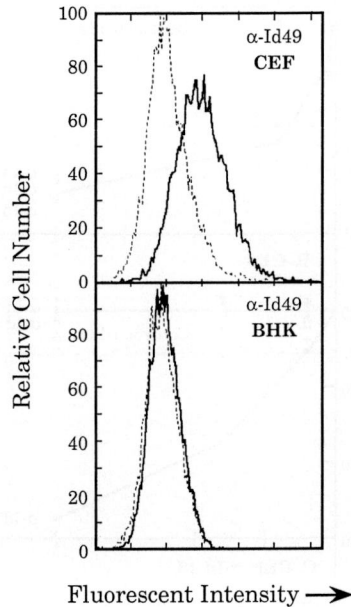

Fig. 2. Binding of α-Id 49 to the surface of cells. Cells were released from culture dishes with EDTA and treated with DNaseI. Ten µg affinity purified α-Id 49 were added per 10^6 cells on ice, followed by incubation with FITC-conjugated goat anti-rabbit IgG. Cells were washed, filtered through a nylon screen and stained with propidium iodide to detect dead cells. FACS analysis was performed using a modified Ortho System 50H flow cytometer. Dashed lines show the distribution of cells treated with preimmune serum, solid lines that with the α-Id 49

Surprisingly, α-Id 49 had no effect upon Sindbis binding to or infection of BHK cells (Fig. 1D). Similarly, the α-Id antibody did not bind to the surface of BHK cells (Fig. 2B), nor did it precipitate a protein from preparations of BHK cells (not shown). It appears that Sindbis uses a protein of 63 kDa as a major receptor to enter chicken cells but the equivalent protein is not present in hamster cells, and thus the major receptor used by the virus to enter mammalian cells must be different from the major receptor used to enter avian cells.

A monoclonal antibody that blocks Sindbis binding to mammalian cells

Since Sindbis appears to use different receptors in avian and mammalian cells, we wished to find an antireceptor antibody for mammalian cells. Mice were immunized with whole BHK cells, hybridomas made, and hybridoma supernatants tested for their ability to interfere with infection of BHK cells by Sindbis. One mAb, called 1C3, was capable of blocking virus binding to BHK cells by up to 80%, as measured either by binding of ^{35}S-labeled virus (Fig. 4B) or reduction of plaques formed in a plaque

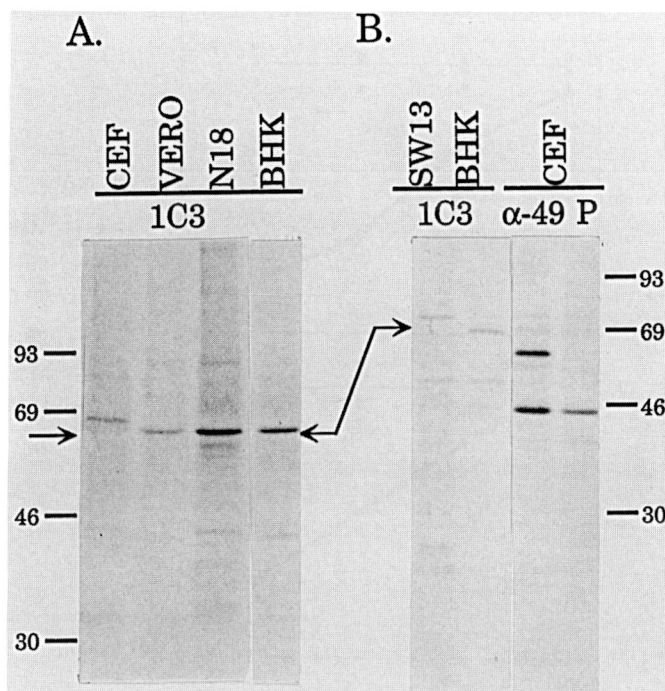

Fig. 3. Immunoprecipitation of cellular proteins by α-Id 49 and mAb 1C3. Membrane preparations from ^{35}S-labelled BHK cells, N18 neuroblastoma cells, Vero cells, or CEF were immunoprecipitated with 5 μg of purified mAb, with polyclonal α-Id 49, or with preimmune serum (P). **A** Immunoprecipitates were collected using rat anti-mouseIgM coupled to Sepharose 4B. **B** Comparison of proteins immunoprecipitated from mammalian cells with mAb 1C3 and chicken proteins precipitated with α-Id 49. The arrow indicates the 67-kDa BHK protein. Sizes of the molecular weight markers in kDa are indicated at the sides

assay (Fig. 4A). This antibody blocked virus binding to three different mammalian cells tested, but did not significantly block virus binding to chicken cells (Fig. 4C). The antibody immunoprecipitated a protein of 67 kDa from several mammalian cell lines and a protein of about 70 kDa from CEF (Fig. 3A). The 70-kDa chicken protein precipitated by 1C3 is clearly different from the 63-kDa chicken protein precipitated by α-Id 49 (Fig. 3B). The 70-kDa protein is presumably the analog of the mammalian 67-kDa protein, but does not appear to function as a significant receptor for entry of virus into CEF. In agreement with the result with α-Id antibodies, it appears that the virus uses different primary receptors for entry into avian cells versus mammalian cells.

mAb 1C3 also interfered with the binding of Sindbis to mosquito cells. At 20 μg 1C3/ml, 59% as much ^{35}S-labeled Sindbis bound to mosquito cells as was bound in the presence of control mAbs (versus 28% for BHK cells). This indicates that a protein reactive with mAb 1C3

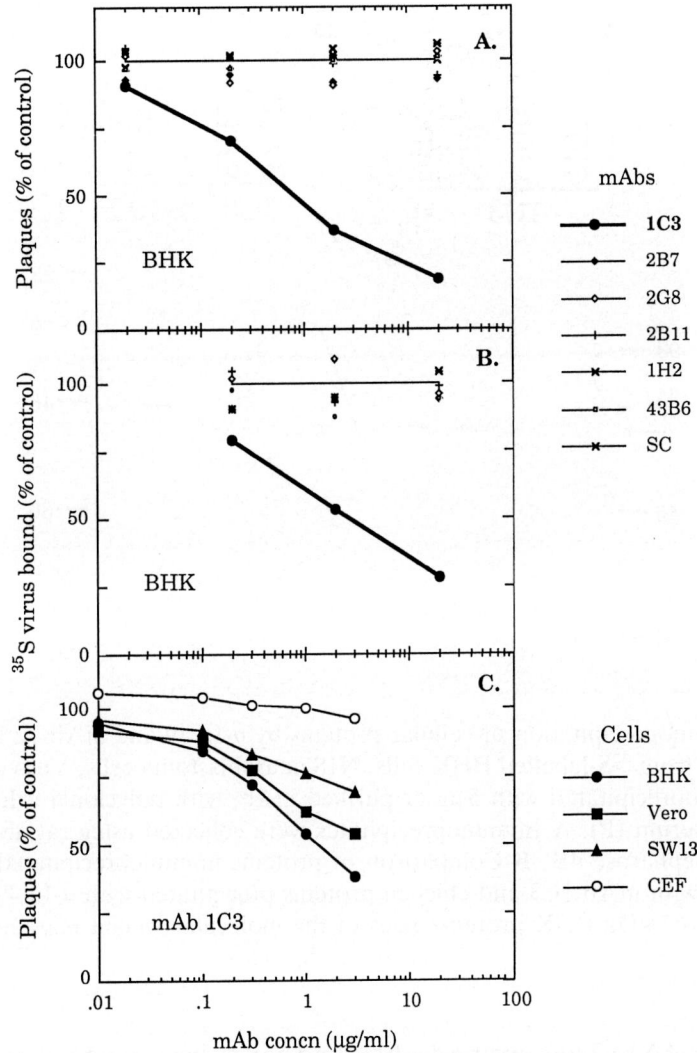

Fig. 4. mAb 1C3 blocks binding of Sindbis to mammalian cells. mAbs 1C3, 2B7, 2G8, 2H11, 1H2 are mAbs from mice immunized with whole BHK cells. 43B6 is a mAb from a mouse immunized with BHK membranes and SC is an anti-rat membrane protein mAb. Antibodies were affinity purified and the assays performed as in Fig. 1. **A** shows the effects of the different mAbs on the number of plaques formed in a suboptimal plaque assay; **B** shows the effects on binding of radiolabeled virus, both in BHK cells; **C** shows the effects of mAb 1C3 on virus binding to various types of cells, as determined by plaque assay

is highly conserved between mosquitoes and mammals and functions as a virus receptor in both.

The high affinity laminin receptor

To identify the BHK receptor protein reactive with mAb 1C3, a λgt11 expression library made with BHK cDNA was screened with this antibody. Several clones reactive with the antibody were identified and sequenced. mAb 1C3 was found to react with the C-terminal domain of a protein of 295 amino acids that has been identified with the high affinity laminin receptor. Laminin is an important component of basement membranes and laminin receptors, of which there are several, are important in the development of multicellular organisms and in metastasis of tumor cells. The high affinity laminin receptor is widely distributed and was originally identified as a 67-kDa protein that bound to laminin during affinity chromatography [17]. Although this 67-kDa protein has not been completely characterized, antibodies to it were found to react with cDNA clones encoding the same 295-residue open reading frame (ORF) identified by us [18]. The primary sequence of the 295-residue protein is highly conserved; there are two amino acid differences between the hamster protein and the human or bovine protein, and the hamster and mouse proteins are identical [19–21]. The exact nature of the high affinity laminin receptor is controversial. The 295-residue ORF produces a product of 37 kDa by translation in vitro or in vivo [19, 22]. A body of evidence has accumulated that the 295-residue protein and the 67-kDa laminin receptor share sequence identity, and it has been proposed that the 295-residue protein is a precursor to the 67-kDa form [23]. The 67-kDa protein is not glycosylated [19, 22], and the modification that might convert the smaller precursor to the 67-kDa protein has been postulated to involve covalent linkage to another protein [23]. A second school of thought has proposed that the 295-residue protein is not in fact a precursor, but rather a cytoplasmic protein associated with ribosomes, and that the mRNA for the 67-kDa protein has not yet been identified [19, 24]. In any event, the 295-residue and the 67-kDa proteins share sequence identity, and the extreme conservation of the 295-residue protein makes it an attractive candidate for a receptor used by a virus with such an extremely broad host range.

Overexpression of the 295-residue ORF makes a cell more susceptible to Sindbis virus

We wanted to know whether the high affinity laminin receptor, or at least a product derived from the 295-amino acid ORF, would serve as a receptor for Sindbis as determined by its ability to cause a transformed cell to bind more virus and become more sensitive to virus infection. Although it would be best to transform a resistant cell to one sensitive to infection by the virus, we could not do this because we have not identified a cell devoid of receptors for Sindbis. However, we reasoned that a cell expressing more receptors would bind more virus and would become infected more rapidly than a cell expressing fewer receptors. We constructed a full length clone of the 295-residue ORF and inserted this into an expression vector in which the ORF is expressed under the control of a high efficiency mammalian promoter. The expression vector also contained a G418 resistence gene. BHK cells were transfected and many independent clones, selected for resistence to G418, were screened for their susceptibility to Sindbis. We found that all clones tested were more sensitive to the virus, with 73% being three- to five-fold more susceptible to the virus (three- to five-fold more plaques were formed on the transfected cell monolayers than on cells transfected with vector not expressing the 295-residue ORF). In a complementary experiment, the 295-residue ORF was inserted in the antisense orientation and drug resistant clones were tested for their sensitivity to Sindbis virus. These cells were less sensitive to the virus, with approximately half of the clones being only one-third to one-half as sensitive to the virus.

Two independent BHK cell lines transfected with plus sense vectors, two transfected with minus sense vectors, and two transfected with vector alone, were compared to the parental BHK cells for their ability to bind mAb 1C3 at their surface, using a FACS assay [14]. The sense transformed lines expressed more 1C3-reactive protein at their surface and the antisense transformed lines expressed less 1C3-reactive protein at the surface. Quantitation showed that the two lines that overexpressed the sense RNA bound about threefold more mAb 1C3 than did the control cells; these cells were four times as sensitive to Sindbis virus by plaque assay. Conversely, the two lines transfected with antisense RNA bound only about 60% as much as cells transfected with the vector alone and were only 40% as susceptible to the virus. It is clear that the level of expression of the 295-residue ORF at the cell surface as determined by the ability of the cell to bind mAb 1C3 is correlated with the susceptibility of the cell to Sindbis virus.

In a second experiment, CHO cells were transfected using a different vector and one transfected clone studied in some detail. Figure 5 shows the binding of ^{35}S-labeled Sindbis to cells transfected with the laminin receptor and to cells transfected with vector alone. Binding in both cases was saturable, and the transfected cells bound 4.6 times as much virus as did the control cells. Moreover, there were 7.7 fold more plaques on monolayers of transfected cells than on the control cells.

Our results indicate that overexpression of the 295-amino acid ORF leads to an increase in the expression of a protein at cell surface that reacts with mAb 1C3 and this leads to increased binding of ^{35}S-labeled virus and to increased uptake of the virus as measured by a plaque assay. Quantitative analysis of our data indicates that binding of virus is directly proportional to the amount of 1C3-reactive protein at the cell surface but the susceptibility of the cell to virus infection as measured by plaque assay varies as the 1.4 power of the concentration of 1C3-reactive pro-

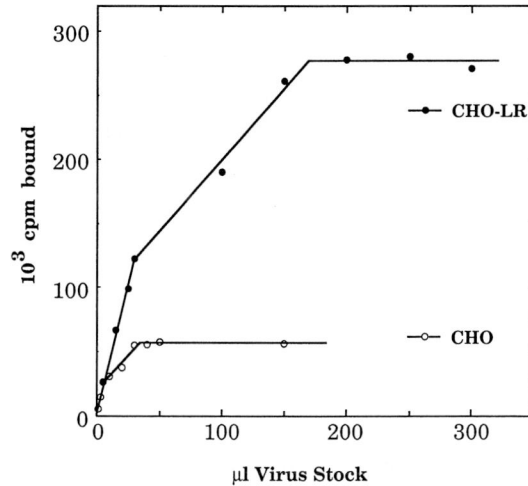

Fig. 5. Binding of Sindbis to CHO cells overexpressing the 295-residue ORF. CHO-LR cells were transfected with vector pEE14 containing the laminin receptor gene. Cell monolayers growing in 96-well plates were incubated with the indicated amounts of purified ^{35}S-labeled Sindbis for 90 min at 8°C. The monolayers were then washed and dissolved for assay of bound radioactivity

tein present on the surface, suggesting that there is some cooperativity in virus entry.

The nature of the protein at the cell surface that reacts with mAb 1C3 and that arises from expression of the 295-residue ORF remains to be determined. We are currently expressing antigenically tagged versions of this 295-residue ORF in an effort to ascertain the nature of this protein.

Conclusions

Our work indicates that Sindbis achieves a very wide host range by two different mechanisms. In one, it utilizes a protein receptor that is highly conserved across the animal kingdom, allowing it to enter both mosquito cells and mammalian cells and presumably to infect cells of many different tissues within these hosts. In a second, the virus utilizes more than one independent protein receptor and the major chicken receptor is different from the major hamster receptor. It is not clear whether the virus has more than one antireceptor domain in the glycoprotein spike allowing it to bind to two or more independent proteins, or whether the multiple protein receptors all interact with a common conformational epitope on the virus.

Other results also indicate that multiple receptors are used by Sindbis. Smith and Tignor [7] used competition binding studies to show that a neurovirulent strain of Sindbis bound to a different set of re-

ceptors than did nonneurovirulent strains, and that the receptors used
by the neurovirulent strain were also used in part by (neurovirulent)
Eastern equine encephalitis virus. In another study, Tucker and Griffin
[25] found that two strains of Sindbis that differed at a single amino acid
in glycoprotein E2 bound equally well to BHK cells but differed in their
ability to bind to cells of neuronal origin. The simplest interpretation of
these data is that the virus uses a different set of receptors on BHK cells
than on neuronal cells, and that these two strains of virus differ in their
ability to bind to these different receptors.

Studies to identify possible receptors for Sindbis also suggest that
multiple receptors are used. We have identified as putative Sindbis
receptors a 63-kDa protein in chicken cells and a 67-kDa protein in
mammalian cells. Maassen and Terhorst [26] used a cross-linking technique to identify a 90-kDa protein as a possible Sindbis receptor or part
of a receptor complex in a human lymphoblastic cell line. Ubol and
Griffin [27] used anti-idiotypic antibodies to identify proteins of 110-
and 74-kDa as putative receptors in neuronal cells. All these results
are consistent with the hypothesis that Sindbis uses multiple receptors
to enter cells, that the constellation of receptors on different cells is
different, and that changes in the virus glycoproteins can alter the binding affinity of the virus for these different receptors.

Acknowledgements

We are grateful to E. Rothenberg, P. Patterson, and T. Rümenapf for their advice and
interest in the project, to R. Diamond for help with the FACS analysis, to S. Ou and
C.-W. Yao for their efforts in obtaining mAb 1C3, and to H. Zhang for technical
assistance. This work was supported by NIH Grant AI 20612 to JHS and by ONR
Contract N00014-84-K-0536 to ALS. KSW was supported by NIH training grant GM
00086, and RJK was supported by NIH fellowship AI 07869.

References

1. Chamberlain RW (1980) Epidemiology of arthropod-borne togaviruses: the role of
   arthropods as hosts and vectors and of vertebrate hosts in natural transmission
   cycles. In: Schlesinger RW (ed) The togaviruses. Academic Press, New York, pp
   175–227
2. Peters CJ, Dalrymple JM (1990) Alphaviruses. In: Fields BN, Knipe DM (eds)
   Virology. Raven Press, New York, pp 713–761
3. Griffin DE (1986) Alphavirus pathogenesis and immunity. In: Schlesinger S,
   Schlesinger MJ (eds) The togaviridae and flaviviridae. Plenum, New York, pp
   209–250
4. Niklasson B, Aspmark A, LeDuc JW, Gargan TP, Ennis WA, Tesh RB, Main AJ
   (1984) Association of a Sindbis-like virus with Ockelbo disease in Sweden. Am J
   Trop Med Hyg 33: 1212–1217
5. Niklasson B (1988) Sindbis and Sindbis-like viruses In: Monath TP (ed) The
   arboviruses, epidemiology and ecology. CRC Press, Boca Raton, pp 167–176

6. Shirako Y, Niklasson B, Dalrymple JM, Strauss EG, Strauss JH (1991) Structure of the Ockelbo virus genome and its relationship to other Sindbis viruses. Virology 182: 753–764

7. Smith AL, Tignor GH (1980) Host cell receptors for two strains of Sindbis virus. Arch Virol 66: 11–26

8. Dalgleish AG, Beverley PCL, Clapham PR, Crawford DH, Greaves MF, Weiss RA (1984) The CD4 (T4) antigen is an essential component of the receptor for the AIDS retrovirus. Nature 312: 763–767

9. Klatzmann D, Champagne E, Chamaret S, Gruest J, Guetard D, Hercend T, Gluckman J-C, Montagnier L (1984) T-lymphocyte T4 molecule behaves as the receptor for human retrovirus LAV. Nature 312: 767–768

10. Staunton ED, Merluzzi VJ, Rothelein R, Barton R, Marlin SD, Springer TA (1989) A cell adhesion molecule, ICAM-1, is the major surface receptor for rhinoviruses. Cell 56: 849–853

11. Greve JM, Davis G, Meyer AM, Forte CP, Yost SC, Marlor CW, Karmarck ME, McClelland A (1989) The major human rhinovirus receptor is ICAM-1. Cell 56: 839–847

12. Mendelsohn CL, Wimmer E, Racaniello VR (1989) Cellular receptor for poliovirus: Molecular cloning, nucleotide sequence, and expression of a new member of the immunoglobulin superfamily. Cell 56: 855–865

13. Wang K-S, Schmaljohn AL, Kuhn RJ, Strauss JH (1991) Antiidiotypic antibodies as probes for the Sindbis virus receptor. Virology 181: 694–702

14. Wang K-S, Kuhn RJ, Strauss EG, Ou S, Strauss JH (1992) High affinity laminin receptor is a receptor for Sindbis virus in mammalian cells. J Virol 66: 4992–5001

15. Schlesinger M, Schlesinger S (1986) Formation and assembly of alphavirus glycoproteins. In: Schlesinger S, Schlesinger MJ (eds) The togaviridae and flaviviridae. Plenum, New York, pp 121–148

16. Gaulton GN, Greene MI (1986) Idiotypic mimicry of biological receptors. Ann Rev Immun 4: 253–280

17. Martin GR, Timpl R (1987) Laminin and other basement membrane components. Annu Rev Cell Biol 3: 57–85

18. Wewer UM, Liotta LA, Jaye M, Ricca GA, Drohan WA, Claysmith AP, Rao CN, Wirth P, Coligan J, Albrechtsen R, Mudry M, Sobel ME (1986) Altered levels of laminin receptor mRNA in various human carcinoma cells that have different abilities to bind laminin. Proc Natl Acad Sci USA 83: 7137–7141

19. Grosso LE, Park PW, Mecham RP (1991) Characterization of a putative clone for the 67-kilodalton elastin/laminin receptor suggests that it encodes a cytoplasmic protein rather than a cell surface receptor. Biochemistry 30: 3346–350

20. Makrides S, Chitpatima ST, Bandyopadhyay R, Brawerman G (1988) Nucleotide sequence for a major messenger RNA for a 40 kilodalton polypeptide that is under translational control in mouse tumor cells. Nucleic Acids Res 16: 2349

21. Yow H, Wong JM, Chen HS, Lee C, Steele GDS, Chen LB (1988) Increased mRNA expression of a laminin-binding protein in human colon carcinoma: complete sequence of a full-length cDNA encoding the protein. Proc Natl Acad Sci USA 85: 6394–6398

22. Castronovo V, Claysmith AP, Barker KT, Cioce V, Krutzsch HC, Sobel ME (1991) Biosynthesis of the 67 kDa high affinity laminin receptor. Biochem Biophys Res Commun 177: 177–183

23. Castronovo V, Taraboletti G, Sobel ME (1991) Functional domains of the 67-kDa laminin receptor precursor. J Biol Chem 266: 20440–20446

24. Auth D, Brawerman G (1992) A 33-kDa polypeptide with homology to the laminin receptor: component of translation machinery. Proc Natl Acad Sci USA 89: 4368–4372

25. Tucker PC, Griffin DE (1991) Mechanism of altered Sindbis virus neurovirulence associated with a single-amino-acid change in the E2 glycoprotein. J Virol 65: 1551–1557

26. Maassen AA, Terhorst C (1981) Identification of a cell-surface protein involved in the binding site of Sindbis virus on human lymphoblastic cell lines using a heterobifunctional cross-linker. Eur J Biochem 115: 153–158

27. Ubol S, Griffin DE (1991) Identification of a putative alphavirus receptor on mouse neural cells. J Virol 65: 6913–6921

Authors' address: Dr. J. H. Strauss, Division of Biology, California Institute of Technology, Pasadena, CA 91125, U.S.A.

Arch Virol (1994) [Suppl] 9: 485–494

Archives
of
Virology
© Springer-Verlag 1994
Printed in Austria

Cell surface receptor for ecotropic host-range mouse retroviruses: a cationic amino acid transporter

M. P. Kavanaugh,[2] **H. Wang,**[1] **C. A. R. Boyd,**[3] **R. A. North,**[2] and **D. Kabat**[1]

[1] Department of Biochemistry and Molecular Biology School of Medicine,
[2] Vollum Institute for Advanced Biomedical Research Oregon Health Sciences,
University Portland, Oregon, U.S.A.
[3] Department of Human Anatomy University of Oxford, Oxford, U.K.

Summary. The cell surface receptor for ecotropic host-range murine leukemia viruses is a sodium-independent transporter for essential cationic acids. Our evidence strongly identifies this receptor as the transporter system y^+, which was previously characterized by transport assays. Mutational analysis indicates that transporter activity is not necessary for viral reception. Infection of cells with ecotropic retroviruses causes only a partial down-modulation of receptor expression on cell surfaces.

Introduction

Several cell surface receptors that mediate virus binding and infection have been identified and molecularly cloned (e.g. [1–11]). Receptors for human immunodeficiency virus, poliovirus, rhinoviruses and the A-59 strain of mouse hepatitis coronavirus are members of the immunoglobulin superfamily of cell surface proteins [1–5]. Others are laminin receptors [6], complement receptors [7], metalloenzymes [8, 9] or transporter proteins [10, 12, 13]. Among the most intriguing is the receptor (ecoR) for ecotropic host-range murine leukemia viruses (ecoMuLVs), retroviruses that infect only mice or rats. A cDNA that encodes ecoR was cloned by Albritton and coworkers [11], and it was then found that ecoR is a Na^+-independent transporter for the basic amino acids arginine, lysine and ornithine [12, 13]. This discovery raised questions about interactions between the dual functions of this receptor/transporter; it is a "Trojan Horse" [14], a bearer of gifts (essential amino acids) and an agent of destruction (retroviral infection).

A structural model of ecoR

The ecoR cDNA sequence showed that the receptor contains 622 amino acids and 14 hydrophobic potential membrane-spanning sequences [11]. Based on this evidence and its lack of an amino terminal signal sequence, it was proposed that both the amino and carboxyl termini of ecoR might be in the cytosol, and that the protein might contain 14 transmembrane sequences and 7 extracellular loops [11].

As an alternative, we propose a model (Fig. 1) that has 12 transmembrane sequences and only six extracellular loops. This model is based on the following considerations. (*i*) Our model places all four N-X-(S/T) sites for potential N-linked glycosylation (these occur at amino acids 223, 229, 373 and 490) on the extracellular side of the membrane in loops 3, 4 and 5, whereas the previous model placed two of these (positions 373 and 490) in the cytosol. Sites for potential N-linked gly-

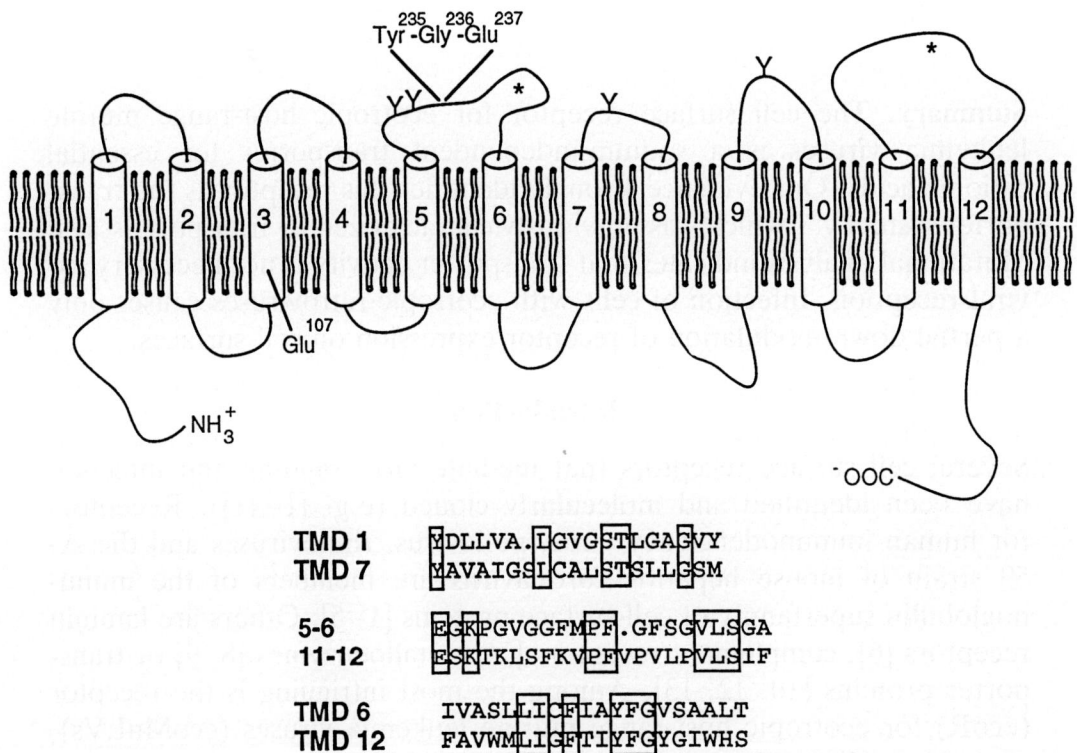

Fig. 1. A structural model of ecoR with twelve transmembrane domains. The positions of Glu[107] in the third TMD and the Tyr[235]Gly[236]Glu[237] sequence in the third extracellular loop are highlighted; these structural features are critical for amino acid transport function and viral binding, respectively (see text). Positions marked with Y are potential sites for N-linked glycosylations. Sequences at bottom illustrate partial homology between TMDs 1 and 7, 6 and 12, and the third and sixth extracellular loops (marked by *), suggesting that the ecoR gene may have arisen by internal duplication

cosylation appear to be at least partially conserved in the homologous transporters of other mammals and in other members of this transporter family in the extracellular loop 3 and 5 regions of our model. The proposed loop 3 sequence must be extracellular because it contains the YGE amino acid sequence at the virus binding site [15]. In addition, species' differences in this region of the encoded proteins affect the Michaelis constant for arginine influx (unpubl. res.). (*ii*) ecoR appears to be related to the family of yeast transporters for arginine (CAN 1), histidine (HIP-1) and choline (CTR), which also contain approximately 600 amino acids [12, 13]. A comparative analysis of this family of transporters suggests that they contain 9–12 transmembrane domains [16]. (*iii*) Features of the ecoR sequence suggest that it may have arisen by duplication of a primordial gene that encoded six transmembrane sequences. For example, in our model the putative transmembrane regions 1 and 7 exhibit some sequence homology, as do transmembrane regions 6 and 12 (Fig. 1). In addition, the loop region between putative transmembrane domains 5 and 6 exhibits apparent homology with the analogous sequence between transmembrane domains 11 and 12. Thus, our model suggests that ecoR may contain two "halves", each with 3 extracellular loops. The previous 14 transmembrane domain model would place these apparently related sequences in opposite topological orientations because each half of the protein would contain an odd number (i.e. [7]) of transmembrane sequences. (*iv*) In our model, the previously proposed transmembrane regions 6 and 13 are instead considered folded into extracellular loops 3 and 6. These sequences exhibit only partial hydrophobic character and contain features (e.g. P, S, T, and C) that could possibly disrupt transmembrane topology. Transmembrane sections of proteins are believed to fold into α-helices or β-sheets in which the carbonyl oxygens are hydrogen bonded to the α-amino nitrogens, thus reducing the polarity of the peptide bond. Breakage of helices (e.g. by prolines) is especially unfavorable in a nonpolar hydrocarbon environment because of the large polarity of the peptide bond and because of the absence of water to replace the broken hydrogen bonds within the protein. Our model is similar to one proposed for voltage-gated ion channels, which is based on a four-domain structure, each domain having six transmembrane sequences and interestingly, a partially hydrophobic segment in the third loop between transmembrane sequences 5 and 6 (see [17]). In voltage-gated potassium channels, residues in this segment of each subunit contribute to the external channel pore.

ecoR as an amino acid transporter

Evidence that ecoR is an amino acid transporter was obtained by injecting synthetic ecoR mRNA into *Xenopus laevis* oocytes and subsequently detecting the transport activity by radiochemical and electrophysiological methods [12, 13]. Figure 2 shows electrophysiological results using the two-electrode voltage clamp method. The application of lysine, arginine, or ornithine to voltage-clamped oocytes expressing ecoR caused inward currents, and the concentration-dependencies of these currents were well fit to the Michaelis-Menten equation (Fig. 2). The inward current correlated with L-[^{3}H]amino acid uptake measurements in a manner suggesting the current results from influx of one positive charge per molecule transported [13]. As shown in Fig. 2C, ecoR also transports arginine reversibly. During prolonged application of arginine the inward current gradually diminishes as the oocyte accumulates substrate and begins to reversibly transport it in both directions. Termination of extracellular arginine application then results in an outward current, corresponding to cation efflux.

A sodium-independent transporter for basic amino acids known as system y^+ has been previously characterized [18]. Like ecoR, it is expressed in almost all cells. Replacement of extracellular sodium with Tris or choline had no significant effect on the lysine, arginine, or ornithine uptake in oocytes expressing ecoR, demonstrating a sodium independence like the cellular y^+ system. Another hallmark of y^+ is that transport of basic amino acids can be inhibited by high concentrations of homoserine and certain other neutral amino acids only in the presence of Na$^+$. Indeed, White obtained evidence suggesting that system y^+ can transport homoserine but only in the presence of sodium [18]. As shown in Fig. 3A, lysine induces inward currents equally in the presence or absence of sodium but inward current caused by homoserine only occurs in the presence of extracellular sodium, implying that it is caused by homoserine-Na$^+$ cotransport. The Km for sodium is dependent on homoserine concentration; at 10 mM homoserine the sodium Km is 59 mM (Fig. 3B). These results provide strong evidence that ecoR is the basic amino acid transport system y^+.

Electrophysiological analysis of system y^+ transport provides a powerful approach for studying its function. Indeed, ecoR is ideal for electrophysiological analysis because it is a Na$^+$-independent system that transports an ionic substrate, allowing real time measurement of transport by voltage clamp current recording. This allows us to make precise measurements of the effect of membrane potential on the kinetics of transport [19]. In addition, rapid voltage-jump studies can be used to gain insight into the molecular mechanisms underlying the

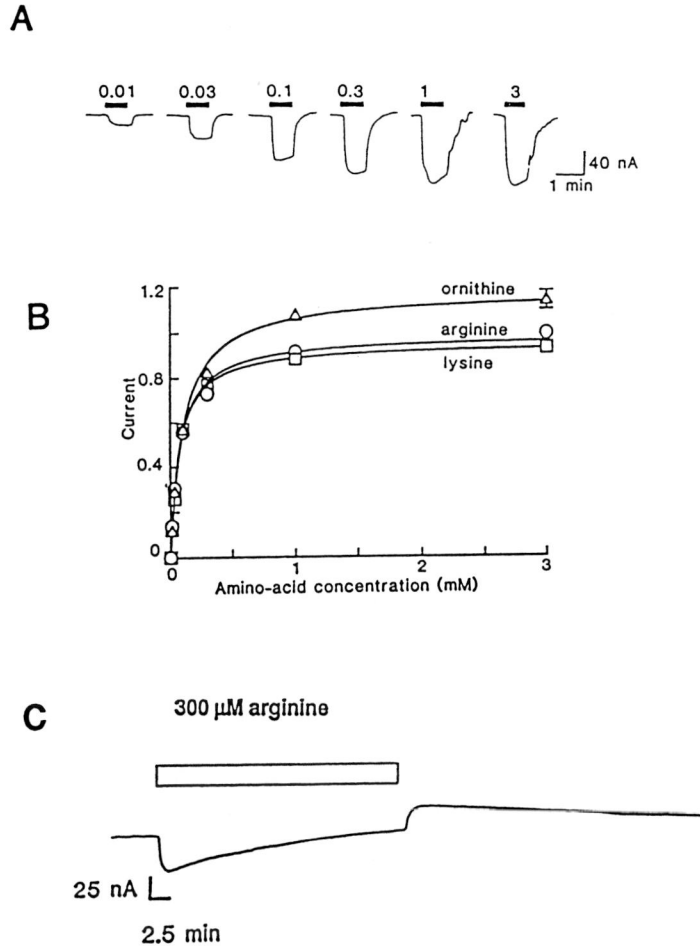

Fig. 2. Electrophysiological analysis of ecoR transporter activity by two-electrode voltage clamp in *Xenopus laevis* oocytes. **A** Voltage-clamped oocyte (-60 mV) expressing ecoR was superfused with various concentrations of lysine (mM) for the duration indicated by bars. Downward deflection represents inward current. **B** Concentration-response curves for the basic amino acids ornithine, arginine, and lysine. Data are fit to the Michaelis-Menten equation; the respective K_M values were 105, 77, and 73 μM. **C** Transport mediated by ecoR is reversible. During extended superfusion of arginine to an oocyte voltage clamped at -20 mV, a decline in influx is seen as a result of intracellular accumulation. Following termination of superfusion, an outward current is measured corresponding to arginine efflux. Data in A and B are from [13]

kinetic behavior by measuring transient ecoR conformational currents using a novel cut-open oocyte fast clamp technique [20]. This transient current reflects the charge transferred in the membrane simply by the conformational change that occurs when the empty transporter moves its substrate binding site from one side of the membrane to the opposite side. We have also begun to study ecoR site-directed mutants and to

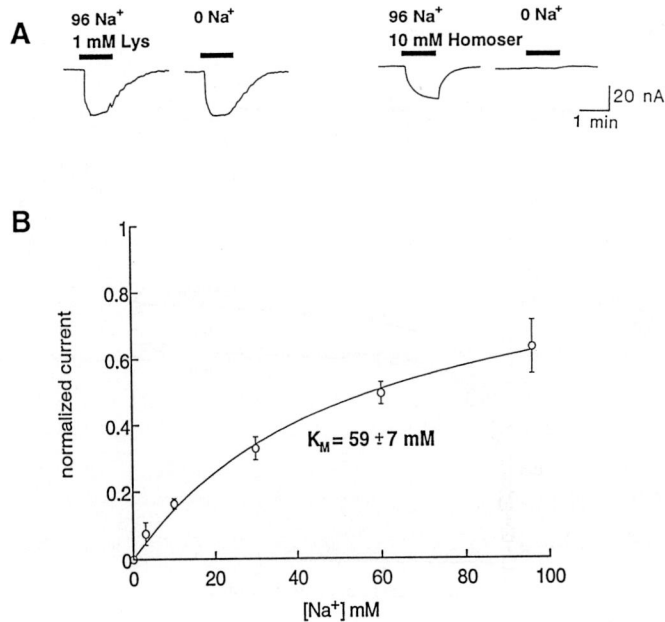

Fig. 3. Na$^+$-dependence of amino acid transport by ecoR. **A** Transport of the basic amino acid lysine is unaffected by removal of external Na$^+$ from the medium (Tris replacement) while transport of the neutral amino acid homoserine is dependent on the presence of Na$^+$. This behavior is identical to that exhibited by the cellular transport system y$^+$. **B** Na$^+$ concentration-response of the current generated by 10 mM homoserine. Data are fit to the Michaelis-Menten equation which yields a K$_M$ for Na$^+$ of 59 mM. Data in **A** are from [13]

correlate effects on specific steps of the transport cycle with effects on viral infection. For example, we have found that a highly-conserved sequence [hydrophobic-hydrophobic-(T/S) (V/I) GEL-X-X-(A/F)] in transmembrane region 3 is essential for y$^+$ transporter function; even a conservative E to D substitution in this sequence at position 107 (see Fig. 1) eliminates transport. This sequence is also the most conserved among the yeast histidine, choline, and arginine transporters. We are now studying the mutant transporters to determine whether changing this residue affects the protein's conformational transition currents. By correlating such analyses with studies of viral binding and infection, we hope to identify aspects of ecoR structure and conformational mobility that may be essential for mediation of infection.

Effects of ecoMuLV on ecoR transporter functions

Ecotropic host-range MuLVs replicate endemically in mice and rats, are inherited in their genomes, and have been associated with diseases. By studying ecoMuLV variants with different abilities to cause specific diseases, *env* glycoproteins have been identified as critical determinants for causation of thymic lymphomas, hemolytic anemias, and neuronal degenerations (e.g. [13, 21, 22]). As with HIV and AIDS [21], it is likely that *env* glycoprotein interactions with receptors are involved in key aspects of ecoMuLV pathogenesis. A critical but poorly understood aspect of retroviral infections is "interference to superinfection", a process whereby persistent infection of a cell causes complete resistance to superinfection by any virus of the same host-range class (i.e. that uses the same receptor for infection). In the cases of HIV and an avian reticuloendotheliosis retrovirus, infection down-modulates receptor expression on cell surfaces. This involves an intracellular interaction of the newly-synthesized receptors and viral glycoproteins that prevents their processing to cell surfaces [21, 23]. Presumably, severe ecoR down-modulation in chronically infected tissues could cause starvation for essential basic amino acids, with attendant abnormalities in cellular functions.

We have detected several effects of ecoMuLV that are specific to mouse y^+ and not to the homologuous mink y^+ transporter [23]. When mouse fibroblasts or mink fibroblasts that express recombinant ecoR are persistently infected with ecoMuLVs, they become completely resistant to superinfection. However, in both of these cell types, infection with ecoMuLV only partially (c.a., 60%) down-modulates mouse-specific y^+ transport activity. For example, as shown in Fig. 4, we compared the initial rates of L-[^{3}H]arginine uptake as a function of arginine concentration into control CCL64 mink lung fibroblasts and into infected and uninfected CCL64 fibroblasts that express recombinant mouse ecoR. As expected, the cells with recombinant ecoR contained substantially more total y^+ transport activity than the control mink cells. This mouse-specific excess y^+ activity was reduced but not completely eliminated by infection with ecoMuLV.

These results and other more detailed studies [23], strongly imply that interference to ecoMuLV superinfection can be accomplished without complete removal of ecoR from cell surfaces. One possibility is that interference involves competitive blockade of ecoR due to adsorption of endogenously synthesized viral *env* glycoprotein onto the cell surface receptors. Alternatively, the ecoR that remains on surfaces of infected cells may have been covalently modified [e.g. by glycosylation (see Fig. 1)] so that it cannot bind extracellular virus. We are currently studying these issues.

We also found that purified ecoMuLV *env* glycoprotein (gp70) acts as a specific weak impermeant inhibitor of mouse y^+ when it is adsorbed onto mammalian cells at 37°C [23]. The amino acid uptake cycle of mouse y^+ is slowed but not blocked by gp70 attachment. Analysis of this

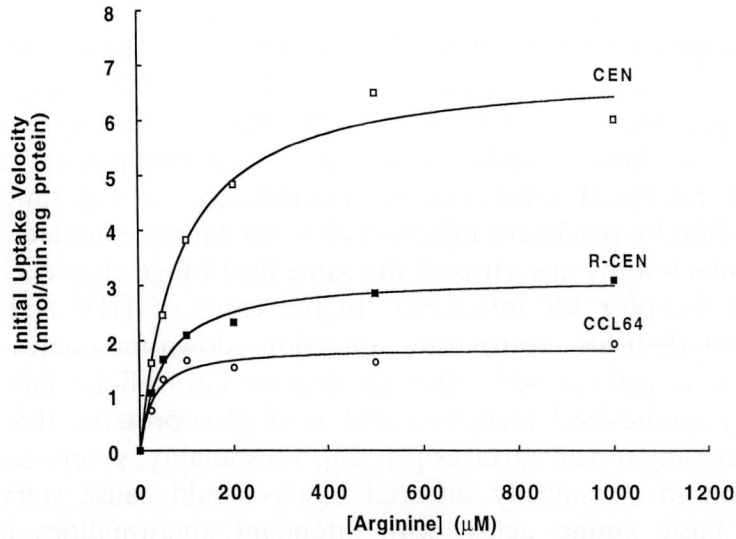

Fig. 4. Partial down-modulation of ecoR transporter activity caused by chronic ecoMuLV infection of cells. Initial rates of uptake of L-[^{3}H]arginine into cultured cells was analyzed as previously described [22]. The cells used were control mink lung CCL64 fibroblasts, a derivative clone (*CEN*) that expresses recombinant mouse ecoR, and a derivative of the latter clone that was chronically and efficiently (greater than 99%) infected with ecoMuLV (Rauscher strain); this infected clone is called *R-CEN*

inhibition suggested that the affinity of gp70 for the viral binding site is affected by the conformational state of y^+. This site apparently changes conformation at rates up to several hundred times per second during the amino acid transport cycle. One implication is that the virus may vibrate at this rate after it has attached to its receptor.

Acknowledgements

We thank E. Dechant, D. Wu and Y.-N. Wu for excellent and enthusiastic assistance. This research has been funded by grants CA25810 and DA03160 from the U.S.A. National Institutes of Health.

References

1. Williams RK, Jiang GS, Holmes KV (1991) Receptor for mouse hepatitis virus is a member of the carcinoembryonic antigen family of glycoproteins. Proc Natl Acad Sci USA 88: 5533–5536
2. Maddon PJ, Dalgleish AG, McDougal JS, Clapham PR, Weiss RA, Axel R (1986) The T4 gene encodes the AIDS virus receptor and is expressed in the immune system and the brain. Cell 47: 333–348

3. Greve JM, Davis G, Meyer AM, Forte CP, Yost SC, Marlor CW, Kamarck ME, McLelland A (1989) The major human rhinovirus receptor is ICAM-1. Cell 56: 839–847

4. Staunton DE, Merluzzi VJ, Rothlein R, Barton R, Marlin SD, Springer TA (1989) A cell adhesion molecule, ICAM-1, is the major surface receptor for rhinoviruses. Cell 56: 849–853

5. Tomassini JE, Graham D, DeWitt CM, Lineberger DW, Rodkey JA, Colonno RJ (1989) cDNA cloning reveals that the major group rhinovirus receptor on HeLa cells is intercellular adhesion molecule I. Proc Natl Acad Sci USA 86: 4907–4911

6. Wang KS, Kuhn RJ, Strauss EG, Ou S, Strauss JH (1992) High-affinity laminin receptor is a receptor for Sindbis virus in mammalian cells. J Virol 66: 4992–5001

7. Moore MD, Cooper NR, Tack BF, Nemerow GR (1987) Molecular cloning of the cDNA encoding the Epstein-Barr virus/C3d receptor (complement receptor type 2) of human B lymphocytes. Proc Natl Acad Sci USA 84: 9194–9198

8. Delmas B, Geli J, L'Haridon R, Vogel LK, Sjostrom H, Noren O, Laude H (1992) Aminopeptidase N is a major receptor for the entero-pathogenic coronavirus TGEV. Nature 357: 417–419

9. Yeager CL, Ashmun RA, Williams RK, Cardellichio CB, Shapiro LH, Look AT, Holmes KV (1992) Human aminopeptidase N is a receptor for human coronavirus 229E. Nature 357: 420–422

10. Johann SV, Gibbons JJ, O'Hara B (1992) GLVR1, a receptor for Gibbon ape leukemia virus, is homologous to a phosphate permease of *Neurospora crassa* and is expressed at high levels in the brain and thymus. J Virol 66: 1635–1640

11. Albitton CM, Tseng L, Scadder D, Cunningham JM (1989) A putative murine ecotropic retrovirus receptor gene encodes a multiple membrane-spanning protein and confers susceptibility to virus infection. Cell 57: 659–666

12. Kim JW, Closs EI, Albritton LM, Cunningham JM (1991) Transport of cationic amino acids by the mouse ecotropic retrovirus receptor. Nature 352: 725–728

13. Wang H, Kavanaugh MP, North RA, Kabat D (1991) Cell surface receptor for ecotropic murine retroviruses is a basic amino acid transporter. Nature 352: 729–731

14. Vile RG, Weiss RA (1991) Virus receptors as permeases. Nature 352: 666–667

15. Albritton, LM, Kim, JW, Tseng, L, Cunningham, JM (1993) Envelope binding domain in the cationic amino acid transporter determines the host range of ecotropic murine retrovirues. J Virol 67: 2091–2096

16. Weber E, Chevlier M, Jund R (1988) Evolutionary relationships and secondary structure predictions in four transport proteins of *Saccharomyces cerevisiae*. J Mol Evol 27: 341–350

17. Durell SR, Guy HR (1992) Atomic scale structure and functional models of voltage-gated potassium channels. Biophys J 62: 238–245

18. White MF (1985) The transport of cationic amino acids across the plasma membrane of mammalian cells. Biochim Biophys Acta 822: 355–374

19. Kavanaugh MP (1993) Voltage-dependence of arginine flux mediated by the system y^+ basic amino acid transporter. Biochemistry (submitted)

20. Taglialatella M, Toro L, Stefani E (1992) Novel voltage clamp to record small, fast currents from ion channels expressed in Xenopus oocytes. Biophys J 61: 78–82

21. Weiss RA (1992) Cellular receptors and viral glycoproteins involved in retrovirus entry. In: Levy JA (ed) The retroviruses. Plenum Press, New York

22. Kabat D (1989) Molecular biology of Friend viral erythroleukemia. Curr Top Microbiol Imunol 148: 1–42

23. Wang H, Dechant E, Kavanaugh MP, North RA, Kabat D (1992) Effects of ecotropic murine retroviruses on the dual-function cell surface receptor/basic amino acid transporter. J Biol Chem 267: 23617–23624

Authors' address: Dr. D. Kabat, Department of Biochemistry, L224, Oregon Health Sciences University, 3181 S.W. Sam Jackson Park Road, Portland, OR 97201-3098, U.S.A.

Virus structure and assembly

Arch Virol (1994) [Suppl] 9: 497–512

Archives
of
Virology
© Springer-Verlag 1994
Printed in Austria

Comparative studies of $T = 3$ and $T = 4$ icosahedral RNA insect viruses

J. E. Johnson, S. Munshi, L. Liljas*, D. Agrawal, N. H. Olson, V. Reddy, A. Fisher**, B. McKinney, T. Schmidt, and T. S. Baker

Department of Biological Sciences, Purdue University, West Lafayette, Indiana, U.S.A.

Summary. Crystallographic and molecular biological studies of $T = 3$ nodaviruses (180 identical subunits in the particle) and $T = 4$ tetraviruses (240 identical subunits in the particle) have revealed similarity in both the architecture of the particles and the strategy for maturation. The comparative studies provide a novel opportunity to examine an apparent evolution of particle size, from smaller ($T = 3$) to larger ($T = 4$), with both particles based on similar subunits. The BBV and FHV nodavirus structures are refined at 2.8 Å and 3 Å respectively, while the NωV structure is at 6 Å resolution. Nevertheless, the detailed comparisons of the noda and tetravirus X-ray electron density maps show that the same type of switching in subunit twofold contacts is used in the $T = 3$ and $T = 4$ capsids, although differences must exist between quasi and icosahedral threefold contacts in the $T = 4$ particle that have not yet been detected. The analyses of primary and tertiary structures of noda and tetraviruses show that NωV subunits undergo a post assembly cleavage like that observed in nodaviruses and that the cleaved 76 C-terminal residues remain associated with the particle.

Introduction

There are a variety of unifying themes in the structure and function of positive strand RNA viruses and these have been used to construct

Present addresses: *University of Uppsala, Biomedical Center, Department of Molecular Biology, Uppsala, Sweden; **Institute for Enzyme Research, University of Wisconsin, Madison, Wisconsin, U.S.A.

detailed evolutionary relationships among these viruses [1, 2]. Most positive strand RNA viruses have icosahedral capsids with $T = 3$ quasi symmetry (primarily plant viruses) or a picornavirus type capsid (primarily animal viruses) [2]. These capsid types are themselves related, with an apparent triplication of the capsid protein gene observed in the $T = 3$ viruses forming the picornavirus capsids [3]. The identification of a particular capsid type with viruses infecting members of a particular kingdom is not universal, however, since there are plant viruses with capsids that are clearly related to the picornaviruses [4] and there are animal viruses with $T = 3$ symmetry [5, 6]. Because there are such limited examples of capsid types the only comparisons to date have been between $T = 3$ and picornavirus capsids [2, 3, 4]. We have been investigating two groups of insect viruses, nodaviruses [7] (displaying $T = 3$ quasisymmetry) and tetraviruses [8] (displaying $T = 4$ quasisymmetry). In this paper we report similarities in structure and function that are apparent from our biochemical and biophysical studies.

The $T = 3$ nodaviruses are among the simplest animal viruses known. Their genomes consist of roughly 4 500 bases that are split between two single-stranded, messenger-sense RNA molecules encapsidated in one virion [9]. The genome encodes only three proteins [10]; a replicase (protein A), the coat protein precursor (protein α) and a protein of unknown function (protein B). Nodaviruses are small non-enveloped viruses that infect insects [7], mice [11], and fish [12]. They can be produced in large quantity and readily crystallized. They undergo a well characterized series of assembly and maturation steps [13]. An infectious clone is available for Flock House nodavirus (FHV) [14], and particles spontaneously assemble and package their own messenger RNA when the FHV capsid protein gene is expressed in a baculovirus system [15]. The structure of the black beetle nodavirus (BBV) has been determined [5] and was found to be similar to all $T = 3$ RNA plant virus structures analyzed, although the β-barrel subunits contained elaborate surface loops and an interior helical domain not observed in the plant viruses. Biochemical studies of FHV showed that 80% of the coat protein subunits α (407 amino acids) underwent a post assembly, autocatalytic cleavage to form protein β (363 amino acids) and protein γ (44 amino acids) in the mature virion and that this cleavage was required for infectivity [16]. This maturation is similar to that observed in picornaviruses [17] and suggests that the nodaviruses, although displaying a different capsid architecture, may share a common biological strategy for particle maturation and possibly uncoating.

The *Tetraviridae* family of viruses, formerly referred to as the *Nudaurelia* β virus family, consists of seven members that from icosahedral particles with $T = 4$ symmetry [8]. All members of this virus group identified to date propagate exclusively in insect hosts. The prototype, *Nudaurelia capensis* β virus (NβV), was originally isolated from the pine emperor moth, *Nudaurelia cytharea capensis* Stoll [18] and it has been the subject of recent studies [19]. This group of viruses form shells that are 350–400 Å in diameter, encapsidating a single-stranded RNA having a molecular mass of 1.8–1.9 × 10⁶ daltons. The particles have sedimentation coefficients between 200S and 210S. Examination of the capsids by SDS-PAGE revealed in all cases a single capsid polypeptide estimated to be 65–68 kD. Compilations of these characteristics for individual members have recently been published [20, 21].

The striking differences between noda and tetraviruses are the size of the capsid subunit and the size of the particle (Table 1). Earlier work on these viruses did not suggest a relationship between them, however, structural studies at the primary and tertiary level of *Nudaurelia capensis* ω virus (NωV), another tetravirus, now suggest that noda and tetraviruses are related [22]. The geometric concepts for generating quasi equivalent shells have been developed on the basis of theoretical and experimental data [23, 24]. The nodaviruses display $T = 3$ symmetry (180 subunits in the capsid) and the tetraviruses $T = 4$ symmetry (240 subunits in the capsid). The first two high resolution virus structures determined were for the $T = 3$ plant viruses, tomato bushy stunt virus [25] and southern bean mosaic virus [26]. Remarkably, they were found to have subunit tertiary and particle quaternary structures that were virtually superimposable in the contiguous shell, although the subunits displayed only limited sequence homology. The $T = 3$ plant and animal viruses have a shape that is strikingly similar to the rhombic triacontahedron [27] and this geometric solid is shown and discussed in Figs. 1a and b. Although structurally similar to the plant viruses in many respects, recent studies have shown that nodaviruses incorporated regions of duplex RNA in the capsid and that the RNA functioned as part of the molecular switch that regulates the formation of the $T = 3$ shell [28].

Comparison of subunit primary structures of noda and tetravirus

The segmented genome of NωV consists of RNA1 (5 kb) and RNA2 (2.5 kb). The sequence of RNA2 was recently reported and the encoded amino acid sequence was compared with nodavirus capsid proteins [22]. Two similarities were observed. First, the NωV coat protein (646 residues) probably undergoes a post translational cleavage. This was apparent when the virus particle was subjected to direct chemical protein sequencing. The 21 "Nterminal" amino acids obtained in this experiment aligned with residues 571 to 591 encoded by the RNA2 gene. The likely explanation for this result is that a posttranslational cleavage occurs

Table 1. A comparison of features between *Nodaviridae*, NωV, and NβV

Virus	Icosahedral symmetry	Number of subunits	Particle size (Å)	Buoyant density (g/cc)	Polypeptide composition	Genomic composition	Encoded products
Nodaviridae	$T = 3$	180	312 (9.4×10^6 mD)	1.33	β (38 kD) γ (5 kD)	RNA1 (3.1 kb) RNA2 (1.4 kb)	protein A, B α (β,γ)
NωV	$T = 4$	240	410	1.285	β (62 kD) γ (8 kD)	RNA1 (5 kb) RNA2 (2.5 kb)	n.d. α (β,γ)
NβV	$T = 4$	240	397 (16.3×10^6 mD)	1.295	β (61 kD) γ (8 kD)	RNA1 (5.4 kb)	n.d.

n.d. Not determined

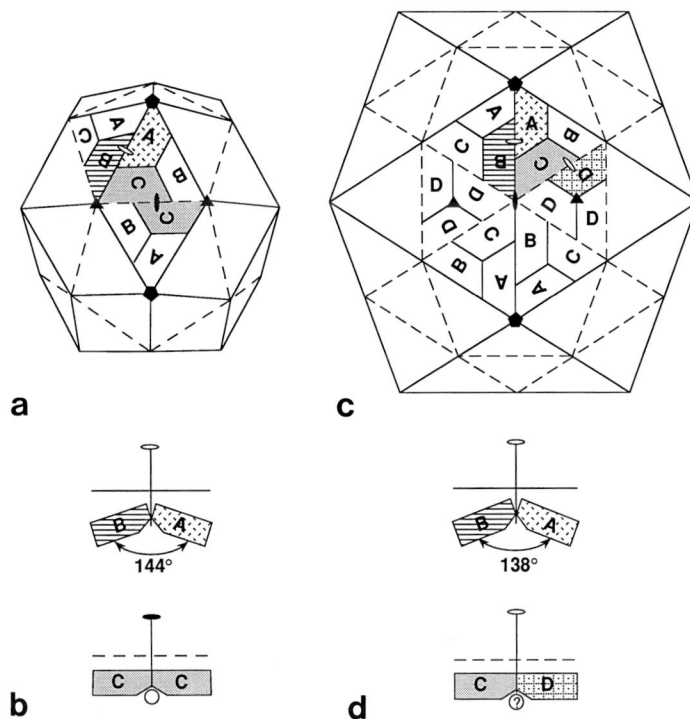

Fig. 1. Diagrammatic representations of the subunit associations accounting for contiguous shells in $T = 3$ (**a, b**) and $T = 4$ icosahedrons (**c, d**). Labeled trapezoids represent the individual capsid subunits. The packing relations, as viewed along an equatorial line within the structure, between C subunits related by the icosahedral twofold (solid ellipse) and between A and B subunits related by the quasi twofold (open ellipse) are shown for the $T = 3$ structure in **b**. Likewise, the C and D, and the A and B subunits of the $T = 4$ structure, related by different quasi twofold axes, are shown in **d**. A peptide and RNA duplex (open circle, **b**) maintains the planar C-C contact as observed in the BBV structure [4]. A similar peptide (**d**) may keep the C-D contact in the $T = 4$ structure planar. Open and closed ellipses, closed triangles, and closed pentagons represent quasi and icosahedral twofold, and icosahedral threefold and fivefold axes, respectively

between residues 570 and 571, making residue 571 an unblocked N-terminus of a polypeptide consisting of residues 571–646 (This would correspond to the γ peptide described for nodaviruses). The true N-terminus of the NωV coat protein translation product is probably blocked like many other viral capsid proteins. Exactly this situation was found with the Flock House nodavirus [13]. When NωV was analyzed on a SDS gel, there was clear evidence for a small polypeptide of molecular weight ~8 kD as well as the larger polypeptide at ~61 kD. Earlier studies had not reported the small protein probably because it ran off the end of the gel under typical electrophoresis conditions. The gene for the NωV capsid protein has been cloned and expressed in *E. coli* and the resultant

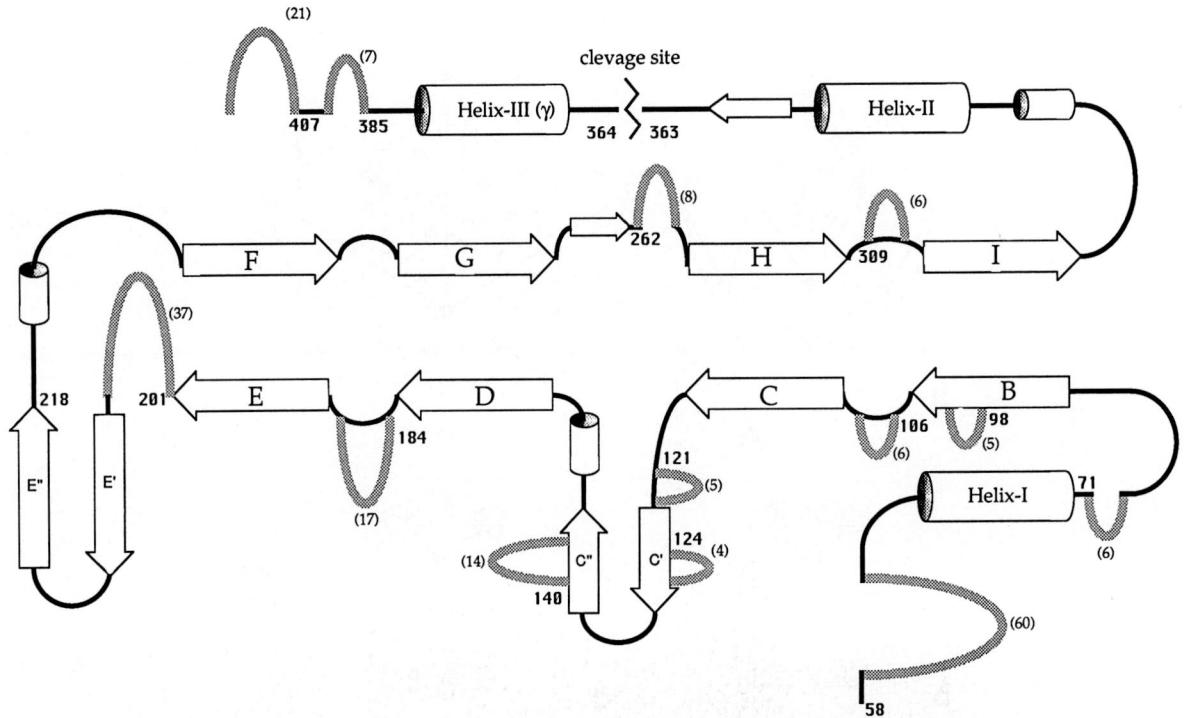

Fig. 2. A schematic representation of secondary structure elements observed in BBV[5] and FHV[28]. Stipled lines show the positions (in the BBV sequence) and the number of residues inserted in the capsid protein of NωV on the basis of the primary structure alignment. Note the alignment places all but one of the insertions between secondary structure elements suggesting that the β-structure is conserved in noda and tetraviruses

gene product is a 71 kD protein that reacts with antibodies to NωV particles in a western blot. The NωV sequence was aligned with the four nodavirus sequences by the program PILEUP. Although the number of amino acids in the nodavirus capsid protein is roughly 400 and the number in the NωV capsid protein is 646, a significant level of similarity was found. Two aspects of the alignment were notable. First, virtually all of the additional amino acids in NωV were accounted for by inserts between strands of the β-barrel in the known structures of the noda-viruses. This suggests that the residues of the β-barrel strands are relatively conserved and that the additional mass of the NωV protein is composed of insertions between these strands. Significant inserts also are located at the beginning and at the end of the β-barrel. The relation between the BBV secondary structures and the NωV inserts is shown in Fig. 2. Another remarkable feature of the alignment is the coincidence of the cleavage sites, located between residues N363 and A364 in BBV and FHV and between residues N570 and F571 in NωV.

Electron microscopy and solution X-ray scattering of tetraviruses

Structures with $T = 4$ symmetry have been found mainly within insect viruses, although both the outer [29] and core [30] structures of the alphaviruses, Sindbis and Semliki forest virus, display $T = 4$ shells. The structure of NβV was determined at 25 Å resolution by both negative stain [31] and cryo-electron microscopy [20]. The striking feature of the NβV structure was its icosahedral shape (Fig. 3) which is in contrast to the more spherical shape predicted for the Sindbis core protein $T = 4$ shell [30]. Although images of negative stained NβV and NωV revealed particles of similar size, the respective three-dimensional reconstructions showed a close relationship between the two structures (Fig. 3). The solution X-ray scattering patterns for NβV and NωV (Fig. 4) are also

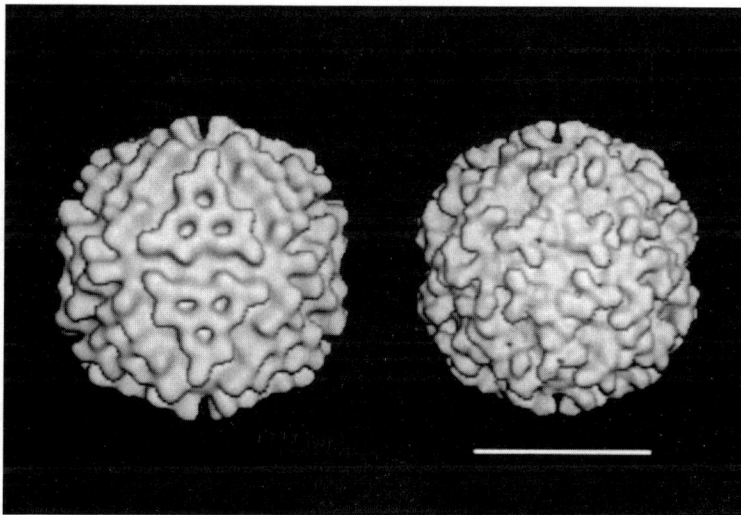

Fig. 3. Shaded-surface representations of three-dimensional reconstructions of NβV (left) and NωV (right) computed from images of frozen-hydrated virions. The bar on the right is 250 Å. Specimens were prepared for cryo-electron microscopy by applying small drops of each sample to perforated carbon films on 400 mesh copper grids. The grids were briefly blotted with filter paper and then rapidly plunged into liquified ethane. They were then transferred to liquid nitrogen before being inserted into a cold, Gatan cryotransfer holder and then into a Philips EM420 transmission electron microscope. The images were recorded at ~1.5 μm underfocus, at 49 000X magnification, and with an electron dose of ~20e⁻/Å². Selected micrographs were digitized, individual particle images from each were boxed, and the background density of each image was subtracted. The initial centers of density of each image were determined by cross correlation methods [36] and the orientation of each particle, with respect to the view axis, was determined with established icosahedral processing procedures [37–39]. Refined data sets of twenty-five images of NβV and 57 of NωV were used to calculate three-dimensional reconstructions of each virus with an effective resolution of 31 Å and 28 Å, respectively

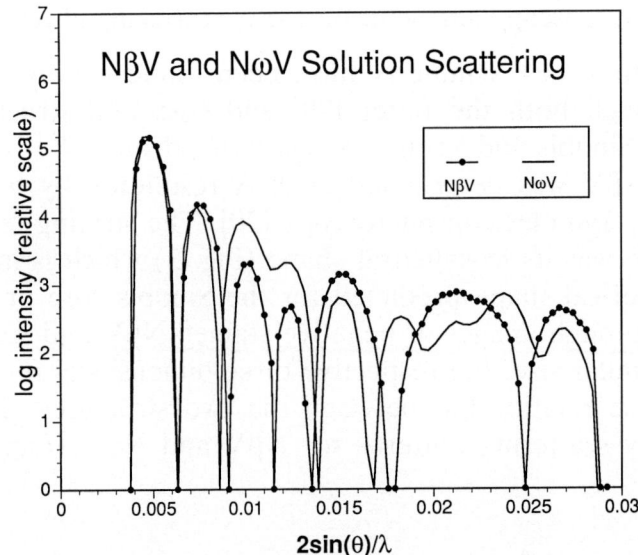

Fig. 4. NβV and NωV solution scattering patterns. Each virus was pelleted, resuspended in buffer at approximately 150 mg/ml, and placed in a 1 mm quartz capillary. Diffraction data were then collected, at 267.5 mm sample-to-film distance, using CuKa radiation, a double-mirror focusing camera [40, 41] with an order-to-order resolution of ~1500 Å, and an Elliott GX-20 rotating anode generator with a 0.15 × 2.5 mm focal spot. Exposure time was about 85 h at an X-ray power of 35 kV and 30 mA. In order to fully record the wide intensity range of each virus's diffraction pattern, three Kodak DEF-5 films were stacked together, with each film both recording and attenuating the scattered radiation. An Optronics C4100 film scanner was used to digitize the film optical densities on a 50 μm raster. The center of each film's diffraction pattern was determined using the circular symmetry of the pattern. The average optical density at each radial distance from the center was then calculated by integrating all optical densities within 50 μm thick circular shells. Each film's final radial intensity distribution was then obtained by subtracting a smooth background curve fitted through the nodes. The three radial intensity distributions (from the three stacked films) were then scaled together to produce the final intensity profile shown for each virus. Additional experimental details can be found in [42]

different, but display an underlying similarity consistent with the electron microscopy reconstructions.

Figures 1c and d represent the tetravirus $T = 4$ structure diagrammatically and show the similarity with the rhombic triacontahedron of the $T = 3$ structure. On the basis of this comparison a detailed model of the NωV $T = 4$ shell was constructed with the subunit tertiary structure observed in the $T = 3$ shell of BBV [5]. The icosahedral asymmetric unit of the $T = 4$ shell (A, B, C, D) (Fig. 1c) was defined as the A, B, C and C2 subunits in the $T = 3$ shell (Fig. 1a). The coordinates of atoms in these four subunits in BBV were adjusted radially until a non-overlapping contiguous shell was formed with atoms in neighboring asymmetric units generated by the icosahedral symmetry operations. Table 2 lists the coordinates and radial distances of well-defined features observed in the subunits of BBV as they were positioned in the $T = 4$ NωV modeled structure. These are compared with features in the electron density map of NωV below.

Electron density of NωV at 8 Å resolution

Crystals of NβV and NωV have been produced and they diffract X-rays to 2.7 Å resolution [32, 33]. Both viruses crystallize in an unusual but similar unit cell. Cell dimensions and crystallization conditions are given in Table 3. The proposed similarity of the noda and tetraviruses is supported by the 8 Å resolution NωV electron density map recently computed. There is clear evidence for a β-barrel core in the electron density for the contiguous shell of NωV and it agrees well with the $T = 4$ model based on the adjusted BBV coordinates. Two additional characteristics found in nodavirus structures that were obvious in the 8 Å electron density map were the helical interior domain that probably contains the C terminal portion of the cleaved capsid protein γ (residues 571 to 646) and a peculiar "doughnut" shaped density feature surrounding the fivefold symmetry axes. Both of these features were clear in the NωV map (Figs. 5a and b) and they served as markers to compare the model of NωV generated with the BBV subunits with the actual NωV structure. The γ peptides of the BBV subunits are placed in helical density in the NωV map (Fig. 5a). In addition the leucine residues on the βD-βE loop of BBV are directly in contact with the "doughnut" shaped density appearing in that structure about the fivefold axes, and the BBV subunits modeled in the NωV map occupy exactly the same position relative to the "doughnut" shaped density in NωV (Fig. 5b). Thus, at 8 Å resolution the electron density supports the similarities between nodaviruses and tetraviruses based on geometric arguments and the sequence alignments.

Discussion

The noda and tetravirus systems offer a unique opportunity to study the factors affecting capsid architecture. Evidence presented here suggests that the subunit building blocks of noda and tetraviruses are similar, particularly with regard to the β-barrel that forms the contiguous protein shell. Other features of the $T = 3$ and $T = 4$ shells, in addition to the β-barrel, are also comparable. These include the helical domain that lies inside the barrel, which is the location of the autocatalytic maturation cleavage in FHV and BBV, and the "doughnut" shaped density that exists on the fivefold symmetry axes in nodavirus structures and NωV.

Table 2. NωV ($T = 4$) model based on BBV

	Residue # (Cα)	X	Y	Z	D (Å)
Farthest distance along Q3	B207	59.	−1.	183.	192.
Farthest distance along I5	A306	103.	−3.	163.	193.
Farthest distance along I3	C207	−2.	1.	183.	183.
Farthest distance along I2	A305	18.	−42.	159.	165.
γ Peptide of A subunit	A379	76.	−3.	119.	142.
	A364	92.	2.	133.	161.
Doughnut along I5	A182	108.	−0.	154.	188.
	A181	106.	1.	150.	184.
Ca^{2+} on Q3 axis	A249	62.	1.	141.	154.
	A251	61.	5.	145.	157.

Residue (measured at)	strand	X	Y	Z	D (Å)
CHEF-sheet of β-barrel of a subunit					
190	E	85.	−6.	150.	173.
299	H	85.	−3.	154.	176.
113	C	84.	1.	157.	178.
277	F	48.	19.	165.	173.
BIDG-sheet of β-barrel of a subunit					
241	G	84.	0.	145.	167.
174	D	87.	6.	147.	172.
314	I	86.	9.	150.	173.
100	B	87.	13.	152.	176.

Loops facing fivefold axis measured at	loop	X	Y	Z	D (Å)
108	βB-βC	96.	−5.	165.	191.
306	βH-βI	103.	−3.	163.	193.
183	βD-βE	105.	1.	156.	188.
236	βF-βG	98.	−0.	149.	178.
135	βC-βC′	80.	−9.	168.	186.

Heavy atom positions (site)	X	Y	Z	D (Å)
A	99.	2.	164.	192.
B	38.	−38.	166.	174.
C	34.	36.	164.	171.
D	24.	−36.	166.	172.

Sites A, B and C are related by quasi threefold symmetry and sites C and D by quasi twofold symmetry

Table 3. Tetravirus crystals

Crystal form (virus)	Space group	Lattice constants	Z
Form III NβV[a]	P1	408.7 Å, 403.14 Å, 409.67 Å, 59.35°, 58.97°, 63.21°	1
NωV[b]	P1	413.55 Å, 410.22 Å, 419.67 Å, 59.13°, 58.9°, 64.04°	1

[a] Growth conditions: Virus (8–14 mg/ml) in 0.07 M acetate buffer pH 5.0 (V) precipitated by vapor diffusion with 0.1 M $Ca(NO_3)_2$, 2.25% PEG 8000 and 0.5% β-octyl glucoside in acetate buffer pH 5.0 (P). Ratio of virus to precipitant solution in drop (V:P = 1:1), vapor equilibrated against P

[b] Growth conditions: Virus (8–10 mg/ml) in 0.07 M acetate buffer pH 5.0 (V) precipitated by vapor diffusion with 2% PEG 8000, 0.25 M $CaCl_2$ and 0.001 M NaN_3 in 0.075 M morpholino propanesulfonic acid (MOPS) buffer pH 7.0 (P). Ratio of virus to precipitant solution in drop (V:P = 1:4), vapor equilibrated against P

Biochemical and structural studies of nodaviruses have suggested a mechanism for the autocatalytic cleavage that involves an acid-catalyzed main-chain hydrolysis event. The cleavage is also required for infectivity in nodaviruses and results in a substantial increase in particle stability. By analogy a similar maturation mechanism can be predicted for NωV. Unfortunately, it has not been possible to propagate NωV in any cell line, so the study of its biochemical and biological properties has not been possible. Schneemann et al. [15] have shown that when the capsid protein gene of FHV is expressed in a baculovirus system the subunits spontaneously assemble and package the messenger RNA for the capsid protein gene. The expression of the wild type sequence of the FHV capsid protein gene results in normal subunit processing and maturation. Mutation of the asparagine(N) at residue 363 to threonine(T), alanine, or aspartic acid result in cleavage inhibition. The N363T mutant expressed in baculovirus is stable as the provirion and this particle type has been crystallized [34]. Currently we are attempting to express the capsid gene of NωV in a baculovirus system and we anticipate that these subunits will assemble to form a $T = 4$ capsid, thus permitting a thorough study of factors affecting the assembly and architecture of a $T = 4$ shell. Such a study is currently being done for the $T = 3$ shell of FHV [35].

The $T = 4$ architecture is intriguing when viewed as the simplified model in Fig. 1. Finch et al. [31] recognized that the icosahedral asymmetric unit of tetraviruses (containing four subunits) does not form a logical assembly unit, and they suggested that the particle may form from preassembled trimers. This assembly hypothesis is consistent with the $T = 4$ organization described here. The detailed model predicts that, prior to assembly, there is probably no difference between trimers related by icosahedral symmetry (DDD) and trimers related by quasi threefold symmetry (ABC) (Fig. 1c) and it is only after the shell starts to assemble that the two classes of trimers begin to differ slightly. The nature of the two classes of quasi twofold axes (bent and flat) in the $T = 4$ particle are nearly identical to the quasi (bent) and icosahedral (flat) twofold axes in the $T = 3$ nodaviruses (Fig. 1). The nodaviruses architecture results from ordered RNA and a 10 amino acid polypeptide that are visible only at the "flat" contact between trimers at the icosahedral twofold axes. At the "bent" quasi twofold axes the RNA and protein polypeptide do not obey icosahedral symmetry and are invisible. Portions of the capsid protein that interact with the RNA and polypeptide at the

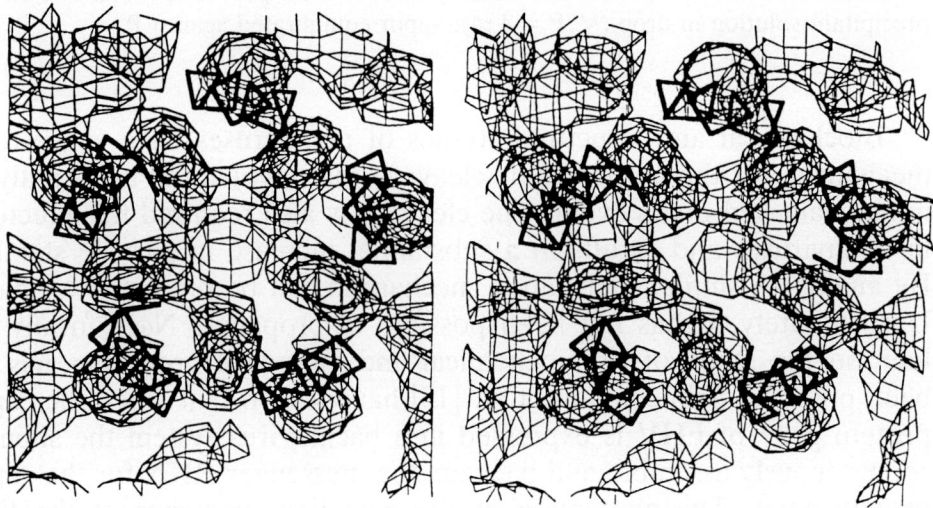

a

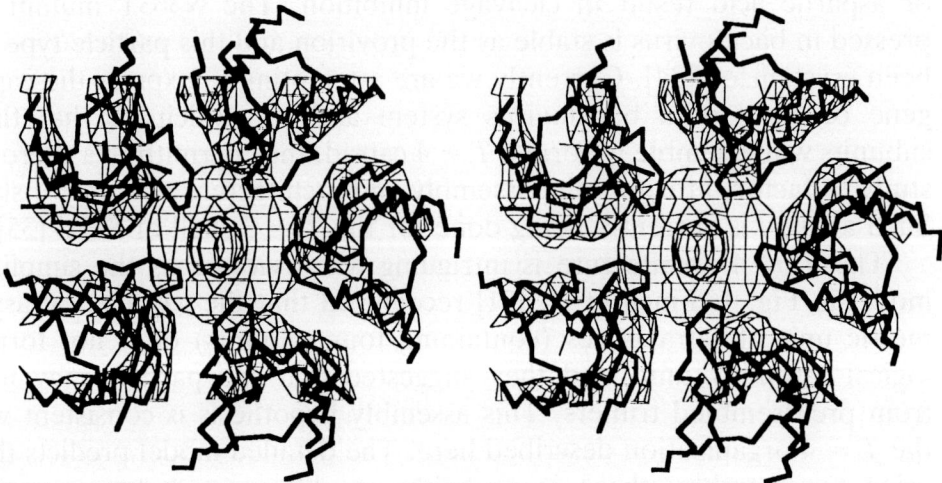

b

flat contact interact with each other at the bent contact. A similar mechanism can be invoked for formation of bent and flat contacts in $T = 4$ structures (Fig. 1b), but at present we have no suggestion for differentiating the two trimers. The difference between $T = 3$ and $T = 4$ structures is apparent at the quasi sixfold axes, which exist in both. In $T = 3$ shells the quasi sixfold axis is coincidental with an icosahedral threefold axis and the actual structure is best described as a trimer of BC dimers. The quasi sixfold axis in the $T = 4$ shell is coincident with an icosahedral twofold axis and thus is probably best described as a dimer of BCD trimers. The 2.8 Å structure of NωV should clearly show what portions of the coat protein or RNA regulate this aspect of the $T = 4$ particle.

Acknowledgements

We thank D. Hendry for supplying samples of *Nudaurelia capensis* ω virus, and S. Fateley for help in preparing the manuscript. This work was supported by a NIH grant (GM34220) to JEJ.

Fig. 5. Features in the electron density map of NωV. The electron density map was computed with structure factors between 20 and 8 Å resolution. Data were recorded on photographic films by oscillation photography at the Cornell High Energy Synchrotron Source. 574 films were indexed and intensities were estimated for 4 411 246 observations measured 3 σ above the background to 2.7 Å resolution. 2 626 720 reflections were unique accounting for 81% of data to 8 Å resolution and 43% of the data to 2.8 Å. The overall scaling agreement factor was 12.3%. An initial set of phases for data between 20 and 15 Å were calculated from the Fourier transform of a spherical shell of inner and outer radii of 139 and 192 Å respectively placed in the NωV unit cell. These phases were refined by real space molecular averaging procedures. The phases were extended to 8 Å resolution in steps of one reciprocal lattice point, followed by 5–6 cycles of density averaging. The handedness of the map was determined by calculating difference Fourier between native data and data collected on the crystals of NωV soaked in 2–5 mM flourecene mercury acetate. The mercury atoms were identified (Table 2) in the negative electron density suggesting the Babinet opposite (phase = phase + 180) solution for the phases. The electron density map at 8 Å was computed with observed amplitudes and phases for the correct hand and compared with the BBV based $T = 4$ model structure identified in bold. **a** A stereoview of the electron density in the helical interior of NωV with the adjusted BBV model superimposed. The γ peptide Cα positions are shown. Note their proximity to the rod-shaped electron density characteristic of a helix. **b** A stereoview of electron density close to the fivefold symmetry axes displaying a characteristic feature observed in both BBV and NωV electron density. The chemical nature of this "doughnut" feature has not been determined in BBV or NωV. Note the proximity of the loops defined by Cα positions from the adjusted BBV coordinates. The same relation between this density feature and the βD-βE loop (closest to ring shaped density) is seen in the BBV structure

References

1. Goldbach R (1987) Genome similarities between plant and animal RNA viruses. Microbiol Sci 4: 197–202
2. Rossmann MG, Johnson JE (1989) Icosahedral RNA virus structure. Annu Rev Biochem 58: 533–573
3. Rossmann MG, Rueckert RR (1987) What does the molecular structure of viruses tell us about viral functions? Microbiol Sci 4: 206–214
4. Chen Z, Stauffacher C, Li Y, Schmidt T, Bomu W, Kamer G, Shanks M, Lomonossoff G, Johnson JE (1989) Protein-RNA interactions in an icosahedral virus at 3.0 Å resolution. Science 245: 154–190
5. Hosur MV, Schmidt T, Tucker RC, Johnson JE, Gallagher TM, Selling BH, Rueckert RR (1987) Structure of an insect virus at 3.0 Å resolution. Proteins Struct Funct Genet 2: 167–176
6. Prasad BVV (1992) The structure of a calcivirus by cryo-electron microscopy (pers. comm.)
7. Hendry DA (1991) Nodaviridae of invertebrates. In: Kurstak E (ed) Viruses of invertebrates. Marcel Dekker, New York, pp 227–276
8. Moore NF (1991) The *Nudaurelia* β family of insect viruses. In: Kurstak E (ed) Viruses of invertebrates. Marcel Dekker, New York, pp 277–285
9. Newman JFE, Brown F (1977) Further physicochemical characterization of Nodamura virus. Evidence that the divided genome occurs in a single component. J Gen Virol 38: 83–95
10. Friesen P, Rueckert RR (1981) Synthesis of black beetle virus proteins in cultured Drosophila cells: differential expression of RNAs 1 and 2. J Virol 37: 876–886
11. Newman JFE, Brown F (1973) Evidence for a divided genome in Nodamura virus, an arthropod-borne picornavirus. J Gen Virol 21: 371–384
12. Mori K-I, Nakai T, Muroga K, Misao A, Mushiake K, Furusawa I (1992) Properties of a new virus belonging to Nodaviridae found in larval striped jack (*Pseudocaranx dentex*) with nervous necrosis. Virology 187: 368–371
13. Gallagher TM, Rueckert RR (1988) Assembly dependent maturation cleavage in provirions of a small icosahedral insect ribovirus. J Virol 62: 3399–3406
14. Dasmahapatra B, Dasgupta R, Saunders K, Selling B, Gallagher T, Kaesberg P (1986) Infectious RNA derived by transcription from cloned cDNA copies of the genomic RNA of an insect virus. Proc Natl Acad Sci USA 83: 63–66
15. Schneemann A, Dasgupta R, Johnson JE, Rueckert RR (1993) Use of recombinant baculoviruses in synthesis of morphologically distinct viruslike particles of Flock House virus, a nodavirus. J Virol 67: 2756–2763
16. Schneemann A, Zhong W, Gallagher TM, Rueckert RA (1992) Maturation cleavage required for infectivity of a nodavirus. J Virol 66: 6728–6734
17. Fernandez-Tomas CB, Baltimore D (1973) Morphogenesis of poliovirus. J Virol 12: 1122–1130
18. Struthers JK, Hendry DA (1974) Studies of the protein and nucleic acid components of *Nudaurelia capensis* β virus. J Gen Virol 22: 355–362
19. du Plessis DH, Mokhosi G, Hendry DA (1991) Cell free translation and identification of the replicative form of *Nudaurelia* β virus RNA. J Gen Virol 72: 267–273
20. Olson NH, Baker TS, Johnson JE, Hendry DA (1990) The three-dimensional structure of frozen hydrated *Nudaurelia capensis* β virus, a *T* = 4 insect virus. J Struct Biol 105: 111–122

21. Reinganum C (1991) Tetraviridae. In: Adams JR, Bonami JR (eds) Atlas of invertebrate viruses. CRC Press, Boca Raton, pp 387–392

22. Agrawal DK, Johnson JE (1992) Sequence and analysis of the capsid protein of *Nudaurelia capensis* ω virus, an insect virus with $T = 4$ icosahedral symmetry. Virology 190: 806–814

23. Caspar DLD, Klug A (1962) Physical principals in the construction of regular viruses. Cold Spring Harbor Symp Quant Biol 27: 1–24

24. Johnson JE, Fisher AJ (1993) Principles of virus structure. In: Webster RG, Granoff A (eds) Encyclopedia of virology. Academic Press, London, in press

25. Harrison SC, Olson AJ, Schutt CE, Winkler FK, Bricogne G (1978) Tomato bushy stunt virus at 2.9 Å resolution. Nature 276: 368–373

26. Abad-Zapatero C, Abdel-Meguid SS, Johnson JE, Leslie AGW, Rayment I, Rossmann MG, Suck D, Tomitake T (1980) Structure of southern bean mosaic virus at 2.8 Å resolution. Nature 286: 33–39

27. Williams R (1979) The geometrical foundation of natural structure. Dover, New York

28. Fisher AJ, Johnson JE (1992) Ordered duplex RNA controls capsid architecture in an icosahedral animal virus. Nature 361: 176–179

29. von Bonsdorff C, Harrison S (1975) Sindbis virus glycoproteins form a regular icosahedral surface lattice. J Virol 16: 141–145

30. Choi H-K, Tong L, Minor W, Dumas P, Boege U, Rossmann MG, Wengler G (1991) Structure of Sindbis virus core protein reveals a chymotrypsin-like serine proteinase and the organization of the virion. Nature 354: 37–43

31. Finch JT, Crowther RA, Hendry DA, Struthers JK (1974) The structure of *Nudaurelia capensis* β virus: the first example of a capsid with icosahedral surface symmetry $T = 4$. J Gen Virol 24: 191–200

32. Sehnke P, Harrington M, Hosur M, Li Y, Usha R, Tucker R, Bomu W, Stauffacher C, Johnson J (1988) Crystallization of viruses and virus proteins. J Crystal Growth 90: 222–230.35

33. Cavarelli J, Bomu W, Liljas L, Kim S, Minor W, Munshi S, Muchmore S, Schmidt T, Johnson J (1991) Crystallization and preliminary structure analysis of an insect virus with $T = 4$ quasi-symmetry: *Nudaurelia capensis* ω virus. Acta Crystallogr B47: 23–29

34. Fisher A, McKinney B, Schneeman A, Rueckert RR, Johnson JE (1993) Crystallization of virus-like particles assembled from Flock House virus coat protein expressed in a baculovirus system. J Virology 67: 2950–2953

35. Zlotnick A, Reddy V, Schneemann A, Dasgupta R, Rueckert R, Johnson J (1993) A buried aspartic acid catalyzes autoproteolytic maturation in a family of insect viruses (in prep.)

36. Olson N, Baker T (1989) Magnification calibration and the determination of spherical virus diameters using cryo-microscopy. Ultramicroscopy 30: 281–298

37. Crowther R, DeRosier D, Klug A (1970) The reconstruction of a three-dimensional structure from projections and its application to electron microscopy. Porch Roe Soc Loaned A317: 319–340

38. Fuller S (1987) The $T = 4$ envelope of Sindbis virus is organized by interaction with a complimentary $T = 3$ capsid. Cell 48: 923–934

39. Baker T, Drak J, Bina M (1988) Reconstruction of the three dimensional structure of Simian virus 40 and visualization of the chromatin core. Proc Natl Acad Sci USA 422–426

40. Franks A (1955) An optically focusing X-ray diffraction camera. Proc Phys Soc Sect B 68: 1054–1064

41. Harrison S (1968) A point focusing camera for single crystal diffraction. J Appl Crystallogr 1: 84–90

42. Schmidt T, Johnson J, Phillips W (1983) The spherically averaged structures of cowpea mosaic virus components by X-ray solution scattering. Virology 127: 65–73

Authors' address: Dr. J. E. Johnson, Department of Biological Sciences, Purdue University, West Lafayette, IN 47907, U.S.A.

Arch Virol (1994) [Suppl] 9: 513–522

Archives
of
Virology
© Springer-Verlag 1994
Printed in Austria

Retroviral RNA packaging: a review

A. Rein

Laboratory of Molecular Virology and Carcinogenesis, ABL-Basic Research Program,
NCI-Frederick Cancer Research and Development Center, Frederick, Maryland,
U.S.A.

Summary. In retroviruses, the "Gag" or core polyprotein is capable of assembling into virus particles and packaging the genomic RNA of the virus. How this protein recognizes viral RNA is not understood. Gag polyproteins contain a zinc-finger domain; mutants with changes in this domain assemble into virions, but a large fraction of these particles lack viral RNA. Thus, one crucial element in the RNA packaging mechanism is the zinc-finger domain. RNA sequences required for packaging ("packing signals") have been studied both by deletion analysis and by measuring encapsidation of nonviral mRNAs containing limited insertions of viral sequence. These experiments show that all or part of the packaging signal in viral RNA is located near the 5 end of the genome. These signals appear to be quite large, i.e., hundreds of nucleotides. Each virus particle actually contains a dimer of two identical, + strand genomic RNA molecules. The nature of the dimeric linkage is not understood. In some experimental situations (including zinc-finger mutants), only a small fraction of the particles in a virus preparation contain genomic RNA. It is striking that the genomic RNA packaged in these situations is dimeric. Because of this important observation, it is speculated that only dimers are packaged, and that the dimeric structure is an element of the packaging signal. It is also suggested that the dimers undergo a conformational change ("RNA maturation") after the virus is released from the cell, and that this change may depend upon the cleavage of the Gag polyprotein, a post-assembly event catalyzed by the virus-coded protease.

Introduction

In all viruses, the genomic DNA or RNA is specifically packaged by the viral proteins during assembly of the particle. Thus, one event in virus assembly is the specific recognition and selection of the viral genome for packaging.

One important approach to understanding the specific interaction between viral proteins and the viral genome is, of course, to identify the sites in both proteins and nucleic acids which are involved in this interaction. The present review will consider the packaging of genomic RNA by retroviruses, and will deal first with the nature of interacting sites in the viral proteins, and then with the location of "packaging signals" in the viral RNA. Much of this material was also recently reviewed by Linial and Miller [1].

Aspects of viral proteins involved in packaging

In retroviruses, a single protein species, the "Gag" or core polyprotein, can assemble into viruslike particles when it is expressed in the cell in the absence of other viral proteins. These particles bud from the cell in the same manner as complete virions, and are morphologically quite similar to the latter. Further, when viral genomic RNA is also present in the cell, the particles assembled from the Gag polyprotein package this RNA. Thus, this protein species is sufficient for virus assembly and RNA packaging.

In the course of normal virion synthesis, the Gag polyprotein is cleaved into a series of distinct proteins by the viral protease (PR). This cleavage occurs during or after the release of the particle from the cell. One of these cleavage products is known to represent a Gag domain which is significant in RNA packaging. This product, derived from the C-terminal region of the Gag precursor polyprotein, is termed the nucleocapsid (NC) protein. Retroviral NC proteins are small, basic proteins which bind single-stranded nucleic acids with no detected specificity in vitro, and are apparently associated with the genomic RNA in the mature virion.

All retroviral NC proteins (except those of the spumavirus group, which is not well characterized as yet) contain either one or two copies of a sequence motif, $C-X_2-C-X_4-H-X_4-C$, which has been termed the "cysteine array", "Cys-His box", or "zinc finger-like sequence" [2–3].

The conserved spacing of cysteines and histidines in the cysteine array resembles those in the "zinc finger" motifs, which are found in many eukaryotic transcription factors and are implicated in nucleic acid sequence recognition by these factors (reviewed in [4]). Indeed, recent evidence shows that zinc ions are indeed present in intact retrovirus particles [5], and are bound to the NC cysteine arrays [6, 7]. In studies to identify the function(s) of the arrays, a number of laboratories have made and analyzed mutants in cysteine arrays of murine leukemia viruses

(MuLVs), avian C-type viruses, and human immunodeficiency virus (HIV) [8–14].

In general, these studies have shown that virtually any change in the cysteine array dramatically reduces tha amount of viral genomic RNA packaged per virion. The degree of reduction can be as low as 2- to 3-fold or, frequently, as much as 100-fold or more (e.g. [10, 13]). Even quite conservative changes, such as the replacement of a single cysteine residue with a serine residue (a change of a sulfur atom to an oxygen atom in the protein), can drastically decrease the efficiency of RNA packaging.

As noted, many retroviral NC proteins have not one but two cysteine arrays. In these viruses, both arrays are required for efficient RNA packaging (e.g. [9, 13]). Indeed, in HIV, if the two arrays are left intact but their positions in the NC protein are exchanged, RNA packaging is reduced 10- to 20-fold [14a].

The results outlined above show clearly that the intact cysteine arrays are required for the normal packaging of viral RNA during retrovirus assembly. What is the nature of this requirement? One possibility is that the array is involved in the specific recognition of the viral RNA; in this case it might be imagined that mutant viruses with altered cysteine arrays would be lacking in specificity, and might package cellular RNAs in place of the viral genome. Alternatively, it is possible that the array is required for binding and packaging *any* RNA, so that cysteine-array mutants would lack RNA altogether.

In fact, a retrovirus particle normally contains a large number of tRNAs and other small RNA molecules. While MuLV cysteine-array mutants still contain these small RNAs, they do not contain high-molecular-weight cellular RNAs in place of the viral genome [11, 15]. Thus, it seems possible that the arrays are required for binding and packaging *any* high-molecular-weight RNA, rather than for specific recognition of the genome.

One spontaneous mutant of an avian type-C retrovirus is unique in that, in stark contrast to the cysteine-array mutants, it *does* package cellular mRNAs [16]. Thus this mutant, termed SE21Q1b, appears to have partially lost its ability to specifically recognize genomic RNA during virion assembly. While the defect responsible for this remarkable phenotype has not been localized, it is somewhere in Gag, but not in the NC protein [17].

Taken together, the results described above suggest the possibility that two separate domains in the Gag polyprotein cooperate in RNA packaging. One of these is outside the NC region and specifically recognizes "packaging signals" in the genomic RNA. The other is the cysteine array(s) (and neighboring residues in the NC region [18]), which may bind the RNA and enable it to be incorporated into the particle. It would obviously be of great importance ot know more about the former domain.

"Packaging signals" in retroviral genomic RNAs

Wild-type retroviral proteins show a very high degree of specificity in packaging retroviral genomic RNA. This specificity must reflect the recognition of signals contained in the sequence of the RNA, since to our knowledge there are no other, qualitative differences between the viral RNA and other cellular mRNAs. That is, the genomic RNA has a cap at its 5' end and a poly A tract at its 3' end; indeed, this RNA is chemically identical to the Gag and Gag-Pol mRNA molecules present in the infected cell. The packaging signals must be present in this full-length viral RNA but absent from the spliced mRNA molecules derived from it.

One approach to the identification of packaging signals is that of deletion analysis. It has been known for many years that deletion of sequences near (but not at) the 5' end of the retroviral RNA molecule renders the RNA unpackageable [19, 20]. These results indicate that retroviral packaging signals involve sequences near the 5' end of the RNA. Such deletions in MuLVs have been termed Ψ mutants [20].

While the experiments just mentioned demonstrate that sequences near the 5' end of the RNA are *necessary* for packaging, they do not show that the packaging signals are restricted to this region. Obviously, one powerful approach to analyzing packaging signals is to identify sequences *sufficient* for packaging, by adding retroviral sequences to nonretroviral RNAs and measuring the packaging of these chimeric RNAs.

One careful study has defined the MuLV sequences sufficient for packaging a nonviral mRNA. Adam and Miller [21] placed fragments of MuLV and a closely related virus, murine sarcoma virus (MuSV), in the 3'-untranslated region of an mRNA encoding a drug-resistance marker. They found that addition of a 325-nucleotide stretch of MuSV sequence, i.e. bases 215 to 539, was sufficient to render the heterologous RNA packageable by MuLV proteins with nearly the same efficiency as authentic MuLV genomic RNA. Interestingly, the corresponding stretch of MuLV sequence, i.e. bases 215 to 563, was only about 1/10 as effective as this MuSV sequence. (This MuLV sequence is clearly a part of the MuLV packaging signal, since its deletion renders MuLV genomic RNA unpackageable [20].) As shown in Fig. 1, the two sequences are extremely similar (>97% identity) except for a highly divergent region, i.e. MuSV bases 408–439, which show no obvious resemblance to MuLV

```
MuSV   215 CCAGCAACTTATCTGTGTCTGTCCGATTGTCTCTAGTGTCTATGTTTGATGT
           ||||||||||||||||||||||||||||||||||||||||||| |||| |
MuLV   215 CCAGCAACTTATCTGTGTCTGTCCGATTGTCTCTAGTGTCTATGACTGATTT

MuSV   265 TATGCGCCTGCGTCTGTACTAGTTAGCTAACTAGCTCTGTATCTGGCGGA
           |||||||||||| ||||||||||||||||||||||||||||||||||||
MuLV   265 TATGCGCCTGCGTCGGTACTAGTTAGCTAACTAGCTCTGTATCTGGCGGA

MuSV   315 CCCGTGGTGGAACTGACGAGTTCTGAACACCCGGCCGCAACCCTGGGAGA
           ||||||||||||||||||||||| ||||||||||||||||||||||||||
MuLV   315 CCCGTGGTGGAACTGACGAGTTCGGAACACCCGGCCGCAACCCTGGGAGA

MuSV   365 CGTCCCAGGGACTTTGGGGGCCGTTTTTGTGGCCCGACCTGAGGAAGGGA
           ||||||||||||| |||||||||||||||||||||||||||||     |
MuLV   365 CGTCCCAGGGACTTCGGGGGCCGTTTTTGTGGCCCGACCTGAGTCCAAAA

MuSV   415 GTCGATGTGGAATCCGACCC..................CGTCAGGATAT
           ||    ||| | |   ||| |                  | ||||||
MuLV   415 ATCCCGATCGTTTTGGACTCTTTGGTGCACCCCCCCTTAGAGGAGGGATAT

MuSV   446 GTGGTTCTGGTAGGAGACGAGAACCTAAAACAGTTCCCGCCTCCGTCTGA
           ||||||||||||||||||||||||||||||||||||||||||||||||||
MuLV   465 GTGGTTCTGGTAGGAGACGAGAACCTAAAACAGTTCCCGCCTCCGTCTGA

MuSV   496 ATTTTTGCTTTCGGTTTGGAACCGAA.....GCCGCGCGTCTTGTCTGC
           |||||||||||||||||||| ||||||       |||||||||||||||
MuLV   515 ATTTTTGCTTTCGGTTTGGGACCGAAGCCGCGCCGCGCGTCTTGTCTGC
```

Fig. 1. Comparison of MuLV nucleotides 215–563 with nucleotides 215–539 of MuSV

bases 408–458. It seems reasonable to speculate that this divergent region is an important part of the MuSV packaging signal.

In the studies by Adam and Miller [21], it was found that the MuLV "Ψ" sequence, i.e. MuLV bases 215 to 563, is "defective" or "incomplete" as a packaging signal, since it is not sufficient to render a heterologous RNA efficiently packageable. It thus appears that this sequence is only a part of the MuLV packaging signal, and that the defect in this sequence is compensated for by sequences elsewhere in the MuLV genome. In fact, there is little loss of packaging efficiency if this sequence is moved from its normal location to a different region in the genome [22]; this observation suggests that this region, while it is only part of the packaging signal, acts autonomously with respect to packaging.

The packaging signal of the avian C-type retroviruses has not been defined in detail, but experiments by Sorge et al. [23] suggest that it is discontinuous, consisting of sequences near the 3′ end of the viral genome in addition to sequences near the 5′ end. There is no evidence that this is true for other retroviruses, however.

Another recent analysis [24] has partially defined the sequences required for packaging in retroviral vectors derived from Harvey sarcoma virus (HaSV). It was striking that relatively efficient packaging was obtained with genomes in which the only viral sequences at the 5′ end were the first 271 nucleotides of HaSV. The first ~210 bases of the HaSV genome are derived from a region of Moloney MuLV which is not required for packaging [20, 21].

The experiments cited above suggest that packaging signals can be quite large, and can be composed of several distinct elements which are all required for efficient packaging. The large size of the required sequences raises the possibility that the signals are actually secondary or tertiary structures, as well as specific sequences, in the viral RNA.

Attempts to define packaging signals are further complicated by the fact that, while these signals may act autonomously in a heterologous "reporter" RNA [21], they are nevertheless sensitive to their context. For example, sequences in the "U5" region of the genome are not part of the MuLV packaging signal identified by Adam and Miller, but Murphy and Goff [25] found that deletions in U5 render MuLV genomic RNA unpackageable. In a somewhat analogous experiment, Donzé and Spahr [26] found that sequences in U5 are essential for the efficient packaging of avian retroviral RNA. These avian sequences are very short open reading frames; the significance of this fact is completely unclear at present. In the case of HIV, Hayashi et al. [27] inserted a 46-base viral sequence from the "leader" region, i.e. between the 5′ long terminal repeat and the beginning of the protein-coding region of the viral genome, into either of two positions in a "reporter" RNA. Surprisingly, they found that this sequence led to efficient packaging in one position, but no detectable packaging in the other. Obviously, it is extremely difficult to identify autonomously acting structural elements in an RNA molecule, since inserted sequences can always interact with regions elsewhere in the "host" or reporter RNA molecule.

Is RNA dimerization required for packaging?

As far as is known, the genomic RNAs in all retrovirus particles are in the form of a dimer. The nature of the linkage between the two monomers, which are both of + strand polarity, is not understood, but presumably consists of hydrogen bonds or other weak, noncovalent bonds, since the dimer is dissociated into monomers under relatively mild conditions.

Where and when is the dimer formed? In an attempt to answer this question, several laboratories isolated RNA from "rapid harvest virus", i.e. very young virus particles. They reported [28–30] that rapid harvest virus contains monomeric RNA, and that this RNA dimerizes upon incubation of the virus. The obvious interpretation of these findings is that virus initially packages monomeric RNA, and that the two mono-mers are only joined together after the virus is released from the cell.

On the other hand, other observations in the literature are very difficult to reconcile with this conclusion. These are the results obtained in situations where the packaging of RNA is very inefficient. Thus, when MuLV-producing cells are treated for several hours with actinomycin D [31], and also in the case of certain cysteine-array mutants of MuLV [11], the level of genomic RNA packaged per particle is drastically reduced. However, the RNA in these particles is still dimeric [11, 31]. The fact that dimers, rather than monomers, are found in particles when the vast majority of particles contain no genomic RNA strongly suggests that the two monomers do not enter the assembling virion independently of each other; rather, it is consistent with the idea that some type of linkage between them is formed *before* they enter the nascent particle.

One hypothesis that would appear to be consistent with all of the results discussed above is as follows. RNA is initially packaged as a dimer, but the monomers in this "young" dimer are held together in a more fragile linkage than that found in mature virus particles. Such a dimer might not be stable under some of the extraction conditions used to isolate dimers from standard virus preparations. Then, at some time after the virus is released from the cell, the dimer undergoes a stabilization or "maturation" event.

Stoltzfus and Snyder [32] analyzed the RNA present in rapid harvest virions of avian C-type virus. They found that rapid harvest virus contains dimeric RNA, but that this RNA dissociates into monomers at a lower temperature than that found in mature virus. Thus, their results lend strong support to the overall hypothesis presented above.

One important implication of this hypothesis is the possibility that only dimers can be packaged. Perhaps the dimeric structure itself is a crucial part of the "packaging signal" in genomic RNA; the genetic experiments defining viral sequences necessary or sufficient for packaging may really be determining sequences required to form the dimer structure.

Darlix and colleagues have recently made the intriguing observation that RNA molecules made by in vitro transcription of retroviral sequences can form dimers spontaneously. The dimerization depends upon the presence of sequences near the 5′ end of the viral genome, and only occurs with molecules of the + polarity (e.g. [33–35]). The requirement for sequences in the Ψ region is, of course, independent support for the idea that dimerization and packaging are intimately connected. The phenomenon may shed light on the mechanism of dimerization, and in addition, may make it possible to generate large enough quantities of dimer for structural studies.

As noted above, the proteins of the retrovirus particle are already known to undergo a maturation event, i.e. cleavage of the Gag precursor polyprotein by the viral protease, after the particle is assembled. In the discussion just concluded, I have proposed that the RNA dimer of the particle also undergoes a maturation event. Recent data from our laboratory (W. Fu and A. Rein, unpubl.) suggest that these two events are connected: the dimeric RNA isolated from mutant, protease-deficient MuLV particles is in a different, less stable conformation than that isolated from wild-type particles (data not shown). Perhaps the maturation of the RNA is catalyzed by a Gag cleavage product, and is thus dependent on the prior cleavage of the Gag precursor. Future experiments will extend the genetic analysis of this phenomenon, as well as analyze rapid-harvest MuLV RNA.

In summary, it seems possible that the dimeric structure of the genomic RNA in retrovirus particles has a crucial significance for packaging. It is likely that structural, as well as genetic and biochemical, progress will be required before we arrive at a real understanding of mechanisms involved in RNA packaging during retrovirus assembly.

Acknowledgements

I wish to thank L. Henderson and J. Levin for helpful comments on this manuscript, and Carol Shawver for assistance in its preparation. Research sponsored by the National Cancer Institute, DHHS, under contract NO. NO1-CO-74101 with ABL. The contents of this publication do not necessarily reflect the views or policies of the Department of Health and Human Services, nor does mention of trade names, commercial products, or organizations imply endorsement by the U.S. Government.

References

1. Linial ML, Miller AD (1990) Retroviral RNA packaging: sequence requirements and implications. In: Swanstrom R, Vogt PK (eds) Retroviruses. Strategies of replication. Springer, Berlin Heidelberg New York Tokyo, pp 125–185
2. Henderson LE, Copeland TD, Sowder RC, Smythers GW, Oroszlan S (1981) Primary structure of the low-molecular-weight nucleic acid binding proteins of murine leukemia viruses. J Biol Chem 256: 8400–8406
3. Covey SM (1986) Amino acid sequence homology in *gag* region of reverse transcribing elements and the coat protein gene of cauliflower mosaic virus. Nucleic Acids Res 14: 623–633
4. Evans RM, Hollenberg SM (1988) Zinc fingers: gilt by association. Cell 59: 103–112
5. Bess JW Jr, Powell PJ, Issaq HJ, Schumack LJ, Grimes MK, Henderson LE, Arthur LO (1992) Tightly bound zinc in human immunodeficiency virus type 1, human T-cell leukemia virus type I, and other retroviruses. J Virol 66: 840–847
6. Summers MF, Henderson LE, Chance MR, Bess JW Jr, South TL, Blake PR, Sagi I, Perez-Alvarado G, Sowder RC III, Hare DR, Arthur LO (1992) Nucleocapsid zinc fingers detected in retroviruses: EXAFS studies of intact viruses and the solution-state structure of the nucleocapsid protein from HIV-1. Protein Sci 1: 563–574

7. Chance MR, Sagi I, Wirt MD, Frisbie SM, Scheuring E, Chen E, Bess JW Jr, Henderson LE, Arthur LO, South TL, Perez-Alvarado G, Summers MF (1992) Extended x-ray absorption fine structure studies of a retrovirus: equine infectious anemia virus cysteine arrays are coordinated to zinc. Proc Natl Acad Sci USA 89: 10041–10045

8. Méric C, Spahr P-F (1986) Rous sarcoma virus nucleic acid-binding protein p12 is necessary for viral 70S RNA dimer formation and packaging. J Virol 60: 450–459

9. Méric C, Gouilloud E, Spahr P-F (1988) Mutations in Rous sarcoma virus nucleocapsid protein p12 (NC): deletions of Cys-His boxes. J Virol 62: 3328–3333

10. Gorelick RJ, Henderson LE, Hanser JP, Rein A (1988) Point mutants of Moloney murine leukemia virus that fail to package viral RNA: Evidence for specific RNA recognition by a "zinc finger-like" protein sequence. Proc Natl Acad Sci USA 85: 8420–8424

11. Méric C, Goff SP (1989) Characterization of Moloney murine leukemia virus mutants with single-amino-acid substitutions in the Cys-His box of the nucleocapsid protein. J Virol 63: 1558–1568

12. Aldovini A, Young RA (1990) Mutations of RNA and protein sequences involved in human immunodeficiency virus type 1 packaging result in production of noninfectious virus. J Virol 64: 1920–1926

13. Gorelick RJ, Nigida SM Jr, Bess JW Jr, Arthur LO, Henderson LE, Rein A (1990) Noninfectious human immunodeficiency virus type 1 mutants deficient in genomic RNA. J Virol 64: 3207–3211

14. Dupraz P, Oertle S, Meric C, Damay P, Spahr P-F (1990) Point mutations in the proximal Cys-His box of Rous sarcoma virus nucleocapsid protein. J Virol 64: 4978–4987

14a. Gorelick RJ, Chabot DJ, Rein A, Henderson LE, Arthur LO (1993) The two zinc fingers in the human immunodificiency virus type 1 nucleocapsid protein are not functionally equivalent. J Virol 67: 4027—4036

15. Gorelick RJ, Nigida SM Jr, Arthur LO, Henderson LE Rein A (1991) Roles of nucleocapsid cysteine arrays in retroviral assembly and replication: possible mechanisms in RNA encapsidation. In: Kumar A (ed) Advances in molecular biology and targeted treatment for AIDS. Plenum Press, New York, pp 257–272

16. Linial M, Medeiros E, Hayward WS (1978) An avian oncovirus mutant (SE21Q1b) deficient in genomic RNA: biological and biochemical characterization. Cell 15: 1371–1381

17. Anderson DJ, Lee P, Levine KL, Sang J, Shah SA, Yang OO, Shank PR, Linial ML (1992) Molecular cloning and characterization of the RNA packaging-defective retrovirus SE21Q1b. J Virol 66: 204–216

18. Fu X, Katz RA, Skalka AM, Leis J (1988) Site-directed mutagenesis of the avian retrovirus nucleocapsid protein pp 12: mutation which affects RNA binding in vitro blocks viral replication. J Biol Chem 263: 2134–2139

19. Shank PR, Linial M (1980) Avian oncovirus mutant (SE21Q1b) deficient in genomic RNA. Characterization of a deletion in the provirus. J Virol 36: 450–456

20. Mann R, Mulligan RC, Baltimore D (1983) Construction of a retrovirus packaging mutant and its use to produce helper-free defective retroviruses. Cell 33: 153–159

21. Adam MA, Miller AD (1988) Identification of a signal in a murine retrovirus that is sufficient for packaging of nonretroviral RNA into virions. J Virol 62: 3802–3806

22. Mann R, Baltimore D (1985) Varying the position of a retrovirus packaging sequence results in the encapsidation of both unspliced and spliced RNAs. J Virol 54: 401–407

23. Sorge JD, Ricci W, Hughes SH (1983) cis-Acting RNA packaging locus in the 115-nucleotide direct repeat of Rous sarcoma virus. J Virol 48: 667–675

24. Torrent C, Wang P, Darlix J-L (1992) A murine leukemia virus derived retroviral vector with a rat VL30 packaging psi sequence. Bone Marrow Transplant 9 [Suppl 1]: 143–147

25. Murphy JE, Goff SP (1989) Construction and analysis of deletion mutations in the U5 region of Moloney murine leukemia virus: effects on RNA packaging and reverse transcription. J Virol 63: 319–327

26. Donzé O, Spahr P-F (1992) Role of the open reading frames of Rous sarcoma virus leader RNA in translation and genome packaging. EMBO J 11: 3747–3757

27. Hayashi T, Shioda T, Iwakura Y, Shibuta H (1992) RNA packaging signal of human immunodeficiency virus type 1. Virology 188: 590–599

28. Cheung K-S, Smith RE, Stone MP, Joklik WK (1972) Comparison of immature (rapid harvest) and mature Rous sarcoma virus particles. Virology 50: 851–864

29. Canaani E, Helm KVD, Duesberg P (1973) Evidence for 30–40S RNA as precursor of the 60–70S RNA of Rous sarcoma virus particles. Proc Natl Acad Sci USA 72: 401–405

30. Korb J, Travnicek M, Riman J (1976) The oncornavirus maturation process: Quantitative correlation between morphological changes and conversion of genomic virion RNA. Intervirology 7: 211–224

31. Levin JG, Grimley PM, Ramseur JM, Berezesky IK (1974) Deficiency of 60 to 70S RNA in murine leukemia virus particles assembled in cells treated with actinomycin D. J Virol 14: 152–161

32. Stoltzfus CM, Snyder PN (1975) Structure of B77 sarcoma virus RNA: stabilization of RNA after packaging. J Virol 16: 1161–1170

33. Prats AC, Roy C, Wang P, Erard M, Housset V, Gabus C, Paoletti C, Darlix J-L (1990) cis elements and trans-acting factors involved in dimer formation of murine leukemia virus RNA. J Virol 64: 774–783

34. Bieth E, Gabus C, Darlix JL (1990) A study of the dimer formation of Rous sarcoma virus RNA and of its effect on viral protein synthesis in vitro. Nucleic Acids Res 18: 119–127

35. Darlix J-L, Gabus C, Nugeyre M-T, Clavel F, Barré-Sinoussi F (1990) Cis elements and trans acting factors involved in the RNA dimerization of HIV-1. J Mol Biol 216: 689–699

Author's address: Dr. A. Rein, Laboratory of Molecular Virology and Carcinogenesis, ABL-Basic Research Program, NCI-Frederick Cancer Research and Development Center, Frederick, MD 21702, U.S.A.

Arch Virol (1994) [Suppl] 9: 523–529

Archives of
Virology
© Springer-Verlag 1994
Printed in Austria

Structural studies of viruses by electron cryomicroscopy

M. F. Schmid, B. V. V. Prasad, and **W. Chiu**

Summary. Electron cryomicroscopy is a unique biophysical technique for studying molecular structures of viruses which are difficult to analyze by x-ray diffraction. The structural information derived from the low resolution reconstructions of viruses has so far been useful to understand various functional properties of the viruses such as antibody neutralization, receptor binding and assembly. Electron cryomicroscopy has enabled the visualization of the four core alpha helices of the coat protein in tobacco mosaic virus. This represents the highest resolution detail of a virus studied by electron cryomicroscopy. The prospects of attaining similar resolution beyond 10 Å for spherical viruses as well are encouraging, with newly available instrumentation, data collection and processing procedures.

Introduction

Electron microscopy has been an essential tool in molecular virology for identification and classification of viruses [25]. The pioneering work of Klug and co-workers introduced computational procedures for retrieving three-dimensional structures from two-dimensional electron micrographs [12–14]. Their work has led to many fundamental insights into the principles of assembly and structural organization for icosahedral and helical viruses. In the early 1980s, another significant technical advance was introduced in electron microscopy. This was to embed virions in a thin matrix of vitreous ice and to observe them at low dose and low temperature conditions [1]. This cryo-technique eliminates the use of negative stain and fixative and allows the study of virus structure at different functional or chemical states (see [9, 17]). The virion structure can also be preserved better, so that higher resolution structural details can be determined. Furthermore, the availability of affordable computer workstations with color graphics capability has made data analysis much faster and easier than before. There have been an increasing number of examples of different kinds and sizes of viruses being studied by electron cryomicroscopy. In the future, it is expected that this technology will provide better contrast and resolution of single virions. This paper will review both the technical aspects and applications of electron cryomicroscopy for studying molecular structures of viruses.

Materials and methods

Several experimental and computational steps are required to determine the three-dimensional structure of a virus by electron cryomicroscopy. In terms of specimen requirements, it has been necessary to have a concentration of the virions in the range of 10^{11}–10^{12}/ml, much more than that used in conventional negative stain procedures. This requirement is imposed in order to have enough virions (over 50) per micrograph, so that they can be analyzed and merged for reconstruction without the need for correction of different electron optical parameters, such as magnification and defocus in different micrographs. However, as the computational technique is being improved to make necessary corrections [8, 22] and scaling between different images [2], this requirement for high concentration of virions may likely be relaxed.

The specimen preparation for imaging has been done by rapidly plunging a grid that has been blotted nearly dry into a bath of liquid ethane cooled by liquid nitrogen. Most laboratories doing this work have built a simple apparatus to carry out this step, though there is a commercial apparatus now available. The disadvantage of the commercial apparatus is its relatively high cost and the large consumption of liquid nitrogen. In order to provide safety precautions for handling pathogens and the potentially explosive cryogen (ethane) in the gas phase, a containment box has been built in which the freezing step can be performed [23]. This cooling is quite straightforward but takes some practice in order to obtain the proper thickness of embedding ice and also to avoid ambient water contamination of the frozen specimen. The subsequent step is to transfer the frozen, hydrated specimen grid onto to cryospecimen holder, which is kept at low temperature ($< -160°C$) during the entire period of observation in the microscope. Because the specimen surface is cold, it easily attracts residual gases which can condense onto the specimen grid inside the microscope. Therefore, it is critical to have an efficient anticontaminator close to the specimen area. This can be installed readily in most microscopes [16].

Once the frozen, hydrated specimen is inserted into the microscope column, one usually waits for 40 minutes or longer to allow it to reach a thermal and mechanical equilibrium prior to carrying out the low dose imaging. The most frequently encountered problems at this point include poor specimens due to ice being either too thick or completely dry. Beginning users may also encounter ice contamination during the preparative or transfer step [17]. The low dose imaging procedure can be easily done with most modern microscopes. Images usually are recorded with a specimen dose less than 8 electrons/$Å^2$ at a magnification of 30 000 to 50 000 [22]. For recording film we use Kodak SO163, which achieves high film speed. A focal pair of images are taken, with the first one at a relatively small defocus value (0.5–0.8 μm), the second one with a higher defocus value such as 1–1.5 μm [19]. The difference between these two images manifests itself in the low contrast and high resolution of the small defocus image, and high contrast but low resolution in the high defocus image. The highly defocused image makes the subsequent image processing easier. It has become routine to use the small defocus images for reconstructions, in order to obtain maximum structural information.

Images of icosahedral viruses are evaluated by visual inspection for specimen contrast, ice thickness, defocus, charging and mechanical drift. So far, this step is quite subjective, and it would be better to quantify this for higher resolution structure determination. In the case of helical viruses, it is relatively easy to evaluate the structural quality of the images by an optical diffraction analysis because the structural data of a helical array are distributed in layer lines of a diffraction pattern [14]. The chosen images are then digitized with a scanner at an interval and scanning spot size according to the Sampling Theorem [13]. The image processing steps are to determine virus orientation, to refine these orientational parameters, to merge them and evaluate the merged data and to reconstruct its three-dimensional mass density [3, 12, 13]. Detailed mathematical formulae in the image processing steps are different for viruses with icosahedral or helical symmetry [11, 12, 15]. So far, virus particles without either of these symmetries have not been studied to the extent of three-dimensional reconstruction, though an algorithm has already been established for images of small single asymmetric particles such as ribosome subunits [18]. Once the three-dimensional map is obtained, the most exciting and challenging task is to visualize it at various angular perspectives using a color graphics computer terminal and to interpret its structural features in terms of known biochemical and functional data. A potential difficulty in interpreting low resolution maps is the lack of stereochemical constraints as present in an atomic model.

Results and discussion

The use of the above techniques has been expanding and numerous studies have been applied to animal, plant and bacterial virions. Some of these studies have provided surprising results. For example, the first example of Fab-virus was done with a rotavirus and Fab from monoclonal antibody against the spike protein [28]. This antibody-bound virus structure has yielded novel and useful information regarding binding sites, conformational variations of the Fab fragments, and the dimeric nature of the spike proteins. Recently, this kind of application in visualizing antibodies or receptors bound to the virions has been used to confirm hypothetical models of the binding sites [27, 32].

Another useful application of the cryomicroscopy technique is to observe structural variation of the virions at different assembly states. Investigations of precursor and the mature particles of P22 bacteriophage have led to a mechanistic model for the exit of the scaffolding protein during genome packaging [29]. Electron cryomicroscopy also has been found useful for probing attachment sites and modes of packing of nucleic acids inside virions [4, 6]. A recently determined three-dimensional density map at 22 Å of calicivirus, a positive-stranded RNA virus with a diameter of 400 Å, suggests an icosahedrally ordered RNA [30]. Although distinct from the structures of other animal viruses, the three-dimensional structure of this virus exhibits remarkable similarities to the structures of T = 3 plant viruses, such as tomato bushy stunt virus [20] and turnip crinkle virus [21].

Though electron cryomicroscopic studies have provided new and insightful contributions to understanding of the structure and function of

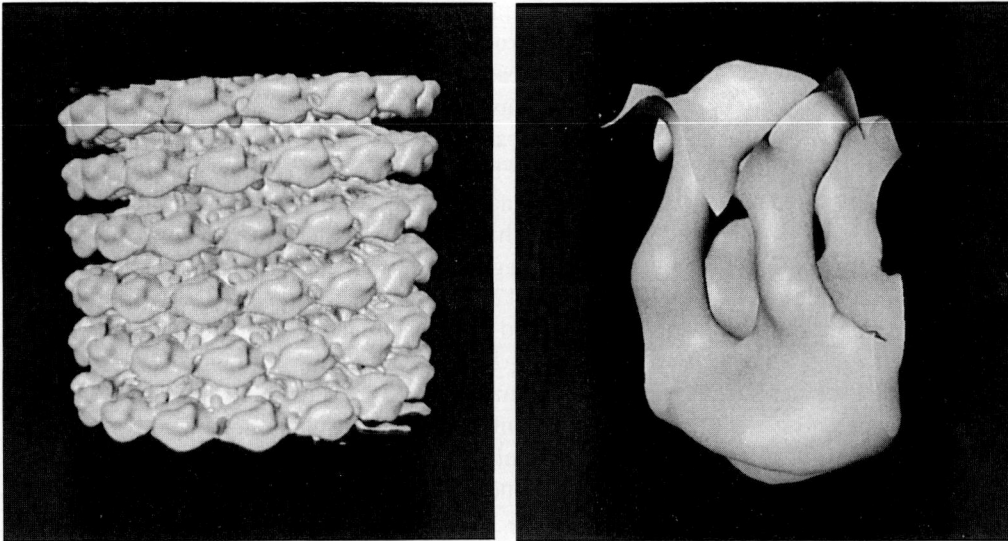

Fig. 1. a Surface rendering of six helical turns of tobacco mosaic virus in a mass density map at ~9 Å resolution, determined from computer reconstruction of electron images of ice embedded TMV particles. Coat proteins are arranged helically with 49 proteins in three helical turns; this display shows two helical repeat in a length of 138 Å from part of a TMV particle. **b** Computer-extracted single coat protein subunit of TMV where the four core alpha helices nearly normal to the TMV helix axis are clearly seen. The density map of this reconstruction was obtained from the previous work of Jeng et al. [22]; this display was done with the EXPLORER software in an Iris Indigo workstation of Silicon Graphics Computer System

viruses, resolution usually is limited to 20–40 Å. The highest resolution so far obtained has been on tobacco mosaic virus at 9 Å resolution [22]. At this resolution, the four nearly parallel core alpha helices can be resolved as shown in Figs. 1a and 1b. Because the TMV structure has been well established by X-ray crystallography of the virus and of the reconstituted assemblies [5, 26], electron cryomicroscopic analysis was used mainly to demonstrate the resolving power of the method. It is relatively easier to obtain and evaluate the structural data of a helical than an icosahedral object. This imaging capability has opened new research possibilities for studying the assembly questions of helical virus, such as TMV, to a reasonably high resolution. For instance, TMV can exist as a disk or as a helix under different chemical conditions [7, 24]. A carefully co-ordinated investigation of both biochemical and cryomicroscopic analyses can lead to a clear interpretation of the structural morphogenesis of the virus in vitro.

Conclusion

Electron cryomicroscopy and computer reconstruction have been used to provide information on different biological processes carried out by viruses including attachment, entry, neutralization and assembly. Though these structural determinations have been carried out at low resolution (20–40 Å), the information has led to other more directed and meaningful studies to advance our understanding of the various system. The feasibility analysis of TMV at 9 Å has suggested the potential of this method for yielding resolution such that the secondary structure of the viral proteins of a highly ordered and rigidly helical virus can be resolved. With further enhancement of the structural and computational tools, there is optimism that this method can resolve even finer detail of complex virus assembly [10, 33]. Moreover, a combination of high resolution X-ray structures of individual proteins and electron cryomicroscopy of a larger complex of virus with receptor/antibody has been shown to provide even greater information than either of these techniques can provide alone [27, 31, 32].

Acknowledgement

This research has been supported by NIH grants RR02250 and GM41064 and by the W. M. Keck Foundation.

References

1. Adrian M, Dubochet J, Lepault J, McDowall AW (1984) Cryo-electron microscopy of viruses. Nature 308: 32–36
2. Aldroubi A, Trus BL, Unser M, Booy FP, Steven AC (1992) Magnification mismatches between micrographs: corrective procedures and implications for structural analysis. Ultramicroscopy 46: 175–188
3. Amos LA, Klug A (1975) Three-dimensional image reconstructions of the contractile tail of T4 bacteriophage. J Mol Biol 99: 51–73
4. Baker TS, Cheng RH, Johnson JE, Olson NH, Wang GJ, Schmidt TJ (1992) Organized packing of RNA inside viruses as revealed by cryo-electron microscopy and x-ray diffraction analysis. In: 50th Ann Electr Micros Soc Amer. Boston, San Francisco Press, pp 454–455
5. Bloomer AC, Champness JN, Bricogne G, Staden R, Klug A (1978) Protein disk of tobacco mosaic virus at 2.8 Å resolution showing the interactions within and between subunits. Nature 276: 362–368
6. Booy F, Trus BL, Newcomb WW, Brown JC, Serwer P, Steven AC (1992) Organization of dsDNA in icosahedral virus capsids. In: 50th Ann Electr Microsc Sco Amer. Boston, San Francisco Press, pp 452–453
7. Caspar DLD, Namba K (1990) Switching in the self-assembly of tobacco mosaic virus. Adv Biophys 26: 157–185
8. Cheng RH (1992) Correlation of cryo-electron microscopic and x-ray data and compensation of the contrast transfer function. In: Proc 50th Ann Meeting of Electr Microsc Soc Amer. Boston, San Francisco Press, pp 996–997

528   M. F. Schmid et al.

9. Chiu W (1986) Electron microscopy of frozen, hydrated biological specimens. Annu Rev Biophys Biomol Struct 15: 237–257

10. Chiu W (1993) What does electron cryomicroscopy provide that X-ray crystallography and NMR spectroscopy cannot? Annu Rev Biophys Biomol Struct 22: 233–255

11. Chiu W, Schmid MF, Prasad BVV (1993) Teaching electron diffraction and imaging of macromolecules. Biophys J 64: 397–402

12. Crowther RA (1971) Procedures for three-dimensional reconstruction of spherical viruses by Fourier synthesis from electron micrographs. Phil Trans Roy Soc London Ser B 261: 221–230

13. Crowther RA, DeRosier DJ, Klug A (1970) The reconstruction of a three-dimensional structure from projections and its application to electron microscopy. Proc Roy Soc London 317: 319–340

14. DeRosier DJ, Klug A (1968) Reconstruction of three-dimensional structures from electrons micrographs. Nature 217: 130–134

15. DeRosier DJ, Moore PB (1970) Reconstruction of three-dimensional images from electron micrographs of structures with helical symmetry. J Mol Biol 52: 355–369

16. Downing KH, Chiu W (1990) Cold stage design for high resolution electron microscopy of biological materials. J Electron Microsc (Tokyo) 3: 213–226

17. Dubochet J, Adrian M, Chang JJ, Homo JC, Lepault J, McDowall AW, Schultz P (1988) Cryo-electron microscopy of vitrified specimens. Q Rev Biophys 21: 129–228

18. Frank J, Radermacher M (1992) Three-dimensional reconstruction of single particles negatively stained or in vitreous ice. Ultramicroscopy 46: 241–262

19. Fuller SD (1987) The T = 4 envelope of Sindbis virus is organized by interactions with a complementary T = 3 capsid. Cell 48: 923–934

20. Harrison SC, Olson A, Schutt CE, Winkler FK, Bricogne G (1978) Tomato bushy stunt virus at 2.9 Å resolution. Nature 276: 368–373

21. Hogle JM, Maeda A, Harrison SC (1986) The structure and assembly of turnip crinkle virus I: X-ray crystallographic analysis at 3.2 Å. J Mol Biol 191: 625–638

22. Jeng TW, Crowther RA, Stubbs G, Chiu W (1989) Visualization of alpha-helices in tobacco mosaic virus by cryo-electron microscopy. J Mol Biol 205: 251–257

23. Jeng TW, Talmon Y, Chiu W (1988) Containment system for the preparation of vitrified-hydrated virus specimens. J Electron Microsc Tech 8: 343–348

24. Klug A (1983) From macromolecules to biological assemblies. Bioscience Reports 3: 395–430

25. Levine AJ (1992) Viruses Scientific American Library. W.H. Freeman, New York

26. Namba K, Stubbs G (1986) Structure of TMV at 3.6 Å resolution: implications for assembly. Science 231: 1401–1406

27. Olson NH, Kolatkar PR, Oliveira MA, Cheng RH, Greve JM, McClelland A, Baker TS, Rossmann MG (1993) Structure of a human rhinovirus complexed with its receptor molecule. Proc Natl Acad Sci USA 90: 507–511

28. Prasad BVV, Burns JW, Marietta E, Estes MK, Chiu W (1990) Localization of VP4 neutralization sites in rotavirus by three-dimensional cryo-electron microscopy. Nature 343: 476–479

29. Prasad BVV, Prevelige P, Marietta E, Chen R, Thomas D, King J, Chiu W (1993) Three-dimensional transformation of capsids associated with genome packaging in a bacterial virus. J Mol Biol 231: 65–74

30. Prasad BVV, Matson DO, Smith AW (1993) Three-dimensional structure of a calicivirus (submitted)
31. Stewart PL, Burnett RM, Cyrklaff M, Fuller SD (1991) Image reconstruction reveals the complex molecular organization of adenovirus. Cell 67: 145–154
32. Wang G, Porta C, Chen Z, Baker T, Johnson JE (1992) Identification of a Fab interaction footprint site on an icosahedral virus by cryoelectron microscopy and x-ray crystallography. Nature 355: 275–278
33. Zhou ZH, Chiu W (1993) Prospects for using an IVEM with a FEG for imaging macromolecules towards atomic resolution. Ultramicroscopy 49: 407–416

Authors' address: Dr. M. F. Schmid, Verna and Marrs McLean Department of Biochemistry, The W. M. Keck Center for Computational Biology, Baylor College of Medicine, One Baylor Plaza, Houston, TX 77030, U.S.A.

Arch Virol (1994) [Suppl] 9: 531–541

Archives
of
Virology
© Springer-Verlag 1994
Printed in Austria

Crystallographic and cryo EM analysis of virion-receptor interactions

M. G. Rossmann[1], **N. H. Olson**[1], **P. R. Kolatkar**[1], **M. A. Oliveira**[1], **R. H. Cheng**[1], **J. M. Greve**[2], **A. McClelland**[2,*], and **T. S. Baker**[1]

[1] Department of Biological Sciences, Purdue University, West Lafayette, Indiana,
[2] Institute for Molecular Biologicals, Miles Inc., West Haven, Connecticut, U.S.A.

Summary. Cryoelectron microscopy has been used to determine the first structure of a virus when complexed with its glycoprotein cellular receptor. Human rhinovirus 16 (HRV16) complexed with the two amino-terminal, immunoglobulin-like domains of the intercellular adhesion molecule-1 (ICAM-1) shows that ICAM-1 binds into the 12 Å deep "canyon" on the surface of the virus. This is consistent with the prediction that the viral receptor attachment site lies in a cavity inaccessible to the host's antibodies. The atomic structures of HRV14 and CD4, homologous to HRV16 and ICAM-1, showed excellent correspondence with observed density, thus establishing the virus-receptor interactions.

Introduction

Human rhinoviruses are one of the major causes of the common cold. They, like other picornaviruses, are icosahedral assemblies of 60 protomers that envelope a single, positive-sense strand of RNA. Each protomer consists of four polypeptides, VP1–VP4. The three external viral proteins (VP1–VP3) each have an approximate molecular weight of 30 000 and a similar folding topology [12, 29]. The external viral radius is ~150 Å and the total molecular weight is roughly 8.5×10^6. A surface depression, or canyon, that is about 12 Å deep and 12–15 Å wide, encircles each pentagonal vertex (Fig. 1c). Residues lining the canyon are more conserved than other surface residues among rhinovirus serotypes [28]. The most variable surface residues are at the sites of attachment of neutralizing antibodies [29, 31]. It has been proposed that the cellular receptor molecule recognized by the virus binds to conserved residues in the canyon, thus escaping neutralization by host antibodies

* Present address: Genetic Therapy Inc., Gaithersburg, Maryland, U.S.A.

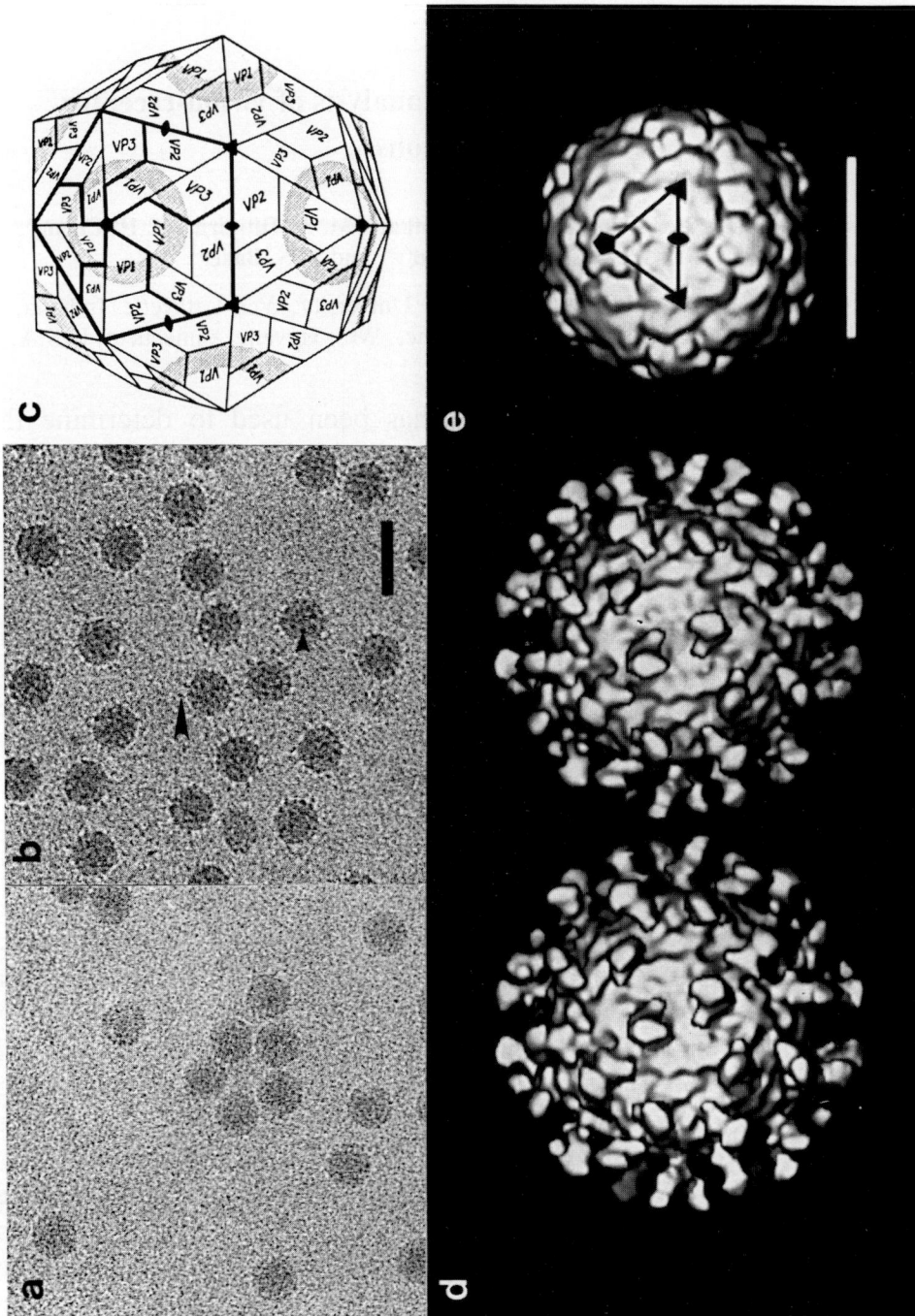

that are too big to penetrate into that region. This hypothesis [27, 29] is supported by site-directed mutagenesis of residues lining the canyon which alters the ability of the virus to attach to HeLa cell membranes [5]. Also, conformational changes in the floor of the canyon, produced by certain antiviral agents that bind into a pocket beneath the canyon floor, inhibit viral attachment to cellular membranes [26]. Conservation of the viral attachment site inside a surface depression has been observed for Mengo [14] and influenza virus [38]. On the other hand, Yeates et al. [40] suggest that for a mouse adapted poliovirus, which has ~36% amino acid identity in VP1, the canyon hypothesis may not be applicable.

There are well over 100 human rhinovirus serotypes, which can be divided into roughly two groups according to the cellular receptor they recognize [1]. The structures of human rhinovirus 14 (HRV14) [29] and HRV16 [25a], which belong to the major group of serotypes, and of HRV1A [13], which belongs to the minor group of serotypes, have been determined. There are at least 78 serotypes [36] that bind to intercellular adhesion molecule-1 (ICAM-1), the major group rhinovirus receptor [10, 34]. The ICAM-1 molecule has five immunoglobulin-like domains (D1 to D5 numbered sequentially from the amino end), a trans-membrane portion, and a small cytoplasmic domain [33]. Domains D2, D3 and D4 are glycosylated. Unlike immunoglobulins, ICAM-1 appears to be monomeric [34]. Mutational analysis of ICAM-1 has shown that domain D1 contains the primary binding site for rhinoviruses as well as the binding site for its natural ligand, lymphocyte function-associated antigen-1 (LFA-1) [20, 23, 35]. Other surface antigens within the immunoglobulin superfamily that are utilized by viruses as receptors include CD4 for human immunodeficiency virus-1 [6, 15], the poliovirus

---

Fig. 1. Cryoelectron microscopy of HRV16 particles and their complex with D1D2. **a** Native HRV16. **b** HRV16:D1D2 complex. D1D2 molecules (the two amino terminal domains of ICAM-1) are seen edge-on at the periphery of the virions (large arrow), or end-on in projection (small arrow). Cryoelectron microscopy was performed essentially as described by Cheng et al. [4] with images recorded at a nominal magnification of 49 000X and with an electron dose of $\sim 20 e^-/\text{Å}^2$. **c** Schematic diagram of HRV showing the icosahedral symmetry, subunit organization and canyon (shaded). Thick lines encircle five protomers of VP1, VP2 and VP3. The fourth viral protein, VP4, is inside the capsid. **d** Stereoview of the reconstruction of the HRV16:D1D2 complex, viewed along an icosahedral twofold axis in approximately the same orientation as in **c**. Sixty D1D2 molecules are bound to symmetry-equivalent positions at the twelve canyon regions on the virion. The reconstruction was modified to correct for defocus and amplitude contrast effects present in the original micrographs (Cheng, submitted). **e** Shaded-surface view of HRV14, computed from the known atomic structure [1], truncated to 20 Å resolution. The triangular outline of one icosahedral asymmetric unit corresponding to that in **c** is indicated. Bar = 500 Å (**a,b**); 200 Å (**d,e**)

receptor [25], and the mouse coronavirus receptor [39]. In ICAM-1, in the poliovirus receptor [8, 16], and in CD4 [2] the primary receptor-virus binding site is domain D1. The structures of the two amino-terminal domains of CD4 have been determined to atomic resolution [30, 37]. Truncated proteins corresponding to the two amino-terminal domains of ICAM-1 (tICAM-1(185)) as well as the intact extracellular portion of ICAM-1 (tICAM-1(453) or domains D1 to D5) have been expressed in CHO cells [11]. The desialated form of tICAM-1(185), which will be referred to hereafter as molecule D1D2, has recently been crystallized [17].

The attachment of rhinovirus to the receptor molecule at the cell surface is only the first step of virus uncoating. Subsequent to binding receptor, virus is apparently internalized by receptor-mediated endocytosis and enters the endosomal compartment. Productive rhinovirus uncoating and infection requires an intracellular low pH step [21]. In vitro, low pH treatment will convert rhinovirus to both 135S (missing VP4) and 80S (missing VP4 and RNA) subviral particles [18]. A number of studies have shown that poliovirus can be conformationally altered to a 135S form upon interaction with its receptor, and rhinovirus can be converted to an 80S empty capsid by incubation in the presence of soluble ICAM-1 [11]. Thus, both virus-receptor binding and low pH (presumably in the endosomal compartment) appear to play active roles in the controlled disassembly of virus during uncoating, although the relative contributions of these two factors and their temporal relationship in vivo are unclear.

A model of the amino-terminal domain D1 of ICAM-1, based on its homology to known structures of the constant domains of immunoglobulins, was reported by Giranda et al. [9]. Guided by mutational studies of HRV14 and ICAM-1, they were able to fit this model into the known canyon structure of HRV14. We have utilized cryoelectron microscopy and image analysis techniques to calculate a three-dimensional reconstruction of the complex of HRV16 and D1D2 to ~28 Å resolution. The reconstruction clearly shows that the receptor binds into the canyon of rhinovirus as predicted [27, 29]. In addition, we use the known structures of HRV14 and CD4 and the predicted structure of D1 of ICAM-1 to identify atomic interactions.

Structure of the virus: receptor model

Complexes between HRV16 and ICAM-1 D1D2 were prepared by incubating a 3.3 mg/ml solution of HRV16 with a 6.6 mg/ml solution of D1D2 for ~16 hours at 34°C. Under these conditions, saturated complexes of HRV16 with D1D2 can be generated, with approximately 60 moles of D1D2 per mole of virus. HRV14 could also form complexes, although these rapidly broke down to empty capsids (Hoover-Litty and Greve, in prep.) and electron micrographs of such specimens revealed severely disrupted particles in a background of protein. HRV16 complexes with D1D2 or with the complete D1 to D5 extracellular domain fragment were both used in the investigation. We present here only the results obtained on the HRV16:D1D2 complex.

Forty-four electron micrograph images for unstained HRV16:D1D2 complex were combined to compute a three-dimensional reconstruction (Fig. 1d) with an effective resolution of ~28 Å [3]. The density value of the D1D2 feature in the reconstruction was roughly the same as the density of the virion capsid, thus indicating that the D1D2 molecules nearly saturated the 60 available sites on the virion. The position of the ICAM relative to the icosahedral symmetry axes of the virus is unambiguous. Each D1D2 molecule has an approximate dumbbell shape, consistent with the presence of a two-domain structure.

A difference map between the EM model and the 20 Å HRV14 model was computed (Fig. 2). The difference map showed that the D1D2 molecule binds to the central portion of the canyon in a manner roughly as predicted by Giranda et al. [9], confirming the predictions inherent in the canyon hypothesis [27]. The D1D2 molecule is closely associated with the "southern" (see Fig. 3) wall and rim of the canyon, extending 10 Å to 12 Å into the canyon. The binding site is near the center of the triangle formed between a fivefold and two adjacent threefold axes (Fig. 3). The ICAM fragment is oriented roughly perpendicular to the viral surface and extends to a radius of ~205 Å. Its total length is ~75 Å as measured in the difference map.

Studies with interspecies chimeras and site-directed mutagenesis of ICAM-1 aimed at identifying the regions necessary for rhinovirus binding indicated that only the first domain is essential, and that the residues most involved in virus binding were concentrated on the outside end of domain D1 [23, 25]. However, it has proven difficult to produce an active form of domain D1 in solution, domains D1 + D2 being the minimal soluble virus-binding species [11]. The inability to produce domain D1 in isolation and the sequence alignment between ICAM-1 and CD4 suggested that domains D1 and D2 of ICAM-1 are intimately associated through a common, extended β-strand as is seen in the structure of CD4 [30, 37]. Thus, it seemed reasonable to use the known structures of CD4 for fitting the reconstructed density map (Fig. 2), although there was slightly too little density for domain D1 and too much density for D2. A better assessment of the fit of domain D1 to the density was obtained by taking the predicted D1 structure of ICAM-1, including all side chains, and superimposing it onto the fitted C_α backbone of CD4. One major difference is that although domain D1 of CD4 resembles a

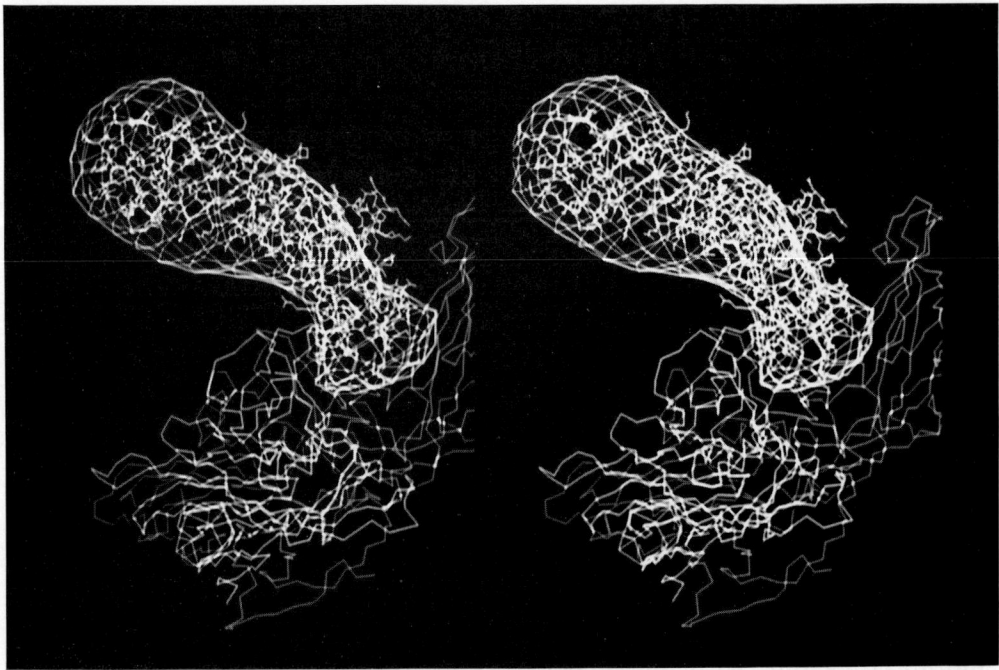

a

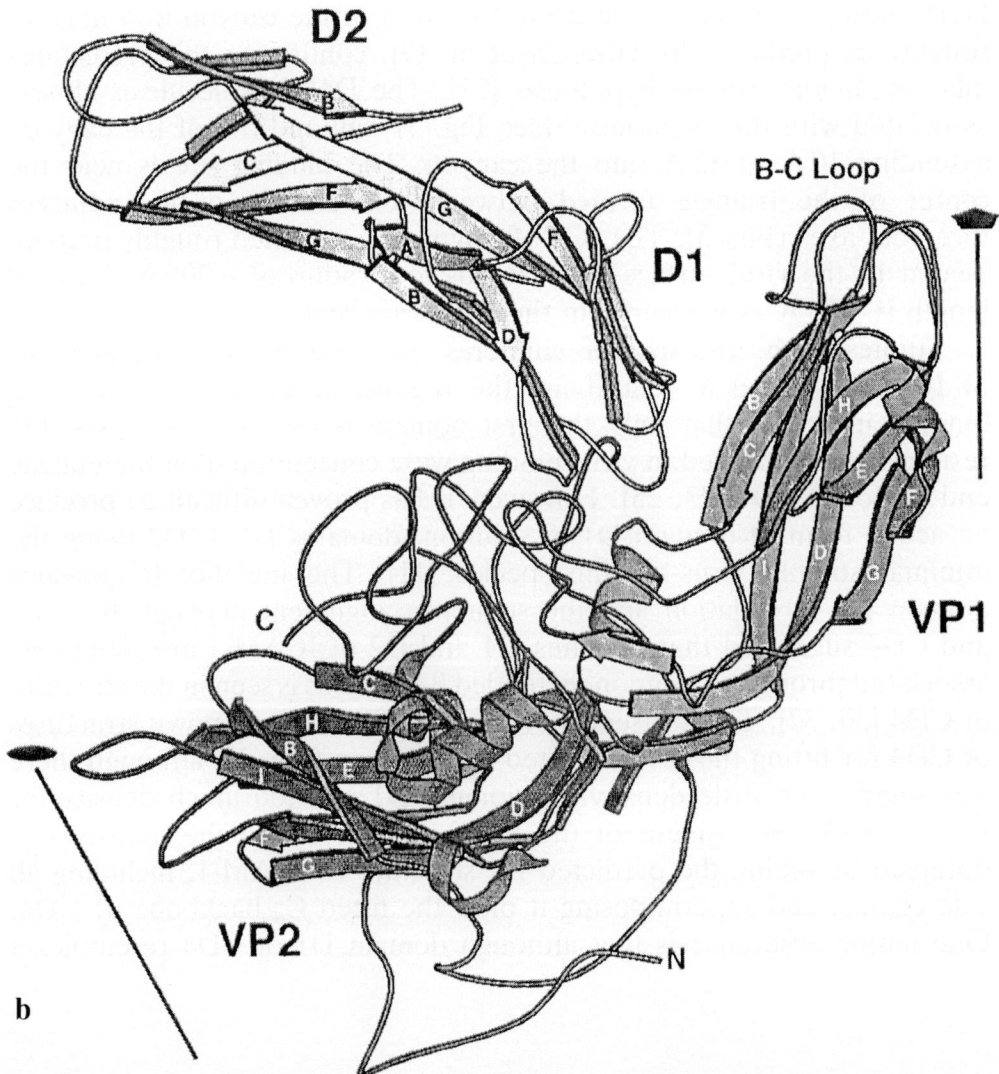

b

variable, immunoglobulin-like domain with two extra β-strands, the ICAM-1 prediction is based on a more likely analogy to an immunoglobulin constant domain. This gives domain D1 of ICAM-1 a sleeker appearance, consistent with the observed difference density (Fig. 2a). The extra density in D2 (in the region furthest away from the virus (Fig. 2a)) compared to D2 of CD4 is probably due to the associated carbohydrate groups that are located in this region.

The atomic structure of HRV14 closely matched the reconstructed density that was not occupied by the D1D2 fragment. The only exception occurred in the BC loop of VP1 of HRV14 (see Fig. 2), which extended about 3 Å outside the reconstructed density on the "northern" rim of the canyon. However, this region of the polypeptide chain is a highly variable structure and is the site of one of the two largest conformational differences between HRV14 and HRV1A [10]. Furthermore, it is also the site of major differences in the structures of homologous poliovirus serotypes 1 and 3 [7].

We describe here the first structure of a virus-receptor complex in which the receptor is a membrane-bound glycoprotein molecule that is used by a virus for recognition of a specific host tissue for attachment and subsequent entry. This receptor molecule belongs to the immunoglobulin superfamily, a class of molecule frequently employed on cell surfaces for the recognition of other molecules (or recognized by viruses) that are subsequently transferred across the membrane.

Since the general nature of the complex described here had been predicted on the basis of the strategy used by HRV to hide its receptor attachment site, perhaps many other viruses use a similar strategy. Poliovirus is clearly homologous to HRV, and both poliovirus [25] and the major rhinovirus group use an immunoglobulin-like molecule as receptor. Thus, it would be expected that the poliovirus receptor binds into the poliovirus canyon in a manner similar to that of the complex formed for rhinoviruses [8]. The structure of a mouse-adapted chimera

Fig. 2. a Stereo diagram showing the fit of the CD4 structure into the difference density between the HRV16:D1D2 reconstruction (Fig. 1d) and the X-ray map of HRV14 (Fig. 1e). A radial scale factor was first determined to slightly adjust the EM model to the accurate dimensions of the X-ray model. The difference electron density has been fitted with the CD4 amino-terminal two-domain structure. A cross-sectional view of the HRV14 structure is also shown outside the ICAM-1 difference density. The additional strands βC′ and βC″ in D1 of CD4 compared to ICAM-1 lie outside the difference density. **b** A diagrammatic drawing [19] of the structure shown in **a**. Secondary structural elements of the CD4 fragment and of HRV14 (homologous structures used to represent ICAM-1 and HRV16, respectively) are identified by the standard nomenclature. The N and C termini of VP1 are also marked

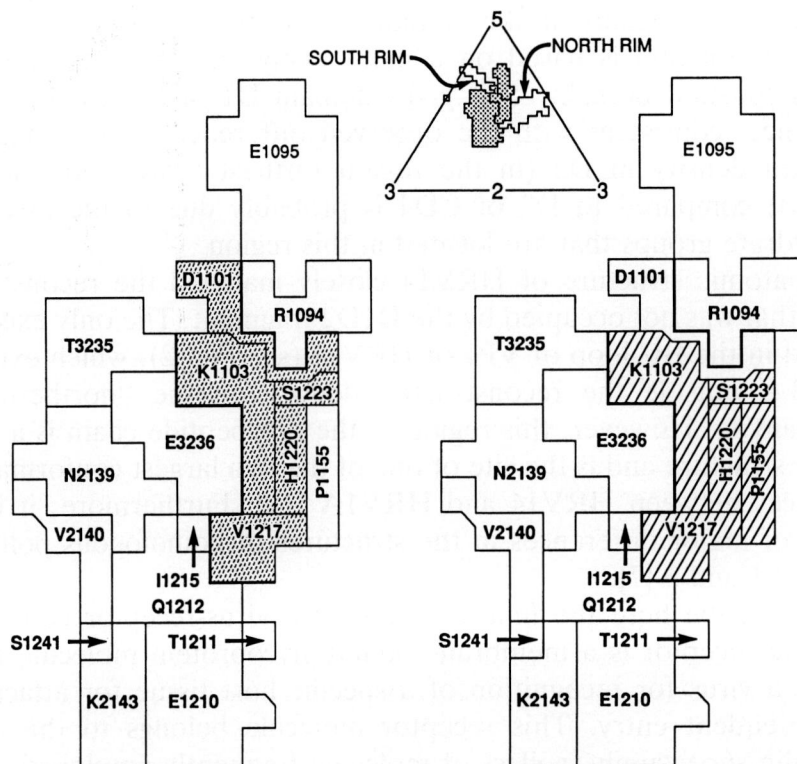

Fig. 3. Top: View of the icosahedral asymmetric unit bounded by adjacent five- and threefold axes, outlining residues on the viral surface. Shown are the limits of the canyon, arbitrarily demarcated by a 138 Å radial distance from the viral center [28], and the ICAM-1 footprint (stippled). Improved resolution of the electron density could only marginally alter the HRV residues at the virus:receptor interface. Left and right: Enlarged view of the residues in the ICAM-1 footprint showing (right) the residues which, when mutated, affect viral attachment [5], and (left) the residues altered in structure by the binding of antiviral compounds that inhibit attachment and uncoating [32]

of human poliovirus 2 has been determined [40]. The major structural change occurs in the chimera in the BC loop, not in the canyon floor. In this instance, therefore, the BC loop might modulate the virus-receptor interaction. Knowledge of the virus-receptor interaction will illuminate various strategies currently being developed to interfere with early stages of rhinoviral and other viral infections [11, 22, 24].

Acknowledgements

We are grateful for many helpful discussions with R. Rueckert (University of Wisconsin) and M. McKinlay, F. Dutko, G. Diana and D. Pevear (Sterling Winthrop Pharmaceuticals Research Division). We are grateful to M. Kremer, C. Forte, and C. Music for technical assistance. We thank W. Hendrickson and S. Harrison for sharing coordinate information. We also thank H. Prongay and S. Wilder for help in preparation of this manuscript. The work was supported by National Institutes of Health grants to M.G.R. and to T.S.B., a National Science Foundation grant to T.S.B. and a Lucille P. Markey Foundation Award.

References

1. Abraham G, Colonno RJ (1984) Many rhinovirus serotypes share the same cellular receptor. J Virol 51: 340–345
2. Arthos J, Deen KC, Chaikin MA, Fornwald JA, Sathe G, Sattentau QJ, Clapham PR, Weiss RA, McDougal JS, Pietropaolo C, Axel R, Truneh A, Maddon PJ, Sweet RW (1989) Identification of the residues in human CD4 critical for the binding of HIV. Cell 57: 469–481
3. Baker TS, Newcomb WW, Olson NH, Cowsert LM, Olson C, Brown JC (1991) Structures of bovine and human papillomaviruses. Analysis by cryoelectron microscopy and three-dimensional image reconstruction. Biophys J 60: 1445–1456
4. Cheng RH, Olson NH, Baker TS (1992) Cauliflower mosaic virus: a 420 subunit (T = 7), multilayer structure. Virology 186: 655–668
5. Colonno RJ, Condra JH, Mizutani S, Callahan PL, Davies ME, Murcko MA (1988) Evidence for the direct involvement of the rhinovirus canyon in receptor binding. Proc Natl Acad Sci USA 85: 5449–5453
6. Dalgleish AG, Beverley PCL, Clapham PR, Crawford DH, Greaves MF, Weiss RA (1984) The CD4 (T4) antigen is an essential component of the receptor for the AIDS retrovirus. Nature 312: 763–767
7. Filman DJ, Syed R, Chow M, Macadam AJ, Minor PD, Hogle JM (1989) Structural factors that control conformational transitions and serotype specificity in type 3 poliovirus. EMBO J 8: 1567–1579
8. Freistadt MS, Racaniello VR (1991) Mutational analysis of the cellular receptor for poliovirus. J Virol 65: 3873–3876
9. Giranda VL, Chapman MS, Rossmann MG (1990) Modeling of the human intercellular adhesion molecule-1, the human rhinovirus major group receptor. Proteins 7: 227–233
10. Greve JM, Davis G, Meyer AM, Forte CP, Yost SC, Marlor CW, Kamarck ME, McClelland A (1989) The major human rhinovirus receptor is ICAM-1. Cell 56: 839–847
11. Greve JM, Forte CP, Marlor CW, Meyer AM, Hoover-Litty H, Wunderlich D, McClelland A (1991) Mechanisms of receptor-mediated rhinovirus neutralization defined by two soluble forms of ICAM-1. J Virol 65: 6015–6023
12. Hogle JM, Chow M, Filman DJ (1985) Three-dimensional structure of poliovirus at 2.9 Å resolution. Science 229: 1358–1365
13. Kim S, Smith TJ, Chapman MS, Rossmann MG, Pevear DC, Dutko FJ, Felock PJ, Diana GD, McKinlay MA (1989) The crystal structure of human rhinovirus serotype 1A (HRV1A). J Mol Biol 210: 91–111
14. Kim S, Boege U, Krishnaswamy S, Minor I, Smith TJ, Luo M, Scraba DG, Rossmann MG (1990) Conformational variability of a picornavirus capsid: pH-dependent structural changes of Mengo virus related to its host receptor attachment site and disassembly. Virology 175: 176–190
15. Klatzmann D, Champagne E, Chamaret S, Gruest J, Guetard D, Hercend T, Gluckman JC, Montagnier L (1984) T-lymphocyte T4 molecule behaves as the receptor for human retrovirus LAV. Nature 312: 767–768

16. Koike S, Ise I, Nomoto A (1991) Functional domains of the poliovirus receptor. Proc Natl Acad Sci USA 88: 4104–4108

17. Kolatkar PR, Oliveira MA, Rossmann MG, Robbins AH, Katti SK, Hoover-Litty H, Forte C, Greve JM, McClelland A, Olson NH (1992) Preliminary X-ray crystallographic analysis of intercellular adhesion molecule-1. J Mol Biol 225: 1127–1130

18. Korant BD, Lonberg-Holm K, Yin FH, Noble-Harvey J (1975) Fractionation of biologically active and inactive populations of human rhinovirus type 2. Virology 63: 384–394

19. Kraulis PJ (1992) MOLSCRIPT: a program to produce both detailed and schematic plots of protein structure. J Appl Crystallogr 24: 946–950

20. Lineberger DW, Graham DJ, Tomassini JE, Colonno RJ (1990) Antibodies that block rhinovirus attachment map to domain 1 of the major group receptor. J Virol 64: 2582–2587

21. Madshus IH, Olsnes S, Sandvig K (1984) Different pH requirements for entry of the two picornaviruses, human rhinovirus 2 and murine encephalomyocarditis virus. Virology 139: 346–357

22. Marlin SD, Staunton DE, Springer TA, Stratowa C, Sommergruber W, Merluzzi VJ (1990) A soluble form of intercellular adhesion molecule-1 inhibits rhinovirus infection. Nature 344: 70–72

23. McClelland A, deBear J, Yost SC, Meyer AM, Marlor CW, Greve JM (1991) Identification of monoclonal antibody epitopes and critical residues for rhinovirus binding in domain 1 of ICAM-1. Proc Natl Acad Sci USA 88: 7993–7997

24. McKinlay MA, Pevear DC, Rossmann MG (1992) Treatment of the picornavirus common cold by inhibitors of viral uncoating and attachment. Annu Rev Microbiol 46: 635–654

25. Mendelsohn CL, Wimmer E, Racaniello VR (1989) Cellular receptors for poliovirus: molecular cloning, nucleotide sequence, and expression of a new member of the immunoglobulin family. Cell 56: 855–865

25a. Oliveira MA, Zhao R, Lee WM, Kremer MJ, Minor I, Rueckert RR, Diana GD, Pevear DC, Dutko FJ, McKinlay MA, Rossmann MG (1993) The structure of human rhinovirus 16. Structure 1: 51–68

26. Pevear DC, Fancher MJ, Felock PJ, Rossmann MG, Miller MS, Diana G, Treasurywala AM, McKinlay MA, Dutko FJ (1989) Conformational change in the floor of the human rhinovirus canyon blocks adsorption to HeLa cell receptors. J Virol 63: 2002–2007

27. Rossmann MG (1989) The canyon hypothesis. Hiding the host cell receptor attachment site on a viral surface from immune surveillance. J Biol Chem 263: 14587–14590

28. Rossmann MG, Palmenberg AC (1988) Conservation of the putative receptor attachment site in picornaviruses. Virology 164: 373–382

29. Rossmann MG, Arnold E, Erickson JW, Frankenberger EA, Griffith JP, Hecht HJ, Johnson JE, Kamer G, Luo M, Mosser AG, Rueckert RR, Sherry B, Vriend G (1985) Structure of a human common cold virus and functional relationship to other picornaviruses. Nature 317: 145–153

30. Ryu SE, Kwong PD, Truneh A, Porter TG, Arthos J, Rosenberg M, Dai X, Xuong N, Axel R, Sweet RW, Hendrickson WA (1990) Crystal structure of an HIV-binding recombinant fragment of human CD4. Nature 348: 419–426

31. Sherry B, Rueckert R (1985) Evidence for at least two dominant neutralization antigens on human rhinovirus 14. J Virol 53: 137–143

32. Smith TJ, Kremer MJ, Luo M, Vriend G, Arnold E, Kamer G, Rossmann MG, McKinlay MA, Diana GD, Otto MJ (1986) The site of attachment in human rhinovirus 14 for antiviral agents that inhibit uncoating. Science 233: 1286–1293

33. Staunton DE, Marlin SD, Stratowa C, Dustin ML, Springer TA (1988) Primary structure of ICAM-1 demonstrates interaction between members of the immunoglobulin and integrin supergene families. Cell 52: 925–933

34. Staunton DE, Merluzzi VJ, Rothlein R, Barton R, Marlin SD, Springer TA (1989) A cell adhesion molecule, ICAM-1, is the major surface receptor for rhinoviruses. Cell 56: 849–853

35. Staunton DE, Dustin ML, Erickson HP, Springer TA (1990) The arrangement of the immunoglobulin-like domains of ICAM-1 and the binding sites for LFA-1 and rhinovirus. Cell 61: 243–254

36. Tomassini JE, Maxson TR, Colonno RJ (1989) Biochemical characterization of a glycoprotein required for rhinovirus attachment. J Biol Chem 264: 1656–1662

37. Wang J, Yan Y, Garrett TPJ, Liu J, Rodgers DW, Garlick RL, Tarr GE, Husain Y, Reinherz EL, Harrison SC (1990) Atomic structure of a fragment of human CD4 containing two immunoglobulin-like domains. Nature 348: 411–418

38. Weis W, Brown JH, Cusack S, Paulson JC, Skehel JJ, Wiley DC (1988) Structure of the influenza virus haemagglutinin complexed with its receptor, sialic acid. Nature 333: 426–431

39. Williams RK, Jiang G-S, Holmes KV (1991) Receptor for mouse hepatitis virus is a member of the carcinoembryonic antigen family of glycoproteins. Proc Natl Acad Sci USA 88: 5533–5536

40. Yeates TO, Jacobson DH, Margin A, Wychowski C, Girard M, Filman DJ, Hogle JM (1991) Three-dimensional structure of a mouse-adapted type2/type 1 poliovirus chimera. EMBO J 10: 2331–2341

Authors' address: Dr. M. G. Rossmann, Department of Biological Sciences, Purdue University, West Lafayette, IN 47907-1392, U.S.A.

Arch Virol (1994) [Suppl] 9: 543–558

Archives
of
Virology
© Springer-Verlag 1994
Printed in Austria

Assembly of tobacco mosaic virus and TMV-like pseudovirus particles in *Escherichia coli*

D.-J. Hwang[1], **I. M. Roberts**[2], and **T. M. A. Wilson**[2]

[1] AgBiotech Center, Cook College, Rutgers University, New Brunswick,
New Jersey, U.S.A.
[2] Department of Virology, Scottish Crop Research Institute, Invergowrie,
Dundee, U.K.

Summary. High-level expression of plant viral proteins, including coat protein (CP), is possible in *Escherichia coli*. Native tobacco mosaic virus (TMV) CP expressed in *E. coli* remains soluble but has a non-acetylated N-terminal Ser residue and, following extraction, is unable to package TMV RNA in vitro under standard assembly conditions. Changing the Ser to Ala or Pro by PCR-mutagenesis did not confer assembly competence in vitro, despite these being non-acetylated N-termini present in two natural strains of TMV. All TMV CPs made in *E. coli* formed stacked cylindrical aggregates in vitro at pH 5.0 and failed to be immuno-gold-labelled using a mouse monoclonal antibody specific for helically assembled TMV CP. TMV self-assembly has been studied extensively in vitro, and an origin of assembly sequence (OAS) mapped internally on the 6.4 kb ssRNA genome. Pseudovirus particles can be assembled mono- or bi-directionally in vitro using virus-derived CP and chimeric ssRNAs containing the cognate TMV OAS, but otherwise of unlimited length and sequence. Studies on plant virus assembly in vivo would be facilitated by a model system amenable to site-directed mutagenesis and rapid recovery of progeny particles. When chimeric transcripts containing the TMV OAS were co-expressed with TMV CP in vivo for 2–18 h, helical TMV-like ribonucleoprotein particles of the predicted length were formed in high yield (up to 7.4 μg/mg total bacterial protein). In addition to providing a rapid, inexpensive and convenient system to produce, protect and recover chimeric gene transcripts of any length or sequence, this *E. coli* system also offers a rapid approach for studying the molecular requirements for plant virus "self-assembly" in vivo. Transcription of a full-length cDNA clone of TMV RNA also resulted in high levels of CP expression and assembly of sufficient intact genomic RNA to initiate virus infection of susceptible tobacco plants.

Introduction

Tobacco mosaic virus (TMV) was the model system of choice for early studies on the spontaneous "self-assembly" of complex biological structures in vitro [1–3]. Historically, the high yield of virus from infected plants (1–5 g/kg leaf) and its remarkable stability during prolonged storage are responsible for the assembly system of TMV being the most extensively studied of any virus infecting eukaryotes.

For use in ssRNA encapsidation in vitro, two simple protocols were developed to prepare assembly-competent TMV coat protein (CP) from isolated virions [4, 5]. During the 1970's the 3-dimensional structure and interchangeable polymorphic forms of TMV CP were studied extensively [6]. Functional assays with 3′-coterminal nested sets of ssRNA genome fragments revealed the internal location [7] and nucleotide sequence [8] of the viral origin-of-assembly sequence (OAS) before the complete 6 395-nt genome of the common (U1 or *vulgare*) strain had been sequenced [9]. Equivalent OAS regions have now been located in several fully-sequenced *Tobamovirus* genomes. Mostly they lie about 1.0 kb from the 3′-terminus, within a gene encoding the viral 30 kDa cell-to-cell movement protein; however, in Sunn-hemp mosaic virus (*alias* the C_c or bean strain of TMV) the OAS resides in the CP gene itself, which creates a useful self-assembling cassette [10].

In 1986 we showed that, in addition to native genomic TMV RNA or suitable 3′-coterminal fragments thereof, a 432 bp cDNA copy of TMV U1 genome co-ordinates 5 112–5 543 could be cloned downstream of almost any foreign gene sequence and that the resulting chimeric transcripts were encapsidated by assembly-competent TMV coat protein (CP) in vitro [11]. Recently it has also been shown that a single stem-loop of only 75-nts from this longer sequence is necessary to direct the efficient assembly of contiguous foreign RNA sequences of unlimited length [12, 13]. Phosphodiester charge neutralization and H-bonding to the 2′-OH moiety determine that TMV CP can only encapsidate single-stranded (ss) RNA, not ssDNA [14]. This allowed in vitro packaging ("chimeric transcript rescue") to be conducted directly after termination of the in vitro transcription reaction [15]. Differences in the optimal divalent cation conditions preclude the possibility of co-transcriptional RNA-packaging in vitro, despite its obvious attractions for increasing product yield and stability. Identification of the prefabricated, oligomeric form(s) of TMV CP used for initiation and bidirectional elongation during virus assembly has been a controversial subject for many years [6, 13, 16–18] and the data still fail to provide a consensus model.

Transcripts containing the 432-nt TMV U1 OAS can also be encapsidated by TMV CP in vivo, during systemic virus infection of transgenic tobacco plants which express a chimeric gene from a constitutive plant promoter [19]. Thus there seems to be no selective subcellular compartment or virus-specific structure (e.g. inclusion body or replicase-dependent complex) where OAS-dependent virus or virus-like particle assembly occurs in vivo. It is also well-documented that chloroplast transcripts, particularly mRNA for the large subunit of ribulose-bisphosphate carboxylase, can be packaged by TMV CP in vivo [20], and that cytoplasmically synthesized TMV CP can enter chloroplasts despite the absence of any recognizable signal transit peptide [21, 22]. Subsequent interactions between TMV CP and the light-harvesting photosystems (especially PSII) in thylakoids largely account for the disease phenotype and symptoms after which the virus was named [23].

Our experimental results indicate that there are few sequence constraints or length restrictions in chimeric ssRNAs made suitable for encapsidation in TMV CP by having a cognate OAS [11, 14, 15]. The exceptional stability of TMV or TMV-like particles to proteases and RNases can therefore be exploited to recover, store and protect otherwise labile ssRNA molecules, and also to deliver them into animal or plant cells as pseudovirus particles for subsequent cotranslational disassembly [24]. To prepare these nucleocapsids in vitro requires tedious, often inefficient protocols and expensive reagents to clone, purify, digest and in vitro transcribe suitable plasmids, the facilities to propagate large quantities of virus, to purify CP in an assembly competent form, to perform in vitro packaging reactions and to recover particles. Thus, to realize fully the technological advantages of encapsidated ssRNA requires a more rapid and efficient production procedure.

Studies on the mechanism(s) and CP structural constraints for virus "self-assembly" in plants would be facilitated by an easily and rapidly manipulated in vivo model system, one that is independent of virus replication, since plant infection is a complex, slow and asynchronous process. Plant protoplasts require some form of wounding to initiate a synchronous infection and often in only a small proportion of the population. Mutational studies using full-length infectious cDNA clones or transcripts in plants or protoplasts are often difficult to interpret since non-viable virus mutants can be created (and lost) for many reasons other than inefficient particle assembly. Conventional prokaryotic expression vectors and transformed bacterial cells offer a homogeneous, inducible and malleable system in which to study plant virus assembly, without the complex and ill-defined requirements for plant virus replication. This paper reports the first example of the "self-assembly" of pseudovirus particles, and an infectious plant virus, in vivo in a convenient prokaryotic system.

High level expression of plant viral coat proteins in *E. coli*

Recently, several groups have reported efficient, high level expression of cloned CP genes from a variety of plant RNA viruses in both *Escherichia coli* and yeast cells [25–28]. However, in no case has the assembly of ribonucleocapsids been detected in vivo.

Johnsongrass mosaic potyvirus (JGMV) CP made in *E. coli* or yeast was reported to form flexuous, stacked-disk particles [28] similar to those obtained by in vitro assembly of potyvirus CP monomers in the absence of RNA [29]. The JGMV CP was purified from crude cell extracts by column chromatography or by sucrose gradient ultracentrifugation and could be polymerized by dialysis in phosphate buffer (pH 7.2 containing 0.1 M NaCl) into protein-only rods of variable length, often with a characteristic striated or stacked-disk appearance [28]. It was unclear whether or not such assemblies also arose in vivo in the induced cell cultures, or during sample preparation of the crude extracts for immune electron microscopy.

In 1986 it was reported that a synthetic TMV U1 strain CP gene with the optimal codon usage for *E. coli* expression could be used to produce high levels of "native" CP or TMV CP with a spacer (Gly-Gly) and a poliovirus-3 VP1 epitope consisting of 8 amino acids attached to the C-terminus [25]. The aim was to purify the hybrid protein and exploit the intrinsic property of TMV CP to form long helical, RNA-free aggregates at or below pH 6.0, in order to increase the effective molecular size and local concentration of the poliovirus epitope for subsequent immunization of rats. The pH 5.0-treated TMV CP-polio 3 elicited a higher neutralizing response than free subunits. However, the precise amino terminal sequence of the TMV CP expressed in *E. coli* was not examined.

More recent work using fast-atom bombardment mass spectrometry has shown that TMV U1 CP expressed in *E. coli* has the amino terminal N-formyl-methionine residue removed, as in virions recovered from plants, but the penultimate Ser residue is not acetylated, in contrast to native virus [27]. The effect of the additional positive charge was to disrupt the normal polymorphic aggregation behaviour of TMV CP at different pH's and ionic strengths, as determined by Durham and colleagues [3, 5]. In particular, formation of the 20S virus assembly nucleation species and of helical aggregates under virus assembly conditions or at pH 5.0, respectively, were retarded. Our results confirmed this behaviour and showed that native TMV U1 CP, and several derivatives mutagenized at the second amino acid position (see below), synthesized in and purified from *E. coli*, failed to assemble with TMV RNA under "standard" in vitro incubation conditions. Most of the protein tended to precipitate during the conventional dialysis steps [5] and, even when this could be avoided, at pH 5.0 we too saw predominantly stacked disk-like structures in the electron microscope (Fig. 1). SDS-PAGE did not show any appreciable proteolytic degradation of the purified TMV CP, which might have accounted for the irreversible stacking and striated appearance [30]. It is significant that Haynes and colleagues [25] also showed truncated, stacked disk-like structures when their *E. coli*-made native TMV CP was dialysed against 0.1 M sodium acetate, pH 5.0 (see Fig. 4A in [25]). When the TMV CP-polio 3 molecules were treated in the same way, longer rods appeared. It is not possible to discriminate between helical or stacked disk forms of CP in their photographs [25]; however, it may suggest that the detrimental effect of the additional amino-terminal positive charge (non-acetylated Ser), could have been neutralized or buried by the ten additional C-terminal amino acids of the polio 3 epitope and spacer. Both the N- and C-termini of TMV CP lie in close proximity at the outer surface of the virion [31]. NMR techniques have shown that there is little flexibility in either end of the TMV CP chain (K. Brierley, personal communication) [32] so the above hypothesis may be correct.

TMV CP expression and mutagenesis

The native TMV U1 strain CP gene, a synthetic *E. coli* codon-optimized U1 CP gene [25], or derivatives of the native CP gene were cloned into pET3a (a colE1 replicon) and expressed following addition of isopropyl-β-thiogalactopyranoside (IPTG) to mid-log cultures (O.D.$_{600}$ ~0.6) of *E. coli* BL21 (DE-3) pLysE [33].

Restriction enzymes were used to isolate the native CP gene from pUC118 containing a full-length TMV cDNA (pTMV210, a generous gift from W. O. Dawson, University of Florida) and from a plasmid containing the synthetic *E. coli* codon-optimized CP gene.

In an attempt to circumvent the problem of poor TMV CP assembly encountered by Shire and colleagues [27], PCR-mutagenesis was used to change only the second amino acid position from the native serine to alanine or proline. The mature N-terminus of U2 strain TMV CP has a non-acetylated proline residue, and the NM strain is claimed to have a non-acetylated alanine in this position [34]. Thus native virus assembly in vivo must tolerate the additional N-terminal positive charge in these cases.

After 2–18 h induction, cleared lysates of cells expressing each TMV CP construct were fractionated on 10% (w/v) polyacrylamide gels in SDS and stained with Coomassie Blue. Most of each TMV CP seemed to remain in the supernatant, and losses in the pellet fraction were due only to contamination. Thus the expressed CPs seemed "soluble" in

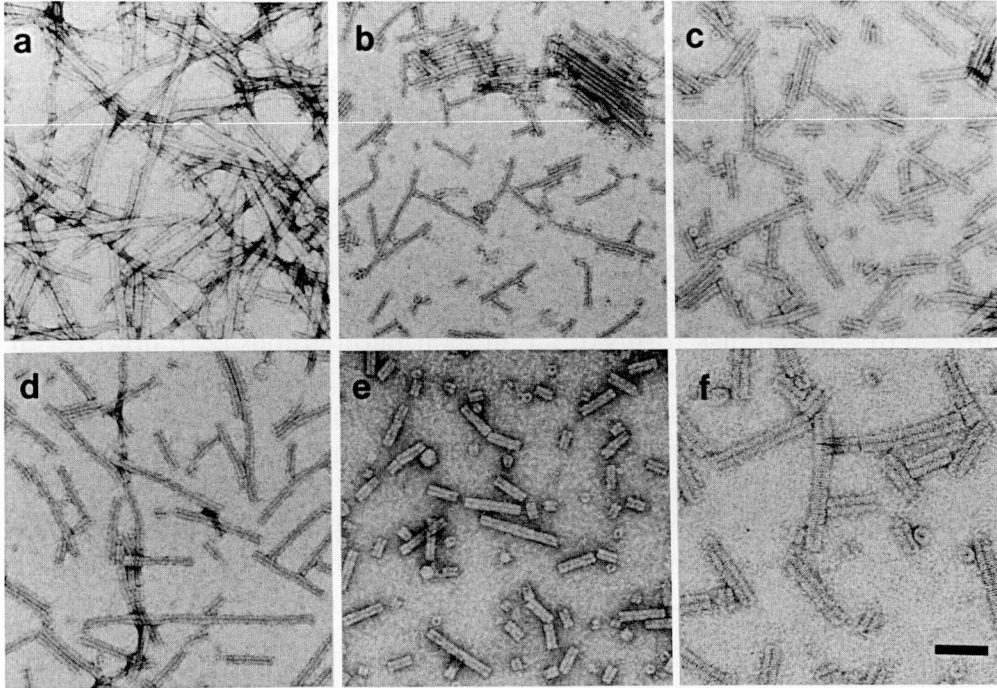

Fig. 1. Immunosorbent electron microscopy of TMV (U1 strain) coat protein produced from virions (**a**), or isolated from transformed cultures of *E. coli* (**b–f**). **a** Helical aggregates of native, virus-derived, TMV coat protein maintained at pH 5.0. TMV CP isolated from *E. coli* expressing the wild-type amino acid sequence encoded by the native (**b**), or *E. coli*-optimized (**c, f**) codons. TMV CP isolated from *E. coli* expressing the alanine (**d**), or proline (**e**) derivatives of the native sequence. All samples were dialysed [5] and stored at pH 5.0. **f** is at twice the magnification of **c**. All samples were negatively stained with 1% uranyl acetate. Scale bar represents 100 nm (**a–e**) and 50 nm (**f**)

vivo. Comparisons with marker tracks, containing known amounts of native TMV CP, showed that each of the three CP gene constructs composed predominantly of native viral codons expressed the 17.6 kDa protein at about 30 µg/ml of original bacterial culture. The prokaryotic codon-optimized construct [25] produced about 60 µg/ml of original culture. The identity and yield of the 17.6 kDa protein species was further confirmed by western immunoblotting using rabbit polyclonal antiserum against native virus-derived CP. The absence of an amino-terminal methionine residue in each CP product was suggested by the lack of detectable radioactivity in each 17.6 kDa species on autoradiograms made from proteins labelled in vivo with L-^{35}S-methionine during T7 polymerase induction with IPTG. TMV CP (U1) has not internal Met residues.

Many attempts were made and different procedures tried [e.g., $(NH_4)_2SO_4$ precipitation or dialysis against sodium acetate, pH 5.0]

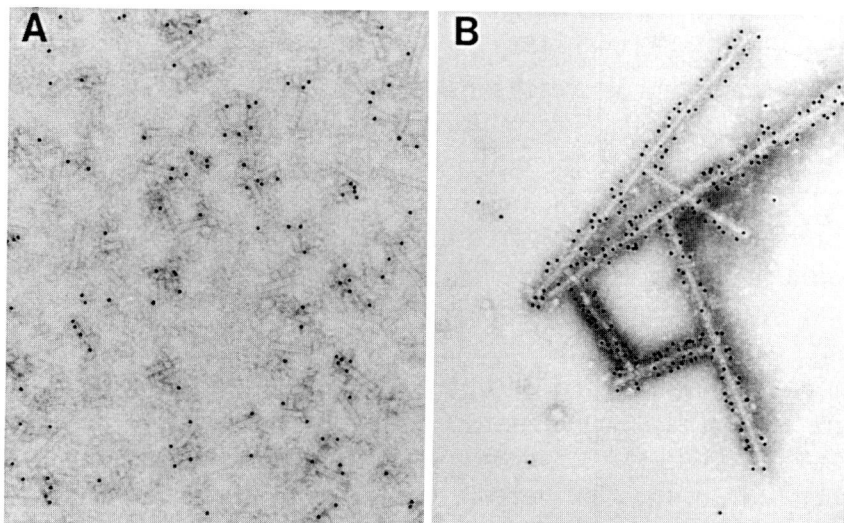

Fig. 2. Immunogold labelling of TMV CP-only aggregates using a mouse monoclonal antibody (MAb) specific for an epitope present only in helically assembled particles (with or without RNA). **A** TMV CP (native sequence expressed from an *E. coli*-codon optimized gene) purified from bacterial cell lysates. **B** CP isolated from TMV harvested from infected tobacco plants. Both samples were prepared by standard methods and stored at pH 5.0 [5]. Following immune trapping on carbon coated EM grids, using rabbit polyclonal antiserum to TMV CP, the material was reacted with the MAb followed by 10 nm-gold conjugated goat anti-mouse IgG

to purify these TMV CP species from cleared bacterial cell lysates for assembly studies with TMV RNA in vitro. Frequently the proteins precipitated during the conventional dialysis procedures [5]. The proline version of the native U1 CP proved particularly insoluble during handling. Although SDS-PAGE showed only a single 17.6 kDa band and no significant proteolytic degradation products, all *E. coli*-made TMV CPs failed to form 20S protohelical or "disk" aggregates, and showed no activity in turbidometric assembly assays with native TMV RNA. Electron microscopy revealed that, at pH 5.0, the purified proteins also failed to form long helical aggregates, but existed as a very heterodisperse population of stacked-disk-like rods (Fig. 1). The non-helical nature of these rods was confirmed by their failure to bind a mouse monclonal antibody (MAb) specific for an epitope (neotope) present only in correctly assembled TMV CP helices or in virions (Fig. 2).

Assembly of pseudovirus particles in vivo

Despite the inability of any purified, *E. coli*-expressed TMV CP to assemble with TMV RNA in vitro, we were still interested to know whether ssRNA encapsidation would occur in vivo. Two approaches were taken to provide suitable ssRNA species for packaging in vivo. First, the cognate TMV OAS (a 432 bp cassette) was added to the 3'-end of the chloramphenicol acetyltransferase selectable marker gene (Cm^r) on the plasmid, pLysE. The multicopy plasmid pLysE is a p15A-based replicon compatible with the colE1 replicon, pET3a which contained the TMV CP gene under a T7 promoter. pLysE also harbours the structural gene for T7 lysozyme which is required to inhibit any IPTG-independent expression of the T7 RNA polymerase structural gene from the *lacUV5* promoter in the chromosome of *E. coli* BL21 (DE3). Alternatively, the cassette T7 promoter-β-glucuronidase (GUS) gene-TMV OAS-T7 RNA pol terminator was inserted into pLysE at a location separate from both the Cm^r and .T7 lysozyme genes. Transcripts, hereafter described as CAT-OAS or GUS-OAS, were therefore produced either constitutively or inducibly from the Cm^r gene promoter or from the IPTG-dependent T7 promoter, respectively.

Co-expression of CAT-OAS mRNA (1.4 kb) and either the native serine or the alanine derivative of TMV CP resulted in short helical ribonucleoprotein rods (Fig. 3) of the predicted length (66 nm). It was necessary to use freeze-thaw or osmotic stress protocols, rather than Triton-X100, in the presence of lysozyme, to achieve bacterial lysis. Triton-X100 seemed to strip the endogenous pseudovirus particles as well as exogenously added TMV in control reconstruction experiments. Similar effects have been reported with poly-L-ornithine and plant protoplast membrane phospholipids [35]. To confirm that these virus-like CAT-OAS rods were helically assembled, they were shown to bind the neotope-specific MAb. In the absence of the TMV OAS few, if any, TMV-like rods could be trapped on polyclonal anti-TMV antibody-coated grids (Fig. 3a). Figure 3 shows CAT-OAS RNA with the alanine version of TMV CP. Both the native (serine) viral and *E. coli* codon-produced versions of the true U1 CP sequence gave identical results. In vivo, the proline version of U1 CP produced the same heterodisperse population of stacked disk-like rods as seen with the purified protein alone in vitro, in the presence or absence of co-expressed CAT-OAS mRNA.

Using the neotope-specific MAb, an ELISA procedure was developed to quantify the yield of assembled particles in bacterial cell lysates. Depending upon the duration of IPTG-induction and the nature of the TMV CP gene, the yield of CAT rods was up to 7.4 µg/mg total bacterial protein. Averaged over several experiments, 20% of the total TMV CP expressed in *E. coli* was assembled into ribonucleocapsids, which is less than in TMV-infected plant cells, but is substantially more rapid, straightforward and less costly than previous methods for producing pseudovirus particles in vitro.

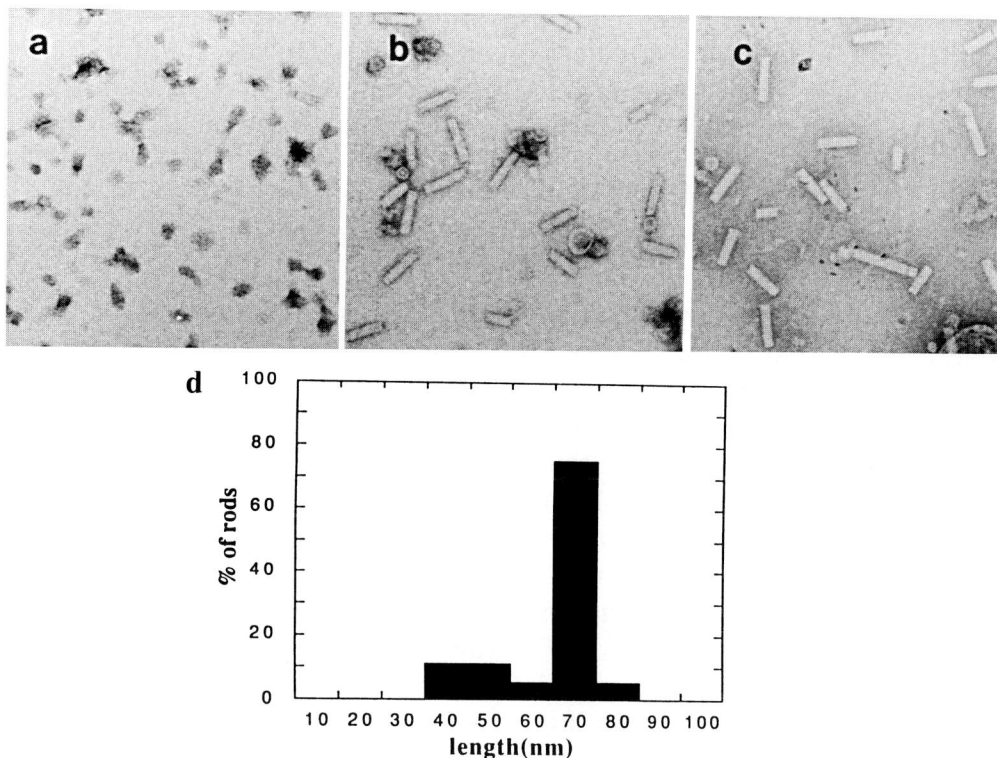

**TMV-like particles assembled *in vivo* in *E. coli* with
TMV CP (Ala) and CAT-OAS RNA**

Fig. 3. Assembly of TMV-like pseudovirus particles in vivo in *E. coli*. Immunosorbent electron microscopy of TMV CP-containing material from IPTG-induced cultures of cells transformed with the alanine-derivative of the native TMV CP sequence alone (**a**) or together with a constitutively expressed CAT-OAS RNA construct (**b, c**). Negative staining was with 1% uranyl acetate (**a, b**) or 2% ammonium molybdate, pH 3.5 (**c**). **d** Histogram showing rod length-frequency distribution with the population mode around 60–70 nm (1.4 kb CAT-OAS RNA predicts a 66 nm rodlet)

To confirm that the purified rodlets contained the expected CAT-OAS mRNA, particles were subjected to phenol-extraction and ethanol precipitation, then were examined for OAS or CAT sequences using specific primers in reverse transcriptase-PCR. Negative controls were included and the expected dsDNA bands were observed only when the CAT mRNA possessed a 3′-OAS. Similarly, northern blotting procedures with OAS (or CAT)-specific probes showed that the expected sequences could be recovered from *E. coli* extracts in a form protected from ribonucleases.

When a GUS-OAS mRNA was expressed synchronously with the TMV CP from a T7 promoter, the 2.8 kb transcripts also resulted in easily identified TMV-like particles in high numbers in cell extracts subjected to immunosorbent electron microscopy. However, in this case, with the Ala-version of the native CP gene (Fig. 4), the rodlets were far shorter (30–60 nm; 640–1 280 nts encapsidated) than predicted (130 nm). The ribonucleoprotein particles had been purified from a crude *E. coli* lysate by low- then high-speed centrifugation through a sucrose cushion. Upon close examination it appeared that most of these truncated GUS-OAS particles had one or more ribosome-like structures attached to one end, predominantly the morphologically distingishable convex end, where the 3′-end of the RNA finally resides [36]. It is proposed that, either because of the greater length or more efficient synchronous induction of the GUS-OAS mRNA transcripts, concomitant with the mRNA for TMV CP, or because of the particular 5′-leader on this construct, GUS mRNA translation commenced efficiently before the 3′-proximal OAS was transcribed. As a result, polysomes and not naked GUS-OAS RNA (as presumably with CAT-OAS RNA), were recruited for pseudovirus particle formation. The particular topography of TMV-based RNA assembly [6, 14, 16], in which the 5′-end of the RNA passes through the central hole, would result in structures depicted in the schematic diagram shown in Fig. 5 – provided 3′-to-5′ nucleocapsid assembly was energetically unable to reverse or dislodge 70S ribosomes actively translocating 5′-to-3′ during mRNA translation. The resulting complex is a sterically blocked, partially assembled, partially translated, polysome – "an assemblosome". The implications of this observation for the functional compartmentalization of progeny plus-strand TMV RNA templates in infected plant cells are obvious. As with CAT-OAS mRNA above, these recovered (truncated) particles gave a strong PCR signal corresponding to the OAS portion of the GUS-OAS mRNA construct, but not for the GUS sequence.

Assembly of infectious full-length TMV particles in *E. coli*

In 1982 it was shown that in vitro translation of full-length (6.4 kb) U1 TMV RNA in *E. coli* S30 extracts resulted in one major 17.5 kDa polypeptide with many of the biochemical and biological properties of TMV CP [37]. Internal initiation at a fortuitous Shine-Dalgarno-like sequence was proposed. To confirm this result in vivo, *E. coli* BH21 (DE3) pLysS cells were simply transformed with the plasmid pTMV212 (another generous gift from W. O. Dawson, University of Florida), which contains a full-length clone of the U1 TMV genome under the control of a T7 promoter but with no T7 terminator. IPTG-induced transcripts of pTMV212 begin pppGG<u>GUAUUU</u> but have no unique 3′-terminus <u>GCCCA</u>$_{OH}$ (native TMV RNA sequences are underlined).

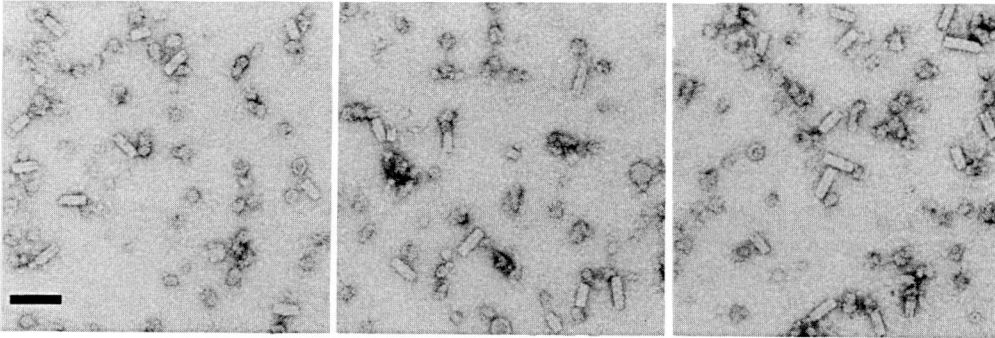

Fig. 4. Incomplete assembly of GUS-OAS RNA into TMV-like pseudovirus particles in *E. coli*. Immunosorbent electron microscopy of TMV CP-containing material from cells induced with IPTG to synthesize, from two separate T7 promoters, the alanine-derivative of the native TMV CP sequence and a GUS-OAS RNA construct (2.8 kb). Negative staining was with 1% uranyl acetate. Scale bar represents 100 nm

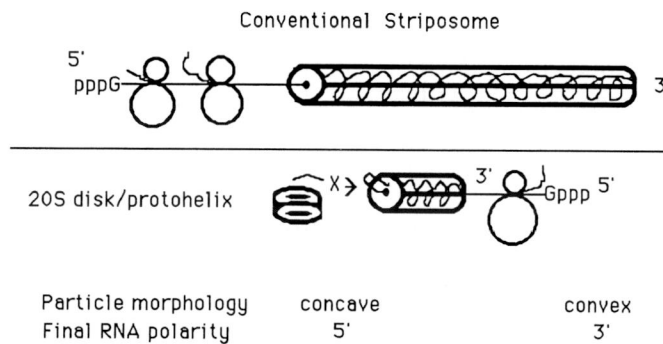

Fig. 5. Schematic diagram depicting how a 70S ribosome, actively translating GUS-OAS mRNA from the free 5'-end (pppG), blocks (X) further 3'-to-5' assembly with prefabricated 20S TMV CP aggregates at the 5'-face (concave end) of the final particle. The free energy lost during ribonucleocapsid assembly is insufficient to dislodge the ribosome, thus movement of the 5'-tail of the RNA through the central hole in the CP helix is sterically blocked

Figure 6 shows the result of only 2 h induction with IPTG in these transformed bacterial cells. Crude extracts (equivalent to 40 μl of original culture) of control *E. coli* cells (lane 1) or of pTMV212-transformed cells (lane 3), or marker proteins including 1 μg TMV CP (lane 2) were fractionated electrophoretically in a 10% polyacrylamide gel with SDS and electroblotted onto nitrocellulose. The blot was probed with rabbit polyclonal antiserum to TMV CP. Parallel blots probed with antiserum to another TMV RNA coded protein (of 126 kDa) gave no signal in any lane (not shown). The level of expression of TMV CP corresponds to about 50 μg/ml original bacterial culture, and is derived from intact or 3'-fragments of the 6.4 kb-or-more T7 transcripts from pTMV212.

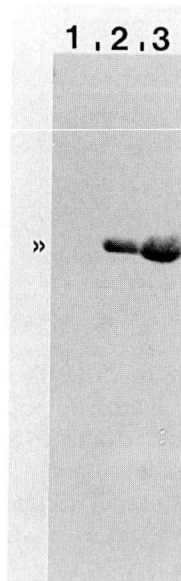

Fig. 6. Expression of TMV (U1) coat protein from "genome-length" transcripts of pTMV212 in transformed *E. coli* BL21 (DE3) pLysS. Bacteria grown at 37°C to mid-log phase were treated with 400 μM IPTG at 27°C for 2 h to induce T7 RNA polymerase and greater-than-full-length transcripts of the TMV genome, initiated with the sequence pppGGGUAU, but with no defined T7 termination signal. Crude extracts of lysed control *E. coli* cells (*1*) or pTMV212-transformed cells (*3*), or standard marker proteins containing 1 μg TMV (U1) CP (*2*), were electrophoresed in a 10% polyacrylamide gel in the presence of SDS and blotted onto nitrocellulose. Blots were probed with rabbit polyclonal antiserum to TMV CP. The position of the 17.5 kDa TMV capsid protein is marked on the left by a double arrowhead. The amount of total protein loaded on *1* and *3* corresponds to approx. 40 μl of each of the original cultures of *E. coli*

Immunosorbent electron microscopy of extracts from pTMV212-transformed *E. coli*, induced with IPTG for 10 h, revealed low numbers of TMV-like particles of varying lengths (Fig. 7). Of course, pTMV212 transcripts carry a functional OAS. To produce fully-infectious RNase-resistant virions (300 nm), 5′-intact, 6.4 kb-long transcripts must be encapsidated. Even after storage for several weeks at ambient temperatures, or at 4°C, the material recovered by high-speed centrifugation from cleared lysates of IPTG-induced, pTMV212-transformed cells could initiate systemic infections on susceptible tobacco cultivars [*Nicotiana tabacum* cvs. Xanthi (genotype nn), or Petite Havana] but was of insufficient titre to produce local lesions on NN-genotype plants. Control *E. coli* extracts to which native TMV virions were added during cell lysis behaved in a similar way, confirming that infectious virus particles could survive the procedures. However, added naked TMV RNA did not survive to produce any infection. Thus, although pTMV212 transcripts

pTMV212-Transformed cells

pUC118-T7 promoter-GG<u>GUAUU....6390nts</u> U1 TMV...pUC118

pLysS-Transformed cells (Control)

20 µg TMV added after harvest, before lysozyme & osmotic shock

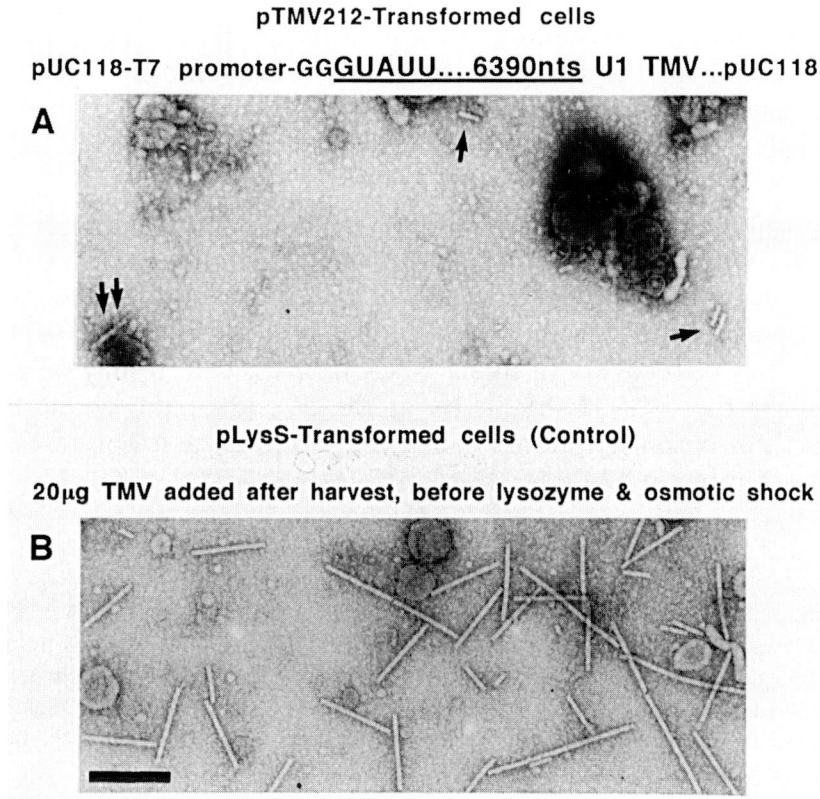

Fig. 7. Immunosorbent electron microscopy of TMV particles harvested from 10 h IPTG-induced (200 ml) cultures of *E. coli* BL21 (DE3) pLysS alone (control) or transformed with pTMV212 (**A**). Purified TMV particles (20 µg; **B**) or TMV RNA (100 µg, not shown) were added to the resuspended pellets (2 ml) of control bacteria during cell lysis, prior to clearing and virus particle concentration by low- and high-speed centrifugation, respectively. Particles were trapped with a rabbit polyclonal antiserum to TMV CP, and stained with 1% uranyl acetate. Arrows in **A** highlight in vivo assembled, TMV-like, pTMV212-derived ribonucleoprotein particles among the cell debris. Both samples shown caused systemic TMV infections on *Nicotiana tabacum* cv. Xanthi nn or Petite Havana

in vivo will be uncapped, carry two extra 5'-G residues and have an undefined 3'-end, some of the packaged molecules must be processed or fragmented correctly and fully encapsidated to initiate plant infection. This is the first report of the production of an infectious plant virus in bacterial cells.

Discussion and future directions

Having established an easily manipulated system for packaging ssRNA in vivo, albeit in bacterial cells, we can now study the sequence requirements in RNA or protein, and a possible role(s) for chaperone proteins in the so-called "self-assembly" of a plant virus. By analogy, we expect that expression of the CP gene from other filamentous or rod-shaped plant viruses may allow their OAS to be mapped by co-expression of cDNA sub-clones of their genome and electron microscopic screening for rodlets.

Expression vectors for blue-green algae are available and we are currently examining the effect(s) of in situ expression of TMV CP (native or mutagenized) on the efficiency of light-harvesting by photosystems I and II and on the structure of cyanobacterial thylakoids. To enhance the efficiency of recovery of intact RNA from *E. coli*, it may be possible to engineer the TMV OAS to be at the 5'-end of the chimeric gene construct, for cotranscriptional packaging, despite the lesser efficiency of 5'-to-3' assembly in TMV [6, 13, 16].

Acknowledgements

We thank J. R. Haynes for supplying the *E. coli* codon-optimized U1 TMV CP gene, W. O. Dawson for pTMV212 and pTMV210, and M. H. V. van Regenmortel for monoclonal antibody 253P, specific for a neotope in helical TMV-like particles. This work was supported by funds from the New Jersey Commission for Science and Technology and the Scottish Office Agriculture and Fisheries Department. The results form part of US Patent Application 07/971, 101 by Rutgers University.

References

1. Fraenkel-Conrat H, Williams RC (1955) Reconstitution of active tobacco mosaic virus from its inactive protein and nucleic acid components. Proc Natl Acad Sci USA 41: 690–695
2. Caspar DLD (1963) Assembly and stability of the tobacco mosaic virus particle. Adv Protein Chem 18: 37–121
3. Durham ACH, Finch JT, Klug A (1971) States of aggregation of tobacco mosaic virus protein. Nature 229: 37–42
4. Fraenkel-Conrat H (1957) Degradation of tobacco mosaic virus with acetic acid. Virology 4: 1–4
5. Durham ACH (1972) Structures and roles of the polymorphic forms of tobacco mosaic virus protein. I Sedimentation studies. J Mol Biol 67: 289–305
6. Bloomer AC, Butler PJG (1986) Tobacco mosaic virus: structure and self-assembly. In: Van Regenmortel MHV, Fraenkel-Conrat H (eds) The plant viruses, vol 2. The rod-shaped plant virusses. Plenum, New York, London, pp 19–57
7. Zimmern D, Wilson TMA (1976) Location of the origin for viral reassembly on tobacco mosaic virus RNA and its relation to stable fragment. FEBS Lett 71: 294–298
8. Zimmern D (1977) The nucleotide sequence at the origin for assembly on tobacco mosaic virus RNA. Cell 11: 463–482
9. Goelet P, Lomonossoff GP, Butler PJG, Akam ME, Gait MJ, Karn J (1982) Nucleotide sequence of tobacco mosaic virus RNA. Proc Natl Acad Sci USA 79: 5818–5822

10. Sacher R, French R, Ahlquist P (1988) Hybrid brome mosaic virus RNAs express and are packaged in tobacco mosaic virus coat protein in vivo. Virology 167: 15–24

11. Sleat DE, Turner PC, Finch JT, Butler PJG, Wilson TMA (1986) Packaging of recombinant RNA molecules into pseudovirus particles directed by the origin-of-assembly sequence from tobacco mosaic virus RNA. Virology 155: 299–308

12. Turner DR, Joyce LE, Butler PJG (1988) The tobacco mosaic virus assembly origin RNA: functional characteristics defined by directed mutagenesis. J Mol Biol 203: 531–547

13. Turner DR, McGuigan CJ, Butler PJG (1989) Assembly of hybrid RNAs with tobacco mosaic virus coat protein. Evidence for incorporation of disks in 5′ elongation along the major RNA tail. J Mol Biol 209: 407–422

14. Gallie DR, Plaskitt KA, Wilson TMA (1987) The effect of multiple dispersed copies of the origin-of-assembly sequence from TMV RNA on the morphology of pseudovirus particles assembled in vitro. Virology 158: 473–476

15. Jupin I, Sleat DE, Watkins PAC, Wilson TMA (1989) Direct recovery of in vitro transcripts in a protected form suitable for prolonged storage and shipment at ambient temperatures. Nucleic Acids Res 17: 815

16. Okada Y (1986) Molecular assembly of tobacco mosaic virus in vitro. Adv Biophys 22: 95–149

17. Caspar DLD, Namba K (1990) Switching in the self-assembly of tobacco mosaic virus. Adv Biophys 26: 157–185

18. Butler PJG, Bloomer AC, Finch JT (1992) Direct visualization of the structure of the "20S" aggregate of coat protein of tobacco mosaic virus. The "disk" is the major structure at pH 7.0 and the *proto*-helix at lower pH. J Mol Biol 224: 381–394

19. Sleat DE, Gallie DR, Watts JW, Deom CM, Turner PC, Beachy RN, Wilson TMA (1988) Selective recovery of foreign gene transcripts as virus-like particles in TMV-infected transgenic tobaccos. Nucleic Acids Res 16: 3127–3140

20. Siegel A (1971) Pseudovirions of tobacco mosaic virus. Virology 46: 50–59

21. Schoelz JE, Zaitlin M (1989) Tobacco mosaic virus RNA enters chloroplasts in vivo. Proc Natl Acad Sci USA 86: 4496–4500

22. Banerjee N, Zaitlin M (1992) Import of tobacco mosaic virus coat protein into intact chloroplasts in vitro. Mol Plant Microbe Interact 5: 466–471

23. Reinero A, Beachy RN (1989) Reduced photosystem II activity and accumulation of viral coat protein in chloroplasts of leaves infected with tobacco mosaic virus. Plant Physiol 89: 111–116

24. Gallie DR, Sleat DE, Watts JW, Turner PC, Wilson TMA (1987) In vivo uncoating and efficient expression of foreign mRNAs packaged in TMV-like particles. Science 236: 1122–1124

25. Haynes JR, Cunningham J, Von Seefried A, Lennick M, Garvin RT, Shen S-H (1986) Development of a genetically-engineered, candidate polio vaccine employing the self-assembling properties of the tobacco mosaic virus coat protein. Bio/Technology 4: 637–641

26. Gal-On A, Antignus Y, Rosner A, Raccah B (1990) Nucleotide sequence of the zucchini yellow mosaic virus capsid-encoding gene and its expression in *Escherichia coli*. Gene 87: 273–277

27. Shire SJ, McKay P, Leung DW, Cachianes GJ, Jackson E, Wood WI (1990) Preparation and properties of recombinant DNA derived tobacco mosaic virus coat protein. Biochemistry 29: 5119–5126

28. Jagadish MN, Ward CW, Gough KH, Tulloch PA, Whittaker LA, Shukla DD (1991) Expression of potyvirus coat protein in *Escherichia coli* and yeast and its assembly into virus-like particles. J Gen Virol 72: 1543–1550
29. McDonald JG, Bancroft JB (1977) Assembly studies on potato virus Y and its coat protein. J Gen Virol 35: 251–263
30. Durham ACH (1974) Cause of irreversible polymerisation of tobacco mosaic virus protein. FEBS Lett 25: 147–150
31. Namba K, Pattanayek R, Stubbs G (1989) Visualization of protein-nucleic acid interactions in a virus. Refined structure of intact tobacco mosaic virus at 2.9Å resolution by X-ray fibre diffraction. J Mol Biol 208: 307–325
32. Jardetsky O, Akasaka K, Vogel D, Morris S, Holmes KC (1978) Unusual segmental flexibility in a region of tobacco mosaic virus coat protein. Nature 273: 564–566
33. Studier FW, Rosenberg AH, Dunn JJ, Dubendorff JW (1990) Use of T7 RNA polymerase to direct expression of cloned genes. Methods Enzymol 185: 60–89
34. Wittmann HG (1965) Die primäre Proteinstruktur von Stämmen des Tabakmosaikvirus. IV. Aminosäuresequenzen (1 bis 61 und 135 bis 158) des Proteins des Tabakmosaikvirus-Stammes U2. Z Naturforsch 20b: 1213–1223
35. Kiho Y, Abe T, Ohashi Y (1979) Disassembly of tobacco mosaic virus by membrane lipid from tobacco leaves and polyornithine. Microbiol Immunol 23: 1067–1076
36. Wilson TMA, Perham RN, Finch JT, Butler PJG (1976) Polarity of the RNA in the tobacco mosaic virus particle and the direction of protein stripping in sodium dodecyl sulphate. FEBS Lett 64: 285–289
37. Glover JF, Wilson TMA (1982) Efficient translation of the coat protein cistron of tobacco mosaic virus in a cell-free system from *Escherichia coli*. Eur J Biochem 122: 485–492

Authors' address: Dr. T. M. A. Wilson, Department of Virology, SCRI, Invergowrie, Dundee DD2 5DA, Scotland.

W. H. Gerlich (ed.)
Research in Chronic Viral Hepatitis

1993. 46 partly coloured figures. XI, 304 pages.
ISBN 3-211-82497-9
Soft cover DM 250,–, öS 1750,–*
(Archives of Virology / Supplementum 8)

O.-R. Kaaden, W. Eichhorn, C.-P. Czerny (eds.)
Unconventional Agents and Unclassified Viruses
Recent Advances in Biology and Epidemiology

1993. 79 partly coloured figures. VIII, 308 pages.
ISBN 3-211-82480-4
Soft cover DM 260,–, öS 1820,–*
(Archives of Virology / Supplementum 7)

P. P. Liberski
The Enigma of Slow Viruses
Facts and Artefacts

1993. 56 figures. XVI, 277 pages.
ISBN 3-211-82427-8
Soft cover DM 250,–, öS 1750,–*
(Archives of Virology / Supplementum 6)

O. W. Barnett (ed.)
Potyvirus Taxonomy

1992. 57 figures. IX, 450 pages.
ISBN 3-211-82353-0
Soft cover DM 290,–, öS 2030,–*
(Archives of Virology / Supplementum 5)

C. De Bac, W. H. Gerlich, G. Taliani (eds.)
Chronically Evolving Viral Hepatitis

1992. 72 figures. XIV, 348 pages.
ISBN 3-211-82350-6
Soft cover DM 260,–, öS 1820,–*
(Archives of Virology / Supplementum 4)

B. Liess, V. Moennig, J. Pohlenz, G. Trautwein (eds.)
Ruminant Pestivirus Infections
Virology, Pathogenesis, and Perspectives of Prophylaxis

1991. 78 figures. VIII, 271 pages.
ISBN 3-211-82279-8
Soft cover DM 220,–, öS 1540,–*
(Archives of Virology / Supplementum 3)

C. H. Calisher (ed.)
Hemorrhagic Fever with Renal Syndrome, Tick- and Mosquito-Borne Viruses

1991. 75 figures. VII, 347 pages.
ISBN 3-211-82217-8
Soft cover DM 258,–, öS 1800,–*
(Archives of Virology / Supplementum 1)

Prices are subject to change without notice

* *10 % price reduction for subscribers to the journal "Archives of Virology"*

Springer-Verlag Wien New York

Sachsenplatz 4–6, P.O.Box 89, A-1201 Wien · 175 Fifth Avenue, New York, NY 10010, USA
Heidelberger Platz 3, D-14197 Berlin · 37-3, Hongo 3-chome, Bunkvo-ku, Tokyo 113, Japan

E. Kurstak (ed.)

Measles and Poliomyelitis

Vaccines, Immunization, and Control

1993. 48 figures. XI, 411 pages. Soft cover DM 198,–, öS 1386,–. ISBN 3-211-82436-7

Elimination of measles and poliomyelitis diseases from the globe is a priority goal of the World Health Organization. For the first time, in a single volume comprising thirty-one well-documented chapters, internationally recognized experts provide a state-of-the-art treatment of these two important viral diseases. The book offers a wide range of new findings and references on the latest advances regarding the measles and poliomyelitis:

- global and molecular genetic epidemiology, characteristics and diseases surveillance
- all available vaccines and research to produce more safe and more potent biotechnology vaccines
- immunization programmes, considering the available vaccines and possibility of vaccinal associations in strategy to eliminate/eradicate these diseases
- immunity to infections and immunogenicity of vaccines
- virus genomes organization and antigenic structures related to vaccine characteristics, stressing their role in immunization strategies
- needs of global cooperation, using all available resources, vaccines and strategies to achieve the global control of the diseases.

It is addressed to all public health professionals concerned with measles and poliomyelitis control, especially in hospitals, clinics, governmental health services, international health organizations, centers of infectious diseases, research institutes, medical schools, vaccine producers and experts in immunization strategies and programmes.

E. Kurstak

Viral Hepatitis

Current Status and Issues

In collaboration with Christine Kurstak, A. Hossain, and A. Al Tuwaijri

1993. 26 figures. X, 217 pages. Soft cover DM 120,–, öS 840,–. ISBN 3-211-82387-5

In the 1990's significant advances in the understanding of viral hepatitis have been observed. In particular, our knowledge of the nature and diversity of viruses causing hepatitis in humans have substantially increased.
"Viral Hepatitis" comprehensively and uniquely presents these valuable information all in a single volume for the utmost benefit of medical practitioners, microbiologists as well as those actively involved in health administration world-wide.
The virological, clinical epidemological, diagnostic, therapeutic, and preventive aspects pertaining to all the types of hepatitis known to date including hepatitis C and E are thoroughly discussed.

Prices are subject to change without notice

Springer-Verlag Wien New York

Sachsenplatz 4–6, P.O.Box 89, A-1201 Wien · 175 Fifth Avenue, New York, NY 10010, USA
Heidelberger Platz 3, D-14197 Berlin · 37-3, Hongo 3-chome, Bunkyo-ku, Tokyo 113, Japan

David H. Walker (ed.)

Global Infectious Diseases

Prevention, Control, and Eradication

With a Foreword by Thomas N. James

1992. 33 figures. XIII, 234 pages.
Soft cover DM 148,–, öS 1036,–
ISBN 3-211-82329-8

Prices are subject to change without notice

The subject of the book is global infectious diseases and includes 12 entities that were carefully selected for diversity of epidemiology, transmission, pathogenesis, and immune mechanisms. Each topic (schistosomiasis, malaria, filariasis, arborviruses, diarrheal diseases, AIDS, hepatitis, fungal diseases, Chagas' disease, rickettsial diseases, Lyme disease, and cysticercosis) is particularly interesting in its own right. Each entity will be reviewed by an expert who perceives the big picture and will cover the epidemiology, ecology, pathogenesis, immunity, and relevant molecular data pertinent to the etiologic agents and their hosts. Each expert will then express his opinion as to the best approach to eradicate, control, or contain the disease in the near future by attacking the most vulnerable point in the interaction of the etiologic agent and the host or the environment. Molecular approaches will be described which point towards novel methods to stimulate effective immunity, eradicate a vector, render it genetically incompetent, or develop a new therapeutic intervention. The ultimate standard and the reasons for its success and the failure of other eradication campaigns will be posed by Dr. D.A. Henderson, leader of the successful campaign to eradicate smallpox and currently a high level advisor in the White House.

Springer-Verlag Wien New York

Sachsenplatz 4–6, P.O.Box 89, A-1201 Wien · 175 Fifth Avenue, New York, NY 10010, USA
Heidelberger Platz 3, D-14197 Berlin · 37-3, Hongo 3-chome, Bunkyo-ku, Tokyo 113, Japan